International Conference on Modeling, Simulation, Optimization and Algorithm (ICMSOA 2019)

Journal of Physics: Conference Series Volume 1419

Sanya, China
9 - 10 November 2019

Editors:

Laura Tan
Liu Yang
Jane Zhang

ISBN: 978-1-7138-1324-8
ISSN: 1742-6588

Printed from e-media with permission by:

Curran Associates, Inc.
57 Morehouse Lane
Red Hook, NY 12571

CURRAN ASSOCIATES INC. proceedings.com

Some format issues inherent in the e-media version may also appear in this print version.

Additional copies of this publication are available from:

Curran Associates, Inc.
57 Morehouse Lane
Red Hook, NY 12571 USA
Phone: 845-758-0400
Fax: 845-758-2633
Email: curran@proceedings.com
Web: www.proceedings.com

TABLE OF CONTENTS

ALGORITHM

MODELING

OPTIMIZATION

SIMULATION

ICMSOA 2019 Preface

Orginazed and sponsored by Yaseen Academy, 2019 International Conference on Modeling, Simulation, Optimization and Algorithm (ICMSOA2019) will be held during Nov. 9-10, 2019 in Sanya, China. This Conference looks for significant contributions to all major fields of Computational Modeling, Simulation, Optimization and Algorithm in theoretical and practical aspects. The aim of the conference is to provide a platform to the researchers and practitioners from both academia as well as industry to meet and share cutting-edge development in the field.

ICMSOA 2019 has received 90 manuscripts, 53 submissions have been accepted by the reviewers. All papers included into this collection have undergone careful peer-review by at least two reviewers before being selected for publication.

We would like to thank the organization staff, the Committee members, all the reviewers and everyone who had worked for and lending a hand in reviewing papers and making valuable suggestions for the improvement of the conference. Special thanks go to IOP Publishing for their sincere help and high efficiency in publishing and indexing for all articles. We'd like to express our gratitude for all authors and participants of this conference.

Jianxin Zhang

Yaseen Academy

ICMSOA 2019 Organizing Committee

www.icmsoa.com

Journal of Physics: Conference Series **1419** (2019) 011002 doi:10.1088/1742-6596/1419/1/011002

ICMSOA2019 Committee

Editors

Laura Tan, Yaseen Academy, Wuhan, Hubei China
Liu Yang, Yaseen Academy, Wuhan, Hubei, China
Jane Zhang, Wuhan University, Wuhan, Hubei, China

Conference General Chair

Prof. Yichuan Zhao, Georgia State University, USA

Program Chair

Prof. Yilun Shang, Northumbria University, United Kingdom

Conference Committee Chair Members

Prof. Zongwei Luo, Southern University of Science and Technology (SUSTech), China
Prof. Ying Ma, Xiamen University of Technology, China
Dr. Hui Sun, Academy of OPTO-Electronics, Chinese Academy of Science, BeiJing, China
Prof. Peng Lu, Central South University, China

International Technical Committee Members

Prof. Bin Liu, Nanjing Medical University, China
Prof. Quanxin Zhu, Hunan Normal University, China
Prof. Zhixiao Wang, Xi'an University of Technology, China
Prof. Yingxue Zhao, University of International Business and Economics, China
Prof. Lulu Wang, Hefei University of Technology, China
Dr. José Joaquim De Moura Ramos, Department of Computer Science, University of A Coruña, Spain
Dr. Pisutti Dararutana, Royal Thai Army Chemical Department, Thailand
Dr. Altan Onat, Eskisehir Technical University, Turkey
Dr. Zul-Atfi Ismail, School of Environmental Engineering, Universiti Malaysia Perlis, Malaysia

International Reviewers

Prof. Kyaw Zaw Ye, Ministry of Defense, Myanmar
Prof. Gorbunov Vladimir, National Research University of Electronic Technology, Russia
Prof. Yu Xue, Nanjing University of Information Science and Technology, China
Prof. Goharik Petrosyan Razmiki, Armenian State Pedagogical University, Armenian
Prof. Ning Chen, Xian Polytechnic University, China
Prof. Nina Popva, St.Petersburg Polytechnic University, Russia
Prof. Anastasios Doulamis, National Technical University of Athens (NTUA), Greece
Dr. Jie Hu, University of Massachuttes, USA
Dr. LO, Man-fung, The Hong Kong Polytechnic University, China
Prof. Peng Lu, Central South University, China
Dr. José Carlos Metrolho, Polytechnic Institute of Castelo Branco, Portugal
Dr. Izhar Mithal Jiskani, China University of Mining and Technology, China

ICMSOA2019 Speeches

Host: Prof. Zhizhang Shen from Plymouth State University, Plymouth, USA

Speech 1: Nazarenko Nikita from Bauman Moscow State Technical University, Moscow, Russia
Speech Title: The method to annotate buildings with images from CCTV cameras with an intersecting field of view
Abstract. Video surveillance and tracking systems help to increase the safety of human activity in a given area and improve detecting the geo-location of an observer (a swarm of drones, autopilot) in modern cities. This article proposes the method for determining the full or partial address (annotation) of buildings by using images from several CCTV cameras with an intersecting field of view. As input parameters, the method takes several images from surveillance cameras, their geographical coordinates and directions of the optical axes and the vector map of the area where the cameras are located. The outputs are annotations of buildings for the initial images. The proposed method allows solving the problem of georeferencing the video analytics system and improving the work of the surveillance and object tracking systems. The resulting annotations of buildings provide an opportunity to determine more accurately the situations with overlapping of the tracked objects with buildings and predict the probable paths of the objects. A lost object on one camera can be found by using annotations from other cameras, which field of view is not overlapped by this building. Besides, the proposed ap0070roach is sustainable to full or partial overlapping of buildings and allows solving to solve the problem in the absence of a detailed height map of the terrain and buildings. Also, the article researches the restrictions of the method. It is necessary to find an angle between two cameras, at which the best result can be achieved. That is, the method should return the correct addresses of all buildings visible in the image.

Speech 2: Yijie Wang from Yunnan University, Kunming, China
Speech Title: Evolutionary dynamics of Chinese tourism market extremely low-priced strategies based on wright-fisher processes
Abstract. Extremely low-priced of package tours is an important cause of Chinese tourism market price disorder. Regulating the price of package tours in Chinese tourism market, can solve the problem of Chinese tourism market price disorder. In order to research the reason of extremely low package tours price, consider the stochastic dynamics of N homogeneous mixed travel agency pricing strategies, and each travel agency could choose Extremely low-pricing or regular pricing. Based on Wright-Fisher processes, we obtain the dominant strategy of Extremely low-pricing and regular pricing under the weak selectivity in the package tours price setting, when the travel agency groups are finite. And then further verify these conditions by numerical simulation.The simulation shows that the Price elasticity of tourism demand, the transparency of the tourism market and the degree of government supervision strength have important effects on the stochastic evolution of package tours price setting. Reducing the demand elasticity of tourism market, increasing the transparency of the tourism market and the degree of government supervision strength can make travel agency choose regular pricing and solve chaotic of Chinese tourism market further. Therefore, this paper puts the suggestion of "two increases and one decrease" for the development of Chinese tourism market.

Speech 3: Yakovleva Olga Viktorovna from Bauman Moscow State Technical University, Moscow, Russia
Speech Title: Predict post spreading in online social networks based on independent cascade model
Abstract. Today, online social networks play an important role in the information spreading. Based on the social influence effect, posts can be spread across the network by repost mechanism. This paper addresses the problem of using independent cascade model for repost dynamics prediction in online social networks. A spreading post probability between two connected users is presented as function of preassigned features and coefficients fitted by the EM algorithm. This approach combines ideas of previously proposed methods. The list of considered features is composed and it includes users, their relationships and posts features. The iteratively solved by the Newton's method system of equation is derived for independent cascade model to obtain coefficients of a probability function. The experiments with the Sina Weibo dataset show that the EM algorithm more accurately estimates unknown coefficients of probability function that the logistic regression method. Since actual values of diffusion probabilities are unknown for a real network, a problem of their estimation is considered as a binary classification problem to determine accuracy of various methods. The independent cascade model tuned by proposed method of diffusion probability estimation was verified. The experimental results show than the model can commendably predict the reposting dynamics for a given post.

Speech 4: Jun He from University of Science and Technology of China, Hefei, China
Speech Title: Financial deepening, inclusive finance and economic growth: Analysis based on endogenous growth theory
Abstract. This article studies the relationship between financial development and economic growth from the perspectives of financial deepening and inclusive finance. The theoretical research expands the financial endogenous economic growth model, constructs the Hamiltonian function, using the first-order optimization conditions to obtain explicit relationship between economic growth and financial development under balanced path. Theoretical research shows that financial deepening and inclusive finance are significant and stable in promoting economic growth.

Speech 5: Murashov Ivan Dmitrievich from Bauman Moscow State Technical University, China
Speech Title: Method of determining vehicle speed according to video stream data
Abstract. This article presents a method of determining the speed of a dynamic object from the data of a video stream. The determination of speed occurs based on the analysis of repeated objects of the environment, the distance between which varies according to a previously known law, such as lamp posts, milestones or bump stops on roads, intermittent road marking lines, emergency exits in tunnels. The method of determining vehicle speed includes three steps: pre-processing of the frame, searching for a pattern in the image, calculating the average vehicle speed obtained on the basis of the number of repeated objects, taking into account the law of their repetition and the distance between them for some fixed time. The advantage of the proposed method is that there is no need to install any additional sensors on the dynamic object except video camera. The development of an additional method for calculating speed has a significant application value for the complexes of control of autonomous traffic, as the existing methods of determining speed have their disadvantages (not high enough accuracy, the requirement for the

presence of additional equipment). This algorithm was studied on regularly recurring objects (lampposts) under different lighting conditions (day, night), the distance between repeated objects and different weather conditions. It was also compared with other methods of determining the speed of the data from the video stream, the results of which the proposed method has a significant advantage in accuracy as compared with analogs in the absence of interference with video.

Speech 6: Zhizhang Shen from Plymouth State University, Plymouth, USA
Speech Title: Faulty node modeling and diagnosis in interconnection networks
Abstract. Faulty processing node analysis, in particular, self-diagnostic paradigm, is an important topic in the area of interconnection network studies. Several diagnostic models have been introduced, with the PMC and MM* models being two of the most popular ones. Researchers have also proposed various extensions and enhancements of fault-tolerance models to better capture the distribution, and identification, of faulty nodes in realistic scenarios. One of them is the g-extra fault-tolerance model, where each cluster in a network with faulty nodes contains at least g+1 fault-free nodes. This paper, following an analytical and constructive approach based on applied graph theory, suggests a general process to identify the maximum number of faulty nodes in a network in terms of the g-extra fault-tolerance model, and as a demonstrative example, provides a specific result for the (n, k)-star graphs.

Speech 7: Cong Ke from Hubei University of Technology
Speech Title: Image Registration of Infrared and Visible Image Based on Geometric Constraint
Abstract. With the wide application of infrared sensors and visible sensors, the registration of two heterogeneous images has caused widespread concern. Multi-angle comprehensive analysis of multi-sensor imaging systems can obtain more accurate image information. However, the main challenge of image registration is the detection of feature points and the acquisition of the correct feature point matching pairs. A Geometric constraint method is proposed to obtain the correct feature matching pairs. Three feature point matching pairs are randomly selected in the two images, and the ratio of the corresponding line segment lengths is obtained according to the coordinates of the three feature points to find the most three similar feature point matching pairs. Compared with the RANSAC algorithm, the method can obtain the correct feature point matching pair better. The algorithm runs on the DSP platform for about 30s, the error is 1 pixel, and the correct rate reaches 80%.

Speech 9: Basim Salem from Bauman Moscow State Technical University, Moscow, Russia
Speech Title: Decision support system for an early-stage keratoconus diagnosis
Abstract. Currently, there is a wide variety of different diseases that exist, a lot of which can be hardly prompt diagnosed even by medical specialists. This paper presents a method for early-stage diagnosis of ophthalmologic disorder keratoconus. Working with medical imagery that was captured by a rotating Scheimpflug camera system for anterior segment analysis, the goal was to create a decision support system to aid ophthalmologists in prompt detection of the disorder to eliminate the chance of further surgical intervention. Given approach uses several steps to achieve that goal, such as find the region of interest on the medical imagery using a Single Shot MultiBox Detector, filter and binarize the image, locate the cornea and approximate the curve with least squares fitting, classify the stage according to the previously labeled dataset by medical specialists

with a random forest method. The suggested approach achieves 76% precision on a dataset containing 500 images of the patients with the first stage of keratoconus and healthy patients. The final result was compared with modern existing medical methods that are usually used in ophthalmologic clinics by medical specialists.

Speech 10: Qijun Zhang from China Communication Microelectronics Technology Co Ltd, Shenzhen, China

Speech Title: Automated Model Generation for Electronic Devices Using Neural Network Approaches

Abstract. This paper describes artificial neural network approaches to convert the conventionally expensive process of electronic modelling into an automated model generation process. Artificial neural networks are trained using machine learning algorithm to learn the electronic behaviour, and the trained neural network becomes a model to help predict the electronic device behaviour. The automated model generation algorithm performs adaptive data sampling to determine the amount and the distribution of training data needed to train neural networks. The algorithm also determines the number of hidden neurons needed to achieve a compact and accurate model. Also incorporated into the automated model generation method is an efficient interpolation approach to make the process much faster. The objective of the described method is to generate a compact neural-network based model with better accuracy and in less time than conventional approach. Examples of automated modeling of radio-frequency and microwave filters used in wireless electronic systems are described showing the advantage of this technique.

Peer review statement

All papers published in this volume of *Journal of Physics: Conference Series* have been peer reviewed through processes administered by the proceedings Editors. Reviews were conducted by expert referees to the professional and scientific standards expected of a proceedings journal published by IOP Publishing.

Journal of Physics: Conference Series **1419** (2019) 012001 doi:10.1088/1742-6596/1419/1/012001

Research on Distributed Target Assignment Based on Dynamic Allocation Auction Algorithm

Haobo Jiang[1], Song Li[2], Chi Lin [3], Chuang Wang[1], Kaiqi Zhong[4], Guannan He[5], Qizhi Zhang[6], Yuanhao Zhao[7] and Jiayi Liu[1]

[1]Graduate School, Air Force Engineering University, Xi'an, Shaanxi, 710051,China
[2]Air and Missile Defense College, Air Force Engineering University, Xian, Shaanxi, 710051, China
[3]Unit 94622, Quanzhou, Fujian,362000, China
[4]Unit 95865, Beijing, Beijing ,102218 ,China
[5]Unit 93015, Harbin, 150036 ,China
[6]Unit 95112, Foshan, 528000, China
[7]Unit 94535, Xuzhou, 221000, China

E-mail: 395049138@qq.com

Abstract. Under the premise of many types and large numbers of targets, how to solve the problem of rational allocation of dynamic multi-objectives has always been a hot issue for experts and scholars. In this paper, a distributed target allocation method based on dynamic allocation auction algorithm is used to solve this kind of problem. This paper establishes a mathematical model based on air defense combat, and uses this method to solve the problem. Finally, the effectiveness and real-time performance are proved by simulation.

1. Introduction

In actual air defense operations, the battlefield environment is dynamically changing. We cannot make decisions based on the battlefield situation at a certain moment. This makes us have to make a target allocation mechanism that can change itself according to the changes in the battlefield situation. However, the centralized centralized target allocation is unified by all the information to the superior command and control center for unified decision-making. When the battlefield environment changes, we pass the battlefield information to the superior command and control center, which will command and decide and distribute it. The results are distributed to the lower-level command and control centers and then to the air defense weapons and equipment. Such a process wastes a lot of time and the real-time performance will be poor. Moreover, all decisions are taken care of by the superior command and control center. If our superior command and control center is hit by the enemy air strike target, so that it can not continue to work normally, then our air defense system will be paralyzed. The subordinate distributed cooperative operations target allocation is supplemented by the upper-level centralized target allocation, and it is necessary to solve the problems existing in the upper-level centralized target allocation.

Many people have put forward their views on the solution of this problem. Jia [1] used a centralized command-distribution autonomous coordination structure to build a model, and used the method based on consistent distributed auction and heuristic based method for decision making. It shows that the algorithm has a good distribution effect. Lei [2] combined the cloud model with the genetic algorithm to improve the convergence speed and verify the feasibility. Sun [3] proposed an improved cuckoo search algorithm, and through simulation experiments, the algorithm can effectively balance the relationship between convergence speed and global exploration ability. Xu [4] used single/double potential well position updating method and particle swarm mutation algorithm to distribute air defense firepower, and verified the effectiveness of the algorithm through simulation. These methods solve this kind of problem quickly, but the real-time performance is poor and cannot be updated in real time. Based on the centralized target allocation in article [5], this paper proposes a distributed target allocation method based on dynamic allocation auction algorithm, which can solve the problem of target allocation well.

2. Description of the problem

In the lower-level distributed cooperative operational target allocation model, each lower-level command and control center can share information, and each lower-level command and control center can receive and send messages to each other. When the target is assigned, the target mobile sensor will be found. The information is sent to the lower-level command and control center, and is coordinated by the lower-level command and control network. The sensor can receive the tracking command sent by the lower-level command and control center, or send the target information to the lower-level command and control center. The air defense weapon can receive the lower-level command and control center. As a result, the target is hit. We use the ellipse to indicate the subordinate accusation center, the circle represents the sensor, and the pentagon represents the air defense weaponry. As shown in Figure 1

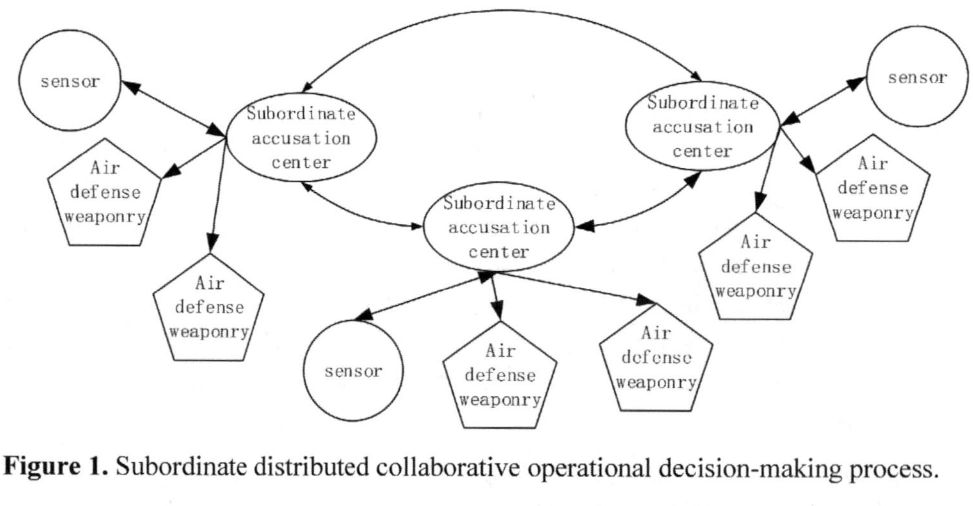

Figure 1. Subordinate distributed collaborative operational decision-making process.

The lower-level distributed cooperative combat target allocation is the last step of the allocation of air defense cooperative combat targets. It has the function of adjusting the centralized target assignment result of the superior, and can deal with the battlefield information in time. Its characteristics are as follows:

(1) High real-time

The air defense battlefield situation information changes rapidly, and the battlefield state will be different at each moment. When the lower-level distributed cooperative combat target changes, we need to adjust it in the lower-level command and control network in time, using the prescribed algorithm. The problem is solved quickly, and finally the target is distributed and hit in time, so that the distribution result has high real-time performance, so as to ensure effective attack.

(2) High synergy

I In the lower level command and control network, information interaction can be achieved, and each lower level command and control center can collect information and send information.

II Multiple air defense weapons can be used to strike a single target if conditions require it.

III When a node is defeated by an attack, other nodes can cooperate to meet the operational requirements.

(3) Rationality of results

By using the lower-level distributed collaborative operational target allocation, the results of the upper-level centralized target allocation can be detected and corrected, and the final allocation result can be rationalized.

(4) High resistance

In the mechanism of lower-level distributed cooperative combat target allocation, each element is deployed in a decentralized manner, making air defense weapons less vulnerable to attack, and no single node is our "critical". When a node is attacked by a target and does not work properly, it can help each other to complete the task that the node should have completed[6].

3. Model establishment

3.1. Target transfer judgment model

When the goal of the upper-level centralized target allocation exceeds the effective strike range of the air defense weapon (The route shortcut is out of range, the target speed is too fast, the incoming target height is higher than the maximum value, the incoming target height is less than the minimum value, or the radar scattering cross-sectional area is less than the minimum value) .Can't continue to strike it.

$$\begin{array}{l} l_i > l_{\max} \quad or \\ h_i > h_{\max} \quad or \\ h_i < h_{\min} \quad or \\ v_i > v_{\max} \quad or \\ RCS_i < RCS_{\min} \end{array} \tag{1}$$

3.2. Combat information transfer model

In the lower-level distributed cooperative combat target allocation, we will speed up the decision-making speed according to the actual situation, in order to save communication time and communication cost, and the firepower unit will notify the lower-level command and control center within a certain communication range when the target unit is transferred. As shown in picture 2.

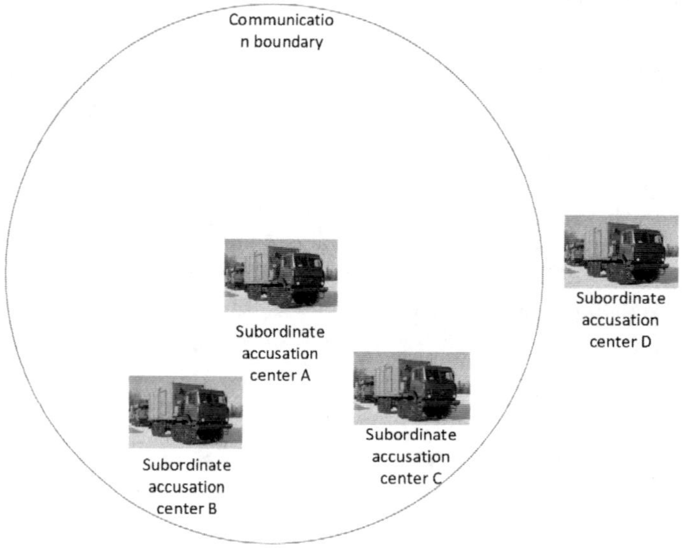

Figure 2. Communication relationship between subordinate units.

$$E = f_E(W, P_s) = \sum_{i=1}^{n} W_i \times \left\{ 1 - \prod_{j=1}^{m} \left[d_{ij} \times (1 - P_{sij}) \right] \right\} \tag{2}$$

The range in which the lower-level command and control center A can communicate is indicated by a blue circle, indicating that the lower-level command and control center A can inform its lower-level command and control center B and the lower-level command and control center C in the lower-level distributed cooperative operations. The set of lower-level command and control centers that satisfy the communication range constraints can be expressed by the formula (2).

$$T = \{x \mid \|P_x - P_A\| \le D_{\text{Communication}},\ x \in Z\} \tag{2}$$

P_x is the location of the lower-level command and control center x, $\|P_x - P_A\|$ is the Euclidean distance from the lower charge center x to A . $D_{\text{Communication}}$ is the communication range, Z is the set of all lower-level command and control centers, and C is the set of all lower-level command and control centers that meet the communication scope constraints.

The subordinate that can attack the target is calculated, as shown in formula (3).

$$N = \{x \in Z \mid l_x \le l_{\max}\ and\ h_{\min} \le h_x \le h_{\max}\ and\ v_x \le v_{\max}\ and\ RCS_x \ge RCS_{\min}\} \tag{3}$$

Among them, N indicates all subordinates that meet the attack limit.

The lower-level command and control center will send the information of the target to the command and control center of the firepower unit that can receive the information and attack the target. The formula is as follows.

$$F=\{x \in Z | x \in (T \cap N)\} \tag{4}$$

F is the collection of subordinate command and control centers that should be sent to.

3.3. *Target priority model*

Target priority refers to the priority order of the current target to confirm the allocation. It can be composed of the following factors: the type of enemy target, the criticality of the incoming target, and the urgency of time. The key types of enemy target types and incoming targets can be judged by the model given above. The urgency of time and the target criticality model are established below.

(1) Urgency of time

The urgency of time refers to the length of time between the current moment and the moment we have to strike. The shorter the time difference, the more urgent we will be on the target. It is expressed by the formula (5).

$$A_{ti}=e^{-\frac{t_i}{t_{\text{definition}}}} \tag{5}$$

A_{ti} is the time urgency function value of the incoming target i, t_i is the time difference between the time of the target i and the current time, and $t_{\text{definition}}$ is the value we set.

(2) Target priority model We use the above mentioned elements to solve the priority model, as shown in equation (6).

$$G_i=A_{ti} \times (\gamma_1 A_{1i} + \gamma_6 A_{6i}) \tag{6}$$

G_i is the priority value of goal i, γ_1 and γ_6 are the weights of the enemy target type factor and the key factors of the incoming target.

(3) Based on dynamic allocation auction algorithm

The auction algorithm is a market mechanism that realizes the rational allocation of resources. The auctioned items are auctioned in an open auction, and finally the one with the highest price is sold. The auction algorithm was first proposed by Bertsekas to solve the single task assignment problem [7]. Xu [8] established a distributed auction algorithm model by considering constraints such as drone capability, and solved the problem by dual decomposition of the problem, and verified the performance and efficiency of the algorithm. There are also many scholars who have analyzed and practiced auction algorithms [9-11].

The auction algorithm has the advantages of simple operation and low complexity. Although these auction algorithms can solve some target allocation problems well, the algorithm is not very suitable for specific situations, and it is easy to conflict. We distribute according to the lower level. The

Journal of Physics: Conference Series **1419** (2019) 012001 doi:10.1088/1742-6596/1419/1/012001

characteristics of the coordinated operation target allocation, and a set of auction algorithms that meet the characteristics of the operation are developed.

This paper proposes an auction algorithm based on dynamic allocation, which can guarantee the high quality of the results and resolve the conflicts. The algorithm is faster. The algorithm steps are as follows:

Step 1: Initialize, sort the priority of the upper-level centralized assignment target by the target priority model.

Step 2: Put the highest priority in the sealed auction algorithm for calculation.

Step 3: According to the target transfer judgment model, it is judged whether the target is not within the attack range of the upper-level centralized allocation of the air defense weapon, or the missile has been allocated by the target with the highest target priority. If yes, go to step 4, if not, keep the original plan and go to step 6.

Step 4: Seal the auction for the target, and send the target information according to the combat information transfer model. Each superior command and control center that receives the information calculates the profit value and killing probability that can be brought by the superior, and transmits it to the auctioneer. The auctioneer assigns a lower-level command and control center with a higher probability of killing in the original target allocation plan (if the accusation center is destroyed, the nearest lower-level command and control center is selected).

Step 5: Select the scheme with the largest absolute return value as the pre-selection scheme.

Step 6: Determine if the killing probability does not exceed our given minimum value or if the selected air defense missile is available, but not enough. If yes, go to step 7, if not, keep the original plan and go to step 8.

Step 7: We reserve a missile with a high probability of killing (if the pre-selection scheme only has one missile to strike, then retain the missile), and the second missile is sealed and auctioned according to step 4.

Step 8: Determine if the minimum kill probability is met. If yes, go to step 9. If not, go to step 10.

Step 9: Find the scheme with the largest absolute return value as a pre-selection scheme if the probability of killing is met. Go to step 11.

Step 10: Select the plan with the highest probability of killing among all the bids as the pre-selection plan.

Step 11: Determine if the target is the target with the lowest priority. If yes, go to step 12. If not, replace the target with the target that the priority is after, and go to step 3.

Step 12: Output and save the assignment result.

Step 13: Determine whether the upper-level centralized allocation is updated. If yes, go to step 1. If not, go to step 2.

The algorithm flow is shown in Figure 3.

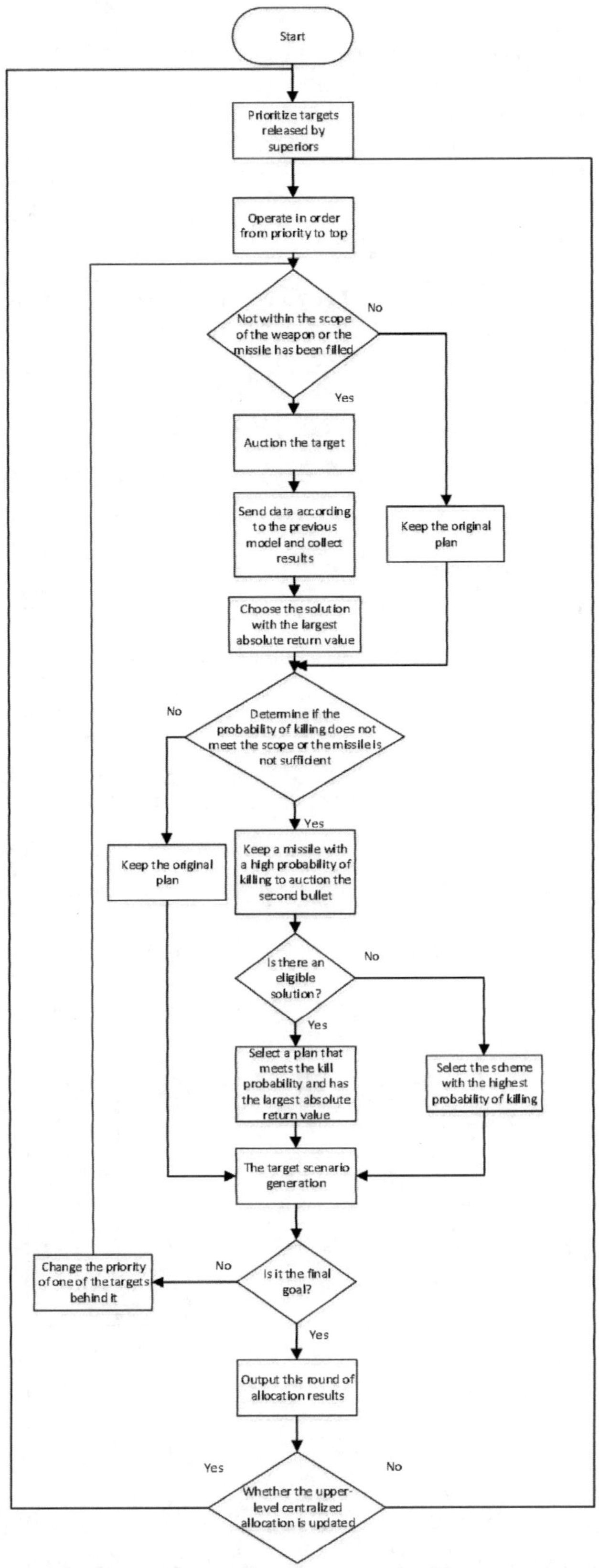

Figure 3. Dynamic allocation auction algorithm flow.

4. Simulation analysis

We use the hypothesis and results of the article [5] to solve the algorithm. We use the result of a centralized assignment of the superior as the initial state of the simulation. The solution of the upper-level centralized target allocation is shown in Table 1.

Table 1. Solution of upper-level centralized target allocation.

Combat unit	Combat unit A	Combat unit B	Combat unit C	Combat unit D
Attack target	10	1、6、9	3、7、8	2、4、5

After the target is distributed, many targets are strongly maneuvered. In this paper, the maneuvering situation is expressed as follows according to the maneuvered time node:

Table 2. shows the situation of the target maneuver.

Target	Time frame	Maneuver
1	0.5	The airplane's flight direction is rotated counterclockwise by 37 degrees
2	0.9	The flight speed is slowed to 250m/s, the flight direction is unchanged, and the flight altitude is increased by 0.3km.
7	1.5	Rotate 5 degrees clockwise for other tasks
8	1.5	Rotate 5 degrees clockwise for other tasks

Set the parameters:

$D_{\text{Communication}}=50\text{km}$, $t_{\text{definition}}=10$, $\gamma_1=\gamma_6=0.5$, The minimum damage probability is 70%.

The problem is solved by the dynamic allocation auction algorithm designed in this paper. The solution results are shown in Table 3.

Table 3. Solution result allocation table.

	Target 1	Target 2	Target 3	Target 4	Target 5	Target 6	Target 7	Target 8	Target 9	Target 10
Combat unit A	0	0	1	0	0	0	0	0	0	2
Combat unit B	0	0	0	1	1	2	0	0	1	0
Combat unit C	0	0	1	1	1	0	1	1	1	0
Combat unit D	2	2	0	0	0	0	1	1	0	0

We compare it with the auction algorithm and the greedy algorithm. The results are shown in Table 4.

Table 4. Comparison of auction algorithm and greedy algorithm.

	Dynamic allocation auction algorithm	Auction algorithm	Greedy algorithm
Time spent in a single algorithm (ms)	186	231	135
Absolute return value	1.568	1.452	1.339
The number of targets that did not reach the minimum probability of killing	0	1	2

The absolute return value of each algorithm at each moment is shown in Figure 4.

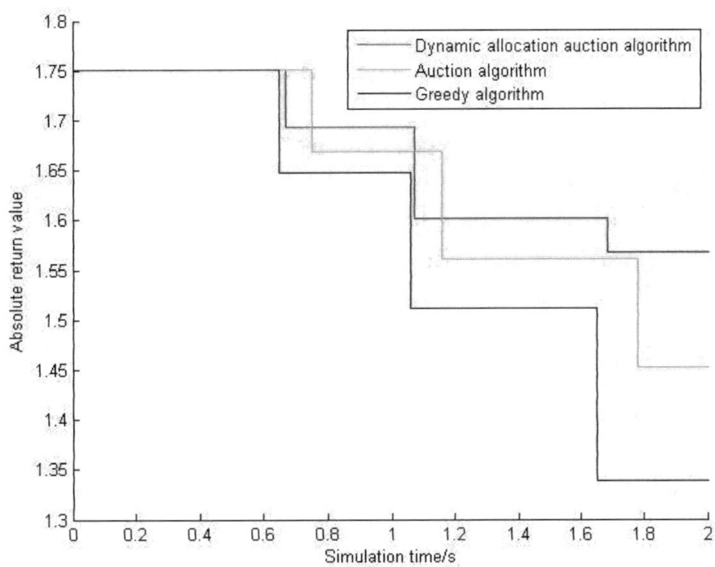

Figure 4. Comparison of the absolute returns of the three algorithms.

It can be seen from the above comparison that the dynamic allocation auction algorithm has the largest absolute return value compared with the auction algorithm and the greedy algorithm, and can also well make the probability of the target target damage to the standard. Although the algorithm is not as good as the greedy algorithm, it is also very fast, which can meet our needs and greatly improve the performance.

5. Summary

Based on the understanding of the auction algorithm, this paper proposes a dynamic allocation auction algorithm, which can solve the problem of insufficient ability to solve dynamic targets in the centralized allocation of superiors. It can achieve the absolute maximum return value, and can reach the target damage probability limit to the maximum extent. Although it takes a long time, it is still not particularly time-saving and can be further improved.

References

[1] Jia J. Research on multi-platform air defense collaborative task assignment. Beijing Institute of Technology, 2016.
[2] Lei M, Xie B. Air defense firepower distribution based on cloud genetic algorithm. Journal of System Simulation, 2018, 30(09): 3533-3537.
[3] Sun HW, Xie XF, Sun T, Pang W. Solution of Air Defense Firepower Optimization Allocation Model for Improved Cuckoo Search Algorithm.Acta Armamentarii, 2019, 40(01): 189-197.
[4] Xu H,Xing QH,Wang W. Air-defense and anti-missile firepower distribution based on fuzzy multi-objective programming[J].System Engineering and Electronics , 2018,40(03):563-570.
[5] Jiang HB. Research on Target Assignment Method Based on Ant Colony-Fish Group Algorithm. 2019 International Conference on Modeling, Simulation, Optimization and Algorithm, 2019.11(in press).
[6] Shang Y . Resilient Multiscale Coordination Control against Adversarial Nodes[J]. Energies, 2018, 11(7).
[7] Bertsekas D P. The auction algorithm: A distributed relaxation method for the assignment problem. Annals of Operations Research, 1988; (14)， 105 — 123.
[8] Xu K, Gong H, Qin XL, Zhang BY. Multi-UAV Grouping Task Assignment Based on Distributed Auction Algorithm.Information and Control,2018,47(03):341-346.
[9] Vartiainen H. Auction design without commitment. Journal of the European Economic Association, 2013, 11(2): 316-342.
[10] Allevi E, Gnudi A, Konnov I V. Combined methods for dynamic spatial auction market models. Optimization and Engineering, 2012, 13(3): 401-416.
[11] Edalat N, Tham C K, Xiao W. An auction-based strategy for distributed task allocation in wireless sensor networks[J]. Computer Communications, 2012, 35(8):916-928.

Research on target assignment method based on ant colony-fish group algorithm

Haobo Jiang[1], Song Li[2], Chi Lin[3], Chuang Wang[1], Guilin Tian[1], Siyuan Wang[1], Kaiqi Zhong[4] and Jiakun Li[1]

[1]Graduate School, Air Force Engineering University, Xi'an, Shaanxi,China
[2]Air and Missile Defense College, Air Force Engineering University, Xian, Shaanxi, China
[3]Unit 94622, Quanzhou, Fujian, China
[4]Unit 95865, Beijing, Beijing,China

E-mail: 395049138@qq.com

Abstract. Target assignment problem has always been the focus of scholars. The solution of target problem belongs to classical mathematical problem. This paper combines ant colony algorithm with artificial fish swarm algorithm by understanding intelligent algorithm. The concept of crowding degree is introduced into fish swarm algorithm. In the early stage of the algorithm, the dispersion is large, avoiding premature maturity. In the later stage, the convergence speed of the algorithm is faster and the time is saved. Compared with other algorithms, the hybrid algorithm has the characteristics of strong optimization ability and fast speed.

1. Introduction

Target assignment problem is essentially the problem of resource allocation. When multiple targets require multiple resources, and each target and resource have a many-to-many relationship, the whole problem will become complicated. How to allocate resources rationally to maximize benefits is a problem we have been trying to solve.

The method of target assignment has been studied by many scholars. Wang [1] improved the genetic algorithm to solve the problem of multi-UAV cooperative combat target assignment in complex environment, improved the population fitness by using heuristic information and random generation, and added penalty mechanism to investigate the scheme that did not meet the conditions. And then inherit the genetics, inserting excellent individuals into the original population and enriching the population. Finally, the feasibility was confirmed by experimental comparison. Yang [2] proposed a new genetic algorithm by introducing the theory of visual fuzzy set to solve the problem of air defense combat target assignment which has slow convergence and easily premature. Wu [3] inserted simulated annealing algorithm into the crossover operator of genetic algorithm, which can not only maintain the breadth of the algorithm, but also enhance the depth of the search. Gao [4] generated random population and artificial population, carried out cross mutation evolution, used new immigration operator to exchange information among populations. Finally, by comparing with artificial immune algorithm, it is found that the algorithm can improve the search efficiency and get the optimal scheme. Sun [5] proposed an improved Cuckoo search algorithm. The simulation results show that the algorithm can effectively balance the relationship between convergence speed and global exploration ability.

All of the above methods are centralized allocation methods, which have the following advantages.

Since the command center has basically fully understood the information of the whole problem, a centralized understanding can be used to obtain a global understanding.

Because superiors and subordinates belong to the relationship between commanded and subordinate, decision-making by superiors can avoid conflicts very well.

However, these algorithms generally have a slow convergence rate, the algorithm is easy to fall into the local optimum, and there are also phenomena such as premature aging and early lag, which greatly affects the speed and quality of decision making.

2. Description of the problem

This paper expounds this kind of problem based on the air defense combat scenario. In the air defense combat, the superior command center acts as the center of command and decision, and collects the incoming target information uploaded from the sensor to the command and control center, It is responsible for collecting the information of attacking targets uploaded from sensors to command and control center, and solves them through the target allocation mechanism, and transmits the results of target allocation to each of subordinate command centers, and they will send the command to the air defense weapon equipment. Its battle process is shown in Figure 1.

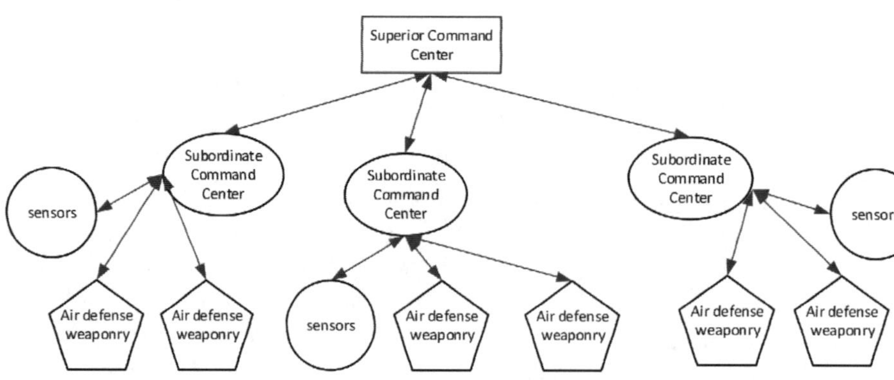

Figure 1. Centralized air defense decision-making process.

Such a decision-making process will bring great communication pressure to the superior command center, and the communication speed will be very slow, which will reduce the decision-making time of the superior command center. The speed and quality of target allocation will directly affect the whole operation, so we need a more reasonable problem solving method in the allocation speed and the quality of the allocation results.

3. Model establishment

3.1. Objective function

In air defense operations, air defense weapons must strike against air strike targets. The significance of target allocation is to exchange a small amount of cost for the maximum benefit. Therefore, we construct its objective function.

$$Y = \mathrm{E} - \mathrm{C} \tag{1}$$

Among them, E is the total return value, which refers to the income from the distribution and attack of the target, and C is the total cost, which refers to the cost of the implementation of the distribution plan, including the cost of the missile, etc. Y is the total absolute return value, which refers to the absolute return of the entire target allocation. Our goal is to have the largest absolute return value. This is also an important criterion for us to measure the effect of target allocation.

The return value can be solved using equation (2).

$$E = f_E(W, P_s) = \sum_{i=1}^{n} W_i \times \left\{ 1 - \prod_{j=1}^{m} \left[d_{ij} \times (1 - P_{sij}) \right] \right\} \tag{2}$$

d_{ij} is the result of the allocation of the air defense weapon j against the target i. The result can only be represented by 0 or 1. 0 is not assigned, 1 is assigned; the W_i is the threat of attacking target i; P_{sij} is the probability of destruction of target i.

It total cost is the sum of the costs of each part. It can be expressed by the formula (3).

$$C = \sum_{i=1}^{m} \sum_{j=1}^{n} (d_{ij} \times \mathrm{c}_j) \tag{3}$$

3.2. Constraints

3.2.1. Space constraintsSpatial

Constraints refer to the fact that certain targets of the target relative to the air defense weaponry are within a certain range to be hit by a certain probability. Formula (4) is a constraint.

$$\begin{aligned} & l_i \le l_{\max} \\ & h_{\min} \le h_i \le h_{\max} \\ & v_i \le v_{\max} \\ & RCS_i \ge RCS_{\min} \end{aligned} \tag{4}$$

Where l_i is the route shortcut of target i, l_{max} is the maximum route shortcut that the target can intercept for air defense weapon equipment; h_i is the height of target i, h_{min} and h_{max} is the minimum height and maximum height that can be intercepted by air defense weapon equipment respectively; v_i is the speed of target , v_{max} is the maximum speed that can be intercepted for air defense weapons. RCS_i is the target radar cross-sectional area , RCS_{min} is the minimum radar cross-sectional area at which the target can be struck.

3.2.2. Target assignment constraint

Target Assignment constraint refers to the target allocation should be carried out according to certain constraint, which are mainly divided into air defense weapon constraint, allocation of target constraint.

Air defense weapon constraint refer to the limitation of the number of air defense and anti-missile missiles and fire passages. We simplified the problem.We assume that each target will be attacked by two missiles with same air defense weapon, which can be expressed by formula (5) (6).

$$\sum_{i=1}^{m} d_{ij} \times 2 \leq n_{\text{missile j}} \tag{5}$$

$$\sum_{i=1}^{m} d_{ij} \times 2 \leq n_{\text{channel j}} \tag{6}$$

$n_{\text{weapon j}}$ is the number of missiles carried by air defense weapon j, and $n_{\text{channel j}}$ is the number of fire channels currently available by air defense weapon j.

We assume that our missiles are sufficient. Target assignment constraint means that each of our targets is allocated at least one air defense weapon to attack, which can be expressed by formula (7).

$$\sum_{j=1}^{n} d_{ij} \geq 1 \tag{7}$$

3.2.3 Time constraint

Time constraint refers to the fact that the command center needs to make decisions on the target within the time from the detection to the launch area. distribute the target reasonably, and pass the distribution result to each weapon. It can be expressed by the formula (8).

$$t_{\text{decision}} \leq t_{\max} \tag{8}$$

$t_{decision}$ represents the decision-making time and t_{max} represents the upper limit of the decision-making time.

4. Ant-fish swarm hybrid algorithm

4.1 Overview of Artificial Fish Swarm and Ant Colony Algorithms

Hybrid algorithms have been studied by many scholars, and many achievements have been made in other fields[6][7]. This paper proposes a hybrid algorithm based on ant colony algorithm and fish colony algorithm, which is introduced below.

4.1.1 Artificial Fish Swarm Algorithm

Artificial fish swarm algorithm was proposed by Li [8] in 2002. Through observation, it was found that the most dense area of fish activity is the place where nutrients are the most abundant in the water area, and then the optimization algorithm of artificial fish swarm algorithm is made.

We assume that the number of fish is A, The current state of the fish is $X = (x_1, x_2, \ldots x_n)$, Each parameter in the state is a optimization variable, We set the nutrient concentration of the fish at its current location as $Y = f_{food}(X)$, The distance between the fish is $j_{ij} = \|X_i - X_j\|$, We assume that the fish's perception range is a spherical area with a radius of $Perceive$, The step size of the fish is $Step$, The algorithm mainly considers three behaviors of fish: foraging, clustering and rear-end. These three behaviors are briefly described below [9].

(1) foraging behavior

Let's assume that our current fish state is X_i, we choose a state of X_j within its range of perception. Expressed by the formula (9).

$$X_j = X_i + Perceive \times Rand \tag{9}$$

Where $Rand$ denotes 0 to 1 random value, Comparing the objective function values Y_i and Y_j of the two, if we are solving the maximum value, when $Y_i < Y_j$, then we move one step to the current range, which can be expressed by formula (10).

$$X_i^{t+1} = X_i^t + \frac{X_j - X_i^t}{\| X_j - X_i^t \|} \times Perceive \tag{10}$$

If $Y_i \geq Y_j$, Then the state is randomly re-selected, and if the appropriate region is not found after repeated repetitions, then the state is randomly moved one step.

(2) Clustering Behavior

Fish will choose the way of group foraging and avoiding attack, which is the survival mode formed by the long-term survival of the fittest. Fish will randomly explore the number of fish in the perceptual range n_k, and find a new location center X_k, expressed by the formula (11).

$$X_k = X_i + Perceive \times Rand \tag{11}$$

The new location center objective function Y_k is compared with the current regional objective function Y_i .If the new objective function is better than the current objective function and is not crowded ($Y_k / n_k > CY_i$), it moves one step to the current location, expressed by the formula (12).

$$X_i^{t+1} = X_i^t + \frac{X_k - X_i^t}{\| X_k - X_i^t \|} \times Perceive \tag{12}$$

Otherwise foraging.

(3)Rear-end behavior

When a few fish find food, the fish in the vicinity and even further will follow. The fish explores the position of the best fish around X_l, and the number of fish around is n_l, when the function value Y_l of the surrounding optimal position is greater than the current position function value Y_i and not crowded($Y_l / n_l > CY_i$).the fish will move one step to it, expressed by the formula (13).

$$X_i^{t+1} = X_i^t + \frac{X_l - X_i^t}{\| X_l - X_i^t \|} \times Perceive \tag{13}$$

Otherwise foraging.

4.1.2 Overview of Ant Colony Algorithm

Ant colony algorithm was proposed by M. Dorigo [10] in 1991 based on the foraging behavior of ants. The main principle of this method is that in order to find a suitable route quickly, the ants search the routes randomly. When the ants encounter the intersection, they will choose the path through the pheromone left by the previous ants. The more pheromones there are, the more likely they will choose the route. The ants will also leave pheromone when they pass through this path, No. Passing pheromones will evaporate over time, and the shortest path pheromones will become more and more, which will form positive feedback, which is the key [11].

Next, we analyze the model.

Suppose that at a certain moment, there is an ant o on the position i, and the probability of selecting the next position as j is as shown in the formula (14).

$$p_{ij}^{o}(t_1)=\begin{cases}\dfrac{\left[\tau_{ij}(t_1)\right]^{\alpha}\left[\eta_{ij}(t_1)\right]^{\beta}}{\sum\limits_{k\in \text{allow}_o}\left[\tau_{ik}(t_1)\right]^{\alpha}\left[\eta_{ik}(t_1)\right]^{\beta}}, \text{j} \in \text{allow}_o \\ 0, \text{j} \notin \text{allow}_o\end{cases} \quad (14)$$

$\tau_{ij}(t_1)$ denotes the current pheromone value from position i to position j at time t_1, and $\eta_{ij}(t_1)$ is a heuristic function, which denotes the expected value from position i to position j at time t_1. $\eta_{ij}(t_1)$ is inversely proportional to the distance between two points, and can be solved by formula (15).

$$\eta_{ij}(t_1)=\frac{1}{j_{ij}} \quad (15)$$

α denotes the importance of pheromones, β denotes the importance of heuristic functions, and allow_o denotes the set of positions that ant o can choose. The pheromones of each location are updated every other unit of time.

$$\tau_{ij}(t_1+1)=(1-\rho)\tau_{ij}(t_1)+\Delta\tau_{ij}(t_1) \quad (16)$$

$$\Delta\tau_{ij}(t_1)=\sum_{o=1}^{q}\Delta\tau_{ij}^{o}(t_1) \quad (17)$$

Q is the total number of ants, $\Delta\tau_{ij}^{o}(t_1)$ refers to the pheromone left by ant o on the path i to j at that time, ρ is the pheromone volatilization coefficient.

4.2 Hybrid algorithm flow

The ant colony algorithm has the advantages of fast solution speed and high efficiency, but its shortcomings are also obvious, and it is easy to fall into local optimum. The artificial fish swarm algorithm has strong global search ability. The main feature of the hybrid algorithm is that it can combine the advantage of two algorithms,and avoiding the shortcomings of the two algorithms, and make full use of their strengths and avoid their weaknesses [12].

Here we embed the congestion directly into the iterative process of the ant colony algorithm, so that when the ant chooses a route with a high congestion degree, it will give up the route and select other routes. Congestion is expressed by formula (18).

$$C_{ij}=\frac{2\tau_{ij}(t_1)}{\sum\limits_{i\neq j}\tau_{ij}(t_1)} \quad (18)$$

In order to expand the search range of the algorithm at the initial time and focus on the search range at the later stage of the algorithm, we calculate the congestion threshold by formula (19).

Journal of Physics: Conference Series **1419** (2019) 012002 doi:10.1088/1742-6596/1419/1/012002

$$\sigma(t) = 1 - e \tag{19}$$

Where b is the congestion threshold coefficient.

The algorithm flow is as follows:

(1) Initialize the parameters, set the initial time, the number of ants, the pheromone, the congestion threshold, the importance of the pheromone, the importance of the heuristic function, and the maximum number of iterations. Put the ants in

(2) According to the pheromone selection path, the selection probability is as shown in formula 13.

(3) After the ant selects the path, the congestion degree detection is first performed. If the congestion degree is greater than the threshold, the ant will select other available paths, and if not crowded, transfer to the route.

(4) Repeat steps 2 and 3 until the ant's selection meets the requirements or the number of iterations reaches a maximum.

(5) The output ends.

The algorithm flow chart is shown in Figure 2.

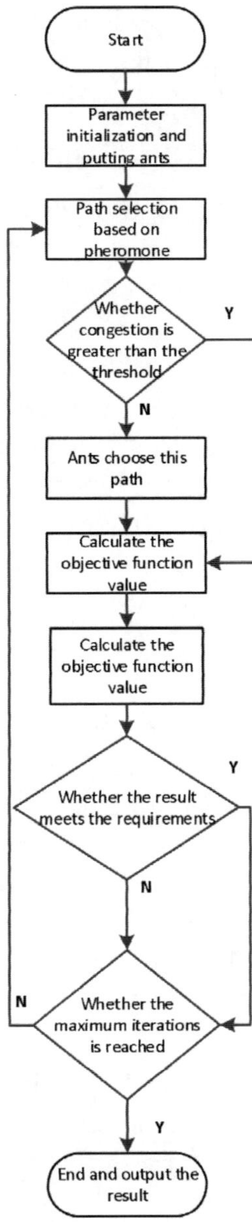

Figure 2. Hybrid algorithm flow chart.

5. Simulation

Assume that the superior command center commands four firepower units, each firepower unit has three launch vehicles, each of which has two available firepower passages, and two missiles. Each firepower unit has only one type of missile, at a certain moment. As for the 10 targets, the target parameters of the 10 incoming targets are shown in Table 1.

Table 1. Target parameter table.

	target1	target2	target3	target4	target5	target6	target7	target8	target9	target10
classification	aircraft	aircraft	UAV	UAV	UAV	UAV	UAV	UAV	UAV	Missile
velocity X (m/s)	116.3	251.8	146.8	146.7	146.8	146.8	146.7	146.8	146.6	1116.6
velocity Y (m/s)	232.6	97.4	104.2	104.2	104.3	104.1	104.2	104.1	104.1	1001.6
velocity Z (m/s)	-0.3	-0.3	-0.2	-0.3	-0.2	-0.2	-0.2	-0.2	-0.2	-0.2
velocity X (km)	42.0	62.8	62.6	63.0	62.7	62.9	62.7	62.9	62.6	107.2
velocity Y (km)	60.0	38.1	55.3	55.4	55.6	55.4	55.6	55.4	55.3	100.8
velocity Z (km)	12.4	13.7	10.2	10.3	10.3	10.5	10.4	10.4	10.6	16.2

The threat degree of the target relative to the combat unit is shown in Table 2.

Table 2. Threat Scale.

	aircraft 1	aircraft 2	UAV 3	UAV 4	UAV 5	UAV 6	UAV 7	UAV 8	UAV 9	Missile 10
CU A	0.670	0.513	0.475	0.449	0.447	0.444	0.455	0.460	0.445	0.974
CU B	0.751	0.574	0.501	0.481	0.475	0.473	0.491	0.497	0.469	0.810
CU C	0.722	0.629	0.671	0.601	0.592	0.571	0.612	0.623	0.584	0.916
CU D	0.604	0.742	0.574	0.523	0.514	0.501	0.539	0.547	0.510	0.717

Each missile launch has its own cost and interception probability, as shown in Table 3.

Table 3. Cost and interception probability of each missile.

	antiaircraft missile A	antiaircraft missile B	antiaircraft missile C	antiaircraft missile D
Cost per missile	0.2	0.16	0.12	0.17
Interception probability	0.7	0.5	0.4	0.6

The location of our combat units is shown in table 4

Table 4. Location of our combat unit.

	Combat unit A	Combat unit B	Combat unit C	Combat unit D
Postion X (km)	0	23	20	40
Position Y (km)	0	14	30	25

We assume that the combat capability parameters of each combat unit are the same, and the same combat unit has different capability parameters for different targets, as shown in table 5.

Table 5. Combat unit operational capability parameter table.

	UAV	aircraft	missile
Height limit (km)	15	15	20
Low limit (km)	1	1	1
Far limit (km)	30	30	40
Close limit (km)	6	6	8
Maximum height and low angle (degrees)	60	60	60
Maximum shortcut (km)	15	20	25

We simplify the problem, and each incoming target is only attacked by two missiles of the same firepower unit. Solving it according to the method described above yields the following results.

We set the various parameters as follows.

$\alpha=1$, $\beta=1$, $\rho=0.5$, m=20, Q=200, c=1

We use hybrid algorithm, ant colony algorithm and artificial fish swarm algorithm to compare and simulate with matlab. The simulation results are shown in Fig. 3.

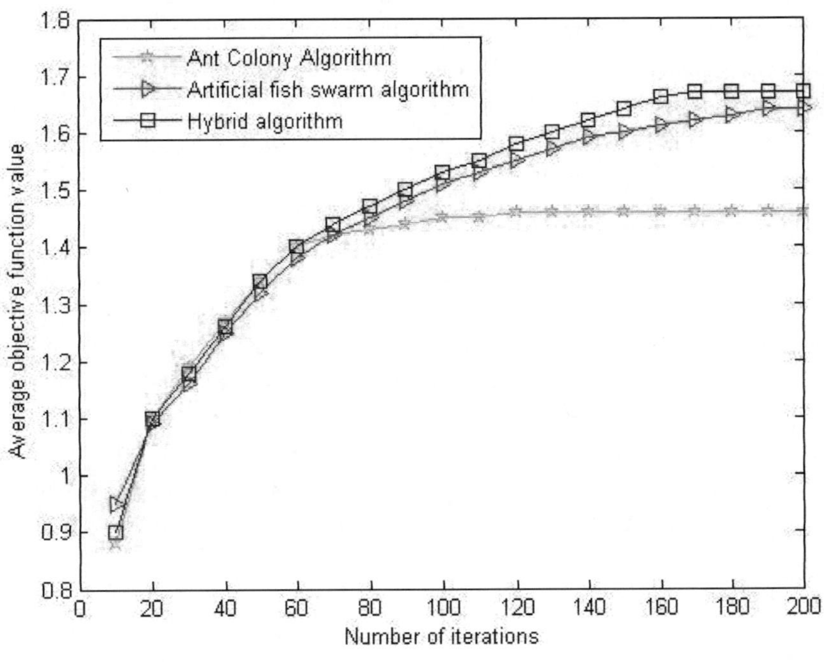

Figure 3. Comparison of the average results of the three algorithms.

By comparing the algorithms, we can draw the following conclusions. Among the three algorithms, ant colony algorithm converges fastest and artificial fish swarm algorithm converges slowest. The results of hybrid algorithm and artificial fish swarm algorithm are far greater than those of ant colony algorithm. Thus, the performance of the hybrid algorithm is better than that of the other two algorithms.

6. Summary

Based on the understanding of ant colony algorithm, this paper quotes the crowding rule of artificial fish swarm algorithm on the basis of ant colony algorithm. Through experiments, it can be found that the crowding rule can avoid the premature problem of ant colony algorithm, and also can converge quickly in the later period of the algorithm. The simulation results show that the algorithm has strong applicability. Although the hybrid algorithm has a good ability, its algorithm speed does not exceed the ant colony algorithm. In the future, it can continue to optimize the algorithm's optimization ability and speed, and propose a new algorithm to solve such problems.

References

[1] Wang QH, Wan G, Chai Z and Li DF. Multiple targets assignment of multiple UAV's cooperation based on improved genetic algorithm. Computer application research, 2018,35(09):2597-2601.

[2] Yang JSi, Li J Wang Y, Wei XH, Wen Tong. Solving Intuitionistic Fuzzy Multi-objection Programming by Improved Genetic Algorithm.Journal of Air Force Engineering University(Natural Science Edition),2017,18(04):93-98.

[3] Wu KH, Zhan SX. Optimization for Target Assignment in Fire Strike Based on Distributed Genetic Simulated Annealing Algorithm. Fire Control & Command Control,2016,41(03):89-92.

[4] Gao YY, Chen SY, Yu MJ, Hai J, Fang RY.A Multi-machine Cooperative Air Combat Target Assignment Method Based on Improved Artificial Immune Algorithm. Journal of Northwestern Polytechnical University,2019,37(02):354-360.

[5] Sun HW, Xie XF, Sun T, Pang W. Improved Cuckoo Search Algorithm for Solving Antiaircraft Weapon-target Optimal Assignment Model. Journal of Military Engineering, 2019,40(01):189-197.

[6] Yilun Shang. "Consensus in averager-copier-voter networks of moving dynamical agents." Chaos: An Interdisciplinary Journal of Nonlinear Science27.2(2017):023116.

[7] Yilun Shang. Hybrid consensus for averager–copier–voter networks with non-rational agents. Chaos, Solitons and Fractals: the interdisciplinary journal of Nonlinear Science, and Nonequilibrium and Complex Phenomena,2018,110.

[8] Li XL, Shao ZJ, Qian JX. An optimizing method based on autonomous animats: fish -swarm algorithm. Theory and practice of system engineering, 2002, 22 (11):32-38.

[9] Li SY, Chen YQ, Li Y. Ant Colony Algorithm and Its Application. Harbin Institute of Technology Press, 2004.

[10] Dorigo M, Maniezzo V, Colorni A. Positive Feedback as a Search Strategy. Technical Report, 1991:91-016.

[11] Shao SJ. Research on weapon target allocation based on intelligent algorithm. Harbin Engineering University, 2019.

[12] Lu SF, Ma K. Application of ant colony-fish group hybrid algorithm in batch

scheduling of differential jobs. Computer System Application, 2018, 27(1):162–167.

Modeling static microstructure of shape memory alloy via Legendre wavelets collocation method

Xuan He[1,2], Haoyuan Du[1], Zixiang Ying[1], Linxiang Wang[1]* and Roderick Melnik[2]

[1]State Key Laboratory of Fluid Power and Mechatronic Systems, Zhejiang University, 310027, Hangzhou, China
[2]MS2Discovery Interdisciplinary Research Institute, Wilfrid Laurier University, Waterloo, ON N2L 3L5, Canada

E-mail: xuanhe@zju.edu.cn

Abstract. The static microstructures of square-to-rectangular phase transformations are modeled in the current paper. Governing equations are described by the static Navier equations. The analysis of the simulation has been finished via Legendre wavelets collocation method. Microstructure evolutions of both 2D and 1D shape memory alloy (SMA) are capture well with a certain boundary condition and given mechanical loadings.

1. Introduction

Shape memory alloys (SMA) are widely applied in industries due to its unique thermo-dynamical properties [1-3]. One of the famous effects of SMA, which is called the shape memory effect, is that the strain of SMA occurred in the martensite phase at low temperature will be recovered in the austenite phase at high temperature. Another famous effect is about elasticity beyond common sense. It is called pseudoelasticity that some phase transformations (austenite → martensite) happen under the external mechanical loadings with a certain range of temperature, and the transformation, as well as strain it caused, will be reversed without the loadings. These two effects are the major topic of the researcher's work. Many related studies are ultimately attributed to the researches of microstructure. The better understanding of microstructure we got, the better application of SMA we did.

One popular way to study microstructure is based on Landau free energy minimization [4-6], and such researches are attributed to the numerical solution of partial differential equations. Legendre wavelets method (LWM) is an emerging spectral method to solve such problems. LWM combines the advantage of both spectral method and wavelet method, and it is good at stability, high precision as well as good efficiency [7].

In the current paper, the static microstructure of SMA model is simulated via Legendre wavelets method at a certain temperature. The analysis of phase combinations in these situations is presented as well. The current paper is organized as follows. An introduction of Legendre wavelets is given in section.2. The model of SMA is presented in section.3. Finally, the simulations and analysis are discussed in section.4 and section.5. Concluding is given in section.6.

2. A mathematical model for SMA

Square-to-rectangular martensitic phase transformations are considered in the current paper, which can be treated as a simplified 2D analog of the cubic-to-tetragonal or tetragonal-to-orthorhombic phase transformations [8]. Considering the fundamental laws, energy balance and so on, the basic functions used in the current paper are illustrated by static Navier equations as follows[8]:

$$\frac{\partial\sigma_{11}}{\partial x}+\frac{\partial\sigma_{21}}{\partial y}+f_x=0, \tag{1.a}$$

$$\frac{\partial\sigma_{12}}{\partial x}+\frac{\partial\sigma_{22}}{\partial y}+f_y=0, \tag{1.b}$$

where σ_{ij} is the stress tensor, and f_i is the mechanical loading in x_i direction.
The stress tensor can be given by the thermos-dynamical equilibrium conditions:

$$\sigma_{ij}=\frac{\delta\Psi(\Delta\theta,\varepsilon)}{\delta\varepsilon_{ij}}, \tag{2}$$

where ε_{ij} is the Cauchy-Lagrangian strain tensor which is given as follows:

$$\varepsilon_{ij}(x,t)=\frac{\frac{\partial u_i(x,t)}{\partial x_j}+\frac{\partial u_j(x,t)}{\partial x_i}}{2}, \tag{3}$$

where u_i is the displacement.
The free energy function $\Psi(\Delta\theta,\varepsilon)$ can be obtained by the Landau theory:

$$\Psi(\Delta\theta,\varepsilon)=\frac{a_1}{2}e_1^2+\frac{a_3}{2}e_3^2+F_l+F_g, \tag{4.a}$$

$$F_l=\frac{a_2}{2}\Delta\theta e_2^2-\frac{a_4}{4}e_2^4+\frac{a_6}{6}e_2^6, \tag{4.b}$$

$$F_g=\frac{k_g}{2}\left[\left(\frac{\partial e_2}{\partial x}\right)^2+\left(\frac{\partial e_2}{\partial y}\right)^2\right], \tag{4.c}$$

where e_1 , e_2 and e_3 are the hydrostatic, deviatoric and shear strain respectively defined as $e_1=\frac{\varepsilon_{xx}+\varepsilon_{yy}}{\sqrt{2}}, e_2=\frac{\varepsilon_{xx}-\varepsilon_{yy}}{\sqrt{2}}, e_3=\frac{\varepsilon_{xy}+\varepsilon_{yx}}{2}$.

3. Numerical algorithm via Legendre wavelets

The simulation of nonlinear wave propagation based on a coupled system of PDEs is not a trivial task to solve the 2D situation. The choice of the algorithm will substantially affect the accuracy of the result. In the current paper, the Legendre wavelets method (LWM) is employed to simulation this hard task.

The wavelet methods can reduce calculation cost-effectively. The Legendre wavelets method(LWM) is originated from applying wavelets method to the mother function Legendre polynomials. The Legendre wavelets family retains the following form[9]:

$$\psi_{n_1,m_1,n_2,m_2}(x,y)=\begin{cases}A\cdot P_{m_1}(n_1^{'})\cdot P_{m_2}(n_2^{'}), & \frac{n_1-1}{2^{k_1-1}}\le x\le\frac{n_1}{2^{k_1-1}},\frac{n_2-1}{2^{k_2-1}}\le y\le\frac{n_2}{2^{k_2-1}}\\ 0, & \text{otherwise}\end{cases} \tag{5}$$

where $n_i^{'}=2^{k_i}x-2n_i+1$ $m_i=0,1,...,M-1$, $n_i=1,2,3,...,2^{k-1}$, $A=\sqrt{(m_1+\frac{1}{2})(m_2+\frac{1}{2})2^{\frac{k_1+k_2}{2}}}$ is the coefficient for orthogonality. P_{m_i} are Legendre polynomials with order m_i on interval [-1,1].
Function $f(x,y)$ defined on interval $[0, 1]\times[0, 1]$ can be expanded as Legendre wavelets as follows:

$$f(x)=\sum_{n_1=1}^{\infty}\sum_{m_1=0}^{\infty}\sum_{n_2=0}^{\infty}\sum_{m_2=0}^{\infty}c_{n_1,m_1,n_2,m_2}\psi_{n_1,m_1,n_2,m_2}(x)=C^T\Psi(x). \tag{7.a}$$

here $c_{n,m}$ denoted the inner product of $f(x)$ and $\psi_{n,m}(x)$. If Eq.(5) is truncated, it can be rewritten as

$$f(x)=\sum_{n_1=1}^{2^{k-1}}\sum_{n_2=1}^{2^{k-1}}\sum_{m_1=0}^{M-1}\sum_{m_2=0}^{M-1}c_{n_1,m_1,n_2,m_2}\psi_{n_1,m_1,n_2,m_2}(x)=C^T\Psi(x), \tag{7.b}$$

The vector C in Eq. (7) is called the spectral space coefficient, and C is the only data needed in the following calculations.

3.1. Operational matrix of derivatives

In the present paper, due to the properties of Legendre polynomials, the calculation of derivatives in this method can be obtained in the matrix form [10]:

$$\frac{d\Psi(x,y)}{dx}=D_x\Psi(x,y), \tag{8}$$

where D_x is a $2^kM2^{k'}M'\times 2^kM2^{k'}M'$ derivative matrix defined as follows:

$$D_x = \begin{pmatrix} F & 0 & \cdots & 0 \\ 0 & F & & 0 \\ \vdots & & \ddots & \vdots \\ 0 & 0 & \cdots & F \end{pmatrix}, \quad (9)$$

where F is a $M2^{k'}M' \times M2^{k'}M'$ matrix which is defined as follow:

$$F_{r,s} = \begin{cases} 2^{k+1}\sqrt{(2r-1)(2s-1)}, & r = 2,\dots,M,\ S = 1,\dots,r-1, \text{ and } (r+s) \text{ odd}, \\ 0, & \text{otherwise.} \end{cases} \quad (10)$$

$$\frac{d\Psi(x,y)}{dy} = D_y\Psi(x,y),$$

$$D_y = \begin{pmatrix} D & 0 & \cdots & 0 \\ 0 & D & & 0 \\ \vdots & & \ddots & \vdots \\ 0 & 0 & \cdots & D \end{pmatrix}$$

$$D = \begin{pmatrix} FF & 0 & \cdots & 0 \\ 0 & FF & & 0 \\ \vdots & & \ddots & \vdots \\ 0 & 0 & \cdots & FF \end{pmatrix}$$

where D is a $M' \times M'$ matrix. FF is a $M' \times M'$ matrix which is defined as follows:

$$FF_{r,s} = \begin{cases} 2^{k+1}\sqrt{(2r-1)(2s-1)}, & r = 2,\dots,M,\ S = 1,\dots,r-1, \text{ and } (r+s) \text{ odd}, \\ 0, & \text{otherwise.} \end{cases}$$

The calculation of derivatives in the spectral space can be easily obtained by multiply spectral space coefficient C and derivative matrix D_x and D_y.

3.2. Numerical algorithm

Above all, the governing system is described as follows:

$$\frac{\partial\sigma_{11}}{\partial x} + \frac{\partial\sigma_{21}}{\partial y} + f_x = 0, \quad (11.a)$$

$$\frac{\partial\sigma_{12}}{\partial x} + \frac{\partial\sigma_{22}}{\partial y} + f_y = 0, \quad (11.b)$$

$$\sigma_{11} = \frac{\sqrt{2}}{2}\left(a_1e_1 + a_2\Delta\theta e_2 - a_4e_2^3 + a_6e_2^5 - k_g\nabla^2e_2\right), \quad (11.c)$$

$$\sigma_{12} = \sigma_{21} = \frac{1}{2}a_3e_3, \quad (11.d)$$

$$\sigma_{22} = \frac{\sqrt{2}}{2}\left(a_1e_1 - a_2\Delta\theta e_2 + a_4e_2^3 - a_6e_2^5 + k_g\nabla^2e_2\right), \quad (11.e)$$

The boundary condition can be described as :

$u_i = 0$, $\frac{\partial u_j}{\partial x_i} = 0$ at the boundary area of x_i direction.

Moving to the time discretization procedure, system(11) can be rewritten as follows:

$$LU + N(x,y,U) = 0, \quad (12)$$

where L is the matrix of the linear operator, N is the matrix of the nonlinear operator and $U = [u_x,u_y]^T$.

This system is solved by using the θ weight formula as follows:

$$\theta LU^{n+1} + (1-\theta)LU^n + N(t,x,U^n) = 0, \quad (13)$$

where n is the current time layer.

Considering the 1-D static situation, the Eq.(11) can be recast as :

$$a_2\Delta\theta\frac{\partial^2u}{\partial x^2} + a_4\left(\frac{\partial u}{\partial x}\right)^2\frac{\partial^2u}{\partial x^2} + a_6\left(\frac{\partial u}{\partial x}\right)^4\frac{\partial^2u}{\partial x^2} - k_g\frac{\partial^4u}{\partial x^4} + f = 0, \quad (14)$$

with the same boundary conditions:

$u(0,t) = 0,\quad u(L) = 0,$

$u_{xx}(0,t) = 0,\quad u_{xx}(L,t) = 0.$

4. Numerical simulation

The numerical experiments data in the current paper were published in [11], which were gotten by

$Au_{23}Cu_{30}Zn_{47}$ materials:

$a_2 = 480 g/(ms^2 cmK), a_4 = 6 \times 10^6 g/(ms^2 cmK)$,

$a_6 = 4.5 \times 10^8 g/(ms^2 cmK)$,

$\theta_0 = 208k$,

$\rho = 11.1 g/cm^3, c_v = 29 g/(ms^2 cmK)$.

All the simulations were taken in the Matlab2015a.

4.1. Dimensionless form of the governing equations

The dimensionless form applied in the current can be referred to as the strategy in Ref. [12]:

4.2. Simulation results of 1-D SMA

In this section, the simulation was performed by the situation with $k = 5$ and $M = 6$ in section 3. The dimensionless form of the governing equations can be gotten from formula(14):

$$\sigma_{\tilde{x}} = \frac{\partial^4 \tilde{u}}{\partial \tilde{x}^4} - \tilde{f}, \qquad \sigma = \widetilde{k_1}\tilde{\varepsilon} + \tilde{\varepsilon}^3 + \tilde{\varepsilon}^5, \tag{15}$$

where $\widetilde{k_1} = 0.192$, $\tilde{f} = 0.4$, $\tilde{\varepsilon} = \frac{\partial \tilde{u}}{\partial \tilde{x}}$.

In this experiment, the temperature was set to $240K$. Thus, the initial condition should begin from the austenitic phase:

$$\widetilde{u_0(x)} = 0. \tag{16}$$

Newton method was applied to complete this 1D simulation with the derivative matrix in section3.1. Fig.1(left) shows the figure of displacement $\tilde{u}$. Fig.1(right) shows the figure of strain $\tilde{\varepsilon}$. Under this mechanical loading, the simulation shows that the rod has finally finished the phase transformation into the martensite phase. And the results are in line with expectations.

4.3. Simulation results of 2-D SMA

In this section, the simulation was performed in an SMA film. The time stepsize in this experiment was set to $\Delta\tilde{t} = 1 \times 10^{-4}$. The initial value is given as a disordered martensite phase.

The mechanical loading which is time-varying is defined as

$$f_x = 500, \tag{17.a}$$

$$f_y = 500, \tag{17.b}$$

Under this mechanical loading, the SMA film should be switching between the martensite phase and the austenite phase. In this section, we set the temperature at $210K$. Both the mechanical and temperature conditions make sure that only the martensite phase exists. And the free energy function here has two local minima which correspond to two martensites. The simulation results are present in Fig.2. It is shown that the initial disordered martensite phase is rearranged into the twins martensite phase ordered. Thus, the phase combinations are caused by the applied mechanical loading. Stress-induced phase combinations are captured well in this section.

5. Conclusion

In the current paper, an isothermal model has been proposed to simulate the static microstructure of SMA. The Microstructure evolutions in both 1D and 2D systems have been simulated via Legendre wavelets method successfully. The numerical algorithm for Legendre wavelets method to solve the partial differential equation has been described in the current paper. The simulation ability of LWM is shown.

References

[1] Mohd Jani J, Leary M, Subic A, et al. A review of shape memory alloy research, applications and opportunities[J]. Materials & Design, 2014, 56:1078-1113.

[2] Humbeeck J V. Non-medical applications of shape memory alloys[J]. Materials Science & Engineering A (Structural Materials, Properties, Microstructure and Processing), 1999,

273-275(none):134-148.

[3] Biesiekierski A, Wang J, Gepreel A H, et al. A new look at biomedical Ti-based shape memory alloys[J]. Acta biomaterialia, 2012, 8(5):1661-1669.

[4] L.X. Wang, R.V.N. Melnik, Simulation of phase combinations in shape memory alloys patches by hybrid optimization methods[J]. Applied Numerical Mathematics, 2008, 58(4):511-524.

[5] L.X. Wang, R.V.N. Melnik, Thermo-mechanical wave propagation in shape memory alloy rod with phase transformations, Mechanics of Advanced Materials and Structures, 2007, 14 (8):665-676.

[6] L.X. Wang, R.V.N. Melnik, Numerical model for vibration damping resulting from the first order phase transformations, Applied Mathematical Modelling, 2007, 31:2008-2018.

[7] Xuan He, Zheng Tong, Haoyuan Du, Dan Wang, Linxiang Wang and Roderick Melnik. Modeling microstructure evolution in shape memory alloy rods via Legendre wavelets collocation method. Journal of Materials Science. (2019) 54:14400-14413.

[8] Wang L X, Melnik R V N. Mechanically induced phase combination in shape memory alloys by Chebyshev collocation methods[J]. Materials Science & Engineering A (Structural Materials, Properties, Microstructure and Processing), 2006, 438-440(none):427-430.

[9] M. Razzaghi, S. Yousefi, Legendre wavelets method for the solution of nonlinear problems in the calculus of variations, Mathematical and Computer Modelling, 34.1–2,(2001), P 45-54,

[10] F. Mohammadi, M.M. Hosseini, A new Legendre wavelet operational matrix of derivative and its applications in solving the singular ordinary differential equations, Journal of the Franklin Institute, 348.8 (2011), p.1787-1796.

[11] L.X. Wang, R.V.N. Melnik, Finite volume analysis of nonlinear thermo-mechanical dynamics of shape memory alloys, Heat Mass Transfer (2007) 43:535–546

[12] Dhote, R. P., Gomez, H., Melnik, R. N. V., & Zu, J. (2015). 3D coupled thermo-mechanical phase-field modeling of shape memory alloy dynamics via isogeometric analysis. Computers & Structures, 154, 48–58.

Acknowledgments
This work has been supported by the National Natural Science Foundation of China (Grant No.51575478 and Grant No.61571007), the National Sciences and Engineering Research Council (NSERC) of Canada, and the Canada Research Chair Program.

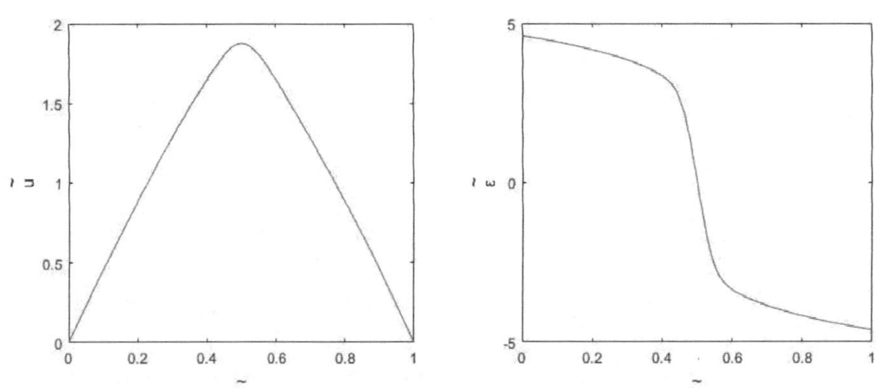

Figure 1. The simulation results of the 1D static system.

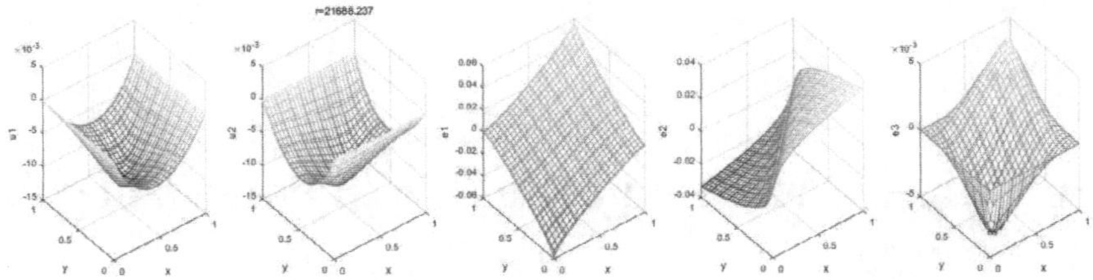

Figure 2. The simulation result of the 2D static system.

Research on netizens' emotion evolution in emergency based on machine learning

Zhongjie Ren[1], Peng Zhang[1], Jing Liu[2] and Yuexin Lan[3]

[1]Department of Fire Commanding, People's Police University of China, Langfang, Hebei, China
[2]Department of Literature and Law, North China Institute of Aerospace Engineering, Langfang, Hebei, China
[3]Department of Border Control, People's Police University of China, Langfang, Hebei, China

E-mail: zp_1981@aliyun.com

Abstract. [Purpose/significance] By analyzing the emotional evolution of Weibo users after emergencies, we can find out the law and potential risks of public opinion evolution and provide instruction for the government to control and guide network public opinion. [Method/process] We proposed an analysis model of public opinion evolution based on Emotional Analysis and GBRT. With the help of Python, a web crawler was developed to collect Weibo comments. After that, a Naive Bayesian Classifier was used for emotional analysis. According to public emotion and the number of comments, we divided the evolution process into fever period, persistence period, incubation period and extinction period. Statistical and visualization methods were used to study the evolution characteristics of word cloud, emotional tendency and age groups. Finally, correlation analysis and GBRT were used to predict each individual's emotions. [Results/conclusion] Taking the dangerous chemical explosion accident in Tianjin as an example, we can validate our model. Results shows that the model can reasonably divide the evolutionary stages, find out the law of public opinion evolution in different stages, and accurately predict users' emotional tendency.

1. Background

Nowdays, the social media platform dominated by micro-blog has gradually become the main space for the evolution of public opinion [1-3]. Weibo users' emotions can be infected by other's quickly and lead to a heated debate on the internet [4-5]. Grasping the law of periodic fluctuation of Weibo users' sentiment and further predicting it can prepare the government for potential risks.

In researches about the evolution of Weibo public opinion, Youyonghua divides the evolution process into rapid diffusion stage, stable development stage and declining stage according to two indicators of attention and attention increment [6]. Wu Xiaojuan divides the development of public opinion of Blue Qianjiang arson case into initial stage, the outbreak stage, the repeated stage and the long tail stage, and researches the themes of each stage [7]. Lan Yuexin and others constructed the logistic model of public opinion propagation process, dividing the incident of "Chengdu men beating women drivers" into five stages [8]. At present, the division is mostly according to micro-blog comment volume, number of followers and number of forwards, without considering emotional characteristics. In this paper, emotional inclination index is introduced into the criterion of stage division of public opinion. There are mainly two kinds of emotional orientation research methods, one is emotional dictionary, the other is machine learning. This paper uses naive Bayesian to conduct emotional analysis.

At present, most of the literature on public opinion emotion prediction is the prediction of group emotion trend. Through time series and gray prediction method, the short-term trend of public sentiment is predicted. Du zhitao et al. predicted online public opinion trend by establishing a grey prediction model [9]. Qian ailing et al. used multi-time series association rules to analyze the BBS trend [10].Zhang heping established an improved Gray Verhulst Markov model to predict the development of public opinion [11].

The current research mostly adopts the time series method, which is used to predict Baidu Index or the average emotional tendency of the public. The time series is simple to use and the system is mature. When the trend of public opinion changes steadily, it can well predict the short-term trend of public average emotion. However, the public opinion is complex, changeable, so the time series is difficult to carry on the long-term forecast; besides, mostly the target of prediction is the average emotion of the public, which cannot reflect the emotion distribution cannot predict the emotion of each member of the group. In addition, users' distinctive features, such as education background, age, gender, number of fans, active status of Weibo, often have a great influence on users' emotional tendency, which cannot be reflected in the methods of time series.

Based on the emotional analysis, statistical analysis, data visualization methods and machine learning, we research the evolution of public sentiment of web users and build a model, which has been validated by Tianjin 8•12 accident cases.

2. Web users' sentiment evolution model

2.1. Research methods

The methods involved in the model are described below.

2.1.1. Web crawler

Web crawlers can bulk collect and download data from web pages. At present, many achievements have been made in studies related to it [12-13].To conduct our research ,we made web crawler Specifically applicable to Weibo with Python, which can simulate users' login and search in Weibo, automatically grab microblog comments on relevant emergencies, and simulate the click action to extract more relevant information.

2.1.2. Emotional analysis

Emotion analysis requires word segmentation to be done first. There are some ways: dictionary-based method, statistical method and rule-based method [14]. This paper adopts SnowNLP, a natural language processing module of Python, to conduct word segmentation. Then Naive Bayes classifier was used for emotion analysis. Naive Bayes classifier has high accuracy, reliable classification. Simply and fast, it can make out each sample's classification probability.

Let's say that there are N categories of problems that we're dealing with, $Y=\{c_1,c_2,c_3,c_4...,c_N\}$.

According to Bayes theorem, that is

$$P(C|X)=\frac{P(C)\,(X|C)}{P(X)} \tag{1}$$

Where $P(C)$ is the prior probability of the class; $P(X|C)$ is the sample's conditional probability

For each sample, we select the category that minimizes the risk of misclassification on that sample

$$h^*=argminR(C|X) \tag{2}$$

Take the lowest misjudgment rate as the target, and the cost function is as follows:

$$\lambda_{ij}=\begin{cases}0, & if \quad i=j\\ 1, & otherwise\end{cases} \tag{3}$$

Journal of Physics: Conference Series **1419** (2019) 012004 doi:10.1088/1742-6596/1419/1/012004

The corresponding conditional risk is

$$R(c|x) = 1 - P(c|x) \tag{4}$$

The classifier whose goal is to minimize error rate is

$$h^* = argmaxP(c|x) \tag{5}$$

Naive Bayes classifier assumes that for a given category, all attributes are independent. Therefore, equation (2) can be written as

$$P(C|X) = \frac{P(C)P(X|C)}{P(X)} = \frac{P(C)}{P(X)}\prod_{i=1}^{d} P(x_i|c) \tag{6}$$

So we can assume that:

$$h(x) = arg\ max\ P(c)\prod_{i=1}^{d} P(x_i|c) \tag{7}$$

2.1.3. Correlation analysis

Correlation analysis can analyze the relationship between two different variables. The simple correlation coefficient is defined as:

$$r = \frac{\sum_{i=1}^{n}(x_i - \overline{x})(y_i - \overline{y})}{\sqrt{\sum_{i=1}^{n}(x_i - \overline{x})^2 \sum_{i=1}^{n}(y_i - \overline{y})^2}} \tag{8}$$

N is the sample size, x_i and y_i are the variable values of the two variables respectively

The test statistic adopts t statistic, which is defined as:

$$t = \frac{r\sqrt{n-1}}{\sqrt{1-r^2}} \tag{9}$$

Simple correlation coefficients, observed values of t-test statistics and corresponding probability values can be calculated.

2.1.4. GBRT model

GBRT is an important algorithm of machine learning. Taking the emotional tendency value obtained in the emotional analysis model as the dependent variable and each attribute of the user as the independent variable ,we can train the GBRT model [15].Gradient promotion tree (GBRT) algorithm, by iterating multiple regression trees to make joint decisions, improves the weak learner to the strong learner and has good generalization characteristics [16].

2.2. Model flow char*t*

The flow chart of the model is shown in figure 1, which includes five stages: data acquisition, data preprocessing, emotion analysis, public sentiment evolution stage division, and GBRT modeling.

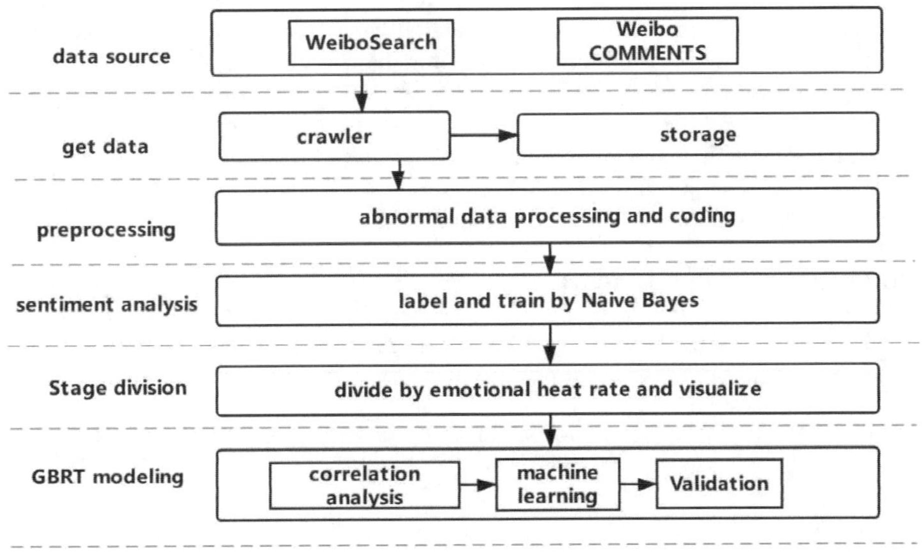

Figure 1. Flow chart of the model.

(1) Data acquisition

Through the analysis of microblog search results, multiple microblog posts with the most comments on the Tianjin incidents were selected, and Python was used to compile crawler for data crawling. The data includes the official microblog of People's Daily, the official microblog of fire department, and other related microblog comments of many official media, as well as the microblog comments of some Internet celebrityies.

(2) Data preprocessing

The data obtained by the crawler may not all conform to the use of modeling, and there may be missing, duplicate and invalid data. Python was adopted to conduct data deduplication, delete missing data, irrelevant data and abnormal data, and store the cleaned data in CSV format.

(3) Emotional analysis

Python language can be used to facilitate the processing of Chinese text data. Firstly, the comment text is processed by word segmentation, and the stop words are removed according to the stop words list to avoid the interference of irrelevant words. We extract some data and manually annotate it , then these samples are regarded as training set, Naive Bayesian classifier was trained to obtains word frequency in each category and further formula out all the comment text emotion inclination degree, the higher the degree of emotional tendency, the more close to the positive emotion, otherwise, the more close to the negative.

(4) Analysis of the evolution stage of public opinion

In various literatures, the number of comments and the number of retweets are often used to measure the popularity of a topic. In general, the more comments, the hotter the topic. However, in order to gain attention, some users use a lot of zombie fan to comment, so the number of comments and retweets cannot effectively and correctly reflect the heat of the event. we introduce the emotion inclination degree here. Considering that the larger the emotional divergence is, the more heated the discussion will be and the hotter the topic will be, the variance of emotion inclination degree can reflect the size of emotional divergence to a certain extent and indirectly indicate the topic heat. In addition, the variance of some dates is relatively large because the number of comments is small, so the variance alone is not enough to fully reflect the fact. After comprehensive consideration, this paper defines emotional heat as follows:

$$\boldsymbol{HOT} = minmax(minmax(num)*q) \tag{10}$$

In formula (3), HOT represents emotional heat, "num" is the number of comments, and q is the variance of emotional tendency. "Minmax" is a standardized treatment. Standardized treatment can remove the influence of dimension and make both factors better reflected in public opinion evaluation.

(5) GBRT model training

Through quantitative data preprocessing and word vectoring process, the user's attributes (including gender, age, location, school/company, the number of fans, the thumb up number, focus on number, the user's first emotional state, average public emotion inclination degree, nickname) information as input data, emotional tendency degree as the output, we build the GBRT model .

3. Tianjin 812 accident emotive analysis

On August 12, 2015, a chemical explosion in Tianjin binhai new area attracted widespread attention on Weibo and became one of the hottest topics that year. People have different opinions and sentiments; the discussions admixture of good and evil, and official information is released improperly, resulting in public opinion crisis [17].

In order to test and verify the model, our model was used to analyze this chemical accident.

3.1. Data acquisition

Search keywords such as "Tianjin 8•12" and "binhai new area" on Weibo, select comments of People's Daily, China fire station and other microblogs, compile crawlers and obtain 15,253 pieces of data. To get each user's information of comments, the crawler is used again to obtain the user's microblog homepage information. The crawling data includes comment content, comment time, commenter ID, nickname, number of thumb up users, number of fans, number of followings, gender, information source, age, gender and province of some users.

3.2. Data preprocessing

Delete duplicate, missing, invalid data. For example, in some data, there are "0 years old", "105 years old", etc., which are inconsistent with cognition and belong to abnormal data. The value of abnormal attributes corresponding to the data is deleted, and the remaining attributes remain unchanged. After the process, there are 15164 pieces data remaining.

3.3. Naive Bayesian classifier training and prediction

In this paper, Naive Bayesian model is used for sentiment analysis. 300 comments of positive emotion and negative emotion were randomly selected, and the positive and negative emotional tendency degree was manually calibrated as training data. After segmentation, we remove the stop words. Naive Bayes classifier was trained to calculate the emotional propensity of all other comments. Taking the comment "we pray for you, and we firmly support you" as an example, the positive emotion is denoted as S, and the negative emotion is denoted as N, then:

$$P(S|\text{"祈福"},\text{"天津"},\text{"坚定"},\text{"支持"}) = \frac{P(\text{"祈福"},\text{"天津"},\text{"坚定"},\text{"支持"}|S)}{P(\text{"祈福"},\text{"天津"},\text{"坚定"},\text{"支持"})} \quad (11)$$

$$P(N|\text{"祈福"},\text{"天津"},\text{"坚定"},\text{"支持"}) = \frac{P(\text{"祈福"},\text{"天津"},\text{"坚定"},\text{"支持"}|N)}{P(\text{"祈福"},\text{"天津"},\text{"坚定"},\text{"支持"})} \quad (12)$$

Then, according to the conditional independence hypothesis of Naive Bayes classifier, we assume that each word is independent, then

$$P(\text{"祈福"},\text{"天津"},\text{"坚定"},\text{"支持"}|S) = P(\text{"祈福}|S)*P(\text{"天津"}|S)*P(\text{"坚定"}|S)*P(\text{"支持"}|S)$$

$$S = \frac{P(S|\text{"祈福"},\text{"天津"},\text{"坚定"},\text{"支持"})}{P(N|\text{"祈福"},\text{"天津"},\text{"坚定"},\text{"支持"})} \quad (13)$$

S is the probability of positive emotion. As S is calculated through the positive and negative word frequency, it can fully reflect the degree to which emotions tend to be positive, and part of the results are shown in figure 2.

[' 祈祷']	0. 829446
[' 看第一幅图就好心疼']	0. 867565
[' 回复:对呀，只能祈福了']	0. 043149
[' 居民楼都炸成这样了？才死这么点人？楼里都没人吗？我只想说，骗子死全家。	0. 004117
[' 这是化学物品爆炸，出动防化部队没错，真不知道这有什么好喷的，怎么救如何	0. 838826
[' 回复:你根本不知道现在消防员的工资有多少，他们连最起码的高温补助都没有	0. 002227
[' 回复:我不满意别人对消防员的不尊重难道有错？你真是狭隘，没有经过了解就	0. 637814
[' 回复:你不满意，所有的人就不满意了？']	0. 190414
[' 为什么每次火灾都会牺牲很多的消防员 他们都那么年轻 到底是一些领头人为	0. 10705
[' 这就是职能部门的职责问题。这样危险极高的场所负责安全的部门尽责了吗？领	0. 000949
[' 看一次眼泪掉一次，真的乞求不要再多的伤亡了，愿一切安好????????????']	0. 807676

Figure 2. Results of Naive Bayes prediction model.

According to the graph, the result of emotion analysis accords with cognition.

3.4. Evolution stage division

3.4.1. Distribution rule and analysis of emotion inclination degree

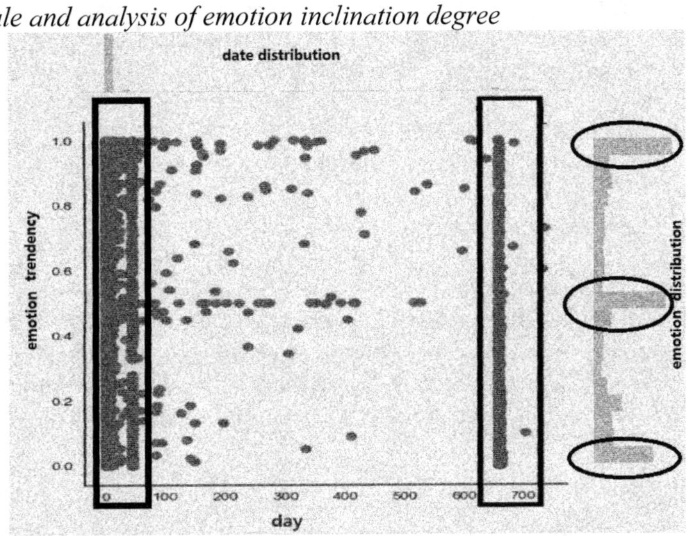

Figure 3. The whole time period comment emotion tendency distribution.

Figure 3 shows the distribution of the emotion inclination degree. The ordinate is the motion inclination degree of each comment, and the larger the value is, the more positive the emotion is, and the closer it is to positive emotion. The abscissa is the number of days since the outbreak of public opinion on August 13.

According to figure 3, the emotional tendency changes greatly with time and comments of different commentators, indicating the characteristics of public opinion on Weibo: Great emotional divergence, changeable, existing explosive and latent periods. By observing the emotional distribution, it can be seen that the emotional distribution has three peaks, which respectively symbolize extremely positive emotion, extremely negative emotion and neutral emotions, indicating that people's emotions fluctuate greatly. From the perspective of time distribution, the emotional tendency tends to become neutral with time after the big divergence in the early stage, but it is the incubation period of public opinion at this time, and occasionally there will be extreme emotional outburst caused by events.

3.4.2. Rule of emotional heat and reason analysis

Calculate the heat according to formula (3) and make the heat -- time curve, as shown in the figure below.

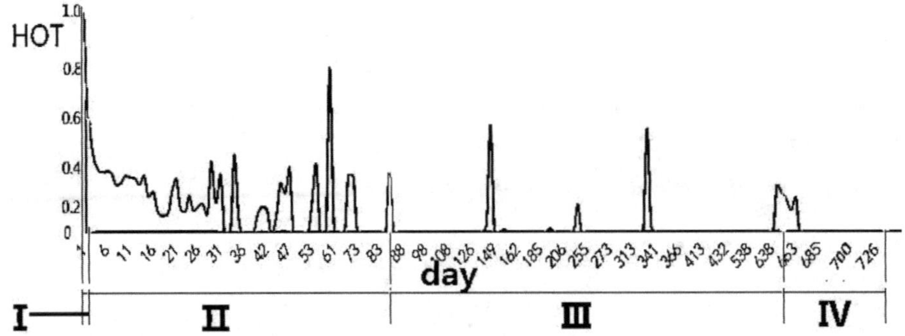

Figure 4. Area diagram of emotional heat changing with time.

According to figure 4, the evolution process of public sentiment can be divided into four stages: hot period, continuous period, incubation period and dying period. The first two days after the outbreak of the accident is regarded as the hot period, and the emotional heat is kept above 0.5.At the beginning of the accident, there are a great deal of comments and people have different emotions towards the accident, information imbalance caused a variety of views among people. From the 3rd to the 85th day after the accident, the emotional heat is stable and high, the information is spread widely, and there are more and more news about the incident. The incubation period of public opinion ranges from 86 to 661 days, netizens' attention is diverted by other things. However, once relevant events occur, the topic will become a hot debate again. Since the 661st days after the incident, the emotional heat maintained at a low value , and it has entered a period of complete extinction.

3.5. Characteristics of the evolution stages

3.5.1. Word cloud in different stages

Figure 5 shows the word cloud of the comments in the four stages. It can be seen that in the whole process of the evolution, the official microblog emphasize firefighters' hero moves and their sacrifices , which aims to guide public opinion in the right way, so that there are words such as "heroes", "pray" in the whole process, but each stage also has its own special words.

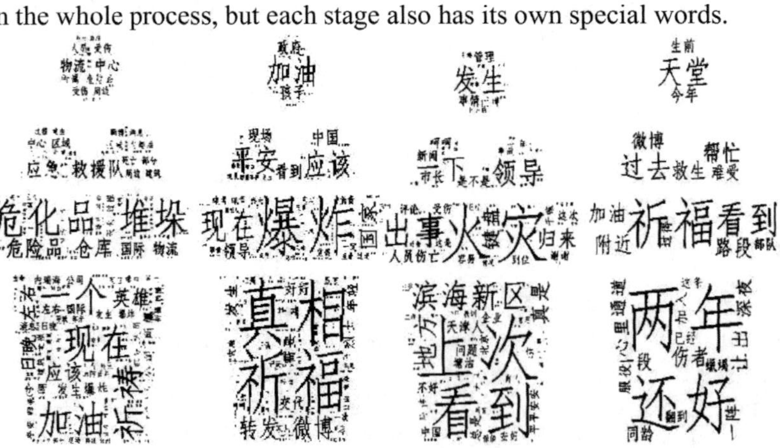

Figure 5. Comment cloud in four stages.

In Stage 1, the comment cloud is mainly a statement of events, such as "warehouse", "hazardous chemicals", "injury", "logistics", etc. It shows people's desire to know and spread the news.

In stage 2, besides positive words such as "pray for good fortune", "peace", "come on", there are words such as "truth" and "leadership", indicating people's different concerns and views on the responsibility of the accident. However, the official media try to be vague about it, which caused the divergence of emotional tendency in public opinion to expand.

In stage 3 ,the words "last time" ,"management" and "leadership" appeared, indicated that people was not satisfied with the leading departments due to the poor management of Tianjin port ,because of the occurrence of another fire in 2016.

In stage 4, the evolution process has been close to the death, some accident witnesses occasionally recall some nostalgic comments in this stage.

3.5.2. Proportion of emotional tendency at each stage

Emotional tendency S was segmented, with 0-0.3 as negative emotion, 0.3-0.6 as neutral emotion and 0.6-1 as positive emotion, and statistical analysis was conducted. The results were shown in table 1.

It can be seen from table 1 that in the first stage, the proportions of the three emotions were equal, which means there were great differences among people's emotions. In stable period, emotions gradually turn into neutral emotions, and negative emotions reduce. During the incubation period, critical events are easy to make it a hot topic again and there are great gaps in people's sentiments. In the dying stage of public sentiments, the proportion of neutral emotion reached a high value, and the evolution process reached dying.

Table 1. Proportion of different emotional types in the four stages.

Period	Proportion		
	Negative	Neutral	Positive
1	0.35	0.31	0.34
2	0.20	0.42	0.38
3	0.57	0.30	0.13
4	0.13	0.63	0.25

3.6. proportion of age groups in different stages

Age groups are divided into five age groups, respectively: [<17], [17-28], [29-37], [38-55] and [>55]. Statistical analysis is conducted on the age of Weibo users at each stage, and the change of the number of different age groups at each stage can be obtained from the curve below, as shown in figure 6.

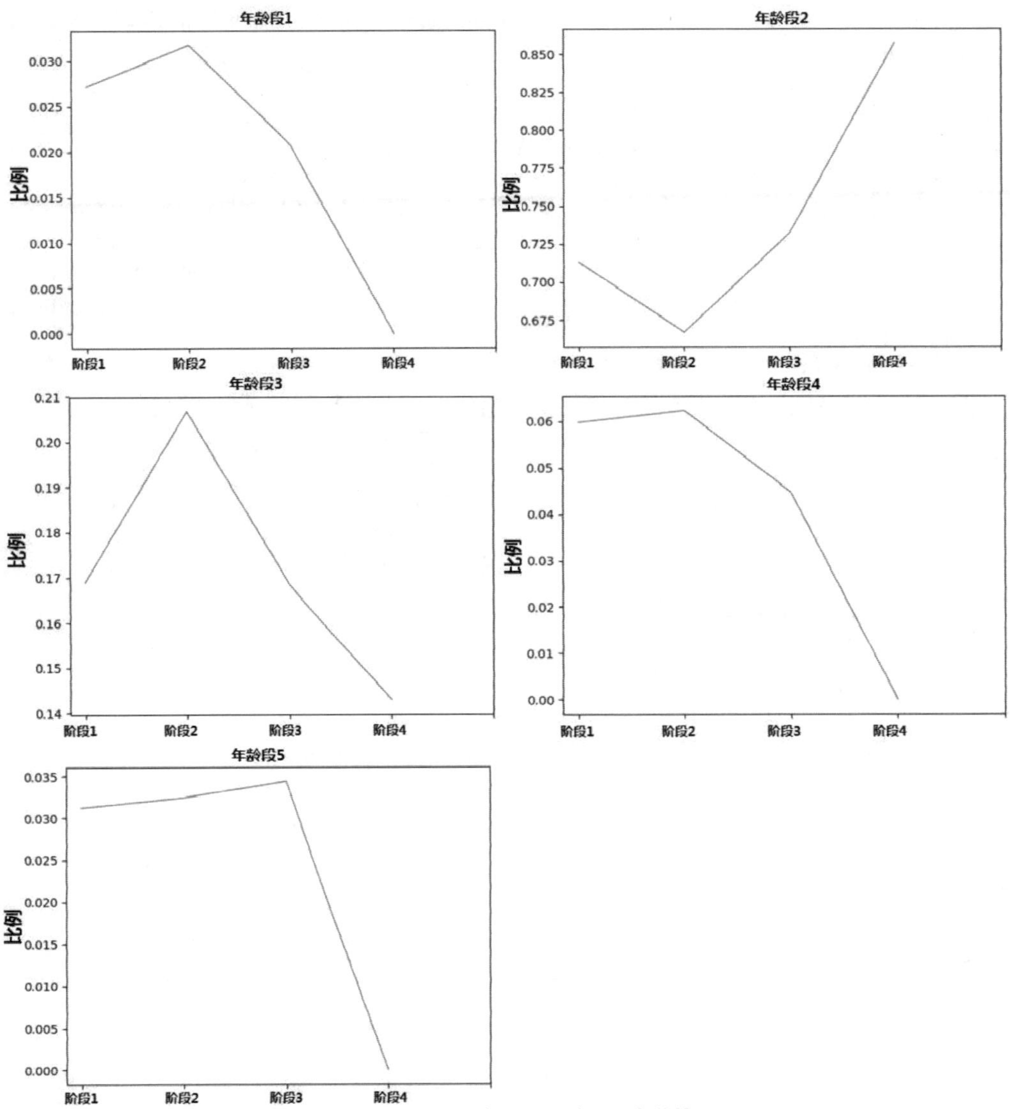

Figure 6. Curve of proportion of different age groups.

As shown in the figure above, group2 (age range [17-28]) and group3 (age range [29-37]) are the dominant groups among the five age groups, accounting for more than 90%. Among them, group2 has the largest proportion, and is always increasing in the four stages. Therefore, our advice for government is that we should focus on young people aged 17-28, especially in the incubation period. Group2 can decide whether the topic will become hot again and when it will die out

4. Tianjin 8•12 accident sentiment prediction model of emotion inclination degree

4.1. Correlation analysis and modelling of emotional propensity

4.1.1. Coding and pre-processing

The gender variable is one-hot coded. Then we use Baidu encyclopedia, sohu news and novels to train Word2Vec model, and the user's personal information, study/work location, nickname and comments release source were transformed into 64-dimensional word vectors with Word2Vec for subsequent analysis and processing.

4.1.2. Correlation coefficient We compute correlation coefficient. To visually display the correlation, a histogram is made, as shown in figure 7.

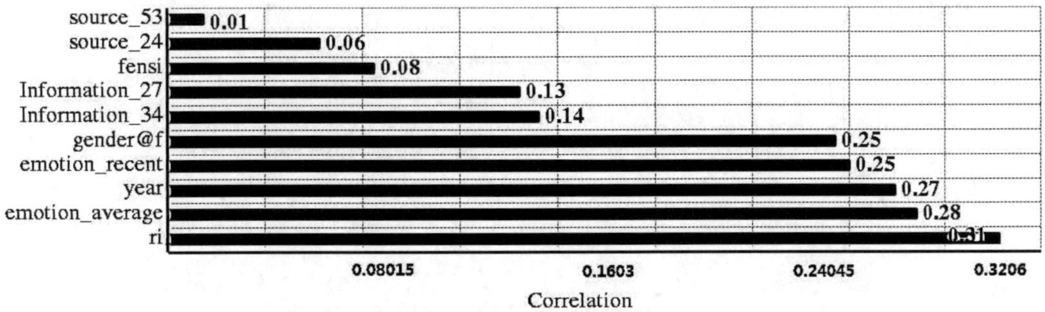

Figure 7. correlation between each variable and emotion inclination degree.

Variables with obvious correlation are "emtion_before" which shows the user's emotional tendency in the previous stage, "ri" which implies the number of days since the event occurred, "emotion_average" which is the group's emotion inclination degree at this stage, etc., these variables have strong positive correlation with a user's emotion inclination degree. In addition, the number of users' followings, gender and personal information are also weakly correlated.

This result shows that emotional tendency changes significantly over time after the event.

For example, there are more negative emotions in the early stage of catastrophic accident, and after that, emotions gradually become neutral over time. So emotion is most affected by time. In addition, due to network empathy and herd mentality, the average emotional inclination of the public also has a positive impact on personal emotional inclination. Besides, even gender can have some influence, and it means woman is more likely to be positive on the issue of emergency.

4.2. Model training

With emotion inclination degree as the prediction target, GBRT was used for regression analysis modeling. Each time, negative gradient of loss function was used to fit the next decision tree. Based on decision integration of multiple regression trees, the prediction model was obtained. The learning rate was set at 0.1, the number of decision trees was set at 100, and the subset of samples accounted for the overall proportion of 0.8.

4.3. Model evaluation

The regression model error can be measured by MSE and MAE.MSE is mean square error, and the calculation expression is:

$$MSE=\frac{1}{N}\sum_{t=1}^{N}(observed_t-predicted_t)^2 \quad (14)$$

MAE refers to the average absolute error. The calculation formula is as follows:

$$MAE=\frac{1}{N}\sum_{i=1}^{N}\text{abs}\,(f_i-y_i) \quad (15)$$

We compute MAE and MSE, MAE value was 0.2672 and MSE was 0.1006, which was within the allowable error range, and the model result was considered reliable.

The model still has a lot of room for improvement. More user portrait attributes can be collected in subsequent work to improve prediction accuracy. On this basis, before and after major disasters, GBRT model is used to predict the emotional tendency of Weibo users on the whole network, which can release targeted guidance information to different groups and guide the correct evolution of public opinions.

4.4. Analysis of model results

After preprocessing and coding user's information, GBRT model is used to predict the emotional tendency of each user in the development of public opinion.

To show the process, information of seven Weibo users in the database were extracted to predict their emotional tendency. To avoid leakage of user information, nickname information is processed here. The sample's personal information is shown in figure 8.

NO.	shijian_review	source	dianzan	gender	guanzhu	fensi	nickname	information	else_info	stage
4	2017/6/26	vivo X5・K歌之王	1	f	239	224	[redacted]洒脱	女 26岁 狮子座 福建	武夷学院	4
10929	2017/6/1	iPhone 6 Plus	2	f	627	179	[redacted]SZ	女 江苏 苏州		3
10857	2017/6/1	柔光自拍vivo X7P	7	f	62	47	[redacted]	女 22岁 水瓶座 四川		3
1877	2015/8/13	三星android智能手	0	f	270	299	[redacted]IN	女 处女座 广东 广州		1
4852	2015/8/13	三星android智能手	1	f	270	299	[redacted]MIN	女 处女座 广东 广州		1
10180	2015/9/26	iPhone 6	0	f	48	728	[redacted]thstarc	女 海外 其他		2
10185	2016/9/26	大神手机	0	m	196	7874	[redacted]某某	男 巨蟹座 福建 厦门	集美大学诚毅	2

NO.	nickname	text	qinggan	type	predict
4	[redacted]洒脱	['深夜翻自己的微博，又看到你们了。心里一阵难受。在天堂的你们还好吗？']	0.373884	2	0.22
10929	[redacted]_SZ	['管理混乱！天津的官员要脚踏实地工作，少一些虚的空的！']	0.005335	1	0.15
10857	[redacted]惓戀	['上次的教训还没完全吸收，这次又来，天津你是想毁了吗']	0.123978	1	0.09
1877	[redacted]N	['不造谣！不信谣！不传谣！这是我们现在能做的！']	0.905739	1	0.77
4852	[redacted]N	['死了很多人、空气里有毒、政府不作为、媒体在隐瞒，键盘侠们的理想状态']	0.074431	1	0.04
10180	[redacted]thstarc	['英雄一路走好，我们不会忘记你们的']	0.996096	2	0.89
10185	[redacted]某某	['事故责任人应该陪葬！']	0.19648	2	0.2

Figure 8. sample user information table.

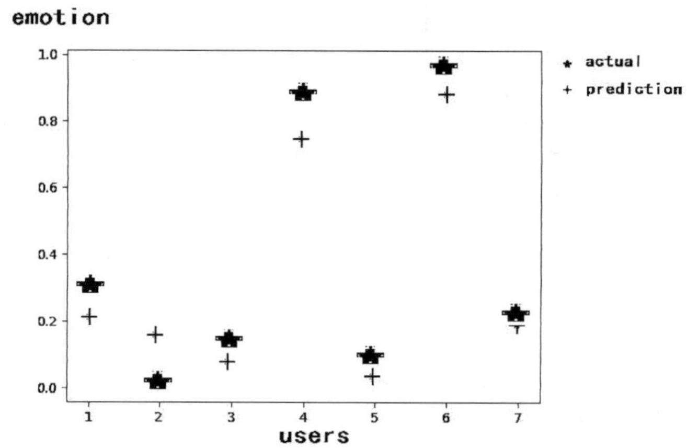

Figure 9. Scatter plots of actual emotion values and predicted values.

Figure 8 shows the value of emotion inclination degree computed by Bayes and the predicted value by GBRT. Figure 9 shows the comparison between them.

The seven predicted values deviated from the actual values, but within the allowable range of error. In the early stage of public opinion, the model can be used to predict the emotional tendency of each person , and government can do some targeted measures according to it. For example, according to word cloud and word frequency in different stages, different age groups, specific guidance should be given to groups with different characteristics and in different stages. Besides, online group identification, offline group guidance can be useful. Identify Internet celebrity users through fan data and microblog data, who could be most the important groups who have the largest influence in the web.

5. Countermeasures and suggestions

According to the analysis results, it is found that the emotional tendency of users is related to the evolution stage of public opinion and the information of user portrait. To some extent, the indicators of user portrait reflect the personality characteristics and knowledge level of users. Therefore, users will have different reactions towards an event. Our model proposed in this paper can be used to predict the emotional tendency of users towards emergencies in advance and guide users' public opinions in a targeted way.

(1) Early warning of public opinion and sentiment.

For example, as shown in section 3.5, user whose ID is 10185 can be predicted in the first stage of the development by model, and combined with the word cloud, the official media can prepare for the potential risks of public opinion in advance.

(2) Public opinion guidance for different user group.

According to the conclusion of correlation analysis in section 3.4, personal emotion is highly correlated with the average public emotion, that is, people will be influenced by others.

Through emotion analysis, positive emotion users are identified and their microblogs are placed at the top of their fans' Weibo page. For users with negative emotions, relevant laws, expert interpretation and other media opinions will be pushed to them to give more reference to him.

(3) Pay attention to the age structure.

In the case of Tianjin 8•12 accident, user group aged 18-37 is the main group of online public opinion. Therefore, after identifying the age characteristics, government should take measures aim to different age group. For example, it is good to adopt more interesting ways for young people. Tik Tok, micro video, kuaishou and other platforms will be useful, as well as the 90's popular cartoon, animation.

(4) Emphasize identity information and Internet celebrity.

Users' information contains variables such as age, gender, educational background, number of fans, number of followings, etc. The number of followers and the number of followings can effectively reflect a person's activeness in Weibo. Among them, users whose number of fans is more than a certain threshold can be identified as Internet celebrity. The emotions of them are often spread through the huge number of fans, which has a significant impact on public opinion. Therefore, the identification of Internet celebrity,targeted emotional analysis and guidance at different stages are conducive to the standardization of cyberspace. In addition, there are school information and workplace information, thus according to it, we can also adopt ways of online identification of the gathering trend of groups and offline guidance.

6. Discussion and conclusion

6.1. Discussion

The research work done in this paper is based on certain assumptions. Based on the following assumptions, the conclusion of this paper is valid:

(1) It is assumed that the number of posts deleted by microblog managers is within the allowable error range and does not significantly affect the analysis of this paper;

(2) It is assumed that the personal information of Weibo users is true and valid;

(3) It is assumed that the comments sample can fully reflect the characteristics of all comments information on Weibo.

6.2. Conclusion

In this paper, crawler is written in Python to obtain microblog comment information. On the basis of data cleaning and preprocessing, emotion analysis modeling is conducted by combining word segmentation technology and Naive Bayes classifier to obtain emotion inclination degree. On this basis, according to the number of comments and emotional inclination degree variance, we define the emotional heat, and then the evolution process of public opinion is divided according to emotion heat ,after that, we use statistical methods to study evolution characteristics, and visualize data Finally, taking Tianjin 8•12 accident as an example, we can research the evolution rules of Tianjin 8•12 accident. It has been proved that the model can operate normally, providing theoretical support for the targeted guidance strategies of public opinion in different stages of emergencies.

References

[1] Li H， Wang LT(2018). Micro-blog Hot Topic Discovery Based on Heat term. *China. J.Information Science*, **36(4)**, 45-50.

[2] Chen J, Liu YP, Deng SL(2018). An Analysis on Factors Influencing the Dissemination Effect of Rumor-refuting Information. *China. J.Information Science*, **36(1)**, 91-95.

[3] Chen CY， Huang XL(2018). A Research on the Generation Mechanism and Propagation Law of Microblog Public Opinion. *China. J.Information Science*， **36(4)**, 32-37.

[4] JI XM(2014). Users'Sentiment Mining and Spreading among Chinese Microblogs in the Context of Specific Events. *China. D. Nankai University.*

[5] Zhang P, Lan YX, Li HQ and Wang JY(2016). Analysis of Internet R umors Classification Based on the Hayshi Quantification Theory. *China. J.Journal of Intelligence*, **35(01)**, 110-115.

[6] Pang T B, Pang B, Lee L(2002). Thumbs up? Sentiment classification using machine learning[J]. *China. J.Empirical Methods in Natural Language Processing*, 79-86.

[7] You YH(2017). The Analysis of Network Public Opinion Development Stage under Big Data Environment. *China. J.Value Engineering*, (35):177-180.

[8] Wu XJ(2018). Analysis of the Theme Evolution of Internet Public Opinion Based On Micro-blog Text —Taking a Fire Case in Blue Qianjiang Community as an example. *China. D. Nanjing University.*

[9] Du ZT, Xie XZ(2013). The Establishment of Public Opinion Forecasting and Early-warning Model with the Methods of Grey Forecasting and Pattern Recognition. *China. J.Library and Information Service*, **57(15)**, 27-33.

[10] Qian AL, Qu BB, Lu YS, Chen PP and Chen GD(2012). Forum Sentiment Trend Prediction Based on Multi Time Series Association Rule Analysis. *China. J.Journal of Nanjing University of Aeronautics & Astronautics*, **44(6)**, 904-910.

[11] Zhang HP, Chen QH(2018). Research on the Prediction of Network Public Opinion Based on Grey Markov Model. *China. J.Information Science*, **36(1)**, 75-79.

[12] Jiang ZY, Ma WR, Zou K and Li L(2018). Research on Emotion Evolution of Network Public Opinion Based on Sentiment Orientation Analysis. *China. J.Journal of Modern Information*, 38(04):50-57.

[13] Pinkerton B(1994). Finding what people want : experiences with the webcrawler. *C.Proc. of the Second International WWW Conference.*

[14] Ahmadi-Abkenari F, Selamat A(2012). An architecture for a focused trend parallel web crawler with the application of clickstream analysis. *J. Information Sciences*, **184(1)**, 266-281.

[15] Li JM , Sun LH, Zhang QR and Zhang CS(2003). Application of native Bayes classifier to text classification. *China. J.*Journal of Harbin Engineering University, **24(1)**, 71-74.

[16] Li H(2012). Statistical learning method [M]. *China. M.*

[17] Sproat R, Emerson T(2003). The first international Chinese word segmentation bakeoff. *C. Sighan Workshop on Chinese Language Processing. Pennsylvania: Association for Computational Linguistics*, 133-143.

Acknowledgments

This work was supported by “National statistical science research key project (no. 2017LZ37)”, “Ministry of Public Security theory and soft science project(No. 2015LLYJWJXY034)”, “Unmanities and social sciences foundation of the Ministry of Education(no.17YJC630214)”, “Hebei province key research and development project(no. 18215601)”, “Hebei statistical research project(no. 2018HY04)”, “Langfang science and technology project (No. 2019013066)”. We are grateful to the anonymous reviewers for their constructive comments that improving the quality of this work.

The methods of placing background image in Rhino

Zichen Ding and Xufeng Cheng

Department of Arts and Design, Beijing Forestry University, Beijing, 100083, China

Abstract. When using Rhino software for computer aided design, designers often place and manipulate background images in the viewports for tracing or design analysis. It is a problem in determining the size and location of the background images. Three methods are provided. The purpose of the paper is to provide a new thought for the designers.

1. Introduction

Rhino is a professional 3D modeling software developed by American Robert McNeel & Assoc. The newest version is Rhino5.0. It has powerful 3D modeling function, simple interface, ease of operation and it has great advantage for fast and accurate performance design. Combined with the corresponding plug-in, the software is widely used in industrial design, architecture design, jewelry design and mechanical design and other fields of design practice and scientific research [1].

In the conceptual design stage of the industrial products, the advantage of Rhino NRBS modeling is obvious. In 3D reconstruction of design sketch, the adaptability for free form surface modeling ability is stronger and the performance of design concept is better, which makes the designer express their design vocabulary more freely [2]. During modeling and according to the idea of "dotting - staying - paving - combined with body" [3], the first step will depend on the reference. Draw the key characteristic curve precisely and make it to be the benchmark of 3D modeling of product. The quality of these lines determines the final quality of 3D model [4]. Therefore, if it is accurate to place background image, directly affect the progress and quality of modeling. It is a problem in determining the size and location of the background images in practice. This article will combine the author's practical experience to introduce different methods aiming at the actual operation of common problems in order to improve the quality of modeling and design efficiency.

2. Common problems of placing background image

Rhino has strong adaptability in free surface modeling, so we need to choice thoughts flexibly according to different objects or different purposes of modeling. No matter how excellent the tool is, it can't complete the design task efficiently without modeler's clear modeling thought and logical thinking [5]. The operation of the Rhino simplicity does not represent the random modeling ideas. It is a necessary guarantee that the user fully understands the final shape and the surface before modeling to map the ideal line. Similar to the structure sketching, we need to target shape form a complete concept and awareness including all the shapes, lines' forms, and the transition relationship between line and line, between surface and surface and so on. The effectiveness of the logical thinking can be reflected when placing bitmap [6].

The basic steps of placing background image are placing the prepared images of all views in the corresponding view as background reference and then adjusting them to the right position and size so that we can observe product modality and interpret the conductions of models visually for analyzing [7] and making models step by step.

Due to different purposes (just like improved design according to the existing or innovative design according to the design draft), the source of image and the complexity of the modeling objects (such as a USB flash disk or a car) are different. Who lacks experience always lacks effective means and methods when placing background image. Two problems often appear. One of them is starting placing reference image without clear thought and adjusting over and over again without baseline. The other is starting modeling by despising the key step of placing reference image which makes it not accurate. And it will cause ambiguity to choose which view. Either case will lead to faults and inefficiency.

3. The introduction to the methods of placing background image

In practice, the writer summarizes that the size and location of the modeling object must be consistent in the reference placed link. According to different requirements, taking proper methods of placing background image is the key to make these two right and efficient. The following three methods are easy-to-use.

3.1. Reference bitmap placed by box

Firstly, if images have exact explanation of sizes, we can build a suitable box. Secondly, the design drafts of the three views or multiple views are imported into the corresponding view window. The design drawings should be precisely cut first, convenient for adjustment later. And then adjust the drawing size and position to make it inscribed on the object in the three side of the box. After the alignment in turn, mark points in the object's key position and examine if it places accurately from different view windows. As shown in Fig 1.

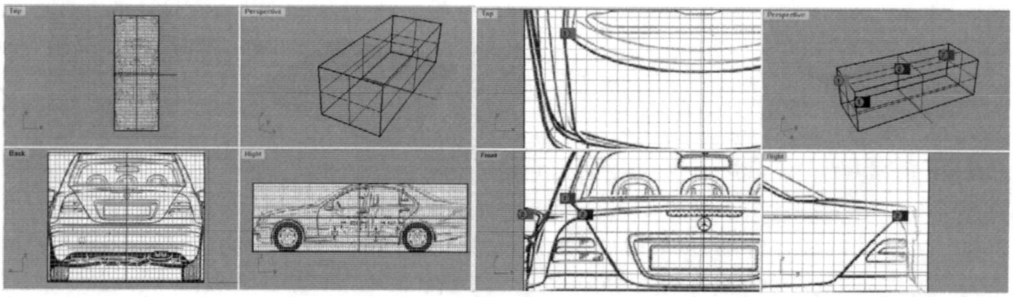

Figure 1. Reference bitmap placed by box.

If there are no sizes provided, we also need to cut the design drawings precisely first. And then lead into the view of reference bitmap and make its size such as length and width and position as a benchmark. Put the corresponding references into the other views according to their length or width. Adjust to the same length or width and the height can be determined naturally good. Next, import other reference bitmaps according to the length, width and height. If it is supposed to be more intuitive, you can build another box after the last step. Finally, examine in the same way.

3.2. Reference bitmap placed by gridding

In product design, designers often need to start modeling from design draft. At this time, it is a problem in determining the size and location of the background images because the design draft may not be so precise. So, there are some experienced designers sketching on grid coordinate paper for later modeling. If there were no grids before, we can add grids to the draft and make the grids and the key points as a benchmark for placing and sampling. The density of grid determines the precision of the model. The greater the density, the higher precision it is. But, this will cut the efficiency of placing image and modeling. So, the density of grid should be determined by the precision needed.

Firstly, import an image to the top view using the main view draft for example. After it is adjusted to a proper size, we draw line A which is based on the grid coordinates with line tool as a reference datum line. Then, draw line B as the medial axis and import the corresponding draft into the front view. Adjust the size and position of the reference image benchmarked against line A. Next, adjust the length of line B in this view according to the draft so that you can ensure the consistency of the position and size. At last, finish adjusting the position and size of the side elevation according to these main lines. As shown in the figure 2, it is a convenient way to placing background image depending on grid and coordinate. Although the outline of the design draft does not fit the three views completely, the sampling and the later modeling won't be influenced.

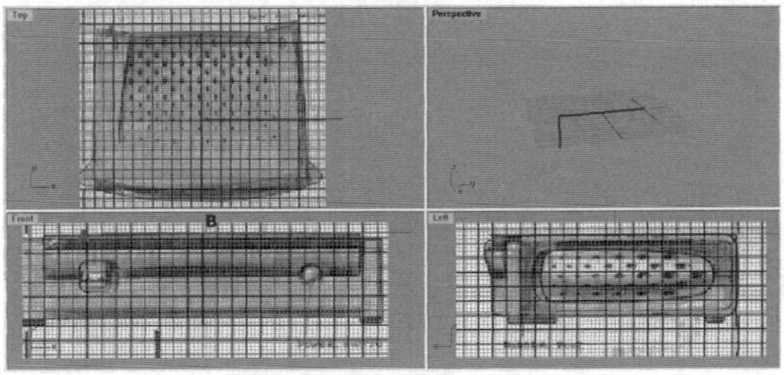

Figure 2. Reference bitmap placed by gridding.

3.3. Reference bitmap placed by photographing

In production practice, more product sample models create directly by using or referring to the product photo. So reference bitmap placed by photographing is the major means of 3D reconstruction. Effectively using of the photo image and getting useful data from it is an important way to develop new product and improve the existing products [9]. But the object which shows three-dimensional space by 2D images will be influenced by high, stadia and vision, focal length and many other factors and then have the deformation on perspective. Therefore, product's pictures of multiple angle of view are needed to be the reference of modeling. Besides, when you take a photo, not only should pay attention to minimizing the distortion deformation effects of perspective and light, but also pay attention to that the body contour and structure line should be clear. In addition, when you take a photo, the object should not be too big or too small and the lens of your camera can't get too close to the object and cannot use wide Angle lens.

When we are taking photos, the selection of reference object have two cases to ensure the minimal perspective deformation of modeling object. First, if there are the regular form of modeling object, we can judge whether these regular form is close to the surface three view drawing according to the standard such as whether the circular deformation for elliptical, rectangular and deformation for the trapezium. Second, if it is the irregular form of modeling object itself, we need to choose relatively simple geometric shapes as a reference. The choice of reference substance is flexible so that we can use local materials. The basic principle is a simple shape and the size of the reference substance is appropriate for the modeling object. If the size of the reference object is relatively smaller, the error of modeling object is bigger. If the reference object is too big, it will also have the perspective error.

The specific method is shown in figure 3. Object modeling is irregular form. In this time, we use the regular form of object as a reference to take photos. After importing the photos of all angles of views, complete the adjustment of the size and location of the reference map in the background with the reference object. The size and location of the object modeling are naturally consistent in this way. Because the perspective distortion brought by deformation is inevitable, object modeling error is inevitable even if the form rule of reference position and size consistent. As shown in figure 3, the right end of the drawn curve in the top view is obviously longer than the modeling object in the front view. It is because of the foreshortening effects and it cannot be avoided. At this point, use various references flexibly and keep the perspective of error in mind and do not operate mechanically. At the same time, we are supposed to think about how to put reference in the different views. Neither it affects counterpoint, also does not produce deformation.

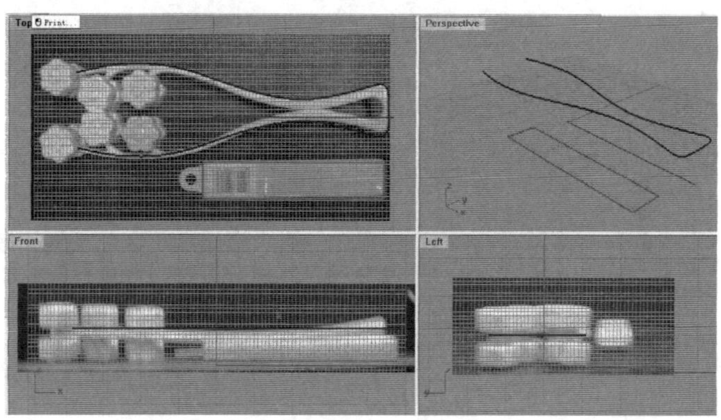

Figure 3. Reference bitmap placed by photographing.

4. The comparison and analysis of the three methods' characteristics

4.1. Reference bitmap placed by box

This method applies to the case which has accurate data and standard reference images. The characteristic of it is placing image fast and clearly. As a result of the quite high reference value of the reference bitmaps, the drawing requirement of this method is very high. The model is loyal to the drawings and it features high precision. And this method is always used for detailed modeling.

4.2. Reference bitmap placed by gridding

This method is mainly aimed at hand-drawn sketches with low accuracy and is suitable for the 3D performance of conceptual design. We should pay attention to selective reference when modeling as a result of the original drafts' low precision. The key outline and line form are which we general refer to. We can adjust the location of the Product's detail parts according to the actual situation of three views.

4.3. Reference bitmap placed by photographing

The biggest advantage of this method is to get the reference image fast. And it is suitable for modeling irregular objects form and modeling without drawings. In the process of taking photos, the effects of the perspective deformation should be reduced as far as possible. When we are making the details of model, photo reference value is not too high regardless of the form or location. It requires people to get a further understanding of the product itself.

These three methods have common ideas. Make sure of the size of the object modeling first and make full use of the grid view and fill the view as far as possible so that the error is relatively low. Then choose or draw the key line (point) flexibly as a benchmark to play an important role [10]. At last, according to the specific situation, make an auxiliary reference line to examine if the location and the size of these two key elements are consistent. The analysis of the three is shown in table 1.

Table 1. The difference among the three methods.

Method	Advantage	Disadvantage	Proper types of reference bitmap
Reference bitmap placed by box	Place image fast and clearly	Need accurate and multiple views design drawings	Precise design drawings
Reference bitmap placed by gridding	Flexible, apply to a wide range	Increase the uncertainty because of the inaccurate drafts	Design drafts
Reference bitmap placed by photographing	Get the reference image fast	Limitation of object size, perspective error	the modeling object with varied form and complex shape

5. Conclusion

Because of the large degrees of freedom of Rhino modeling, we need more reasonable planning and clear ideas. The logic thinking ability shows up as the preparing work (placing background images) before modeling. The above three methods have their own characteristics and the work for modeling provides a logical train of thinking. Only when you choose the proper methods can you model handily and do not be massy at first step. Following is the core of these three methods. Avoid form complex modeling object itself and make the drawn related reference line (or points) as the benchmark in order to make the background reference in all views are consistent with the size and location of the modeling objects. Although the reference image matters to the determination of the structure line of modeling in its infancy, the reference bitmap is only a reference role and we cannot rely too much on it.

Especially when using hand-painted sketch and entity pictures for reference, we need to process the details flexibly. The ability all we need to develop is to have an in-depth understanding of the form and have a well-thought-out plan.

References

[1] Ai Ping, Han Jun. Rhino &V-Ray Product Design and Creative Expression [M].Post and Telecom Press，2011(7):8.

[2] Zhang Quan. The Application CAID Technology in Industrial Design[J].Packaging Engineering ,2007,28(5):123-125.

[3] Ruan Wenjie, Jiang Shan. The Application of Rhino in Digital Product Modeling Design Teaching [J]. Journal of Yangzhou Education Institute, 2009, 27 (1):93-94.

[4] Yan Huijun. The Application of Rhino in the Passenger Train Modeling Design [J]. Bus Technology and Research, 2006(4):15-17.

[5] Lan Shuyuan, Luo Congmin, Xu Nifeng. Learning and Using Rhino [J]. The Computer Age, 2008(11):67-69.

[6] Li Qiguang. The Discussion between 3D Modeling Ideas of Computer Aided Industrial Design——The Comparison on 3DMAX and Rhino this Two Different Modeling Thinking [J].Inner Mongolia Science Technology & Economy, 2007(13):60-61.

[7] Ren Gongchang, Cun Yanling, Yang Junshun. The Industrial Product Modeling Design Based on Rhino 3D [J].Packing Engineering, 2008, 29(1):124-126.

[8] Li Xin. A Modeling and Animation System Based on the Technology of Hand-drawn Sketches [D]. Communication University of China, 2007:29.

[9] Yang Yanyan,Lu Changbiao,Gou Bingchen,Yu Suihuai. The Product Key Techniques of 3D Reconstruction Research Based on Perceptual Factors [J]. Application Research of Computer, 2007, 24(3):275-276,279.

[10] Zhu Yunfeng. The Vacuum Cleaner Shell Design Based on Rhino [J]. Manufacturing Information Engineering of China, 2010, 39(21):57-60.

Acknowledgments

Graduate Program Construction Project of Beijing Forestry University (HXKC18014).

Journal of Physics: Conference Series **1419** (2019) 012006 doi:10.1088/1742-6596/1419/1/012006

Application of Multi-variable Grey Model Method in Power Load Forecasting

Linghua Tong

State Grid Zhejiang Cixi Power Supply Co., Ltd., Ningbo, China

Abstract. Electric power has become one of the most important energy sources in today's society. It is of great significance for power grid enterprises and people's livelihood to accurately grasp the historical data of power load and to mine the potential value of data. In this paper, by making using of the multi-variable grey model, the power load forecasting is studied. Firstly, this paper establishes the multi-variable grey model with grey system theory. Secondly, the correlation between four temperature factors and load is analyzed, which shows that four temperature factors should be considered as the influencing factors of load. Finally, multi-variable grey model is applied to forecast the summer load of some region in Ningbo. The results show that the prediction has better fitting and prediction accuracy.

1. Introduction

Load forecasting [1,2] refers to the process of determining the future load in a certain precision sense, taking into account some important relevant factors and social influences, using a systematic approach. Accurate load forecasting can economically and reasonably arrange the start and stop of the generators of the power grid, maintain the safety and stability of the grid operation, reduce the unnecessary rotating reserve capacity, reasonably arrange the equipment maintenance plan, and ensure the normal production and life of the society. The level of load forecasting has become one of the notable signs of measuring whether a power company's management is modernizing.

For a long time, domestic and foreign scholars have conducted in-depth research on power load forecasting problems, and proposed many effective forecasting methods, such as regression analysis method [3], expert system method, fuzzy prediction method [4], time series method [5], and so on. Based on the continuous development of artificial intelligence methods, in recent 20 years, a large number of artificial intelligence methods have emerged and applied in power load forecasting. Artificial intelligence methods have strong processing capacity for non-linear problems, which can overcome the shortcomings of traditional models. Modern prediction methods are mainly as follows: artificial neural network model [6], support vector machine model [7], wavelet analysis [8], group intelligence optimization method [9], etc.

Grey system theory has many advantages, such as fewer sample data, simple theory, convenient operation, high short-term prediction accuracy and verifiability, and has been widely used [10-13]. For the power load, it is affected by many factors, such as the development of the national economy, the temperature, wind speed, economic structure and production scale changes, population growth, geographical environment and so on. Because there are many factors of electric power load, many of them are difficult to quantify directly, and it is difficult to find a clear corresponding relationship between various factors and load. Therefore, the influence factors of power load can be described by a grey space, which can be modeled and predicted by using grey system theory. In this paper, the multi-variable grey model will be used to forecast the power load in summer.

2. Multi-variable Grey Model

2.1. Grey system theory

Some information is known, some information is unknown, then this system is a gray system. For the electric load system, many factors are difficult to know exactly, so the electric load system is a gray system. The phenomena displayed in the grey process are random and chaotic, but after all, they are ordered and bounded. Therefore, this data set has potential laws. Grey prediction is a method of establishing grey model for grey system and making prediction by using this law.

The actual social and economic system is complex and often contains many variables. These variables interact with each other, interrelate with each other and develop together, forming a complex system. In this case, the development of each variable is not isolated, and the changes of each variable are affected by other variables as well as other variables. From this point of view, this paper adopts a multi-variable grey model, which is denoted by $MGM(1,n)$, aiming at describing all variables from a systematic point of view, explaining the relationship between power load, climate factors and other variables, and revealing the changing law of power load.

2.2. Establishment of Multivariable Grey Model

As a generalization of the model $GM(1,1)$ and the model $GM(1,n)$, the multi-variable grey model $MGM(1,n)$ [14] is a first-order linear differential equations consisting of n variables, which can better reflect the influence of variables on the overall model.

Let $x_i^{(0)}(k)(i=1,2,\cdots n)$ be n time series, and let $x_i^{(1)}(k)(i=1,2,\cdots n)$ be n new sequences which are generated by an accumulation as follows:

$$x_i^{(1)}(k)=\sum_{j=1}^{k}x_i^{(0)}(j) \qquad (k=1,2,\cdots m;\ i=1,2,\cdots n)$$

The first-order ordinary differential equations of $MGM(1,n)$ can be expressed as:

$$\begin{cases} \dfrac{dx_1^{(1)}}{dt}=a_{11}x_1^{(1)}+a_{12}x_2^{(1)}+\cdots+a_{1n}x_n^{(1)}+b_1 \\ \dfrac{dx_2^{(1)}}{dt}=a_{21}x_1^{(1)}+a_{22}x_2^{(1)}+\cdots+a_{2n}x_n^{(1)}+b_2 \\ \vdots \qquad \vdots \qquad \vdots \qquad \vdots \\ \dfrac{dx_n^{(1)}}{dt}=a_{n1}x_1^{(1)}+a_{n2}x_2^{(1)}+\cdots+a_{nn}x_n^{(1)}+b_n \end{cases} \tag{2.1}$$

Let

$$A=\begin{bmatrix} a_{11} & a_{12} & \cdots & a_{1n} \\ a_{21} & a_{22} & \cdots & a_{2n} \\ \vdots & \vdots & & \vdots \\ a_{n1} & a_{n2} & \cdots & a_{nn} \end{bmatrix}$$

$$B=(b_1,b_2,\cdots,b_n)^T$$

and take

$$X^{(0)}(k)=\left(x_1^{(0)}(k),x_2^{(0)}(k),\cdots x_n^{(0)}(k)\right)^T$$

$$X^{(1)}(k)=\left(x_1^{(1)}(k),x_2^{(1)}(k),\cdots,x_n^{(1)}(k)\right)^T$$

Then, (2.1) can be rewritten as:

$$\frac{dX^{(1)}}{dt}=AX^{(1)}+B \tag{2.2}$$

Here A and B are parameters to be identified. Thus, the continuous time response of equation (2.1) is

$$X^{(1)}(t)=e^{A(t-t_0)}X^{(1)}(t_0)+A^{-1}\left(e^{A(t-t_0)}-I\right)B \tag{2.3}$$

In order to identify the parameters A and B, the differential quotient in the equation (2.1) is replaced by the difference quotient with equal time interval. Then， equation (2.1) can be discretized to get

$$x_i^{(0)}(k)=\sum_{j=1}^{n}\frac{a_{ij}}{2}\left(x_j^{(1)}(k)+x_j^{(1)}(k-1)\right)+b_i$$

where we note that

$$x_i^{(1)}(k)-x_i^{(1)}(k-1)=x_i^{(0)}(k)$$

Let $a_i=(a_{i1},a_{i2},\cdots,a_{in},b_i)^T, i=1,2,\cdots,n$. The estimated value $\hat{a}_i$ of a_i can be obtained by the least square method:

$$\hat{a}_i=\begin{bmatrix}\hat{a}_{i1}\\ \hat{a}_{i2}\\ \vdots\\ \hat{a}_{in}\\ \hat{b}_i\end{bmatrix}=\left(L^TL\right)^{-1}L^TY_i,\qquad i=1,2,\cdots n$$

Among them,

$$L=\begin{bmatrix}\frac{1}{2}\left(x_1^{(1)}(2)+x_1^{(1)}(1)\right) & \cdots & \frac{1}{2}\left(x_n^{(1)}(2)+x_n^{(1)}(1)\right) & 1\\ \frac{1}{2}\left(x_1^{(1)}(3)+x_1^{(1)}(2)\right) & \cdots & \frac{1}{2}\left(x_n^{(1)}(3)+x_n^{(1)}(2)\right) & 1\\ \vdots & & \vdots & \vdots\\ \frac{1}{2}\left(x_1^{(1)}(m)+x_1^{(1)}(m-1)\right) & \cdots & \frac{1}{2}\left(x_n^{(1)}(m)+x_n^{(1)}(m-1)\right) & 1\end{bmatrix}$$

$$Y_i=\left(x_i^{(0)}(2),x_i^{(0)}(3),\cdots,x_i^{(0)}(m)\right)^T$$

At the same time, the estimated values $\hat{A}$ and $\hat{B}$ of parameters A and B can be obtained:

$$\hat{A}=\begin{bmatrix}\hat{a}_{11} & \hat{a}_{12} & \cdots & \hat{a}_{1n}\\ \hat{a}_{21} & \hat{a}_{22} & \cdots & \hat{a}_{2n}\\ \vdots & \vdots & \vdots & \vdots\\ \hat{a}_{n1} & \hat{a}_{n2} & \cdots & \hat{a}_{nn}\end{bmatrix},\hat{B}=\left(\hat{b}_1,\hat{b}_2,\cdots \hat{b}_n\right)^T$$

Thus, by (2.3), the multi-variable grey model $MGM(1,n)$ can be calculated as

$$\begin{cases} \hat{X}^{(1)}(k) = e^{\hat{A}(k-1)}X^{(0)}(1) + A^{-1}(e^{\hat{A}(k-1)} - I)B,\ k = 2,3... \\ \hat{X}^{(1)}(1) = X^{(1)}(1) = \hat{X}^{(0)}(1) = X^{(0)}(1) \\ \hat{X}^{(0)}(k) = \hat{X}^{(1)}(k) - \hat{X}^{(1)}(k-1), k = 2,3... \end{cases} \quad (2.4)$$

where $\hat{X}^{(0)}(k)$ is the predicted value.

3. Correlation Analysis of Ningbo Power Data

The variation of power load is closely related to meteorological conditions. Among the meteorological factors, temperature has the greatest impact on load. How to inspect the impact of meteorological factors on power load and how to inspect the comprehensive impact of these factors on power load are the urgent problem to be solved in current load characteristic analysis and load forecasting. In this paper, the data of power load and temperature in summer in Ningbo are used to carry out correlation analysis, and the results are as follows.

Table 1. Correlation analysis of load at 11 am.

Maximum temperature	minimum temperature	Average temperature	8 am Temperature
0.8615	0.8656	0.8550	0.8542

As can be seen from Table 1, the four indicators of maximum temperature, minimum temperature, average temperature and 8 am temperature are highly correlated with the load of 11 am in summer. In the load forecasting of this paper, these four temperature factors should be considered as the influencing factors of load.

4. Power load forecasting

4.1. Establishment of $MGM(1,5)$

In the following, this paper will use multi-variable grey model to forecast the summer load of some region in Ningbo at 11 a.m. Through the above correlation analysis, maximum temperature, minimum temperature, average temperature and 8 am temperature should be considered as four variables in the prediction model. In addition, with load as a variable, the forecasting model in this paper should has 5 variables. Thus, $n = 5$. Let $x_1\ x_2\ x_3\ x_4\ x_5$ denote load, maximum temperature, minimum temperature, average temperature, and 8 am temperature. The multi-variable grey model $MGM(1,5)$ to forecast the load at 11 am can be established as follows

$$\begin{cases} \frac{dx_1^{(1)}}{dt} = a_{11}x_1^{(1)} + a_{12}x_2^{(1)} + a_{13}x_3^{(1)} + a_{14}x_4^{(1)} + a_{15}x_5^{(1)} + b_1 \\ \frac{dx_2^{(1)}}{dt} = a_{21}x_1^{(1)} + a_{22}x_2^{(1)} + a_{23}x_3^{(1)} + a_{24}x_4^{(1)} + a_{25}x_5^{(1)} + b_2 \\ \frac{dx_3^{(1)}}{dt} = a_{31}x_1^{(1)} + a_{32}x_2^{(1)} + a_{33}x_3^{(1)} + a_{34}x_4^{(1)} + a_{35}x_5^{(1)} + b_3 \\ \frac{dx_4^{(1)}}{dt} = a_{41}x_1^{(1)} + a_{42}x_2^{(1)} + a_{43}x_3^{(1)} + a_{44}x_4^{(1)} + a_{45}x_5^{(1)} + b_4 \\ \frac{dx_5^{(1)}}{dt} = a_{51}x_1^{(1)} + a_{52}x_2^{(1)} + a_{53}x_3^{(1)} + a_{54}x_4^{(1)} + a_{55}x_5^{(1)} + b_5 \end{cases}$$

By making using of the matrix operation method in Section II, the following estimations $\hat{A}$ and $\hat{B}$ of A and B can be obtained

$$\hat{A}=\begin{bmatrix} -0.0983 & -0.0182 & -0.0407 & -0.0023 & -0.0063 \\ 15.0939 & 0.4892 & 0.6413 & 0.5940 & 0.5010 \\ -0.8054 & 0.2136 & 0.1705 & 0.1037 & 0.1872 \\ -1.1687 & 0.1260 & 0.0486 & 0.1280 & 0.0366 \\ -12.9034 & -0.7629 & -0.6641 & -0.8363 & -0.7320 \end{bmatrix} \quad \hat{B}=\begin{bmatrix} 111.7176 \\ 24.0888 \\ 25.4516 \\ 22.7369 \\ 24.4804 \end{bmatrix}$$

Thus, the $MGM(1,5)$ model that was finally established in this application is as follows:

$$\begin{cases} \frac{dx_1^{(1)}}{dt}=-0.0983x_1^{(1)}-0.0182x_2^{(1)}-0.0407x_3^{(1)} \\ \qquad -0.0023x_4^{(1)}-0.0063x_5^{(1)}+111.7176 \\ \frac{dx_2^{(1)}}{dt}=15.0939x_1^{(1)}+0.4892x_2^{(1)}+0.6413x_3^{(1)} \\ \qquad +0.5940x_4^{(1)}+0.5010x_5^{(1)}+24.0888 \\ \frac{dx_3^{(1)}}{dt}=-0.8054x_1^{(1)}+0.2136x_2^{(1)}+0.1705x_3^{(1)} \\ \qquad +0.1037x_4^{(1)}+0.1872x_5^{(1)}+25.4516 \\ \frac{dx_4^{(1)}}{dt}=-1.1687x_1^{(1)}+0.1260x_2^{(1)}+0.0486x_3^{(1)} \\ \qquad +0.1280x_4^{(1)}+0.0366x_5^{(1)}+22.7369 \\ \frac{dx_5^{(1)}}{dt}=-12.9034x_1^{(1)}-0.7629x_2^{(1)}-0.6641x_3^{(1)} \\ \qquad -0.8363x_4^{(1)}-0.7320x_5^{(1)}+24.4804 \end{cases}$$

According to formula (2.4), we can forecast the load of some region in Ningbo at 11 am in summer. The predicted and measured values are listed in Table 2.

Table 2. Predicted and measured values at 11 am of in summer.

Date	Measured value(MW)	Fitted value (MW)	Date	Measured value(MW)	Fitted value (MW)
1	114.78	122.01	13	183.41	191.57
2	116.13	118.24	14	186.14	188.30
3	116.12	118.29	15	181.44	182.01
4	124.65	122.01	16	174.95	173.32
5	133.68	129.01	17	173.86	163.09
6	154.66	138.53	18	192.53	152.27
7	168.05	149.62	19	180.44	141.85
8	160.01	161.16	20	152.77	132.73
9	165.05	172.05	21	150.31	125.67
10	169.73	181.25	22	138.94	121.19
11	172.96	187.92	23	122.84	119.52
12	171.29	191.45	24	136.99	120.57

The fitting diagram is shown in the following figure.

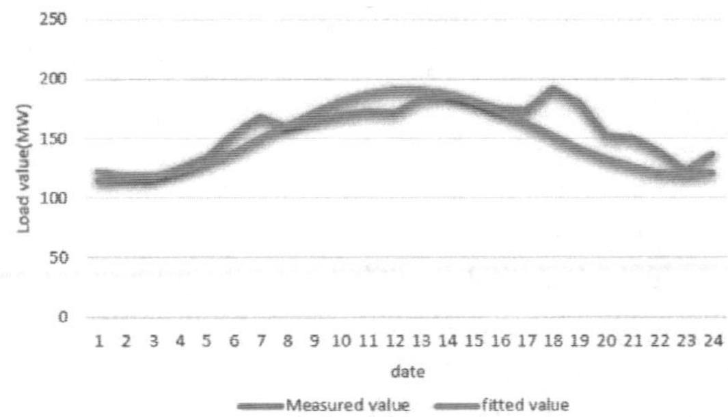

Figure 1. Comparison of measured values and fitted values at 11 am.

4.2. Model prediction

By making use of the multi-variable grey model $MGM(1,5)$, the power load as 11 am in the next 5 days in summer can be predicted. The results are compared with the actual measured values, as shown in Table 3 and Figure 2, respectively.

Table 3. Forecast of power load at 11 am in the next 5 days.

Date	Measured value	Predictive value	Relative error
25	128.16	130.99	2.21%
26	143.64	139.13	3.14%
27	146.14	143.15	2.05%
28	150.77	146.06	3.12%
29	150.40	148.85	1.03%

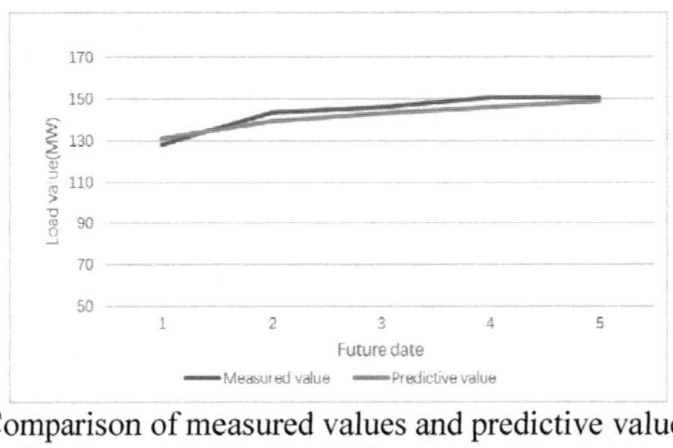

Figure 2. Comparison of measured values and predictive values at 11 am.

The above results show that the prediction has high accuracy. In the same way, we can predict the load at other time through the multi-variable grey model $MGM(1,5)$. Due to space limitations, this article will not list them here.

5. Summary and outlook

In this paper, the multi-variable grey model is used to forecast the power load in summer. The model takes into account the effects of the highest temperature, the minimum temperature, the average temperature and the 8 am temperature on the load, and better reflects the interaction and correlation between the load and the temperature. The results show that the method has good accuracy. This method can be used to predict power load, and provide reliable basis for the scientific and rational implementation of DSM in power sector, so as to improve the efficiency of terminal power consumption and optimize the allocation of resources.

References

[1] Kang C, Xia Q. Review of power system load forecasting and its development[J]. Automation of Electric Power Systems, 2004, 28(17):1-11.
[2] Hippert H S, Pedreira C E, Souza R C. Neural networks for short-term load forecasting: a review and evaluation[J]. IEEE Transactions on Power Systems, 2001, 16(1):0-55.
[3] Song K B, Baek Y S, Hong D H, et al. Short-term load forecasting for the holidays using fuzzy linear regression method[J]. IEEE Transactions on Power Systems, 2005, 20(1):96-101.
[4] Papadakis S E, Theocharis J B, Bakirtzis A G. A load curve based fuzzy modeling technique for short-term load forecasting[J]. Fuzzy Sets & Systems, 2003, 135(2):279-303.
[5] Parrmann L D, Mohamed D N. Adaptive online load forecasting via time series modeling[J]. Electric Power Systems Research, 1995, 32(32):219–225.
[6] Li L, Huang W. A Short-Term Power Load Forecasting Method Based on BP Neural Network[J]. Applied Mechanics & Materials, 2014, 494-495:1647-1650.
[7] Fan Z, Zhong Q, Hong T, et al. The studying of combined power-load forecasting by error evaluation standard based on RBF network and SVM method[C]. International Conference on Chinese Control & Decision Conference. 2009.
[8] Zhai M Y. A new method for short-term load forecasting based on fractal interpretation and wavelet analysis[J]. Power System Technology, 2015, 69(1):241-245.
[9] Wei S, Ying Z. Short Term Load Forecasting Based on BP Neural Network Trained by PSO[C]. International Conference on Machine Learning & Cybernetics. 2007.
[10] Hsu C C, Chen C Y. Applications of improved grey prediction model for power demand forecasting[J]. Energy Conversion & Management, 2003, 44(14):2241-2249.
[11] Yin W, Meng,P, Li,Y. Application of improved grey prediction model in Jilin Province of the software Industry[J]. Advanced Materials Research, 2014, 998-999:1079-1082.
[12] Li M, Sha X. Application of improved GM(1,1) grey prediction model[J]. Computer Engineering & Applications, 2016.
[13] Lei M , Feng Z . A proposed grey model for short-term electricity price forecasting in competitive power markets[J]. International Journal of Electrical Power & Energy Systems, 2012, 43(1):531-538.
[14] Zhai J, Sheng J, Feng Y. The grey model MGM(1,n) and its application[J]. Systems Engineering Theory & Practice, 1997, 17(5):109-113.

Acknowledgments

This work was supported by State Grid Zhejiang Electric Power Co., Ltd. Science and Technology Project of Collective Enterprises in 2019: Power Emergency Assessment System for Bay Area Based on Vulnerability.

Journal of Physics: Conference Series **1419** (2019) 012007 doi:10.1088/1742-6596/1419/1/012007

Assessment of Environmental Degradation Costs: A Case Study of Beijing and Tianze Agricultural Parks

Liufen Zhang, Zebin Zhang, Yanxin Ma and Xiangdong Liu

Jinan University, Guangzhou,GuangDong, China

E-mail: 1426161942@qq.com, tliuxd@jnu.edu.cn

Abstract. Nowadays, the community of human destiny has gradually become the consensus of people. The pursuit of economic benefits and the neglect of ecological benefits will bring serious consequences. Therefore, predicting the cost of environmental degradation for managers is an urgent problem we need to solve.

We establish a mathematical model to predict the cost of environmental degradation. And we ignore some insignificant factors and choose seven factors as indicators in our model under the consideration of actuality. To make the model more intuitive, we transform environmental degradation costs into ecosystem services values and calculate the value of each service in the ecosystem separately. At the same time, we use Analytic Hierarchy Process to calculate the weight of each service according to the judgment matrix. Then we introduce a concept in economics " Weighted Average Cost of Capital (WACC) ". Hence, we can derive the final cost of environmental degradation. Next, we take Beijing's construction as an example of a large national project and Tianze agricultural park as an example of a small community-based project to estimate their respective degradation costs.

As for the effectiveness analysis, we use polynomial 7-th approximation to fit curve and prove its effectiveness successfully. According to the data we get from fitting curve, we analyze the ideal environment level of 2020. The result shows that land use projects exist many shortcomings and we give land use project planners and managers some advice to help them reach the ideal environment level in 2020.

Since ecosystem service valuation is pretty complex and some indicators are influenced by humanity factor, we cannot ensure the accuracy of our calculation. We improve our model and add fuzzy mathematic method into AHP. By using Fuzzy Analytic Hierarchy Process, we can let the weight change automatically over time. Therefore, we success in improving our model and adding time factor into our improved model.

Finally, we analyze our works and then discuss strengths and weaknesses of our model.

1. Introduction

We gradually realize that our environment has degraded, and that we have lost the function that environment brings to us and the world——ecosystem services. Ecosystem services are the many and varied benefits that humans freely gain from the natural environment and from properly-functioning ecosystems.

People tend to ignore that land use projects may cause environmental degradation and limit or remove ecosystem services. Especially in small-scale land use projects, these impacts easily seem negligible. However, the accumulation of these small projects must be the negative impact on ecosystem and lead to environmental degradation. If we take these ecosystem service functions into consideration of economic costs, we can reduce the negative impact on the biosphere caused by land use. First of all, our task is to establish an ecological services valuation model to account for environmental degradation costs into land use development projects. Then we can use this model to calculate the true economic cost of land use projects. Next, we need to perform a cost benefit analysis of land use projects of varying sizes through our model. Thirdly, we need to evaluate the effectiveness of our model. Finally, we add time parameters to the model to predict future land use project costs and use it to suggest land use project planners and managers.

In fact, the concept of ecosystem services is not a new one nowadays. As early as 1974, Holdren and Ehrlich proposed it. However, until 1997, Costanza estimated the value of ecosystem services which really arouse the world's attention to this topic [1] .In 2005, the concept of ecosystem services gained broader attention when the United Nations published its Millennium Ecosystem Assessment (MEA). On the basis that Costanza has defined the connotation of ecosystem service functions, researchers in different countries have proposed various methods for the value of ecosystem services, including the equivalent method, market value method and shadow price method. On the other hand, different researchers have various classifications of ecosystem service functions. Hence, we make categories of ecosystem service ourselves according to one of numerous papers [2], simplifying and modifying as we need. Then we integrate several methods for the value of ecosystem services above in our model to estimate ecosystem services value.

2. Analysis

Natural environmental resources such as the atmosphere, water and land are non-exclusive resources that have long been used freely by people [3] In the past, environmental values have not received enough attention. Nowadays, with the development of society, we have gradually realized the importance of the value of ecosystem services. We have begun to demand a corresponding price for some projects that consume natural resources. For this purpose, we establish a model for quantifying the cost of environmental degradation.

The value of ecological services is a relatively abstract concept and is hard to quantify. When we define this concept and refer to the methods used in the World Bank report[4] and other articles, we decide to use economic methods such as environmental cost value to quantify.

The cost of environmental degradation is measured by estimating the value of the original ecosystem and the possible consequences of development. First, we divide ecosystem services into three types. Among the three types, we ignore the services with less economic value and consider the specific seven services. Secondly, we introduce the concept of "Weighted Average Cost of Capital" to further refine our ecological environment value. Furthermore, instead of simply using the analytic hierarchy process, we use the fuzzy analytic hierarchy process to obtain the proportion of these four types in the evaluation of ecosystem service value. At the same time, through methods like the shadow price method and market assessment method, we quantify the value of these specific services and combine them with their respective weights to arrive at the final environmental degradation cost calculation formula. With the WACC formula, we can calculate the cost of environment degradation.

3. Assumptions and Justifications

- The ecological environment value can be quantified, and we use this to estimate the cost of environmental degradation.
- We assume that the functions of the ecosystem have three broad categories, and the three categories can be subdivided into seven indicators.
- We assume that each ecosystem could produce their own products. They are independent of each other and do not affect each other.
- To simplify the calculation, we assume that the land's ability to dispose of waste is the ability to digest nitrogen and phosphorus in the land. At the same time, we consider that the value of water conservation in the land is the cost of storing water.

- We believe that the various functions of the soil can be transformed by methods such as market value method and shadow price method because they are abstract.
- We assume that the relevant data in the model will not change in the short term, which means that the price and other indicators are sticky.
- We assume that the value of the main elements of the main nutrient is approximately equal to the soil nutrient value in the ecosystem for the convenience of calculation.
- We ignore some parts of ecosystem services value because of their negligible influence on the final result.

4. Notations

Symbols	***Definition***	***Unit***
E	The ecosystem services value	yuan
A_i	The indicator of the i-type of ecosystem services	
B_j	The indicator of specific services j	
β_{Ai}	The weight of A_i	
β_{Bj}	The weight of B_j	
E_j	The value of services j	yuan
W_c	The amount of water conservation	t
P_w	The cost of storing unit water	yuan/t
S_p	The area of the plant of the project	hm2
P_c	The price of carbon	yuan/t
P_o	The price of oxygen	yuan/t
B_n	The annual net productivity of the plant	t/(hm2*age)
R_c	The rate of carbon content of carbon dioxide	
V_c	The annual value of plant carbon sequestration	yuan/age
V_o	The annual value of plant oxygen release	yuan/age
S_l	The area of the cultivated land of the project	hm2
M	Cultivated land erosion modulus	t/(hm2*age)
W_l	The reduction amount of cultivated land soil erosion	t
T_l	The average thickness of the land tillage layer	m
W_a	The reduction area of land resource	hm2
I	The annual average income of cultivated land	yuan/hm2
V_e	The value of cultivated land reduction soil erosion	yuan
V_s	The value of reduce siltation	yuan
R_N	The nitrogen content in unit erosion	g/kg
R_P	The phosphorus content in unit erosion	g/kg
R_K	The potassium content in unit erosion	g/kg
P_N	The market price of nitrogen	yuan/kg
P_P	The market price of phosphorus	yuan/kg
P_K	The market price of potassium	yuan/kg
V_f	The fixed fertility value	yuan
R_m	The organic matter content in unit erosion	g/kg
P_f	The average price of fuel wood	yuan/kg
V_m	The value of organic matter	yuan
S_r	The area with cultural research ecosystem value	hm2
P_r	The cultural research value per unit area	yuan/hm2
H	The total tourism revenue in the project area	yuan

R_t	The proportion of ecological attractions to the total number of tourist attractions	
V_n	The functional value of nutrient cycling in cultivated land	yuan
V_b	The value of the biobank involved in the circulation of nutrients	yuan

5. The Basic Model of Ecosystem Service valuation

In our model, the value of ecological services is divided into three categories: regulation service, cultural service, and support service. These three types are divided into air quality regulation value, erosion regulation and soil formation value, culture and research value, recreation and ecotourism value, nutrient cycling value, land consumption of manure value and water conservation. We calculate the value of each of these seven categories separately [5], and then calculate the final value based on the weight. Finally, we can arrive at the cost of land degradation.

Goal	Criteria	Indicator
Ecosystem Service Valuation ***E***	Regulating Services A_1	Air quality regulation B_1
		Erosion regulation and soil formation B_2
	Cultural Services A_2	Culture and research B_3
		Recreation and ecotourism B_4
	Supporting Services A_3	Nutrient cycling B_5
		Land consumption of manure B_6
		Water conservation B_7

Figure 1. Hierarchy structure.

According to our classification, we make a hierarchy structure above (Figure 1).
First of all, we use Analytic Hierarchy Process (AHP) method to calculate the weight of each indicator.

5.1. The process of AHP

Analytical Hierarchy Process is a methodology developed by Saaty[6] which can be used in any situation with many integrating, interrelating and interdependent factors. By using this method, we can give each valuation a weight.
The AHP algorithm has four major steps:

5.1.1. Establish hierarchical structure model

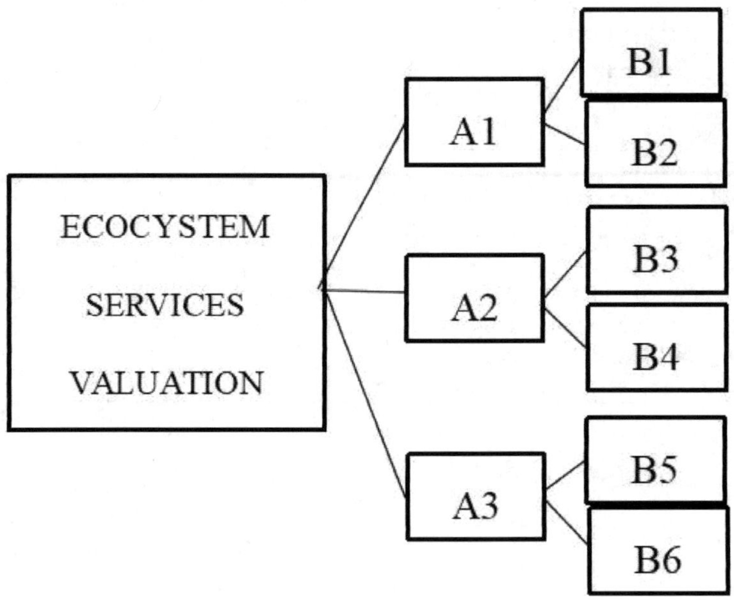

Figure 2. Hierarchy structure.

5.1.2. Construct judgement matrix

Since the hierarchy structure is established, we decision-maker should derive "1-9" ratio-scale [6] priorities reflecting the relative importance of objectives via pairwise comparisons with respect to the goal of the problem. Following this rule, we decision-makers can derive ratio-scale which reflects the priority of each ecosystem service valuation.

By comparing the effect of indexes and add our judgement from data we have into consideration, we construct the pairwise comparison matrix E-B_i and B_i-C_j.

where i=1,2,3,4 and j=1,2, … ,8

While choosing the definition of scales, we employ the research result of Satty[6].

In psychology, the more gradients of scale, the more difficulties we will meet while judging. Finally,we choose "1-9" ratio-scale chart.

Intensity of importance	Definition
1	Equal importance
3	Weak importance
5	Strong importance
7	Demonstrated importance
9	Absolute importance
2,4,6,8	Intermediate values

Chart 1: Ratio-scale chart of AHP

5.1.3. Hierarchical sorting and consistency testing

As we all know, N-th order positive reciprocal matrix A is uniform matrix if and only if the largest eigenvalue cis equal to n, else the largest eigenvalue λ max is greater than n. Therefore, we can utilize the λ max to test the consistency of each judgement matrix.

Firstly, calculate the coincidence indicator CI:

$$CI = \frac{\lambda_{\max} - n}{n-1} \tag{1}$$

Secondly, find out the corresponding mean random consistency index RI from the following chart given by Saaty [6].

n	1	2	3	4	5	6	7	8	9
RI	0	0	0.58	0.90	1.12	1.24	1.32	1.41	1.45

Chart 2: The value of *RI*

Thirdly, run out the consistency ratio CR

$$CR = \frac{CI}{RI} \tag{2}$$

If CR $<$ 0.10, we can consider the consistency is acceptable, else we should go back to Step2 and reconstruct the judgement matrix.

5.1.4. Calculate the weight of each indicator

As to each judgement matrix, find out the eigenvector corresponding to the maximum eigenvalue λ max and standardize it. Then we have the weight of each indicator.

5.2. The seven kinds ecosystem services value

5.2.1. The air quality regulation value

In this paper, two indicators of carbon fixation and oxygen release are used to reflect this function.
Carbon sequestration value V_c:

$$V_c = 1.63 \times S_p \times P_c \times R_c \times B_r \tag{3}$$

V_c is the annual value of plant carbon sequestration; S_p is the area of the plant of the project(hm2); P_c represents the price of carbon(yuan/t); R_c represents the rate of carbon content of carbon dioxide; B_n is the rate of carbon content of carbon dioxide(t/(hm2*age)).
Oxygen release value V_o:

$$V_o = 1.19 \times S_p \times P_o \times R_c \times B_n \tag{4}$$

V_o is the annual value of plant oxygen release; Po represents the price of oxygen(yuan/t).
Therefore, the air quality regulation value E_1:

$$E_1 = V_c + V_o \tag{5}$$

5.2.2. The erosion regulation and soil formation value

The soil-fixing and fertilizing functions of the land have three aspects: maintaining soil fertility, reducing sedimentation in rivers and lakes, and fixing fertility (fertilizer and organic fertilizer). We use the alternative market method[5] and shadow engineering method[5] to calculate the value of these three aspects of land development. And the following is the equation.

$$E_2 = V_e + V_s + V_f + V_m \tag{6}$$

where

- E_2 represents the value of erosion regulation and soil formation;
- V_e represents the value of cultivated land reduction soil erosion;
- V_s represents the value of reduce siltation;
- V_f represents the fixed fertility value;
- V_m represents the value of organic matter.

The formula for calculating the total amount of soil erosion in cultivated land is

$$W_1 = S_1 \times M \tag{7}$$

The area W_a of farmland to reduce the loss of land resources is obtained by the total reduction amount of cultivated land soil erosion divided by the average thickness of the cultivated layer of the land.

$$W_a = \frac{W_1}{T_1} \tag{8}$$

where

- S_l represents the area of the cultivated land of the project;
- M represents cultivated land erosion modulus;
- W_a represents the reduction area of land resource(hm2);
- W_l represents the reduction amount of cultivated land soil erosion(t);
- T_l represents the average thickness of the land tillage layer(m).

The annual average income of cultivated land is used as the opportunity cost of reducing the abandoned land to calculate the value of the total amount of soil erosion. V_e is the value of reducing the total amount of soil erosion in cultivated land. I is the annual average income of cultivated land. The value of land erosion can be obtained by substituting above numbers.

$$V_e = W_a \times I \tag{9}$$

The value of reducing siltation

According to the laws of sediment movement in China's main river basins, 24% of the sediments lost by soil erosion redeposited in reservoirs, rivers and lakes. This part of the sediment directly causes the decrease of water storage of reservoirs, rivers and lakes, which increase the opportunities for drought and flood disasters, and another 33% have been stranded and 37% have entered the sea [7]. This paper only considers the siltation of 24% of reservoirs and rivers and lakes, which is the annual economic value of sedimentation. According to the shadow engineering method, we calculate the value of reducing sedimentation in rivers and lakes:

$$V_s = W_1 \times 24\% \times P_W \tag{10}$$

The value of fixed fertility

The value of fixed fertility is mainly calculated based on the value of soil fertility in reducing soil loss. Among them, the value of organic matter loss is determined by the cost of fuel wood that can be increased by cultivated land.

$$V_m = \sum (R_j \times P_j) \times W_1 \tag{11}$$

$$V_m = W_1 \times R_m \times P_f \tag{12}$$

where

- R_j is the content of the j-th nutrient element in the unit erosion (g/kg);
- P_j is the market price of the j-th nutrient element (yuan/kg);
- R_m is the organic matter content in unit erosion;
- P_f is the average price of fuel wood.

5.2.3. The value of cultural and research

The value of cultural research functions of various ecosystems mainly includes related basic scientific research, teaching internships, and cultural propaganda. To accurately calculate the value of cultural research functions often requires data such as relevant research investment and actual research costs, only the general value can be referred to since the relevant information cannot be obtained at present.

$$E_3 = S_r \times P_r \tag{13}$$

where

- Sr represents the area with cultural research ecosystem value;
- Pr represents the cultural research value per unit area.

5.2.4. The value of recreation and ecotourism

The value of leisure and entertainment can be calculated according to the proportion of ecotourism in the value of tourism.

$$E_4 = H \times R_t \tag{14}$$

where

- H represents the total tourism revenue in the project area;
- R_t represents the proportion of ecological attractions to the total number of tourist attractions.

5.2.5. The nutrient cycling value[5]

The nutrient cycling of ecosystems is mainly carried out between biobanks, litter banks and soil banks. Our model only considers soil banks and biobanks. In this paper, only the relatively high levels of nitrogen, phosphorus and potassium are considered for the nutrients in the biobanks and soil banks involved in the evaluation.

$$E_5 = V_n + V_b \tag{15}$$

$$V_n = \sum_{1}^{m} NPP_J (C_{nj} P_n + C_{pj} P_{nk} + C_{kj} P_k) V_n = \sum_{1}^{m} NPP \tag{16}$$

$$V = \sum_{1}^{m} M_J (S_{nj} P_n f_n + S_{pj} P_p f_p + S_{kj} P_k f_k) V \tag{17}$$

E_5 is the functional value of the nutrient circulation of the land; V_n is the value of the bio-library involved in the circulation of nutrient elements; C_{nj} is the percentage of nitrogen in the biomass of the j-type agricultural products. C_{pj} is the percentage of phosphorus in the biomass of the j-type agricultural products; C_{kj} is the percentage of potassium in the biomass of the j-type agricultural products; P_n, P_p and P_k correspond to the market price of nitrogen, phosphorus and potassium respectively. V_b is the value of nutrient cycling in the soil bank. M_J is the total soil pool of the j-type agricultural products; S_{nj} is the percentage of nitrogen in the soil bank of the j-type agricultural products; S_{pj} is the percentage of phosphorus in the soil bank of the j-type agricultural products; S_{kj} is the j-th percentage of potassium in the soil bank of agricultural products.

5.2.6. The value of land consumption of manure.

The land consumption of livestock manure can reduce the pollution of livestock and poultry manure. Our model mainly considers the ability of land to absorb nitrogen and phosphorus. We determine the calculation method for arable land consumption waste:

$$E_6 = S_1 \times (\sum V_j \times P_j) \tag{18}$$

V_j is the amount of nitrogen and phosphorus in the year of cultivation. P_j is the price of j fertilizer, and S_l is the area of cultivated land.

5.2.7. The value of land conservation water source

$$E_7 = W_c \times P_w \tag{19}$$

where

- E_7 is the value of water conservation.
- W_c is the conservation water source.
- P_w is cost of storing unit water.

5.3. Calculation of the cost of Environmental Degradation

We combine the cost of environmental degradation and economic method WACC together, by defining parameters in this method we can transform this cost into economic category, so that we success to quantify this cost.

$$W_i = \beta_{B_i} \tag{20}$$

$$K_i = E_i \tag{21}$$

$$WACC = \sum_{i=1}^{n} W_i \times K_i \tag{22}$$

6. Application of models in land use projects

6.1. Calculate the weights

Following the process of AHP, we consider different degrees of indicators and get these judgement matrixes.

① Judgement Matrix E~A:

$$E \sim A = \begin{pmatrix} 1 & 5 & 3 \\ \frac{1}{5} & 1 & \frac{1}{3} \\ \frac{1}{3} & 3 & 1 \end{pmatrix}$$

	A_1	A_2	A_3	βA
A_1	1	5	3	0.6370
A_2	$\frac{1}{5}$	1	$\frac{1}{3}$	0.1047
A_3	$\frac{1}{3}$	3	1	0.2583

CI=0.0193; CR=0.0332<0.1000

So, matrix E~A through the consistency examination.

② Judgement Matrix A1~B:

$$A_1 \sim B = \begin{pmatrix} 1 & 3 \\ \frac{1}{3} & 1 \end{pmatrix}$$

	B_1	B_2	βB
B_1	1	3	0.7500
B_2	$\frac{1}{3}$	1	0.2500

CI=0; CR=0<0.1000

So, matrix E~A through the consistency examination.

③ Judgement Matrix A2~B:

$$A_1 \sim B = \begin{pmatrix} 1 & \frac{1}{7} \\ 7 & 1 \end{pmatrix}$$

	B_3	B_4	βB
B_3	1	$\frac{1}{7}$	0.1250
B_4	7	1	0.8750

CI=0; CR=0<0.1000

So, matrix E~A through the consistency examination.

④ Judgement Matrix A3~B:

$$A_3 \sim B = \begin{pmatrix} 1 & 2 & \frac{1}{5} \\ \frac{1}{2} & 1 & \frac{1}{7} \\ 5 & 7 & 1 \end{pmatrix}$$

	B_5	B_6	B_7	βB

B_5	1	2	$\frac{1}{5}$	0.1666
B_6	$\frac{1}{2}$	1	$\frac{1}{7}$	0.0938
B_7	5	7	1	0.7396

CI=0.0193; CR=0.0332<0.1000

So, matrix E~A through the consistency examination.

After calculating all the weights, we can get the total sequencing:

Table 1. Total sequencing.

	A1 ***0.6370***	***A2*** ***0.1047***	***A3*** ***0.2583***	***Sequencing***
B1	0.7500	0	0	**0.4778**
B2	0.2500	0	0	**0.1593**
B3	0	0.1250	0	**0.0131**
B4	0	0.8750	0	**0.0916**
B5	0	0	0.1666	**0.0430**
B6	0	0	0.0938	**0.0242**
B7	0	0	0.7396	**0.1910**

$$CR = \frac{0.0193 \times 0.2583}{0.5800 \times 0.2583} = 0.0333 < 0.1000$$

6.2. Application in small community-based projects

Using the basic model we establish above, we take "Tian-ze agricultural park construction project" [8]as an example. By combining the data[8] and WACC together, we finally work out the cost of this project.

Table 2. The data of the small project.

Indicator	**Amount**
C(t)	377561.72
O(t)	275643.22
Soil Erosion Modulus ($t/(km^2a^1)$)	1940
Reduction in Soil Erosion(t)	9113.78
Soil Water Storage(m^3)	12492.18
Area(m^3)	15188.93
Soil Fertility (yuan)	33800

WCAA(S. project) = 2.3721×10⁷ yuan

6.3. Application in large national projects

Similarly, we take "Beijing urban and rural planning standardized land use project" [5]as an example. Finally, we work out the cost of this project.

Table 3. The data of the large project.

Indicator	**Amount**
Soil turnover of *N*	0.08
Soil turnover of *P*	0.01
Soil turnover of *K*	0.01

Average content of elements in biological library-*N*	3.09%
Average content of elements in biological library-*P*	0.74%
Average content of elements in biological library-*K*	3.28%
The cost of the reservoir (yuan/m^3)	6.48
Mean soil density(g/cm^3)	1.3
Mean soil thickness(m)	0.5
Average land income(yuan/hm^2)	32000
Annual consumption of *N*(kg/hm^2)	60
Annual consumption of *P*(kg/hm^2)	12
Price of *N* (yuan/t)	3152
Price of *P* (yuan/t)	8823
Price of *K* (yuan/t)	3455
Price of Oxygen (yuan/t)	1000
Price of Carbon (yuan/t)	1200
The carbon content of carbon dioxide	27.27%
Scientific research value of unit area of wetland(yuan/(a×hm^2)	382

$$WCAA(L.\ project) = 7.6913 \times 10^9\ yuan$$

7. The effectiveness of the model

As the ecosystem service valuation is hard to quantize, we apply AHP method into our model and it is used to measure the importance between these indicators and allocate weights of each one. However, AHP is a method with too many humanity factors, that is to say, it has poor objectivity. Comparing to linear addition method, giving each indicator a weight is more important. Because each indicator plays different role in different ecosystem, while using AHP we judge each importance according to data we get from specific projects so that we can reduce some error led by subjectivity.

7.1. Polynomial 7-th Approximation

To test the effectiveness, we decide to apply our model to China Environment Data from 2000-2015. After working out all the valuation, we get a set of data and we fit curve through tendency of increase and decrease these data. Using the curve we have, we calculate the data of 2016 and compare it to valuation we calculate by real data.

We get a scatter plot of Data, and then we use polynomial approximation to fit the curve. Finally, we get the plot of Fitted Curve.

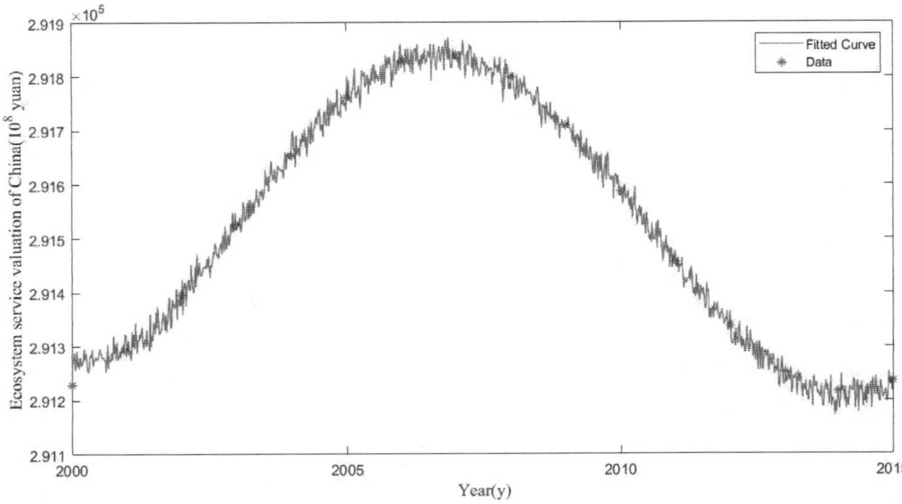

Figure 5. The plot of fitted curve.

7.2. The feasibility of curve fitting

By using Polynomial 7-th Approximation, we get the curve above. The Adjusted R-square of this approximation is 0.9872, which is highly approach to 1. Hence, we can think this curve is relatively scientific.

7.3. The effectiveness examination

According to the data we get from China Statistical Yearbook 2016, we calculate the data of 2016. Then we go back to the fitted curve we get and compare the data of 2016 on this curve to the data we calculate.

Fitted data	2.913
Real data	2.9128957

We can see that the data we fitted is highly approach to the real data. Hence, the model is effective and scientific.

8. Implications on land use project planners and managers

For project planners and managers, the cost of environmental degradation is a non-negligible project when calculating costs. In our model, we could calculate the value weights which are contained in each part of the ecosystem. Although it includes subjective factors, it can adjust the proportion of factors with environmental changes, so this is also an advantage for measuring value.

We analyze the data we calculate while testing the effectiveness of model, and we find that the ecosystem service valuation of China shows a letter "U" tendency. In the plot, the valuation arrives the maximum in 2007-2008 when the whole country was preparing the Olympics. After that, land development is dominant. The most significant change is the valuation decreased over time.

For land use project planners and managers, they should realize that the rapid urbanization has seriously endangered the ecological quality of ecosystem. Especially for land plan bureau in government, it should pursue new ecological projects and consider the cost of environment degradation when developing land. It is necessary to aware that we should not only consider the current profits, but also consider the long-term profits according to the development purpose.

Take data from the 13th five-year plan - statistics of major indicators of resources and environment into our consideration and apply our model. We work out the hypothetical ecosystem service valuation of 2020.

Hypothetical ideal data	2.91687
Fitted data	2.91572

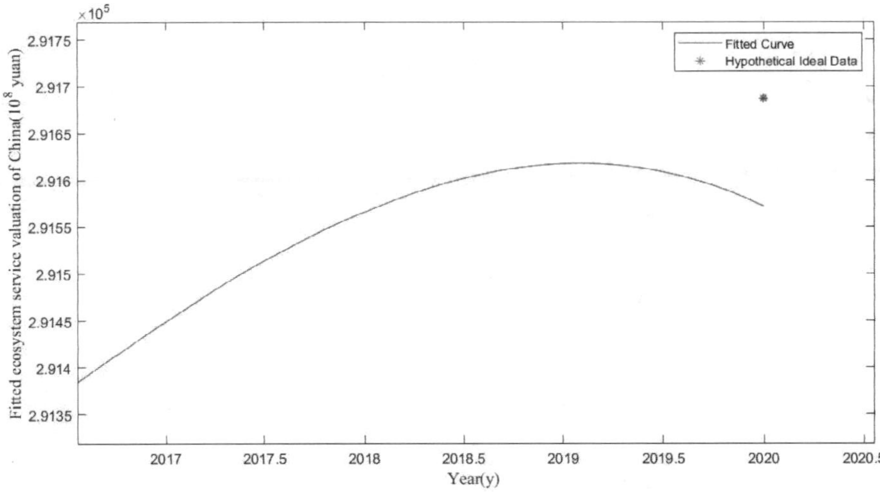

Figure 6. The fitted curve from 2016 to 2020.

From this figure, we can know that if land use planners and managers use the old way, the indicators of 2020 cannot be reached. It's high time that they should devote greater effort to optimize their land use projects.

9. The improved model about time factor

The ecological environment will change with natural factors such as storms, mudslides and other unpredictable factors over time. It is important to note that our model is applicable in the short term under the assumption we design. But in the long run, we should properly modify the weight according to changes in the environment. At the same time, the market value of many elements will change with time. In a certain period of time, we need to recalculate the ecological environment value to estimate the corresponding environmental degradation cost. But the basic framework of the calculation remains the same.

As time going by, the ecosystem will change. The weight of each indicator we calculate before will not be applicative. Hence, we should reconstruct the judgement matrix. While in this improved model, we combine fuzzy mathematical method and AHP together, which we call "FAHP". It has a new type of ratio-scale chart.

For fuzzy judgement matrix R, we can let

$$r_{ij}(\alpha) = \log_{\alpha} a_{ij} + 0.5$$
$$0 \le r_{ij}(\alpha) \le 1 \tag{23}$$

Intensity of importance	Definition
0.5	Equal importance
$\log_{\alpha} 3 + 0.5$	Weak importance
$\log_{\alpha} 5 + 0.5$	Strong importance
$\log_{\alpha} 7 + 0.5$	Demonstrated importance
$\log_{\alpha} 9 + 0.5$	Absolute importance
$\log_{\alpha} i + 0.5$， i=2,4,6,8	Intermediate values

Chart 3: Ratio-scale chart of FAHP

According to the research of Xu Yang etc. [10], we can define the parameter α as a bridge of time and ratio-scale.

$$\alpha = 81 \times \psi$$
$$\psi = \left[\frac{year}{10} \right] \tag{24}$$

With time going by, the improved model can change weight of each indicator. This is what parameter α in this model.

10. Strengths and Weaknesses

10.1. Strengths

- Using the fuzzy analytic hierarchy process to derive the proportion of each value component in the ecosystem, which could give the planners and managers a more direct value assessment.
- Water resources, land resources, the atmosphere, etc. considered into the model, the deviation between the ecological environment value estimation and the actual value is acceptable. It means that our model could be used to predict the true economic cost of environmental degradation.
- Using a variety of analytical calculation methods such as analytic hierarchy process, shadow price method, market value method to convert the abstract ecological value cost into economic cost.
- After testing, the model used is applicable to both large projects and small projects. The model is effective.

10.2. Weaknesses

- The analytic hierarchy process is subjective, and the actual weight may be slightly different from the calculation.
- The calculation method will have errors, so the calculated value may be not so accurate.
- The weights are averaged, so the weights in the model and the exact values corresponding to each year may be biased.
- The subtle effects generated in small projects are not taken into account in the model. When multiple small projects are combined, the combination of these subtle effects has a large impact, which could not be calculated in our model.

References

[1] Costanza R, D'Arge R, Groot R D, et al. The value of the world's ecosystem services and natural capital[J]. *World Environment*, 1997, 25(**1**):3-15.

[2] Ken J. Wallace. Classification of ecosystem services: Problems and solutions[J]. *Biological Conservation*,2007,139(**3**).

[3] Lin Peiying, Yang Guodong, Pan Shumin. *Case Study on Environmental Problems* [M].

[4] World Bank report. *Huaihe River Basin Water Pollution Control Project Evaluation Report.* Beijing: World Bank Beijing Office. 2000.3: 2-17

[5] Tang Xiumei, Pan Yuchun, & Liu Yu. Analysis of ecological value assessment and spatial-temporal change of cultivated land in Beijing. *China Agricultural Resources and Regionalization*, 2018,39(**3**):132-140.

[6] Zhao Zhongbao, Li Keguo, Zeng Guangjuan. Evaluation of Forest Ecosystem Service Function in Qinhuangdao City[J]. *Journal of Arid Land Resources and Environment*, 2012, 26(**2**): 31-36.

[7] Saaty T L , Vargas L G . *Models, methods, concepts & applications of the analytic hierarchy process*[M]. Springer US, 2001.

[8] Wu Gang, Feng Zongqi, Wang Xiaoke. Nutrient cycling of N, P and K in the agro-forest ecosystem of the Huang-Huai-Hai Plain: A case study of the paulownia-wheat and maize intercropping system[J]. *Chinese Journal of Applied Ecology*, 1993, 4(**2**): 141-145.

[9] Pang Ruyue. Ecological value assessment and comprehensive benefit evaluation of ecological

construction projects——Taking Tianze Agricultural Ecological Park as an example. (Doctoral dissertation,2016).
[10] Lan Jibin, Xu Yang, Huo Liangan. Research on Weights of Fuzzy Analytic Hierarchy Process[J]. *Systems Engineering - Theory & Practice*, 2006, 26(**9**): 107-112.

Appendix

Indicators	2015	2020
Cultivated land ownership (100 million mu)	18.6 5	18.65
New construction land scale (ten thousand mu)	-	-
Water consumption per ten thousand yuan of GDP decreased (%)	-	-
Energy consumption per unit of GDP reduced (%)	-	-
Proportion of non-fossil energy in primary energy consumption (%)	12	15
Carbon dioxide emissions per unit of GDP reduced (%)	-	-
Forest development	21.6 6	23.04
	151	165
Air quality	76.7	>80
	-	-
Surface water quality	66	>70
	9.7	<5

Chart 1. Statistics of major resources and environment indicators in China's 13th five-year plan.

Year	**Hectares**
2000	65336.2
2001	65331.6
2002	65660.7
2003	65706.1
2004	65701.9
2005	65704.7
2006	65718.8
2007	65702.1
2008	65687.6
2009	64777.5
2010	64728.0
2011	64686.5
2012	64646.6
2013	64616.8
2014	64574.1
2015	64545.7
2016	64512.7

Chart 2. Statistics of land use in China over the years (2000-2016).

Year	**Total water resources (billion cubic meters)**	**Rainfall (billion cubic metres)**

2000	27701	60092
2001	26868	58122
2002	28261	62610
2003	27460	60416
2004	24130	56876
2005	28053	61010
2006	25330	57840
2007	25255	57763
2008	27434	62000
2009	24180	55959
2010	30906	65850
2011	23257	55133
2012	29529	65150
2013	27958	62674
2014	27267	62602
2015	27963	62569
2016	32466	68672

Chart 3. Statistics of water environment in China over the years (2000-2016).

Year	Total industrial emissions (billion cubic metres)	Total nox emission (tons)
2000	138145	1995.1
2001	160863	1947.2
2002	175257	1926.6
2003	198906	2158.5
2004	237696	2254.9
2005	268988	2549.4
2006	330990	2588.8
2007	388169	2468.1
2008	403866	2321.2
2009	436064	2214.4
2010	519168	2185.1
2011	674509	2217.9
2012	635519	2117.6
2013	669361	2043.9
2014	694190	1974.4
2015	685190	1859.1

Chart 4. Statistics on China's annual exhaust emission and treatment (2000-2015).

Year	Total investment in environmental pollution control
2001	1166.7
2002	1456.5
2003	1750.1
2004	2057.5
2005	2565.2
2006	2779.5
2007	3668.8
2008	4937.0
2009	5258.4
2010	7612.2
2011	7114.0
2012	8253.5
2013	9037.2
2014	9575.5
2015	8806.4
2016	9219.8

Chart 5. Statistics on China's annual exhaust emission and treatment (2000-2015).

Image Registration of Infrared and Visible Image Based on Geometric Constraint

Juan Wang[1,2], Kaiwen Cheng[1], Cong Ke[1], Min Liu[1,*], Linkang Cai[1] and Hao Shi[1]

[1]Hubei Key Laboratory for High-Efficiency Utilization of Solar Energy and Operation Control of Energy Storage System, Hubei University of Technology, Wuhan 430068, People's Republic of China
[2]Post-doctoral Research Workstation, Wuhan Huaan Science and Technology Co., Ltd., Wuhan 430068, People's Republic of China

E-mail: happywj@hbut.edu.cn

Abstract. With the wide application of infrared sensors and visible sensors, the registration of two heterogeneous images has caused widespread concern. Multi-angle comprehensive analysis of multi-sensor imaging systems can obtain more accurate image information. However, the main challenge of image registration is the detection of feature points and the acquisition of the correct feature point matching pairs. A Geometric constraint method is proposed to obtain the correct feature matching pairs. Three feature point matching pairs are randomly selected in the two images, and the ratio of the corresponding line segment lengths is obtained according to the coordinates of the three feature points to find the most three similar feature point matching pairs. Compared with the RANSAC algorithm, the method can obtain the correct feature point matching pair better. The algorithm runs on the DSP platform for about 30s, the error is 1 pixel, and the correct rate reaches 80%.

1. Introduction

Image registration originated in the US military project in the 1970s. It is mainly used in the aircraft navigation assistance system, guidance system and photoelectric pod system, greatly improving the missile hit rate. At present, image registration technology has been widely used in 3D reconstruction [1], satellite remote sensing, medical image analysis, security monitoring and other industries. Image registration of infrared images and visible images belongs to heterogeneous image registration, that is, image registration of different sensors for imaging the same scene. Because of different imaging mechanisms, different resolutions, and different physical imaging conditions, imaging in two different modes has linear transformations such as rotation, translation, and scaling [2]. Image registration is to find an optimal geometric transformation to achieve geometric alignment of the two images.

Image registration algorithms can be roughly classified into the following categories: gray information-based method, image registration based on transform domain, feature-based image registration, and manual calibration registration algorithm [3]. The gray information-based method is used in the early registration field. The algorithm consists in selecting a similarity measure function and searching according to a certain path to find the geometric registration parameters when the maximum similarity is obtained [4]. Transform domain registration method is mainly to transform two-dimensional images from spatial domain to frequency domain. Common transform domain methods include wavelet transform, fast Fourier transform, discrete cosine transform, and so on [5]. Using the mathematical law of spatial domain transformation to the frequency domain, the registration image can be subjected to frequency domain transformation, and then the phase and frequency mathematical relationship between the images to be registered is obtained according to the transformed result, and then inversely transformed into the spatial domain to become an image. The translation, rotation, and scaling operations result in two image geometric transformation model parameters [6]. Image registration method based on transform domain since the fast Fourier transform algorithm is quite mature, it has certain advantages in calculation. However, the disadvantage is that the image registration method based on the transform domain is also insufficient for complex geometric transformations [7].

Image features are the hotspot research direction in the field of image processing. The feature-based registration algorithm utilizes the geometric positional relationship of the image features to be registered to obtain geometric transformation parameters [8]. Image feature selection is related to the specific application scenario, and often determines the final effect of the entire registration algorithm. Selecting appropriate image features is necessary for image feature-based registration algorithms. Excellent features not only raise the registration effect but also improve the efficiency of the registration algorithm. Common image features generally have a few features, line features and contour features [9]. In addition to the above automatic registration algorithm, there is also an image registration method based on manual calibration. The image registration method based on manual calibration needs to artificially treat the same points in the registration image for coordinate calibration, and then combine the corresponding points in the two images into a pair of matching pairs. This manual calibration registration method [10] is simple to calculate and easy to operate. It is widely used in many low-end industrial binocular vision systems.

In this paper, a feature-based registration algorithm is used. The SURF algorithm [11] is used to extract visible image and infrared image feature points, and three feature points are respectively extracted to form a triangle template. The feature point matching uses the template matching constraint to eliminate the wrong match. The correct three pairs of feature point matching pairs are obtained to solve the affine transformation model parameters between the images with registration.

2. Preliminaries

2.1. SURF algorithm

2.1.1. Hessian matrix.

The Hessian matrix is mainly used to detect feature points in an image. The Hessian matrix is mathematically defined as a square matrix of second-order partial derivatives of a multivariate function. Its formula is as follows

$$H(P,\delta) = \begin{vmatrix} L_{xx}(P,\delta) & L_{xy}(P,\delta) \\ L_{xy}(P,\delta) & L_{xy}(P,\delta) \end{vmatrix} \quad (1)$$

Where P is a pixel on the two-dimensional image, and $L_{xx}(P,\delta)$ represents a Gaussian second-order partial derivative of the image in the x direction at the pixel point P, and the rest are similar. The parameter δ represents the variance of the two-dimensional Gaussian function, which is called the scale in the field of image processing. The final calculation is the determinant of the Hessian matrix. Each pixel of the image has its own Hessian determinant value. The Hessian value of all pixels constitutes the Hessian matrix of the image. When the determinant obtains the region maximum value, the pixel point P considered as an image feature point.

The SURF algorithm uses a box filter to approximate the Gaussian second-order derivative filter. This method can not only solve the signal aliasing phenomenon of multiple Gaussian convolution but also use the integral image to improve the calculation speed of the Hessian matrix. The Hessian determinant using the approximated box filter can be approximated by the following formula:

$$H(P,\delta) = D_{xx} * D_{yy} - (0.9D_{xy})^2 \quad (2)$$

Where $D_{xx}D_{xy}D_{yy}$ represents the second-order partial derivative of the image in the x direction, the xy direction and the y direction after the pixel point P is approximated, and a weighting coefficient of 0.9 is multiplied by D_{xy}in order to balance the error.

2.1.2. Build scale space. The SURF algorithm forms the image order according to the size of the box filter. Different sizes of the three box filters are obtained to obtain Hessian matrix diagrams of different sizes. The image is calculated by four kinds of three kinds of box filters of different directions ($D_{xx}D_{xy}D_{yy}$ three types of templates) to obtain four Hessian matrix diagrams and form a first-order image. Due to the different sizes of the box filters, the scale of the filtered images is different, which constitutes the three-dimensional scale space of the SURF.

2.1.3. Feature point location.

Non-maximum suppression is performed on the region of the Hessian matrix in the three-dimensional scale space. Each pixel is not only compared with the surrounding 8 pixels at the same scale, but also with the upper and lower Hessian matrix of the same order 19 pixels are compared, and the feature point is recognized only when the Hessian value of the pixel is greater than the other 26 Hessian values in the 3*3*3 region. In order to obtain the feature points of sub-pixel point precision, a second-order Taylor expansion of the scale space function is needed to fit the feature point position [12]. At this point, the position and scale information (x, y, δ) of the feature points can be known.

2.1.4. Characteristic point main direction analysis.

In order to make the feature points have rotational invariance, in addition to the above position and scale information, the feature points also need to have direction information. In the field of image processing, the direction θ of a pixel can be calculated from its horizontal x direction and the first derivative of the vertical y direction. The definition of θ is as follows:

$$\theta = \frac{dy}{dx} \quad (3)$$

The SURF algorithm is used to determine the direction of the feature points [0]. It is necessary to calculate the Harr wavelet response of the horizontal direction and the vertical direction of all the pixels in the area by using the position of the feature point as the center and the radius of the image area of 6δ. A sector with a central angle of $\pi/3$ is used to wrap around the 6δ circular region around the feature point, and the sum of the Harr wavelet responses of the x and y directions of all the pixels in the sector is calculated to obtain the x direction and y. The total response in both directions of the direction constitutes a vector, and the modulus length of the vector is obtained. Record all the vectors obtained by rotating the fan one round, then take the largest of them. This vector can be used as the direction of the feature points.

2.1.5. Feature point vector descriptor construction.

The descriptor [13] vector of the feature points allows the visual features of the SURF feature points to be specifically quantized as "digital features". With the feature point description [0] sub-vector, it is possible to compare the "similarities and differences" with different features to determine whether the feature is the same feature. When constructing the feature point to describe the sub-vector, the feature point is taken as the origin. The feature point direction obtained in the above section is the positive direction of the y-axis, and the coordinate system is established. The square area of 20δ*20δ centered on the coordinate origin is selected, and the square is obtained. The area is evenly divided into 4*4 sub-areas. Each sub-region uses a Harr wavelet template with a size of 2δ*2δ to obtain the total response of the region in the x-direction and the y-direction. The response of each pixel in the region is obtained by weighting the Gaussian distribution centered on the feature points. In addition to the total response dx, dy in the horizontal and vertical directions of each region, the absolute values $|dx|,|dy|$are also calculated. So each region has 4 quantities $dx,dy,|dx|,|dy|$to characterize the region. Since the feature points have 16 sub-regions, the SURF algorithm has a total of 64 numbers to characterize the feature points [0], which constitutes a 64-dimensional feature description sub-vector of the SURF feature.

2.2. RANSAC algorithm eliminates mismatch matching

The RANSAC algorithm [14] eliminates mismatched feature point matching pairs. This algorithm for rejecting mismatched pairs mainly uses known data to estimate unknown mathematical model parameters. This method can eliminate obvious erroneous data. Firstly, a set of data is selected from the data points from the feature point rough matching to estimate the mathematical model parameters. After obtaining the mathematical model parameters, the specific model is obtained. Then use this model to verify other data in the dataset. If the verified data conforms to the model, then the confidence of the model is increased by one. After all the data is verified, the total confidence of the model is obtained. Then we use another set of data to get the model parameters, then find the overall confidence of the model. Repeat this step until you have traversed all the data in the data set. According to this iterative method, the confidence of all possible mathematical models can be obtained, and finally the mathematical model parameters with the highest confidence are selected as the final model.

3. Template matching

3.1 Template matching eliminates mismatch matching

In view of the shortcomings of the RANSAC algorithm, this section proposes a feature point matching pair extraction method based on template matching. Imaging of the imaging sensor after camera calibration can be considered as undistorted, eliminating the nonlinear transformation that may be caused by image distortion, which means that there are a large part of similar information in the two images of the same scene, and this part is only a linear geometric transformation relationship in the image. Then the geometrical positions of the image feature points extracted in the similar parts of the two images should also be similar. That is, if three feature points are extracted from the visible image to form a triangle template, the corresponding feature points in the infrared image can also be constructed. A similar triangle template with only translation, rotation, and scaling relationships between the two templates. If there are three pairs and above correct feature point matching pairs for the two images, then the two triangles formed by the pair of feature point matching pairs are also necessarily the most similar among all matching pairs. The constrained condition of the feature point matching pair plus the triangle similar template can be used to eliminate the mismatching pair and the correct three pairs of feature point matching pairs can be obtained, and the affine transformation model parameters between the registered images can be solved.

Whether the two template triangles are similar can be based on the ratio of the length of the line segments. How to find the optimal similarity in the algorithm, we can find a similar quantitative expression to facilitate programming. Suppose the side lengths of the first triangle are n1, n2, and n3, respectively. The side length of the second triangle is m1, m2, and m3, and the ratio of each side is

$$k_1 = \frac{n1}{m1},\ k_2 = \frac{n2}{m2},\ k_3 = \frac{n3}{m3} \tag{4}$$

The length of each line segment is the length of the Euclidean distance [0] in the pixel coordinate system. In theory, the ratio of the corresponding triangle segments should be equal, but the triangles in the program cannot be judged to be similar under the condition of complete equality. So when designing the algorithm, you can set a threshold to ensure that the images are similar. The threshold is set as follows:

$$abs\left(1-\frac{k_1}{k_2}\right) < 0.02\&\&abs\left(1-\frac{k_2}{k_3}\right) < 0.0 \tag{5}$$

When both are less than the threshold 0.02, the above two conditions are satisfied to prove that the ratios of the three line segments are close to the same. If it is greater than the threshold, the triangles are judged to be dissimilar. Under similar conditions, set the dust value to select the most similar triangle. The definition of dist is as follows:

$$\text{dist} = abs\left(1-\frac{k_1}{k_2}\right) + abs(1-\frac{k_2}{k_3}) \tag{6}$$

The smaller the dist. value, the more similar the two triangles are. In the feature point rough matching pair, three pairs of feature point matching pairs are arbitrarily selected, and the three pairs of feature point matching pairs will constitute two triangles. According to the coordinates [0] of the three feature points, the length of each line segment can be obtained, and then the ratio of the length of the corresponding line segment of the triangle is calculated, and three ratios are obtained. Finally, the three similar pairs of feature point matching pairs are found according to the above two conditions. The algorithm flow chart is shown in figure 1.

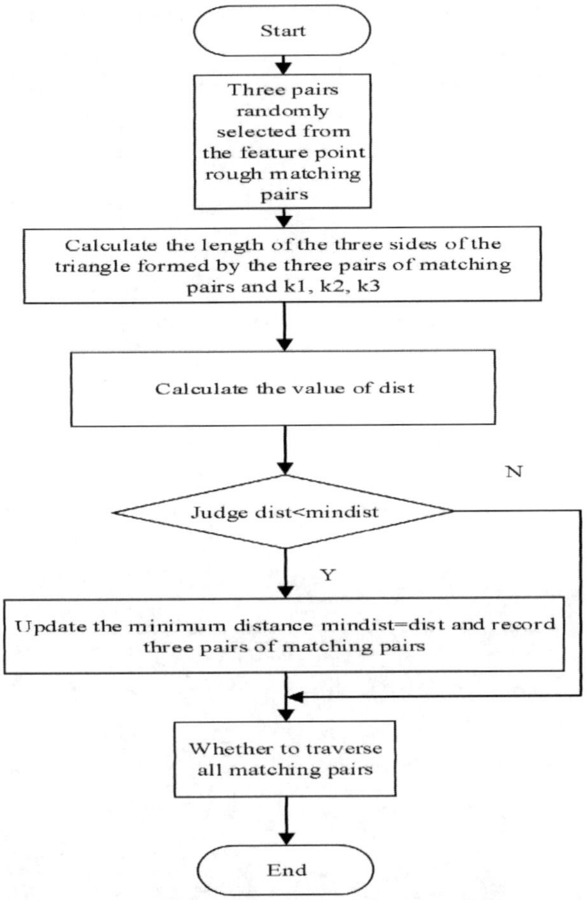

Figure 1. Geometric Constraint algorithm flow chart.

3.2 Edge coincidence degree-based registration evaluation method

This section mainly proposes a registration evaluation method based on edge coincidence degree. Firstly, the Canny algorithm[14] is used to extract the edge image of the registration image separately, and the edge of the image can be approximated to calculate the edge coincidence degree. The edges obtained by the Canny edge extraction of the visible light image or the infrared image are binary images.

Each pixel has an intensity value of 0 or 255, and the pixel with an intensity of 255 is an edge pixel. For the two edge images of the registration image, select thc image with fewer edges in the two edge images as the reference image, traverse the pixel points in the reference image as edges, and then observe whether the corresponding pixel in the registration image is also the edge pixel. If both are edge pixels, the registration confidence is increased by one. By traversing all the edge pixels in the reference image as described above, the total confidence of the registration can be obtained. The objective confidence rate is then obtained by dividing the total confidence by the number of edge pixels in the reference image. The steps of the registration evaluation method based on the edge coincidence degree are shown in figure2.

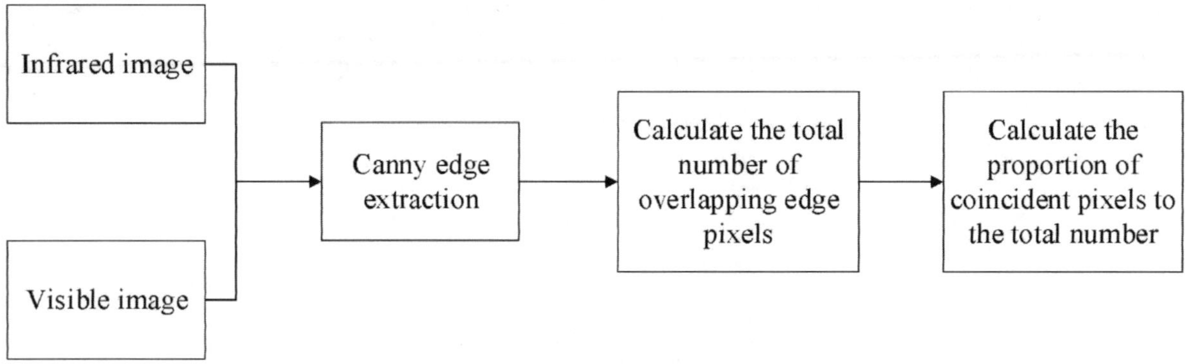

Figure 2. Registration evaluation method based on edge coincidence.

4. Experiment analysis

4.1 Feature point rough matching pair extraction

The feature points extracted by the SURF algorithm are then used to describe the nearest Euclidean distance of the sub-vectors according to the image feature points. The coarse matching of the feature points is as shown in figure 3.

Figure 3. Pairs of visible and infrared image feature points.

It can be seen from Fig. 3 that there are about 40 pairs of feature point matching pairs formed, and the matching of the feature points formed by the infrared image and the visible image through the preliminary screening is mostly mismatched. The main reason for the large number of mismatching pairs here is the difference in the imaging mechanism of the imaging mechanism of the two heterogeneous image sensors. The visible light imaging sensor is imaged according to the brightness of the reflected light of the object, and the intensity values of the pixels in the infrared image sensor are Heat related to object radiation.

4.2 Template matching method for extracting feature point matching pairs

From the feature point rough matching pair, the template matching algorithm and the RANSAC algorithm are used to extract the correct feature point matching pairs respectively. The experimental results are shown in figure 4.

It can be seen from Fig. 4 that the feature point matching pair selected by the template matching method is obviously superior to the feature point matching pair selected by the RANSAC algorithm. It can be seen from the figure that there are 2 pairs of matching pairs in the feature point matching pair selected by the RANSAC algorithm belonging to the mismatch matching pair. It can be found that as long as the feature point matching pair has three pairs and above correct feature point matching pairs, the template matching method can be used to find the correct feature point matching pair.

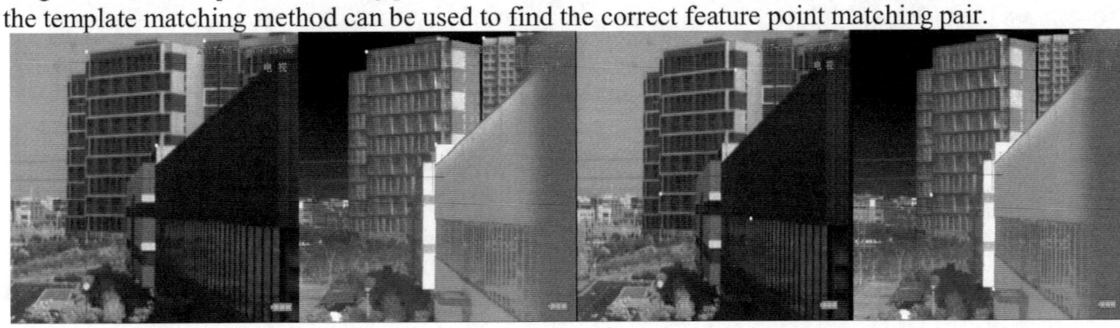

a b c d

Figure 4. (a,b) shows the feature points of the visible image and the infrared image obtained by the Geometric Constraint method, respectively. (c,d) respectively represent feature points of the visible image and the infrared image obtained by the RANSAC algorithm.

4.3. Registration evaluation results based on edge coincidence

The registered infrared image edge and visible image edge obtained by Canny edge extraction algorithm are shown in figure 5.

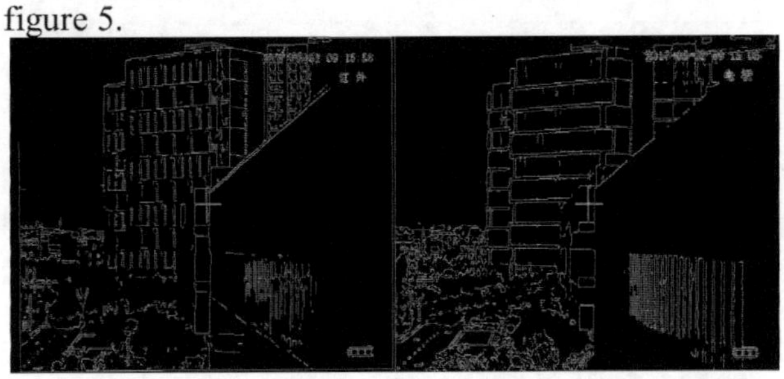

a b

Figure 5. (a) shows the edge of the infrared image (b) shows the edge of the visible image.

It can be seen from figure 5 that the two edges have both a similar portion and a plurality of different portions. For different portions of the edges, matching with other edge pixels of another image may increase the registration confidence without registration. This section mainly demonstrates from the two levels of translation and rotation that the registration rate obtained by this method has a maximum value attribute. For translation, consider the translation in the horizontal and vertical directions, translate one of the two edge images, and then recalculate the coincidence of the two images after translation. For rotation, the image center area is selected for edge coincidence calculation. The results of experimental are shown in Table 1 and 2.

Table 1. Ten-pixel translational registration rate.

vertical \ horizontal	0	1	2	3	4	5	6	7	8	9	10 (/pixel)
0	0.2919	0.2822	0.2841	0.2875	0.2823	0.2783	0.2761	0.2765	0.2690	0.2676	0.2662
1	0.2916	0.2827	0.2837	0.2883	0.2790	0.2734	0.2759	0.2748	0.2692	0.2663	0.2665
2	0.2897	0.2803	0.2831	0.2857	0.2795	0.2735	0.2746	0.2744	0.2656	0.261	0.2617
3	0.2873	0.2781	0.2782	0.2839	0.2748	0.2703	0.2705	0.2689	0.2659	0.2610	0.2605
4	0.2853	0.2778	0.2754	0.2802	0.2736	0.2684	0.2697	0.2669	0.2591	0.2581	0.2578
5	0.2795	0.2701	0.2728	0.2770	0.2709	0.2620	0.2641	0.2646	0.2561	0.2541	0.2560
6	0.2762	0.2673	0.2684	0.2737	0.2641	0.2621	0.2612	0.2600	0.2544	0.2526	0.2514
7	0.2719	0.2652	0.2661	0.2686	0.2635	0.2566	0.2563	0.2568	0.2515	0.2480	0.2505
8	0.2722	0.2615	0.2606	0.2665	0.2568	0.2549	0.2564	0.2541	0.2460	0.2467	0.2566
9	0.2646	0.2579	0.2551	0.2633	0.2553	0.2488	0.2517	0.2496	0.2470	0.2431	0.2447
10	0.2618	0.2538	0.2545	0.2568	0.2499	0.2481	0.2476	0.2468	0.2429	0.2384	0.2406

Table 2. Rotating registration rate within 30 degrees.

angle	**0**	**1**	**2**	**3**	**4**	**5**	**6**	**7**	**8**	**9**	**10**
rate	0.331	0.320	0.304	0.285	0.276	0.276	0.258	0.249	0.235	0.231	0.225
angle	11	12	13	14	15	16	17	18	19	20	21
rate	0.226	0.217	0.223	0.224	0.230	0.231	0.233	0.225	0.240	0.239	0.232
angle	22	23	24	25	26	27	28	29	30		
rate	0.230	0.237	0.236	0.236	0.237	0.235	0.234	0.224	0.231		

Table 1 shows that horizontal and vertical translation amount is 0, the registration ratio is at most 0.2919. According to the experimental data, it can be known that the registration rate obtained by this evaluation method can maintain the maximum value attribute for the translation transformation. Table 2 when the rotation angle is 0, the registration ratio of the two registration images without the rotation plane is up to 0.331. This registration evaluation method can maintain the maximum value attribute for the rotation transformation. The experimental data can prove that the registration evaluation method based on edge coincidence degree is reliable. In practical applications, the registration of the registered images can be manually calibrated, and then the edge coincidence registration ratio of the two images is calculated, and the registration rate is used as a priori condition to detect the final registration effect.

4.4. DSP platform algorithm implementation

TI's Da Vinci series TMS320DM6437 processor is a 32-bit fixed-point multimedia DSP processor with 128MB of Nand Flash ROM and 256MB of DDR2 RAM, clocked at up to 600MHz, peak processing capacity of 5600MIPS, low power consumption. Widely used in high-performance, low-cost vision development systems. Figure 6 is a graphical representation of the experimental results of the entire registration algorithm running on the DSP platform.

a b

Figure 6. DSP platform test results.

The experimental results were taken by the mobile phone. As can be seen from the figure, the gray value is high, the lack of detail is the infrared image, the detail is fine, and the lower gray level is the visible image. Visible images and infrared images basically meet the registration requirements. The running time of the whole algorithm is about 30s, the registration error is 1 pixel, and the correct rate is about 80%.

5. Conclusions

This paper mainly presents a template matching method for infrared image and visible image registration to obtain the correct feature point matching pair. This method uses the triangle template by matching pairs of coarse feature points to obtain the correct feature point matching pairs based on their similarity. At the same time, this algorithm is also implemented on the DSP platform, but it can be found that the execution time of the algorithm on the DSP platform is relatively long, and the FPGA platform or GPU can be used later to reduce the processing time of the algorithm. The entire algorithm process also has a part that is worth optimizing. You can consider the optimization of the calculation to reduce the running time of the program and improve the performance of the program. This type of algorithm uses a fixed threshold, which means that this is not an adaptive algorithm. However, with the wide application of deep learning in the field of image, we can consider the application of deep learning algorithm [15] in the field of image registration.

References

[1] D. G. Lowe, "Local feature view clustering for 3D object recognition," Proceedings of the 2001 IEEE Computer Society Conference on Computer Vision and Pattern Recognition. CVPR 2001, Kauai, HI, USA, 2001, pp. I.

[2] B. S. Reddy and B. N. Chatterji, "An FFT-based technique for translation, rotation, and scale-invariant image registration,"IEEE Transactions on Image Processing, vol. 5, no. 8, pp. 1266-1271, Aug. 1996.

[3] D. G. Lowe, "Object recognition from local scale-invariant features," Proceedings of the Seventh IEEE International Conference on Computer Vision, Kerkyra, Greece, 1999, pp. 1150-1157 vol.2.

[4] H Bay, A Ess, T Tuytelaars, et al. Speeded-Up Robust Features (SURF)[J]. Computer Vision and Image Understanding, 2008, 110(3):346.

[5] E. Rublee, V. Rabaud,K. Konolige and G. Bradski, "ORB: An efficient alternative to SIFT or SURF," 2011 International Conference on Computer Vision, Barcelona, 2011, pp. 2564

[6] D G .Lowe Distinctive Image Features from Scale-Invariant Keypoints[J]. International Journal of Computer Vision, 2004, 60(2):91.

[7] S. Leutenegger, M. Chli and R. Y. Siegwart, "BRISK: Binary Robust invariant scalable keypoints,"2011 International Conference on Computer Vision, Barcelona, 2011, pp. 2548-2555.

[8] K. Mikolajczyk and C. Schmid, "A performance evaluation of local descriptors," in IEEE Transactions on Pattern Analysis and Machine Intelligence, vol. 27, no. 10, pp. 1615-1630, Oct. 2005.

[9] J MMorel, GYu. ASIFT: A New Framework for Fully Affine Invariant Image Comparison[J]. SIAM Journal on Imaging Sciences, 2009, 2(2):438-469.

[10] E. Rosten, R. Porter and T. Drummond, "Faster and Better: A Machine Learning Approach to Corner Detection," in IEEE Transactions on Pattern Analysis and Machine Intelligence, vol. 32, no. 1, pp. 105-119, Jan. 2010.

[11] Fischler, Martin & C. Bolles, Robert. (1981). RandomSample Consensus: A Paradigm for Model Fitting with Applications To Image Analysis and Automated Cartography. Communications of the ACM. 24. 381-395. 10.1145/358669.358692.

[12] M. C. Lin and J. F. Canny, "A fast algorithm for incremental distance calculation," Proceedings. 1991 IEEE International Conference on Robotics and Automation, Sacramento, CA, USA, 1991, pp. 1008-1014 vol.2.

[13] Flusser, Jan. (2011). Moment Invariants in Image Analysis. Transactions on engineering, Computing and Technology. VII.
[14] Canny, J. A computational approach to edge-detection. IEEE Trans. Pattern Anal. Mach. Intell 1986, 8, 679–698.
[15] Shuxiu Wang, Sheng Lei and Faliang Chang, "Image registration based on neural network," 2008 International Conference on Information Technology and Applications in Biomedicine, Shenzhen, 2008, pp.

Acknowledgments
This work was supported by Hubei Provincial Department of Education Scientific Research Project Funded Project Young Talents Program (Q20181401), Hubei University of Technology Ph.D. Research Startup Fund Project (BSQD2015023), Key projects for the development and research of key laboratories in Hubei Province for efficient solar energy utilization and energy storage operation(HBSEES201701). Major Technological Innovation Projects of Hubei (No. 2018AAA028).

A High-Speed Serial Digital Interface Video Signal Jitter Measurement Method

Difei Li, He Chen, Hongying Gao, Kejia Zhao, Zhaochun Wu and Bo Li

Electronics and Information Technology Division, National Institute of Metrology, China.No.18 Bei San Huan Dong Lu, Beijing, China

E-mail: {lidf & chenhe & gaohy & zhaokj & wuzc & libo}@nim.ac.cn

Abstract. This paper designs a measurement method for high-speed serial digital interface video signal jitter. Firstly, we introduce the features of SDI signal and the jitter measurement methods of traditional high-speed serial signals. A jitter model is built for the SDI signal while the properties of the components are analysed. An adjusted dual-Dirac model is adopted to measure the jitter in the signal. This method can accurately analyse different jitter components in SDI signals. It is directly performed on oscilloscope, which is beneficial to the traceability of the jitter. The feasibility of the method is verified by experiments. The measurement results show that the method has better measurement accuracy than the direct spectrum method.

1. Introduction

Serial digital interface (SDI) is a kind of digital video interfaces first standardized by SMPTE (The Society of Motion Picture and Television Engineers) in 1989 [1, 2]. SDI video signal transmission belongs to the mode of baseband serial signal communication. With the development of digital TV broadcasting network technology and related consumer electronics industry, digital video has already moved from full HD to 4K-HD or even 8K-HD. Utra-HD video will become the mainstream development direction of digital TV in the future that require greater fidelity and resolution than standard full HDTV can provide. 6G-SDI and 12G-SDI standards have been published which are prepared for transmission of uncompressed, high capacity and high bit rate digital video signals within video facilities [3, 4].

In the transmission of high-speed SDI signals, the performance of the whole system is often greatly affected due to the existence of jitter. Therefore, jitter has become a key factor restricting the design of high-speed digital systems [5]. As can be seen from the current various high-speed serial bus specifications, the jitter components of the SDI transmission system are extremely explicit. In order to ensure the stability of the high-speed digital system, it is necessary to accurately determine the jitter component of its existence and separate the different jitter components to determine whether the system performance meets the design requirements of bit error rate and stability while minimize the system due to jitter error.

At present, the measurement of serial signals mainly adopts an eye diagram analysis method that only used for qualitative measurement. An eye diagram is a graph of a series of digital signals accumulated on an oscilloscope. From the eye diagram, the effects of crosstalk and noise between codes can be observed which reflects the overall characteristics of the digital signal. That is, directly observing the shape of the eye to judge the quality of the signal. It is impossible to quantitatively measure the jitter component contained in the total jitter. From the eye diagram, we can only obtain the peak-to-peak value of jitters at the current moment [6].

Rising bit-rate result in smaller unit intervals and increasingly tighter timing budgets, thus, can even distinguish the difference in picosecond magnitude. In the past 10 years, the technology used to measure jitter in high-speed serial data signals has made great progress. The goal of all researches is to accurately measure the jitter value when the signal having the extremely low bit error rate (the BER) (e.g. 10^{-12}). Determining jitter at very small bit error ratios on a real-time oscilloscope requires algorithms that extrapolate the measured data rather than by simply calculating peak-to-peak or RMS values directly from the acquired measurements. An important reason for using extrapolation is that if the total jitter is measured directly, the measurement time required is very long and may even take a whole day. Therefore, extrapolation method was used.

The method of spectral measurement (jitter and signal quality specification methodology) which is used in the "MJSQ" in a dual-Dirac jitter model of a communication channel technology has become an industry standard [7]. This paper provides a complete method to measure jitters in high speed SDI video signal by using adjusted dual-Dirac model. This model illustrates the process of converting time interval error (TIE) measurements into jitter values and describes how to represent the view of the jitter components.

2. SDI signal jitter

As shown in Figure 1. In a digital communication link, jitter is the offset between the predicted position of the signal transition and the actual position of the transition. In serial data communication, jitter is a core issue to be aware of because excessive jitter in the transmitter data signal can cause data recovery errors at the receiver. To prevent the bit error rate from being too high, jitter margins are specified in many standards so that the transmit and receive circuits can be designed to operate within jitter budgets and tolerances. To ensure that the equipment works within these budgets, it is necessary to accurately measure jitter. Measurements not only quantify jitter, but also help designers examine the causes and sources of jitter to help reduce or eliminate value of jitter. The deviation faster than 10HZ is defined as jitter, and the deviation over 10Hz is defined as wander. Ideally, the time interval in SDI signals should as wide as the integer multiple of the unit interval (UI) [8]. UI is serial digital signal clock unit corresponds to the crystal oscillator frequency of SDI generator instrument. According to the SMPTE standard specification SDI signal bit rate amount to the clock frequency used to create SDI signal. During the actual transmission of the SDI signal, the transition position of the rising and falling edges of the signal will deviate from the ideal position [9].

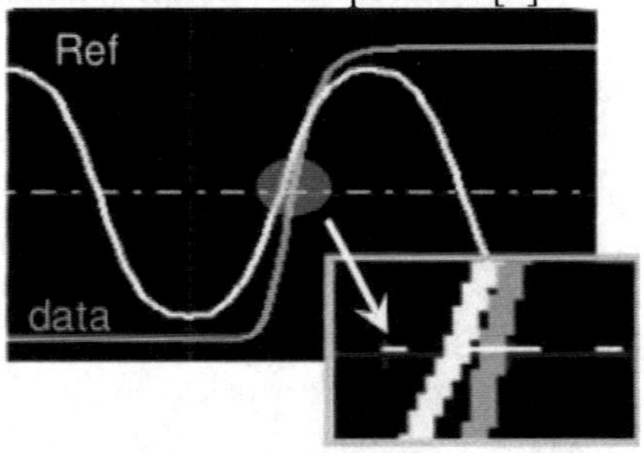

Figure 1. Jitter performance.

There are two types of jitter: deterministic jitter (D_j) and random jitter (R_j). The way these two types of jitter accumulate during serial data communication is different. R_j is considered to be an unbounded component and generally obeys a Gaussian distribution and therefore conforms to certain statistical rules. D_j is considered a bounded component and consists of data-dependent jitter (DD_j), duty-cycle distortion (DCD), periodic jitter (P_j), intersymbol interference (ISI).

D_j is deterministic jitter. It can be predicted and can express specific causes as well. D_j consists of *ISI*, *DCD* and P_j, and the amplitude is non-Gaussian and always bounded. $D_j$depends on the bounded peak to peak.

ISI is inter-symbol interference, which is a data-dependent deterministic jitter that is typically caused by channel scatter or filtering. Inter-code interference, also known as data-dependent jitter

(DD_j), occurs when signals arrive at receiver thresholds at different times, starting at different positions of the bit sequence (symbol).

DCD is the duty cycle distortion. It is the difference between the average pulse width of the positive pulse and the average pulse width of the negative pulse in the clock class code sequence. Amplitude bias error, turn-on delay, and saturation can cause duty cycle distortion.

P_j is periodic jitter, which is repeated in cycles that are independent of data. The typical cause of $P_j$is the power feedthrough of a switched-mode power supply. The P_j model can be built using one or more sine waves and their harmonics.

R_j is random jitter. R_j R is subject to Gaussian distribution, and theoretically its amplitude is unbounded. The characteristics of the Gaussian distribution depend on the root mean square (RMS) value or standard deviation. It can be easily assumed that, in general, any Gaussian random variable will have a span of 14 times more than the standard deviation in 10^{12} samples. If this span is exceeded in the data communication system, the corresponding bit error rate (BER) is 10^{-12}. R_j is mainly caused by thermal noise in electrical components.

Jitter hierarchy is shown in Figure 2.

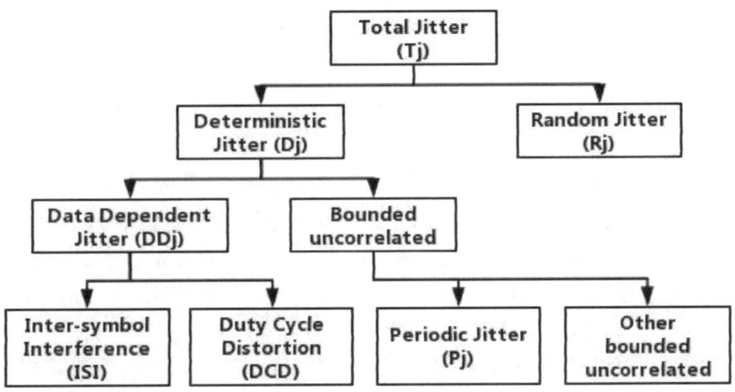

Figure 2. Jitter hierarchy.

3. The dual-Dirac model theory

A common method of viewing jitter has traditionally been to look at a sampling scope eye diagram. By setting a window around the crossing point and plotting the number of hits during a measurement. The more samples recorded in a particular position, the more frequently the edges were in that position. This leads to the idea of the histogram being a measure of the probability of an edge falling at a particular location, or time — a Probability Density Function (PDF). The longer the sampling period, the greater the ability to represent rare events such as edges that are further away from the ideal transition timing. When the SDI signal has both deterministic jitter and random jitter, the statistical histogram appears as a bimodal form, shown as Figure3. The probability density function of jitter can be approximated by the Dual-Dirac model.

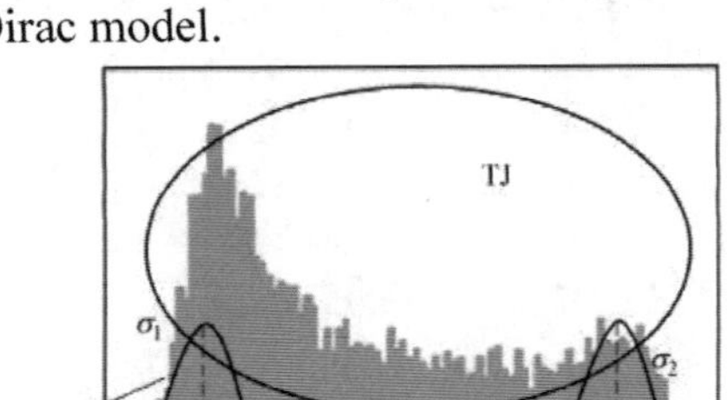

Figure 3. Jitter statistical histogram and dual-Dirac model schematic.

The dual-Dirac model is widely used for its capability in rapid estimating total jitter defined at specified bit error ratio, T_j (BER), and for providing an approach for integrating T_j (BER) from different elements [10]. According to the assumption in [7], total jitter T_j (BER) is the sum of deterministic jitter (D_j) and random jitter (R_j), with R_j weighted by a multiplier α (alpha) that is determined from the bit error ratio[11], represented as (3.1). (α = 14.07 for a typical BER value of 10^{-12}):

$$T_j(BER) = \alpha(BER) * R_j + D_j \quad (3.1)$$

The system corresponds to different coefficient values α in different BER states. The specific numerical correspondence is shown in the following table.

Table 1. Values of $\alpha(BER)$ for different BERs.

BER	$\alpha(BER)$	BER	$\alpha(BER)$
10^{-6}	9.507	10^{-11}	13.412
10^{-7}	10.399	10^{-12}	14.069
10^{-8}	11.224	10^{-13}	14.698
10^{-9}	11.996	10^{-14}	15.301
10^{-10}	12.723	10^{-16}	16.444

R_j and D_j is uncorrelated so the total jitter distribution can be described as the convolution of PDF of R_j and D_j. The jitter distribution can generally be divided into three portions: at the crossing-point, the influence of D_j is dominant; the effect of the intermediate portion R_j is gradually enhanced; and at the asymptotic limit, the shape of the curve is close to R_j distribution. The asymptotic tails of the distribution usually cause errors at the level of BER < 10^{-8}.

The crossing-point is separated into two Dirac-delta functions positioned at μ_1 and μ_2, the D_j dominated region, followed by an artificially abrupt transition to the R_j dominated tails. To implement the dual-Dirac model, we estimated $T_j(BER)$ by describing the tails of the jitter distribution as two Gaussians of width σ separated by a fixed amount $D_j(\delta\delta) \equiv |\mu_1 - \mu_2|$. The model is a Gaussian approximation to the outer edges of the jitter distribution displaced by $D_j(\delta\delta)$. $T_j(BER)$ and its PDF can be represented as (3.2) and (3.3).

$$T_j(BER) = \alpha(BER) * R_j + D_j(\delta\delta) \quad (3.2)$$

$$f_{Tj} = f_{Dj} * f_{Rj} \quad (3.3)$$

In the formula (3.3), f_{Tj}, f_{Dj} and f_{Rj} represent the PDF of T_j, D_j and R_j respectively. It is certainly possible for the components of deterministic jitter to be larger than $D_j(\delta\delta)$, which, as described previously, will be less than D_j (peak-peak).

f_{Rj} is given by a Gaussian distribution whose width is σ and x is the horizontal axis of the eye diagram. Therefore f_{Rj} can be represented as (3.4).

$$f_{Rj}(x) = \frac{1}{\sqrt{2\pi}\sigma}\exp\left(-\frac{x^2}{2\sigma^2}\right) \quad (3.4)$$

Then f_{Tj} can be represented as (3.5).

$$f_{Tj}(x) = \frac{1}{\sqrt{2\pi}\sigma}\int [\delta(x' - \mu_1) + \delta(x' - \mu_2)]\exp\left(-\frac{(x - x')^2}{2\sigma^2}\right)dx' \quad (3.5)$$

T_j, R_j and D_j are calculated via a fit to the dual-Dirac model, as explained above. P_j, DD_j, ISI and DCD are calculated directly from time interval error measurements. We do not need to fit the middle part of the jitter histogram, because the centre part of the histogram does not contribute to outliers that create bit errors. Which cause bit errors is the tails of jitter histogram. We fit the model to the tails with two parameters: $D_j(\delta\delta)$ and $R_j(\delta\delta)$.

$$R_j(\delta\delta) = \sigma \quad (3.6)$$

$$D_j(\delta\delta) = \mu_2 - \mu_1 \quad (3.7)$$

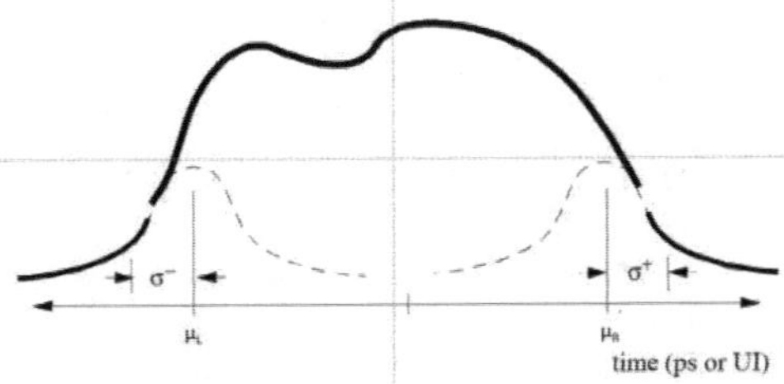

Figure 4: Illustration of a distribution of TIE jitter values overlaid onto the dual-Dirac jitter model. The outer portions beyond the breaks in the curve represent extrapolated values.

4. Application of the dual-Dirac model

4.1 Time interval error measurement

The algorithms all begin with the acquisition of serial data waveform which include a sufficient number of unit intervals and transitions. Acquiring waveforms should contain more than 10^5 unit intervals and at least 100 iterations of a repeating pattern. It is the way to make the jitter measurements statistical.

First step, we use golden PLL to achieve clock recovery process, which calibrates the expected arrival time of the transition edges of the waveform. The actual arrival time of each edge of the input signal is determined by the method which is cubic interpolation of the samples, shown as Figure 5.

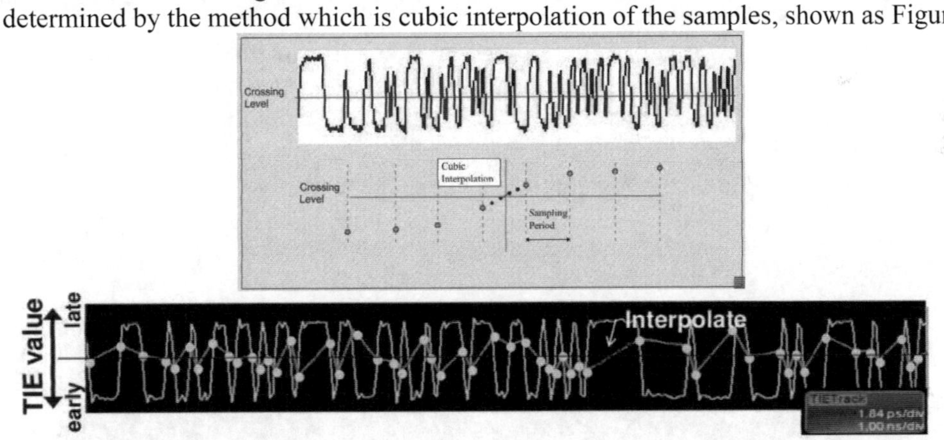

Figure 5. Using interpolation method to identify the arrival time of each edge.

Next, comparing these actual arrival times obtaining from cubic interpolation results to expected arrival times which is represented as the reference clock recovering from the golden PLL outputs. The time interval error (TIE) is obtained by subtracting the actual measured transition time from the expected transition time of the data edge.

4.2 Data-dependent jitter (DD_j) measurement

Pseudo-random sequence of data is commonly used to measure jitter test. In order to locate data-dependent jitter (including *ISI* and *DCD*), we first need to find repeat patterns from the signal. TIE measurements determine an average TIE value for each bit in the repetition pattern. Corresponding

TIE measurements of each iteration of the pattern are averaged resulting in a waveform that contains only data-dependent jitter (DD_j).

The TIE value after a plurality of average calculations is plotted as a histogram. The horizontal axis of the histogram is the moment when the jitter occurs, and the vertical axis is the number of times the jitter occurs, as shown in Figure 6.

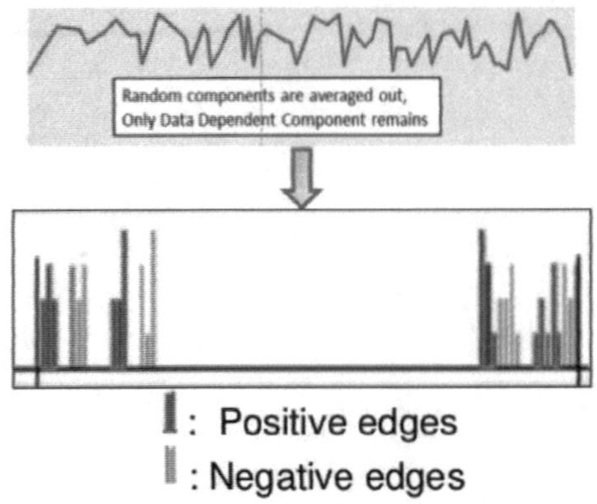

Figure 6. Determine DD_j from TIE.

DD_j, ISI and DCD are determined as follows:

(1) DD_j value is the range of the complete DD_j distribution which is expressed as peak-peak of DD_j histogram, including both positive and negative edges.

(2) ISI value is the larger of the range (max-min) of the individual histograms formed from the DD_j histogram measurements of positive and negative edges.

(3) DCD value is the difference in the means of the histograms for positive and negative edges.

4.3 Pattern Uncorrelated Extraction

To obtain the uncorrelated elements, we shall remove DD_j from TIE. When DD_j is subtracted from full TIE trend, only R_j and bounded uncorrelated jitter whose main component is P_j.

After subtraction, FFT method shall be used to convert waveform to frequency domain. A threshold is set based on the obtained spectral data. Spectral components that exceed the threshold are considered to be caused by periodic jitter. The spectral data information below the threshold is removed leaving only the peak line. Then an inverse FFT is calculated from these isolated P_j peaks. The value for P_j displayed in the jitter measurement table is the peak-to-peak value of the inverse FFT. Figure 8 shows this process.

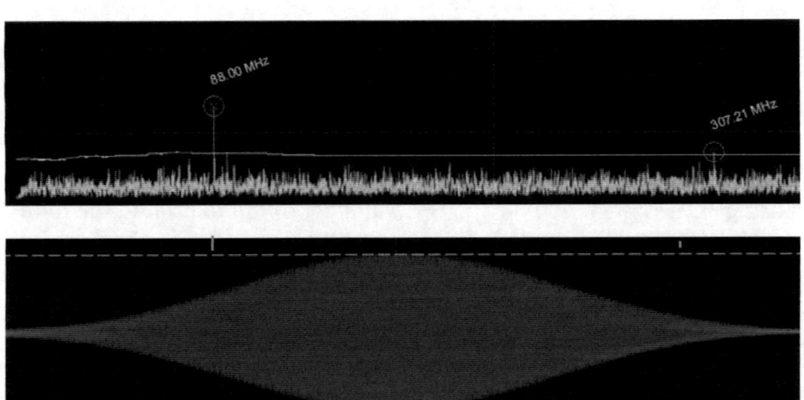

Figure 7. Using FFT to obtain P_j.

Journal of Physics: Conference Series **1419** (2019) 012009 doi:10.1088/1742-6596/1419/1/012009

4.4 Determine R_j and tail extrapolation

Through the FFT transform we determine P_j and remove it from the spectrum. The remaining spectral components contain random jitter and bounded uncorrelated jitter. In general, the remaining spectral components are integrated and the standard deviation σ value is obtained. The σ value is approximately equal to the value of the random jitter, which is the standard deviation of the Gaussian distribution in the dual-Dirac model, as (3.6) presented.

However, it is important to note that the σ result will also include the effects of bounded, correlated jitter (apart from periodic jitter), including crosstalk. In cases where there is high crosstalk or other bounded uncorrelated jitter, the crosstalk will manifest itself as broadband spectral energy over the frequency range and masquerade as R_j when separated using the spectral method. Therefore, when there is crosstalk or other bounded uncorrelated jitter that raises the jitter spectrum's noise floor, the spectral method can fail to give an accurate result. An adjusted method should be used instead.

In the spectrum method above, the extrapolated tail of total jitter is obeyed by the same Gaussian distribution defaulted, as Figure 4 presented. Here we respectively estimate the extrapolated tails on both sides. Therefore Equation (3.5) will be rewritten as follows:

$$\lim_{x\to-\infty} f_{Tj}(x) \to \frac{1}{\sqrt{2\pi}\sigma_1}\left(-\frac{(x-x')^2}{2\sigma_1^2}\right) \tag{4.1}$$

$$\lim_{x\to\infty} f_{Tj}(x) \to \frac{1}{\sqrt{2\pi}\sigma_2}\left(-\frac{(x-x')^2}{2\sigma_2^2}\right) \tag{4.2}$$

The next step in the process is the extrapolation of the tails of the jitter distribution. The uncorrelated elements obtained from section 4.3 is used to form a histogram. The tails of this histogram are extrapolated using the σ value found from the spectral analysis.

4.5 Total jitter reconstruction

We convolve the extrapolated histogram with DD_j histogram calculated in section 4.2. The resulting distribution is the overall jitter probability density function (PDF), which looks similar to the TIE Histogram, but with extrapolated tails as described above. The PDF formed is integrated from the outsides to the center to form the cumulative distribution function (CDF). The CDF is plotted with time on the X-axis, and probability on the Y-axis; the width of the CDF at a particular Y-value (or BER) is the T_j at that BER value, as shown in Figure 8.

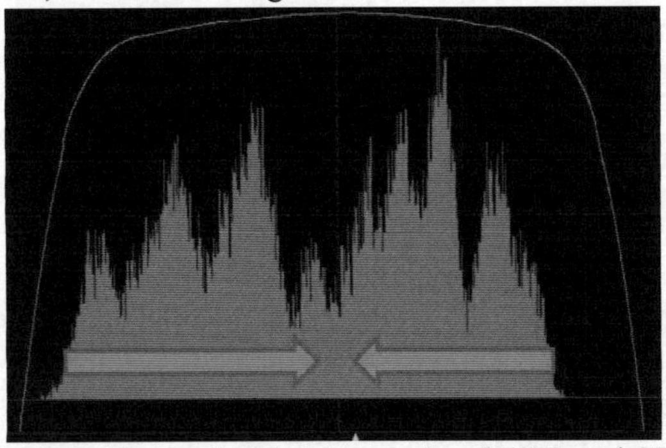

Figure 8. Determine the extrapolated PDF and CDF.

T_j is the width of the CDF at specified BER value. To determine R_j and D_j when using the $R_j + D_j$ CDF Fit method, the CDF is fitted to the Dual Dirac model equation $T_j(BER) = \alpha(BER) * R_j + D_j(\delta\delta)$, where α(BER) is the confidence interval at a confidence level of 1-BER for a single "normal" Gaussian, as shown in (4.3) and (4.4). The points on the CDF that are used in the fit include the selected BER, 1 point above, and two below.

$$T_j(BER) = \alpha(BER) * R_j + D_j(\delta\delta) \quad (4.3)$$

$$\alpha(BER) = 2\left|CDF_{Gaussian}^{-1}\left(\frac{BER}{2\rho_{tx}}\right)\right| \quad (4.4)$$

ρ_{tx} is the transition density which means the ratio of the number of logic transitions to the total number of bits.

5. Experiments

In our experiment, we use Tektronix TG700 signal generation platform to output an SDI signal with jitter which can add known deterministic jitter and random jitter to the signal, and use Teledyne LeCroy oscilloscope to measure the signal in time domain. Using 4K SDI video color bar signal as test signal.

In the experiment, test serial data measuring window in oscilloscope should include a sufficient number of unit intervals and transitions. So that the clock recovery process can accurately determine the expected arrival time of the edges, and the jitter calculation, which includes extrapolation, can have the best statistical significance. We use waveforms with a length of more than 100 iterations of a repeating pattern.

We recover the clock signal from SDI using golden PLL software algorithm. The bandwidth of the loop filtering in PLL will directly affect the measurement of the jitter value. If the bandwidth is set too high, the low-frequency periodic component of the jitter in the actual signal may be filtered out and leave only the jitter component caused by the high-frequency phase noise. It will cause errors in the experimental measurement results. In general, the loop filter bandwidth of the PLL in the experimental system needs to be consistent with the filter bandwidth of the PLL circuit in commercial television signal receivers. We set it to 1KHz in measuring process.

Using the signal generation platform to add deterministic jitter and random jitter to the signal and use this signal as the measured signal for jitter analysis experiments. The experimental results are shown in the table below.

Table 2. Jitter measurement data.

	T_j	D_j	R_j
Generator output	1.760ns	360ps	100ps
dual-Dirac spectral R_j direct method	1.913ns	347.5ps	111.3ps
Adjusted dual-Dirac method	1.699ns	358.5ps	95.3ps

The dual-Dirac spectral R_j direct method is used to calculate σ value of the Gaussian distribution and consider the σ value equals to R_j. However, the σ result will also include the effects of bounded, correlated jitter (apart from periodic jitter), and the extrapolated tail of total jitter may not obey the same Gaussian distribution. Therefore the spectral R_j direct method can fail to give an accurate result.

Through the measurement of jitter, on the one hand, it can help us to judge whether the device under test meets the design specifications, and on the other hand, it can help us find the cause of the problem. This is the traceability analysis method of jitter. From the analysis of the cause of jitter and combined with our actual experience and system design, the analysis of the possible increase in jitter lies in the following aspects:

1) interconnection channel;
2) Power supply ripple and noise;
3) Reference clock.

By separating the components of jitter, if the ISI is large, the main source may cause by interconnect channel, and the impedance continuity and loss of the interconnect channel can be further tested to determine the source of the problem. If the P_j is large, it may be caused by power supply ripple (for non-spreading systems). Then further testing of the power supply ripple is necessary to determine the source of the problem. If the R_j is found to be large, the main reason is heat dissipation and power supply noise.

6. Conclusions

This paper designs a measurement method for high-speed serial digital interface video signal jitter. Jitter measurement and jitter component analysis were performed using an improved dual-Dirac model. The measurement does not depend on BER analysis instrument or other instruments. The measurement and calibration are directly performed by the oscilloscope, which is beneficial to the traceability of the jitter. The experimental results show that the method has better accuracy and the measurement accuracy is higher than the direct spectrum method. The results show that extrapolated tails of total jitter at the asymptotic limit may obey the different Gaussian distribution and the bounded correlation jitter is the part that is easy to be ignored by the measurement, which may have influence on the measurement result.

References

[1] P. Charles, "Digital Video and HDTV Algorithms and Interfaces," *Morgan Kaufmann Publishers, San Francisco,* vol. 260, p. 630, 2003.

[2] J. Hudson, "3Gb/s SDI for Transport of 1080p50/60, 3D," *UHDTV1/4k and Beyond,* 2013.

[3] S. Standard, "2160-Line and 1080-Line Source Image and Ancillary Data Mapping for Single-Link 6G-SDI," 2015.

[4] S. Standard, "2160-line Source Image and Ancillary Data Mapping for 12G-SDI," 2015.

[5] M. P. Li, *Jitter, noise, and signal integrity at high-speed.* Pearson Education, 2007.

[6] R. Stephens, "Analyzing jitter at high data rates," *IEEE Communications Magazine,* vol. 42, no. 2, pp. S6-10 %@ 0163-6804, 2004.

[7] B. Ham, "Fibre channel-methodologies for jitter and signal quality specification-MJSQ," *Proc. Int. Committee Inf. Technol. Standards (INCITS),* pp. 29-38, 2004.

[8] H. Chen *et al.*, "Serial digital interface video signal jitter generator and its calibration method," *Measurement,* vol. 106, pp. 116-120 %@ 0263-2241, 2017.

[9] K. K. Kim, J. Huang, Y.-B. Kim, and F. Lombardi, "Analysis and simulation of jitter sequences for testing serial data channels," *IEEE Transactions on Industrial Informatics,* vol. 4, no. 2, pp. 134-143 %@ 1551-3203, 2008.

[10] R. Stephens, "Jitter analysis: The dual-Dirac model, RJ/DJ, and Q-scale," *Agilent Technical Note,* 2004.

[11] R. Stephens, "What the dual-Dirac model is and what it is not," *Tektronix 360 Knowledge Series,* 2006.

Acknowledgments

This work is supported by National Key R&D Program of China (No. 2018YFF0212103) and Program of NIM (27-AKYCX1907).

The evaluation of the damping control in DFIG wind turbine based on impedance analysis method

Chuan Xiang[1], Lei Yang[2] and Chang Chang[3]

[1]Electric Power Research Institute, Yunan Power Grid CO., LTD, Kunming, China
[2]Electric Power Research Institute, Yunan Power Grid CO., LTD. Kunming, China
[3]College of Electrical Engineering, Southwest Jiaotong University, Chengdu, China

E-mail: 1091930966@qq.com, 15911577929@139.com, chris_e_mail@163.com

Abstract. In this paper, the impedance analysis method is utilized to evaluate the performance of the damping control designed for doubly fed induction generator (DFIG) wind turbine. Two categories of damping control are compared and assessed, parameter-optimization-based damping and direct-addition-based damping. First, the impedance analytical model of the control methods are deduced. Then, the scanning results of the impedance based on Matlab/Simulink are given. Hence, the analysis by synthesis can be implemented. It can be concluded that the damping section eliminates the peak of impedance amplitude, thus finally the impedance of the DFIG can be reshaped by adjusting the key parameters of the damping control.

1. Introduction

Along the rapid development of renewable resources in traditional power system, the permeability of the power electronic devices also becomes more significant. During recent years, frequency oscillation problems occur on many occasions and lead to some severe accidents as devices burnt, wind turbine blockade. At present, the phenomena have been investigated to some extent and several measures are also proposed to resolve the problems. The results of [1], [2], and [3] indicate that the oscillation has close relation with the negative damp at specific frequency points caused by the mismatch of the parameters between wind turbine control and grid. Hence, different damping methods are put forward to enhance the operation stability. An integrated PSS controller with virtual inertia and damping control is proposed by [4]. In [5], a damping control strategy considering active and the reactive power is designed for permanent magnet synchronous generator (PMSG)-based wind turbines. But the influence of the key parameter is abbreviated. Similar to some other researches, they are based on the parameter optimization of the tracking curve. Relatively, the damping control is implemented by the addition of the damping components, as in [6]. The general emulating inertia control and drop control based on the structure of direct addition are briefly analyzed and compared. But the performance differences between the damping based on parameter optimization and damping based on the direct addition are still unclear.

Considering the impedance analysis method has the advantage that can simplify the modeling process and improve the stability analysis compared to the state space method [7-8]. In this paper, the impedance analysis method based on the harmonic linearization is utilized to evaluate the two categories of the damping control. The impedance analytical model is deduced, and the amplitude-frequency curves and the phase-frequency curves generated by impedance scanning in

Matlab/Simulink are given. After the comprehensive and comparative analysis, the property differences of the two categories of damping control will be concluded, some optimizing proposal will also be presented.

The structure of this paper is as follows. The simplification and construction in the model of the DFIG wind turbine is presented in section 2. Section 3 introduces and compares the damping control based on the addition of damping components and the damping control based on the parameter optimization. In section 4, the detail impedance analytical models of DFIG are deduced. Section 5 shows and compares the simulation results of impedance scanning among the different damping methods. Finally, the conclusions and suggestions are given in section 6.

2. DFIG Wind Turbine and Control

2.1. DFIG Wind Turbine Modeling

The paper focuses on the damping assessment of the DFIG wind turbine. the DFIG wind turbine is modeled by three portions, wind turbine, Asynchronous induction generator, and AC-DC-AC converter. The crowbar circuit and chopper circuit are neglected, and the wind turbine is constructed as a dual mass block model. The structure of the whole DFIG wind turbine is shown in Figure1.

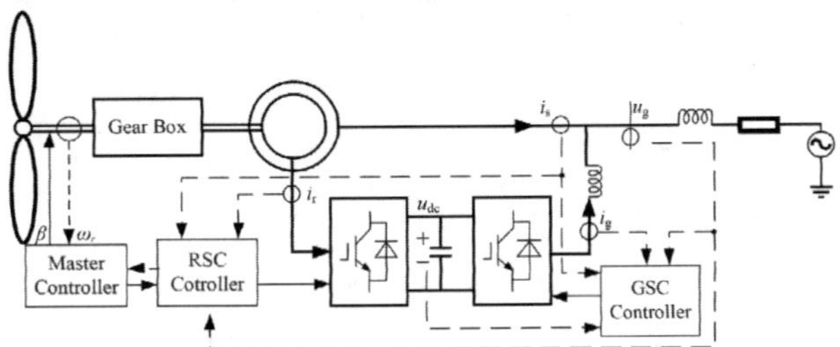

Figure 1. The structure of the DFIG wind turbine.

2.2. Control Structure

The control system is mainly divided into three sections, the master controller, the rotor side converter(RSC) controller and the grid side converter(GSC) controller, as is shown in Fig.1. The master control is adopted to achieve the optimal power calculation and pitch angle adjustment. The structure of the pitch control is shown in Figure 2, and the optimal power calculation is depicted in (1).

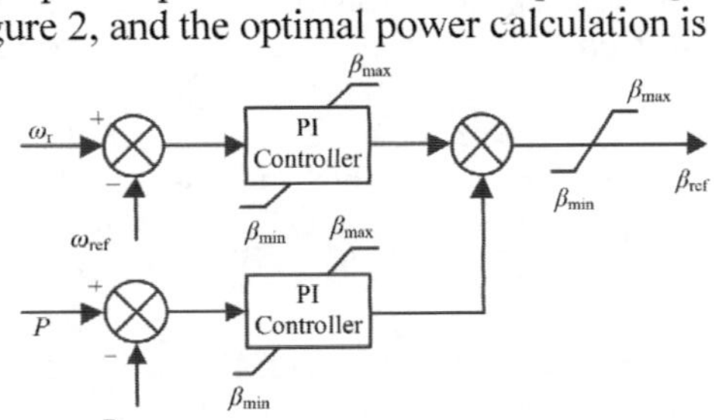

Figure 2. The diagram of the pitch control.

$$P_{out} = \begin{cases} k\omega_r^3, & \omega_0 < \omega_r < \omega_1 \\ \dfrac{P_{max} - k\omega_1^3}{\omega_{max} - \omega_1}(\omega_r - \omega_{max}) + P_{max}, & \omega_1 < \omega_r < \omega_{max} \\ P_{max}, & \omega_r > \omega_{max} \end{cases} \tag{1}$$

where k denotes the coefficient of the tracking curve, ω_r is the rotating speed, ω_0 is the accessing speed, ω_1 is the value of the constant speed period, ω_{max} is the limit speed, and the P_{max} denotes the limit of the output power.

The structure of the RSC controller is dual-closed loop, the outer power loop and the inner current loop in d-q frame, as shown in Figure 3.

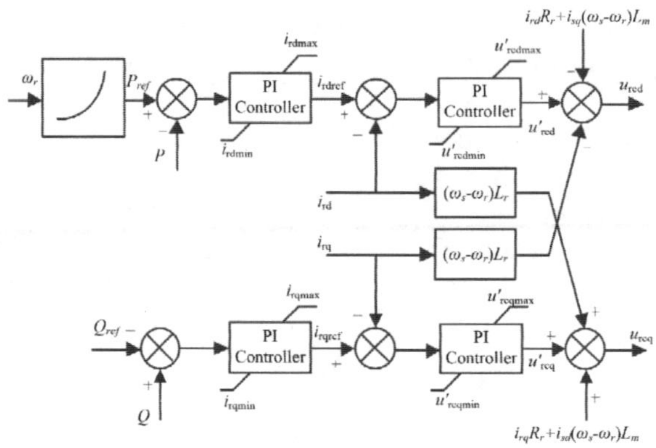

Figure 3. The diagram of the RSC controller.

The GSC controller is also dual-closed loop with the inner current loop in d-q frame, but the outer loop in d axis is the direct voltage control. as depicted in Figure 4.

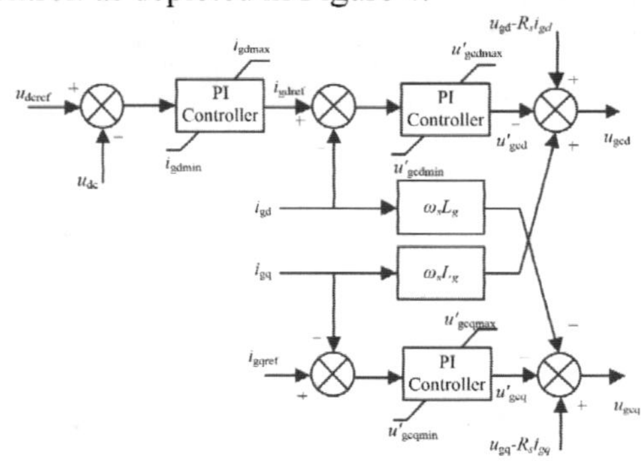

Figure 4. The diagram of the GSC controller.

3. Damping Control

For DFIG wind turbines, since the performance of maximum wind energy capture and the decoupling of the rotating speed and grid frequency, the change of grid frequency will not be responded. Hence the DFIG wind turbines are incapable of inertia response and primary frequency regulation. To make the wind farm participate in the frequency regulation in the power system, the control improvements must be implemented. Considering the RSC has the direct influence on DFIG, the effects will be more remarkable if the work is conducted in the RSC controller, hence the researches of the damping control are generally focus on RSC controller. The studies mainly divided into two sections, the damping controls on the outer loop and the inner loop.

3.1. Damping Control Based on the Optimization

On the active power loop, the improvement can be conducted to fast release or store rotational kinetic energy according to the frequency mutation. Hence the inertia can be stimulated to damp the frequency mutation. The damping based on the tracking curve optimization is depicted in Figure 5, and the optimizing formula is shown in (2).

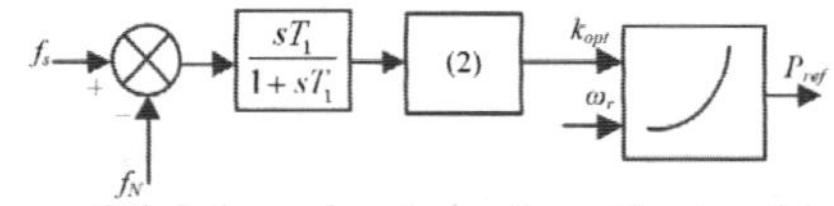

Figure 5. The diagram of the damping based on the tracking curve optimization.

$$k_{opt} = \frac{\omega_{r0}^3}{[\omega_{r0} + 2\pi k_f (f_s - f_N)]^3} k \tag{2}$$

where k_f denotes the coefficient of frequency difference, f_s is the grid frequency, f_N is the nominal frequency, and ω_{r0} denotes the initial rotative speed.

3.2. Damping Control Based on the Optimization

There is another way to achieve the damping on the active power loop, it is the direct adding of frequency difference, as shown in Fig.6. There is no need to acquire the maximum power tracking curve, for most cases, the information is unknown for users.

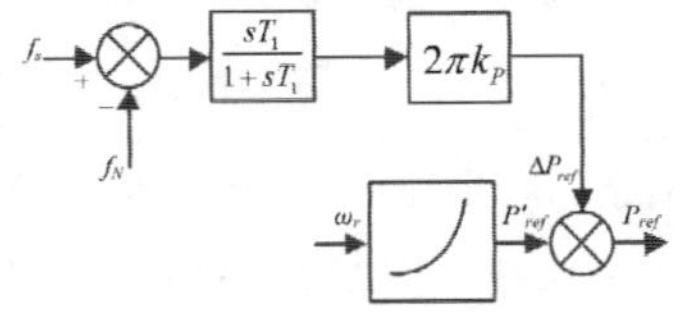

Figure 6. The diagram of the damping based on the direct adding.

Furthermore, if there is enough reactive power margin when the DFIG wind turbine is in operation. The damping effect will be enhanced by dynamically regulating the reactive power. the damping on the reactive power loop is also based on the direct addition of damping components.The structure of the damping on the outer reactive power loop is shown as Fig.7. When the grid voltage difference is below the threshold and the grid frequency difference is beyond the threshold, the function is put into use.

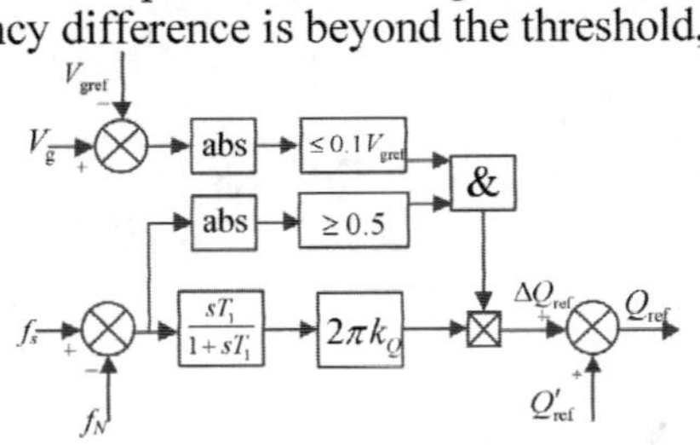

Figure 7. The diagram of the damping on the reactive power loop.

4. Impedance analytical models

Assuming that the capacity of the DC capacitor between the RSC and the GSC is large enough, the voltage can be regarded as a constant value. Hence the DFIG impedance model can be decoupled into two portions, the RSC with induction generator and the GSC, and are defined as Z_1 and Z_2 respectively. The impedance model of DFIG can be depicted by the parallel of Z_1 and Z_2, as shown in (3).

$$Z_{DFIG}=Z_1//Z_2=\frac{Z_1Z_2}{Z_1+Z_2} \quad (3)$$

4.1. The model of the RSC with induction generator

For the induction generator, in frequency domain, the phasor representation of the stator-side equivalent circuit is shown in Figure 8, and the stator voltage equation is depicted as (4).

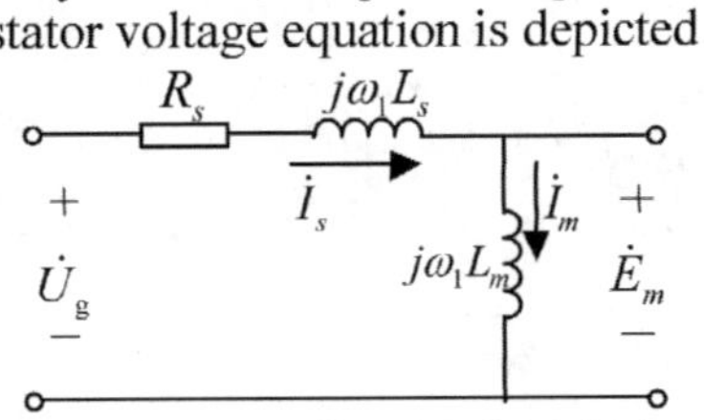

Figure 8. Stator-side equivalent circuit of the induction generator.

$$\begin{cases} \dot{U}_g=\dot{I}_s(R_s+j\omega_1 L_s)+\dot{E}_m \\ \dot{E}_m=j\omega_1 L_m \dot{I}_m \end{cases} \quad (4)$$

where $\dot{U}_g$ is the grid voltage, R_s and L_s are the leakage resistance and leakage inductance of the stator winding, L_m is the inductance of the rotating magnetic field of the stator winding, $\dot{I}_s$ and $\dot{I}_m$ are the stator winding current and excitation current respectively, $\dot{E}_m$ is the induced electromotive force, ω_1 is the stator angular frequency.

Similarly, The equivalent circuit of the rotor-side is shown in Figure 9, and the stator voltage equation can be depicted as (5).

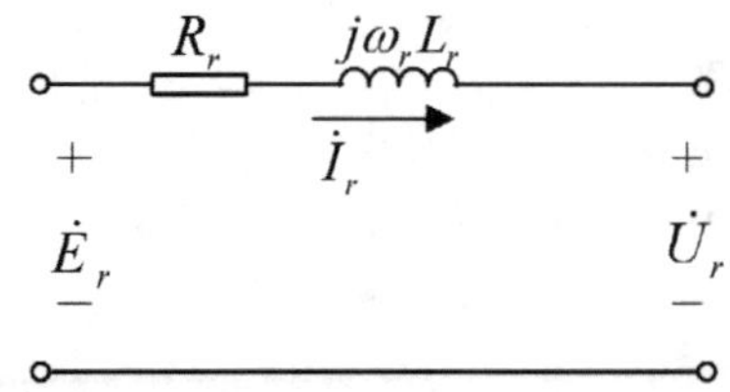

Figure 9. Rotor-side equivalent circuit of the induction generator

$$\dot{E}_r = \dot{I}_r(R_r + j\omega_r L_r) + \dot{U}_r \tag{5}$$

where $\dot{E}_r$ is the induced electromotive force generated by the rotor winding, $\dot{U}_r$ is the rotor voltage, R_r and L_r is the leakage resistance and leakage inductance of the rotor winding, $\dot{I}_r$ is the current of the rotor winding, $\omega_r = \omega_1 - \omega_m$ is the rotor angular frequency, ω_m is the mechanical angular frequency. Defining slip rate as (6).

$$\sigma_1 = \frac{\omega_1 - \omega_m}{\omega_1} \tag{6}$$

Rotor and stator induced electromotive force ratio can be depicted as (7).

$$\frac{E_r}{E_m} = \frac{4.44\sigma_1 f_1 N_2 k_{w2}\varphi_m}{4.44 f_1 N_1 k_{w1}\varphi_m} = \frac{\sigma_1 f_1 N_2 k_{w2}}{f_1 N_1 k_{w1}} = \sigma_1 \alpha \tag{7}$$

where α is the effective turns ratio.
The rotor-side equivalent circuit is converted to the stator side, and the equivalent circuit of induction generator can be acquired and shown in Figure 10.

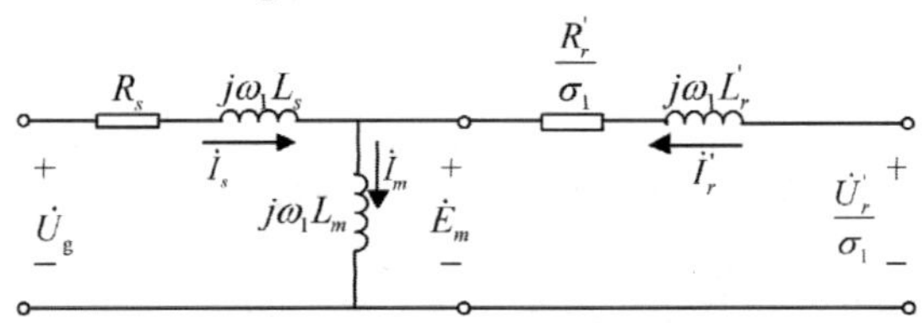

Figure 10. The equivalent impedance circuit of the induction generator.

The parameters on the rotor side is depicted in (8).

$$\begin{cases} \dot{U}_r' = \dfrac{\dot{U}_r}{\alpha}, \dot{I}_r' = -\alpha \dot{I}_r \\ R_r' = \dfrac{R_r}{\alpha^2}, L_r' = \dfrac{L_r}{\alpha^2} \end{cases} \tag{8}$$

As shown in Figure 10, the voltage-current equations of stator side and rotor side can be depicted as (9).

$$\begin{cases} u_{\mathrm{gabc}} = R_s i_{\mathrm{sabc}} + \boldsymbol{L}_{\mathrm{ss}} \dfrac{di_{\mathrm{sabc}}}{dt} + \boldsymbol{L}_{\mathrm{sr}} \dfrac{di_{\mathrm{rabc}}'}{dt} \\ \dfrac{u_{\mathrm{rabc}}'}{\sigma_1} = \dfrac{R_{\mathrm{r}}'}{\sigma_1} i_{\mathrm{rabc}}' + \boldsymbol{L}_{\mathrm{rr}} \dfrac{di_{\mathrm{rabc}}'}{dt} + \boldsymbol{L}_{\mathrm{rs}} \dfrac{di_{\mathrm{sabc}}}{dt} \end{cases} \tag{9}$$

Assuming the system is in symmetrical operation, and the small disturbance signal imposed at the PCC is also symmetric. After the transformation of symmetrical component method, the positive sequence representation of disturbance can be obtained, as shown in (10)

$$\begin{cases} u_{gp} = R_s i_{sp} + (\dfrac{3}{2}L_m + L_s)\dfrac{di_{sp}}{dt} + \dfrac{3}{2}L_m \dfrac{di_{rp}'}{dt} \\ \dfrac{u_{rp}'}{\sigma_p} = \dfrac{R_r'}{\sigma_p} i_{rp}' + (\dfrac{3}{2}L_m + L_r')\dfrac{di_{rp}'}{dt} + \dfrac{3}{2}L_m \dfrac{di_{sp}}{dt} \end{cases} \tag{10}$$

where $\sigma_p = \frac{\omega_p - \omega_m}{\omega_p}$.

The voltage and current containing the disturbance can be depicted in phase domain, as shown in (11), (12) and (13).

$$\boldsymbol{U}_g[f] = \begin{cases} \boldsymbol{U}_{s1} = \frac{U_{s1}}{2} e^{\pm j\varphi_{s1}}, f = \pm f_1 \\ \boldsymbol{U}_{sp} = \frac{U_{sp}}{2} e^{\pm j\varphi_{sp}}, f = \pm f_p \end{cases} \quad (11)$$

$$\boldsymbol{I}_s[f] = \begin{cases} \boldsymbol{I}_{s1} = \frac{I_{s1}}{2} e^{\pm j\varphi_{is1}}, f = \pm f_1 \\ \boldsymbol{I}_{sp} = \frac{I_{sp}}{2} e^{\pm j\varphi_{isp}}, f = \pm f_p \end{cases} \quad (12)$$

$$\boldsymbol{I}_r[f] = \begin{cases} \boldsymbol{I}_{r1} = \frac{I_{r1}}{2} e^{\pm j\varphi_{ir1}}, f = \pm f_1 \\ \boldsymbol{I}_{rp} = \frac{I_{rp}}{2} e^{\pm j\varphi_{irp}}, f = \pm f_p \end{cases} \quad (13)$$

The disturbance voltage will result in the disturbance of the phase-locked loop (PLL). The disturbance from the PLL will inject into the system through the d-q transformation of rotor current and d-q inverse transformation of the control voltage. The structure of PLL is shown in Figure 11.

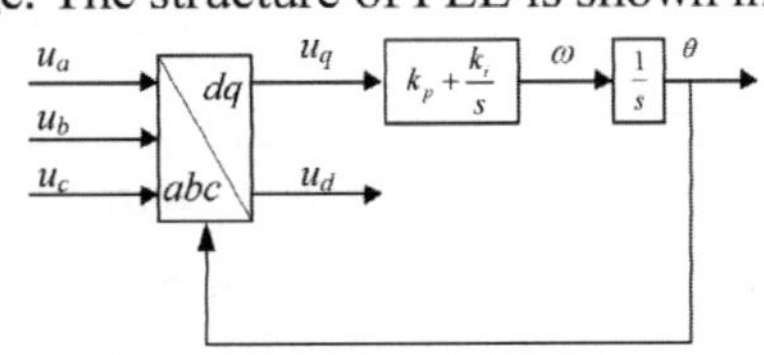

Figure 11. The structure of PLL.

Through the approximating of trigonometric functions, and eliminating the steady state values relationship between practical d-q voltage and ideal d-q-voltage, the PLL output angle can be derived as depicted in (14).

$$\begin{bmatrix} u_{pd}^p \\ u_{pq}^p \end{bmatrix} \approx \begin{bmatrix} u_{pd} + U_q\theta_p \\ -U_d\theta_p + u_{pq} \end{bmatrix} = \begin{bmatrix} u_{pd} \\ -U_{s1}\theta_p + u_{pq} \end{bmatrix} \quad (14)$$

The superscript p means the voltage components in practical conditions, and the subscript p means the components are the disturbances.

The regulator transfer function following the u_q is depicted in (15).

$$H_{pll}(s) = \frac{k_p + \frac{k_i}{s}}{s} \quad (15)$$

The overall transfer function can be deduced as (16)

$$F(s) = \frac{jH_{pll}(s)}{1 + U_{s1}H_{pll}(s)} \quad (16)$$

The rotor current in frequency domain and d-q frame can be depicted as (17).

$$\begin{cases} \boldsymbol{I}_{rpd}[f] = \boldsymbol{I}_{rp} \mp j\sin\varphi_{ir1} I_{r1} F(s)\boldsymbol{U}_{sp}, f = \pm(f_p - f_r) \\ \boldsymbol{I}_{rpq}[f] = \boldsymbol{I}_{rp} \pm j\cos\varphi_{ir1} I_{r1} F(s)\boldsymbol{U}_{sp}, f = \pm(f_p - f_r) \end{cases} \quad (17)$$

The output control voltage in d-q frame can be acquired.

$$\begin{cases} \boldsymbol{U}_{rpd}[f] = jI_{r1}F(s)\boldsymbol{U}_{sp}[\sin\varphi_{ir1}H_{ri}(s) - \cos\varphi_{ir1}k_{rd}] \\ \qquad \pm \boldsymbol{I}_{rp}[-H_{ri}(s) \pm jk_r], f = \pm(f_p - f_r) \\ \boldsymbol{U}_{rpq}[f] = jI_{r1}F(s)\boldsymbol{U}_{sp}[\cos\varphi_{ir1}H_{ri}(s) + \sin\varphi_{ir1}k_{rd}] \\ \qquad \mp \boldsymbol{I}_{rp}[\pm \mathrm{j}H_{ri}(s) + k_r], f = \pm(f_p - f_r) \end{cases} \tag{18}$$

where H_{ri} denotes the transfer function of current control of RSC, k_{rd} is the cross coupling coefficient of RSC.

Ignoring the modulation effect of the converter and the delay action of the switch, the control voltage in d-q frame can be transformed to the three-phase coordinate system, then the result can be converted to the stator-side, as shown in (19).

$$\begin{aligned} \boldsymbol{U}'_{rp}[f] = jI'_{r1}F(s \mp j2\pi f_1)\boldsymbol{U}_{sp}H_{ri}(s \mp j2\pi f_1) + \\ \boldsymbol{I}'_{rp}[-H_{ri}(s \mp j2\pi f_1) \pm jk_{rd}], f = \pm f_p \end{aligned} \tag{19}$$

Combining (10) and (19), The impedance model of the RSC with induction generator can be derived.

$$Z_{1p}(s) = \frac{U_{gp}}{I_{sp}} = \frac{R_s + (\frac{3}{2}L_m + L_s)s - \dfrac{\frac{9}{4}L_m^2 s^2}{\frac{H_{ri}(s \mp j2\pi f_1) \mp jk_r}{\alpha^2\sigma_p} + \frac{R'_r}{\sigma_p} + (L'_r + \frac{3}{2}L_m)s}}{1 - \dfrac{\frac{3}{2}sL_m I'_{r1}F(s \mp j2\pi f_1)H_{ri}(s \mp j2\pi f_1)/\sigma_p}{\frac{H_{ri}(s \mp j2\pi f_1) \mp jk_r}{\alpha^2\sigma_p} + \frac{R'_r}{\sigma_p} + (L'_r + \frac{3}{2}L_m)s}} \tag{20}$$

4.2. The Model of the GSC

Defining $L_a = L_b = L_c = L_g$, the state space equation can be derived.

$$L_g \frac{di_{gabc}}{dt} = u_{gabc} - u_{cabc} \tag{21}$$

where i_{gabc} denote the three phase currents of the GSC, u_{gabc} denote the three phase grid voltages, u_{cabc} denote the three phase output voltages of GSC.

Similar to the section 4.1, the positive sequence representation of disturbance can be obtained, as shown in (22)

$$L_g \frac{di_{gp}}{dt} = u_{gp} - u_{cp} \tag{22}$$

The output control voltage can be deduced.

$$\begin{aligned} \boldsymbol{U}_{cp}[f] &= F(s \mp j2\pi f_1)\boldsymbol{U}_{gp}I_{g1}H_{gi}(s \mp j2\pi f_1) + \\ &\quad \boldsymbol{I}_{gp}[-H_{gi}(s \mp j2\pi f_1) \pm jk_g], f = \pm f_p \\ &= sL_g\boldsymbol{I}_{gp} + \boldsymbol{U}_{gp} \end{aligned} \tag{23}$$

Hence, the impedance model of the GSC with induction generator can be derived.

$$Z_{2p} = -\frac{\boldsymbol{U}_{gp}}{\boldsymbol{I}_{gp}} = \frac{H_{gi}(s \mp j2\pi f_1) \mp jk_g + sL_g}{1 - F(s \mp j2\pi f_1)I_{g1}H_{gi}(s \mp j2\pi f_1)} \tag{24}$$

When the damping control is utilized in the control system of DFIG, it will reshape the equivalent impedance of the DFIG. Some modification can be implemented in original model to obtain the new impedance model with the damping control. The analysis will benefit from the corresponding impedance model.

5. Simulation and Analysis

In this section, a DFIG detail simulation model is constructed in Matlab/Simulink, and the major electrical parameters and the key control parameters are respectively listed in Table 1 and Table 2. The *ω-P* curves under different wind speed with β=0 are shown in Figure 12, thus the optimal wind power tracking curve can be acquired. The symmetric voltage disturbances with different frequency and small amplitude are imposed to the steady-state voltage. Then, the current component of corresponding disturbance frequency shall be measured and utilized to calculate the equivalent impedance.

Table 1. Major Electrical Parameters.

Region	Item	Description	Value	Unit
Supply voltage	V_s	Root-mean-square value	575	V
	Fnom	Nominal frequency	50	Hz
Genarator	Pnom	Nominal power	1.5/0.9	MVA
	R_s/ L_s	Stator resistance and inductance	0.023/ 0.18	p.u.
	R_r/ L_r	Rotor resistance and inductance	0.016/ 0.16	p.u.
	L_m	Mutual inductance	2.9	p.u.
	p	Pole-pairs	3	-
Wind turbine	P_{mec}	Nominal mechanical power	1.5	MW
	H_w	Innertial time constant	4.32	-
	D_w	Damping coefficient	1.5	-
	K_w	Elastic coefficient of shafting	1.11	-
GSC inductance	L_g	GSC output inductance	0.3	p.u.

Table 2. Key Control Parameters.

Region	Item	Description	Value
Pitch Control	K_{p_pitch}	Speed loop	1.5
	K_{p_l}/ K_{i_l}	Power limit loop	3/30
RSC control	K_{p_P}/ K_{i_P}	Outer active power loop	0.3/20
	K_{p_Q}/ K_{i_Q}	Outer reactive power loop	0.5/20
	K_{p_ird}/ K_{i_ird}	Inner d-current loop	0.6/8
	K_{p_irq}/ K_{i_irq}	Inner q-current loop	0.6/8
GSC control	K_{p_dc}/ K_{i_dc}	Outer dc voltage loop	8/400
	K_{p_igd}/ K_{i_igd}	Inner d-current loop	0.83/5
	K_{p_igq}/ K_{i_igq}	Inner q-current loop	0.83/5

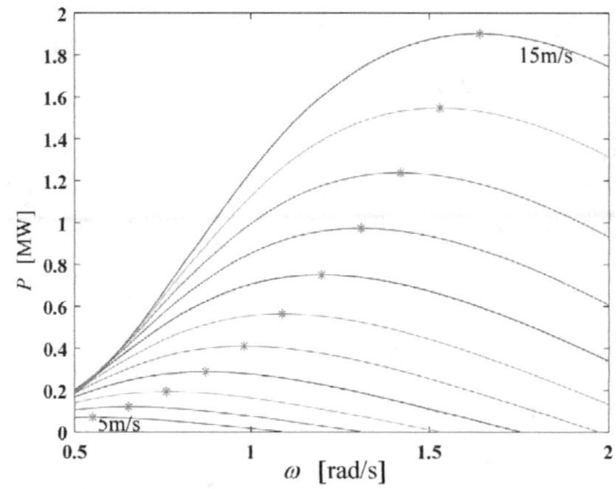

Figure 12. The ω-P curves of the wind turbine.

The impedance scanning results without damping control, with the damping control based on the optimization and the damping control based on addition are implemented, and the corresponding amplitude-frequency curves and phase-frequency curves are acquired, as shown in Figure 13. Here, for the control based on addition, only active power loop is analyzed.

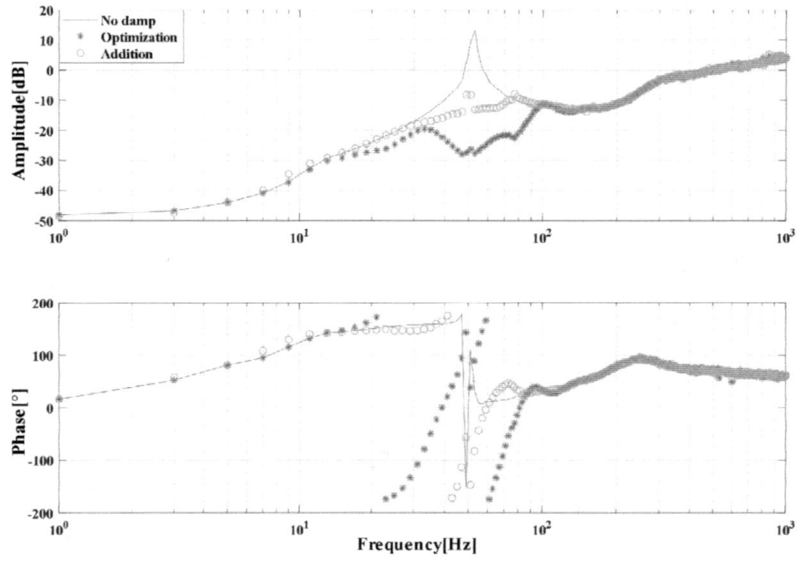

Figure 13. The Bode diagram of the impedance.

It can be seen that in original operation, the amplitude of impedance has a peak. The peak will result in the crossing of the equivalent impedance of DFIG and the impedance of the grid. Generally, the phase of the grid impedance is inductive, if the phase of the DFIG impedance is capacitive, the difference of the DFIG impedance and grid impedance may be greater than 180° . If the situation is at the crossing point, the super-synchronous oscillation or sub-synchronous oscillation may occur.

For the damping based on optimization, it significantly eliminates the peak of impedance amplitude near the nominal frequency point, thus the risk of oscillation can be reduced. It should be noted that there is a big change of the phase, two segments negative phase will prejudice the stability.

For the damping based on direct addition, it also has the capability of reducing the peak. The effect is not as significant as the damping based on optimization, but its negative influence in phase is very small.

The equivalent impedances under different damping coefficients are also obtained to show the influence. The bode diagram of the impedance of damping based on the optimization is shown in Figure 14, and the bode diagram of the impedance of damping based on addition is shown in Figure 15.

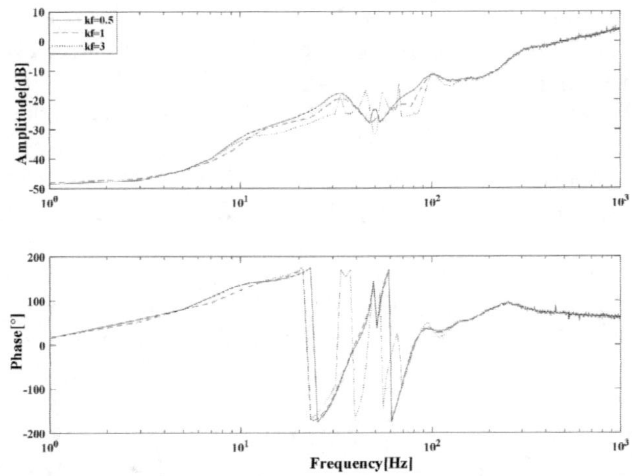

Figure 14. The Bode diagram of the impedance of damping based on the optimization.

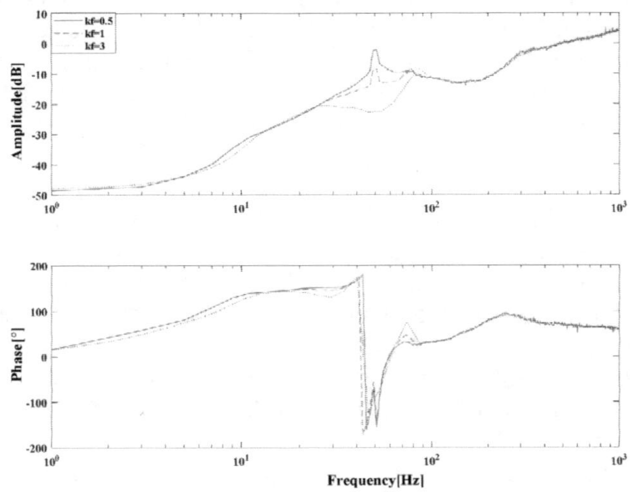

Figure 15. The Bode diagram of the impedance of damping based on addition.

For the change of the damping coefficient k_f in the damping based on the optimization, as shown in Figure 14, it can be seen that from k_f =0.5 to k_f =1,the amplitude of the impedance is further reduced with little change in phase. But when k_f =3, the amplitude has obvious oscillation near the nominal frequency, and three segments of negative phase with large value occurs. Hence the impedance characteristic is deteriorated when k_f=3.

For the change of the k_f in the damping based on addition, as shown in Figure 15, it can be seen that from k_f =0.5 to k_f =3,the amplitude of the impedance is further reduced with little change in phase. The coefficient can be further reduced to get the more suited value.

6. Conclusions

In this paper, the impedance analysis method based on the harmonic linearization is utilized to evaluate the damping based on parameter optimization and the damping based on the direct addition. The impedance analytical model of the DFIG is preliminarily deduced. Based on the simulation model constructed in Matlab/Simulink, the amplitude-frequency curves and the phase-frequency curves generated by impedance scanning are obtained. After the comprehensive and comparative analysis, some conclusion can be acquired.

The amplitude of impedance without the damping has a peak, which will result in the crossing of the equivalent impedance of DFIG and the impedance of the grid. The risk of instability is high.

The damping based on optimization can significantly eliminate the peak of impedance amplitude near the nominal frequency point, thus the risk of oscillation can be reduced. But the big change of the phase results from this method may prejudice the stability.

The damping based on direct addition also has the capability of reducing the peak. Though the effect is not as significant as the damping based on optimization, its negative influence in phase is very small.

For the damping based on optimization, an optimized damping coefficient can be searched in the range from 0.5 to 3. For the damping based on addition, the optimized damping coefficient should be searched over value 3.

In the future work, the analytical impedance considering the damping section will be deduced to further analyze and validate its performance. In addition, the asymmetric condition will be considered in detail by the construction of corresponding positive and negative impedance.

References

[1] Z. Miao, "Impedance-Model-Based SSR Analysis for Type 3 Wind Generator and Series-Compensated Network," in IEEE Transactions on Energy Conversion, vol. 27, no. 4, pp. 984-991, Dec. 2012.

[2] Y. Zhang, Y. Gao, K. Gao, Y. Wang and J. Yang, "Research on Analysis and Suppression Strategy of Subsynchronous Oscillation in Wind Power Grid-Connected System," 2018 Chinese Control And Decision Conference (CCDC), Shenyang, 2018, pp. 4937-4942.

[3] C. Chen, W. Du, H. Wang and T. Littler, "Sub-synchronous oscillations in power systems caused by grid-connected wind farms — A survey of mechanism studies," in CSEE Journal of Power and Energy Systems, vol. 4, no. 4, pp. 495-503, Dec. 2018.

[4] X.Y. Zhang, Y. Fu, Y. Wang, H. Wang and C. Fu,"Integrated PSS Controller of Variable Speed Wind Turbines withVirtual Inertia and Damping Control," Transctions of China Electrotechnical Society, vol.30, no.1 ,pp. 159-169, Jan. 2015.

[5] Y. Wang, X.Y. Zhang, M.H. Li and X.R. Zhu, "Damping Control of PMSG-Based Wind Turbines for Power System Oscillations," Transctions of China Electrotechnical Society, vol.27, no.12 ,pp. 162-171, Dec. 2012.

[6] J.X. Zhao, W. Gao, M.X. Shangguan, X.B. Zha, S. Yue and Y.H. Liu, "Review on frequency regulation technology of power grid by wind farm," Power System Protection and Control, vol.45, no.21, pp. 157-169, Nov. 2017.

[7] Y. Wang, X. Wang, F. Blaabjerg and Z. Chen, "Harmonic Instability Assessment Using State-Space Modeling and Participation Analysis in Inverter-Fed Power Systems," in IEEE Transactions on Industrial Electronics, vol. 64, no. 1, pp. 806-816, Jan. 2017.

[8] J. Sun, "Impedance-Based Stability Criterion for Grid-Connected Inverters," in IEEE Transactions on Power Electronics, vol. 26, no. 11, pp. 3075-3078, Nov. 2011.

Analysis risk of dynamic stability of power network with multiple small motor groups equivalent to each other

Lei Yang, Chuan Xiang, Xin He, Chao Xing, Xinze Xi, Shixian Sheng, Can Xiao, Jiasi Liao

Electric Power Research Institute of Yunnan, Kunming, Yunnan, China

Southwest Jiaotong University, School of Electrical Engineering, Chengdu, Sichuan, China

E-mail: 1373591113@qq.com

Abstract. The BPA data of the regional power grid is usually only used for the detailed modeling of the 110kV grid connected substation, and the 35kV and the following grid station generally use the equivalent parameters of the equivalent capacity of the unit, and then be incorporated into the 110kV power grid for simulation analysis. In order to analyze the accuracy of power system stability simulation after equivalence, a detailed model of the grid connected small hydropower units with 35kV and the following is built. The results show that, using a plurality of small hydropower units group area grid equivalence, dynamic stability analysis of differences, there are regional power grid internal oscillation mode equivalent and cannot reflect the regional oscillation modes in small turbine.

1.Introduction

With the strong demand for environmental protection, hydropower, as a clean energy source, has been valued by the state and the localities. Many small hydropower stations have been established in areas with abundant hydropower resources. In the northwestern part of Yunnan, the number of small hydropower units in the municipal-level regional power grid has reached hundreds. The characteristic of these small hydropower units are that the capacity of a single machine is small, the number is large, and the grid voltage level is low. According to the enterprise standard regulations of China Southern Power Grid Corporation [1], the 50MW and above hydro-generator units and pumped-storage power generation/motor units that are integrated into the southern power grid shall be tested and modeled for the parameters of the excitation system, prime mover and its regulating system. However, there is no clear regulation for small hydropower units below 50MW. Usually, considering that if each small hydropower unit established a detailed model, there may be a problem of "dimensionality disaster" in the calculation, so the regional grid BPA data is only used for the detailed modeling of the 110kV grid-connected plant , and grid-connected plants of 35kV and below are generally equivalent to the equal capacity units with typical parameters, which is incorporated in the 110kV power grid to simulate and analyze, or it is equivalent to the regional load. The characteristics of the small power are difficult to fully reflect.

In recent years, research on the equivalent of power system has been increasing [2]~[18]. In [5], it proposes a homology equivalent method, which is generators that oscillate at a similar speed after a large disturbance in the system are classified as a coherent group, and the generators in each group are combined into an equal value machine. Because this method is based on the coherent characteristics of generators after large disturbances, it is generally applied to the external network equivalent in transient stability analysis. The literature [6] combines the characteristics of reginal small hydropower units and the dynamic stability analysis of power systems' requirements to dynamic equivalents. It improves the traditional coherence equivalence, and proposes a dynamic equivalence modeling method for small hydropower groups based on homology aggregation and network simplification. These methods mainly tend to treat the small hydropower unit as a whole, and analyze the influence on the dynamic characteristics of the main network, and it does not take into account the accuracy of the stability analysis of a small number of small hydropower units within a regional power grid.

In this paper, by establishing a detailed model of grid-connected power stations of 35kV and below in a reginal power grid, the simulation and comparative analysis of the dynamic stability and transient stability of the regional power grid before and after modeling are carried out. The results show that in the regional power grids with multiple small hydropower units, there are internal oscillation mode of the regional power grid and regional oscillation mode with small motor groups that are not reflected by the equivalent. The research results of this paper have certain reference significance for the actual power system stability analysis of engineering.

2.Modeling and calculation standards

2.1. Detailed modeling of 35kV power grid

Detailed models are used for all 35kV grid-connected units in the regional grid. Governor model GH, excitation model FV/F+, PSS model SI/SI+/SP, generator model M/MF.

The "Motor Saturation Factor at Rated Voltage" and "Motor Saturation Factor at 1.2x Rated Voltage" in the generator model's MF card are filled in typical values. Fill in the kinetic energy of the generator:

(1) The T_J parameter in the reported data is calculated according to formula (1) to obtain the kinetic energy of the generator.

$$E_{MWS} = \frac{T_J}{2} S_{N\_GEN} \tag{1}$$

(2) There is no T_J parameter. If a power station is split into N equal capacity units, the generator kinetic energy is filled in by the original 1/N.

(3) There is no T_J parameter. Approximate T_J value is 6(Reference book: turbine rotor time constant is 4~8; governor parameter modeling test experience value:1~10MW turbine group, value

4.2~4.5; considering T_J influence on the dynamic stability ,leaving a certain degree of predetermination, value 6).

In the GH card of the governor model, the R adjustment coefficient is filled 4%, the manual dead zone of 20MW or more for a single machine is filled in as 0.001(the value of the standard), and the dead zone below 20MW is filled in as 0.002(the value of the standard). The remaining paraments are filled in according to typical parameters.

Photovoltaic power generation model is updated to PV, BC, BC+ model.

2.2. Stable calculation of relay electrical protection and safety automatic device action time

According to the requirements of the 《Guidelines for the Analysis and Analysis of Safety and Stability of China Southern Power Grid》, combined with the actual situation of the regional power grid, the operational time regulation for the stable calculation of relay protection and automatic device are shown in Table 1.

Table 1 Operating times of protection relays and automation devices.

Line fault type	Fault removal time (s)		Reclosing time (s)	Coincidence of trip time after fault (s)
	Proximal	remote		
Three-phase short circuit of 220kV line	0.12	0.12		
Single-phase short circuit of 110kV line (optical fiber differential protection)	0.12	0.12	1.2	-
Single-phase short circuit of 110kV line (distance protection)	-	-	2	-
110kV a line	0.65	0.65	2	-
110kV b line	0.6	0.6	2	-
110kV c line	0.65	0.65	2	-
110kV d line	0.4	0.4	2	-
110kV e line	0.3	0.3	2	-
110kV f line	0.6	0.6	2	-
110kV g line	0.3	0.3	2	-
110kV h line	0.65	0.65	2	-
110kV i line	0.3	0.3	2	-

In the calculation process, considering the 110kV single-phase transient and 220kV three-phase permanent fault. Due to the different 110kV line protection configuration of the power grid in the region, the fault is set according to the specific conditions of the protection: 110kV line configured the fiber-based differential protection as the main protection adopts the fault removal time of 0.12s and the reclosing time of 1.2s; 110kV line configured the distance protection as the main protection adopts that the distance II section action time is the fault removal time(mainly considering the distance II section protection line length), and the fault removal time is set according to the distance of the specific line distance II section setting time + circuit breaker trip time. The fault removal time is the same at the near end and the far end (Due to the distance I fault removal time is short and consider the most serious situation, the actual fault end is set to be consistent), unified use of reclosing time 2s.

2.3. Stability criterion

According to the 《Guidelines for the Analysis and Analysis of Safety and Stability of China Southern Power Grid》, the specific requirements are as follows.

(1) Transient stability: After the system failure, the relative angle rocking curve of any two units in the same system is synchronously decreasing oscillation.

(2) Dynamic stability: If any component in the system fails, the protection and the switch operate correctly. The damping ratio of the oscillation mode related to the small and medium power station group in the system is not less than 3.0%, and the system damping ratio is not less than 4.5% when the system disturbance is small.

2.4. Calculation boundary

Hydropower is as full as possible: the hydropower units of the regional power grid are fully loaded according to the installed capacity.

The ultimate method of abundance: the maximum load limit mode of the abundance according to the stability limit of the section.

Calculate according to the way that hydropower is as full as possible, and check according to the ultimate method of abundance.

3.Dynamic stability analysis results

3.1. Hydropower is as full as possible

In the case where multiple small hydropower units are equivalent in the area, there is an oscillation mode inside the grid. The oscillation mode has an oscillation frequency of 1.342 Hz and a damping ratio of 5.5% (Oscillation mode with damping less than 7%).

Tab. 2 Participation factors and eigenvectors of relevant generators in different system oscillation modes of area A when small hydros are taken as several equivalents

Power plant's name	Right mode value	Right angle	Participation factor
A	0.16	0.00	1.00
B	0.12	12.33	0.56
C	0.14	11.30	0.46
D	0.13	-1.06	0.40
E	0.09	156.45	0.21
F	0.04	148.60	0.07
G	0.04	148.60	0.07
H	0.03	178.87	0.04

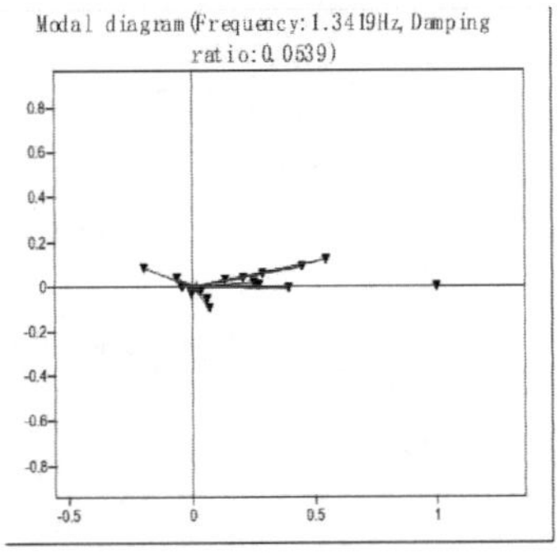

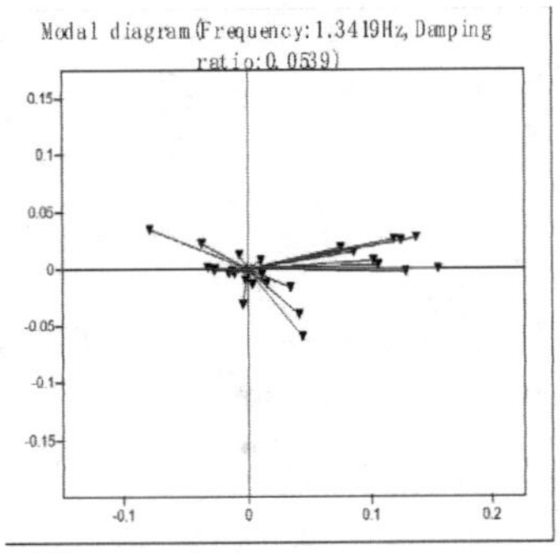

Figure 1. Plots of participation factors and right eigenvectors in different system oscillation modes of area A when small hydros are taken as several equivalents.

The small disturbance analysis results of the oscillation mode of the hydropower in the full-scale mode of the grid after detailed modeling of the 35kV grid are shown in Table 3. Only the oscillation modes with a damping ratio 7% or less are listed in the Table 3. Each oscillation mode participation factor modal diagram and the right eigenvector modal diagram are shown in Figure 2.

Table 3. Results of small disturbance analyzation in different system oscillation modes when small hyrdos are at their maximum power output.

Serial number	Real	Imaginary	Frequency (Hz)	Damping ratio (%)	Oscillation mode
1	-0.346	8.859	1.410	3.9	Mode 1
2	-0.320	7.350	1.170	4.3	Mode 2
3	-0.492	10.016	1.594	4.9	Mode 3
4	-0.726	10.349	1.647	7.0	Mode 4
5	-0.749	10.526	1.675	7.0	Mode 5

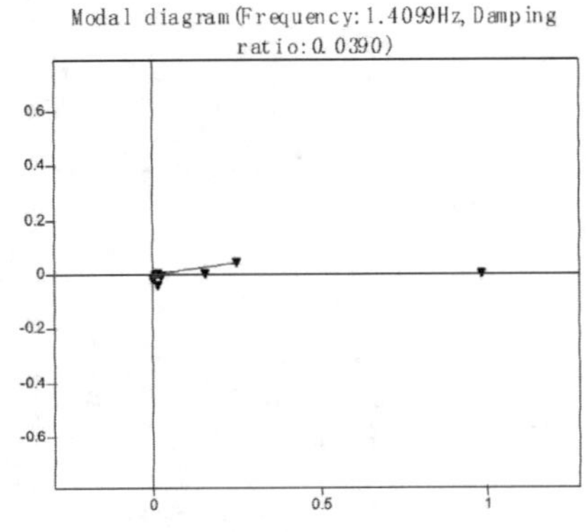

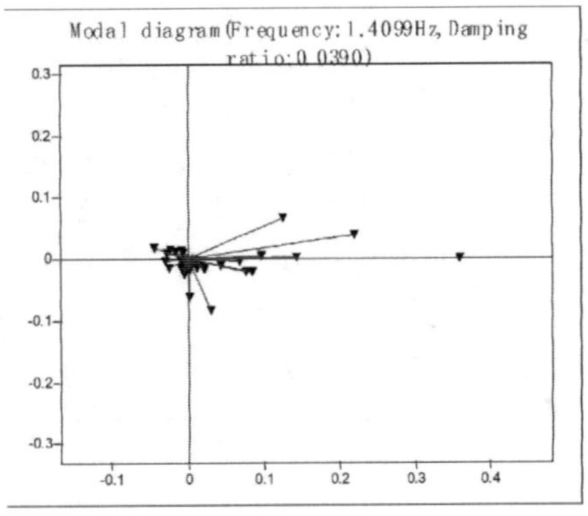

(a) Mode one

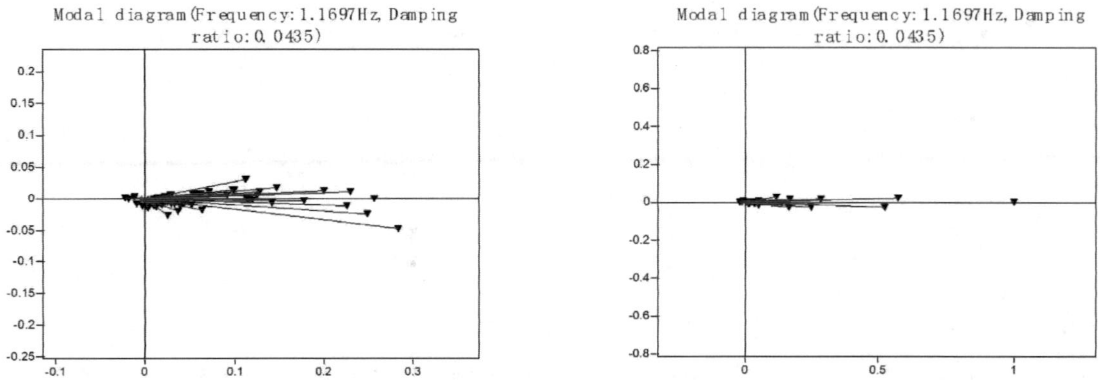

(b) Mode two

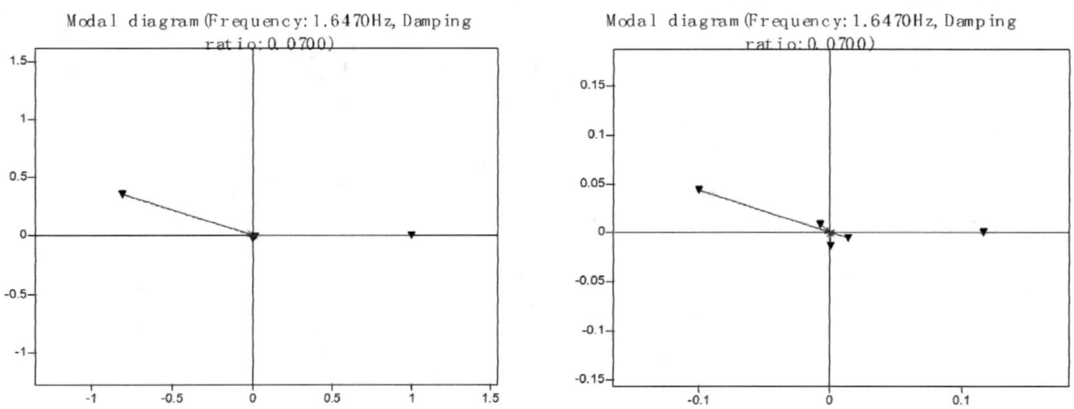

(c) Mode three

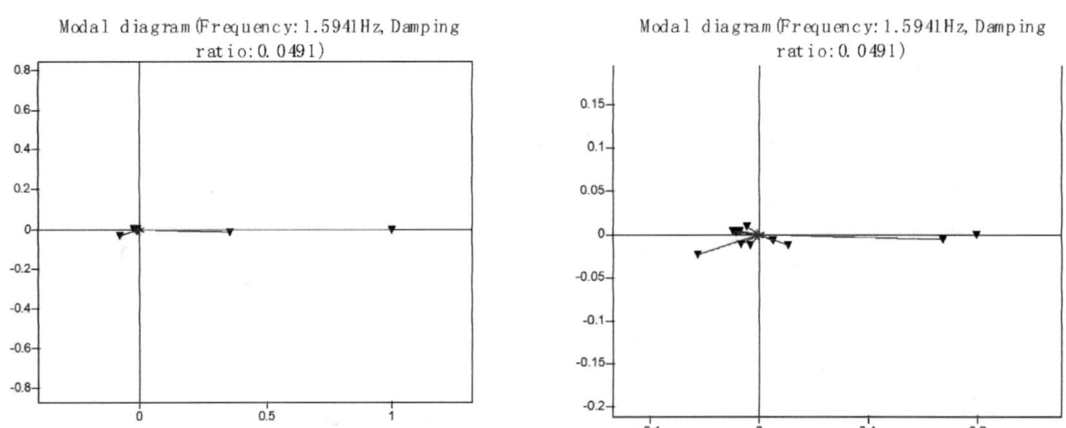

(d) Mode four

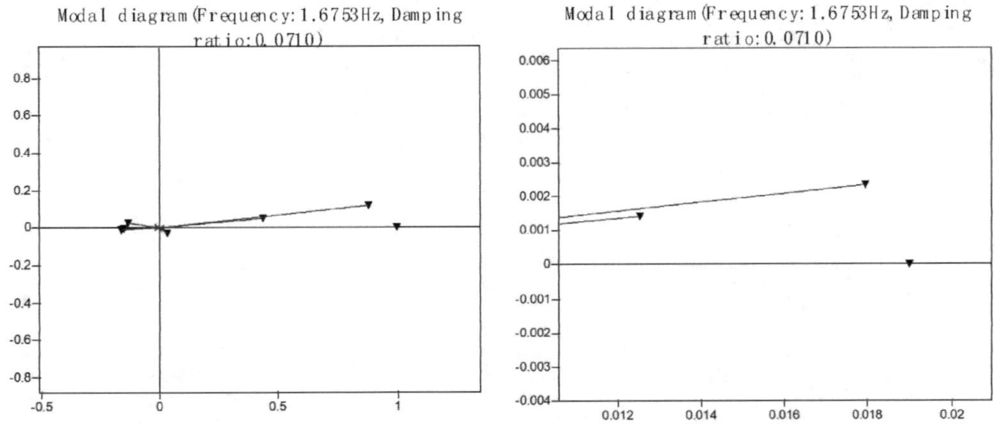

(e) Mode five

Figure 2. Plots of participation factors and right eigenvectors in different system oscillation modes after building a more accurate mode.

It can be seen from Table 2 that the units with large participation factors in the oscillation mode have A、B、C、D、E、F and so on. The oscillation mode participation factor modal diagram and the right eigenvector modal diagram are shown in Figure 1.

In the full-water mode of hydropower, the 35kV power grid in the region is modeled in detail and compared with the oscillation modes of multiple small hydropower groups. The comparison results are as follows:

(1) There is no 1.342 Hz oscillation mode in detailed model.

(2) Other new oscillation modes appear. After the detailed modeling of the 35kV power grid in the region, mode 1 and mode 2 (the main participating units are C、I、J)and mode 3 , mode 5.

(3) Damping ratio reduction. Before the model of the 35kV power grid in the region, the oscillation mode damping ratio associated with the grid unit was greater than 4.5%, which satisfies the stability standard of the southern grid. However, after the model of the 35kV grid in the region, the oscillation mode with damping ratio less than 4.5% appeared. The model 1 damping ratio is 3.9%. There is an oscillation mode between the regions, which has a damping ratio of 4.3%.

The preliminary analysis produces the above three results for:

(1) Units A, B, C, D, E, F and son on, were split, and the 1.342Hz oscillation mode mainly participated by these equivalent units disappears.

(2) The detailed modeling of the 35kV grid-connected unit has appeared to be accurate to the oscillation mode of each unit. For example, mode 1, it is mainly involved in K, L, M, N power plants. Before the 35kV grid is detailed modeled, the M and N stations are equivalent in O, and the K and N stations are equivalent in P. After the detailed modeling of the 35kV grid, the oscillation mode between the split single units is highlighted.

(3) Compared with the model after the detailed modeling of the 35kV grid, the single unit capacity decreased, and most of the small units were not equipped with PSS, and the damping showed a downward trend.

3.2. The ultimate method of abundance

Under the ultimate method of abundance, after detailed modeling of the 35kV power grid in the region, the oscillation modes of the units participating in the area are shown in Table 4. Only the oscillation modes with a damping ratio of less than 8% are listed in the table.

Table 4. Results of small disturbance analyzation in different system oscillation modes when the system is operating under the limit of high flow period.

Serial number	Real	Imaginary	Frequency (Hz)	Damping ratio (%)	Oscillation mode
1	-0.379	8.859	1.410	4.3	Mode one
2	-0.440	7.085	1.128	6.2	Mode two
3	-0.536	9.946	1.583	5.4	Mode three
4	-0.739	10.298	1.639	7.2	Mode four
5	-0.748	10.518	1.674	7.1	Mode five

Consistent with the conclusion of the full-scale hydropower system, under the ultimate method of abundance, the comparison of the oscillation mode of the 35kV power grid in the region with the equivalent mode has the following three results:

(1) The original 1.342Hz oscillation mode disappears.

(2) Other new oscillation modes appear.

(3) Damping is reduced, and the oscillation mode 1 does not meet the dynamic statbility standard of southern power grid.

4.Transient stability analysis

The transient stability of the grid in this area mainly analyzes the stability of the power angle of the three-phase fault in 220kV line and the single-phase fault in 110kV line (protective correct action), which is consistent with the conclusions of the detailed modeling of the 35kV power grid. Under the ultimate method of abundance and the full power of the hydropower, the power angle can maintain stability.

5.Conclusion

In this paper, the dynamic and transient stability analysis of region power grid before and after the equivalent of several hydropower units in a certain area of Yunnan are analyzed. The results show that the conclusions on the transient stability of multiple small hydropower groups before and after equivalence are consistent; the dynamic stability difference is large, and the oscillation mode inside the regional power grid cannot be highlighted after the equivalence, especially the oscillation before the equivalent small hydropower units, and the oscillation mode between the regions cannot be obtained. The conclusions of this paper have certain warning significance for the stability analysis of power system with multiple small water machine groups.

References

[1] 2010 *Regulations for Measurment and Modeling of Parameters for Excitation System, Prime Motive and Its Regulation System of Sychronous Generators of China Southern Power Grid*

[2] Yang Fu, Liang Jun and Niu Qing-da 2012 A practical method of parameter aggregation for power system dynamic equivalence *Power System Protection and Control* vol 40(16) pp 70-6

[3] Li Xiao-hui, Luo Min and Liu Li-xia 2010 A new dynamic equivalence method and its

application in tianjin grid *Power System Protection and Control* vol 38(3) pp 61-6

[4] Zou Gen-hua, Huang Wei and Yao Zhu-xiang 2009 An implementation scheme of model integration considering external network equivalence for Jiangxi power networks *Power System Protection and Control* vol 37(13) pp 94-7

[5] Ju Ping 2010 *Theory and Method of Power System Modeling* (Beijing: Science Press)

[6] Wang Min, Wen Jinyu and Hu Wenbo 2013 A dynamic equivalent modeling for regional small hydropower generator group *Power System Protection and Control*, vol 41(7) pp 1-9

[7] Wang Jiali, Liu Dichen and Liao Qingfen 2016 Generator tripping strategy based on transient energy of power system *Power System Protection and Control* vol 44(10) pp 73-7

[8] Ni Yi-xin, Chen Shou-sun and Zhang Bao-lin 2002 *Theory and Analysis of Dynamic Power System*(Beijing: Tsinghua University Press)

[9] Geng Qingshen, Lu Yu and Fan Hairong 2016 Comparative analysis of operation losses of UHV AC and EHV AC transmission systems *Power System Protection and Control* vol 44(16) pp 52-5

[10] Hu Jie and Yu Yi-xin 2006 A practical method of parameter aggregation for power system dynamic equivalence *Power system Technology* vol 30(24) pp 26-30

[11] M L Ourari 2006 Dynamic equivalent modeling of large power systems using structure preservation technique *IEEE Transactions on Power Systems* vol 21(3) 1284 - 95

[12] Yang Jing-ping 2007 Study on dynamic equivalents methods for large-scale power systems (Hangzhou: Zhejiang University)

[13] Wu Yan, Yang Shuqiang and Jiang Zhixun 2015 Research on the equivalence of large-scale power system using node sensitivity analysis approach *Power System Protection and Control* vol 43(10) pp 83-6

[14] Liu Ju, Sun Haishun and Wen Jinyu 2013 A Method for Analyzing Large Disturbance Impact Characteristics on Tie-line in AC Interconnected Power Grid *Automation of Electric Power Systems* vol 37(21) pp 17-22

[15] Gu Lihong, Zhou Xiaoxin and Tao Hongzhu 2010 Analysis on mechanism of inter-area forced oscillation caused by local weak interconnection in interconnected power grid *Power System Technology* vol 34(12) pp 36-43.

[16] Wu Cheng, Xu Zheng and Han Song 2010 Approximate method to analyze low frequency oscillation of power systems *High Voltage Engineering* vol 36(10) pp 2594-9

[17] Miao You-zhong, Tang Yong and LI Dan 2007 Tentative study of inter-area large power oscillation mechanism caused by the local mode *Proceedings of the CSEE*, vol 27(10) pp 73-7

[18] Liu Xu-bin 2010 Application research of the power system's external network equivalence(Jinan: Shangdong University)

A Low Complexity Robust Adaptive Beamforming Method with Orthogonality Constraint

Bin Zhao and Jia Yang

63891 Troops of People's Liberation Army, Luoyang, Henan, China

E-mail: passionhzy@sina.com

Abstract. Performance of many adaptive beamforming methods degrades when antenna array steering vector mismatch exists and popular optimization-based methods suffer from heavy computation complexity. In this paper, we investigate robust adaptive beamforming technique with low complexity. An optimization problem is constructed based on orthogonality constraint instead of conventional norm constraint; therefore, semidefinite programming (SDP) relaxation is avoided. We present the approach to obtain closed-form solution by imposing Lagrange multiplier methodology. Numerical experiments demonstrate that proposed low complexity method outperforms other previously developed beamforming methods and reduces computation time significantly.

1. Introduction

Adaptive beamforming has been widely utilized in many fields, i.e., wireless communications, radar, astronomy and medical imaging [1-2]. If the knowledge of desired signal steering vector is known exactly, conventional adaptive beamforming methods can suppress interference effectively; otherwise, desired signal may be suppressed as an interference, resulting in drastically reduced antenna array output signal-to-interference-plus-noise ratio (SINR) [3-5]. Steering vector mismatch including pointing error and array geometry perturbations is the common case in practical situations, hence many approaches have been proposed for improving robustness against steering vector mismatch.

Eigenspace-based beamforming [6] is a prevalent robust adaptive technique, however, it doesn't work well at low signal-to-noise ratio (SNR) because of subspace swap phenomenon. With the presumed steering vector and the prior information that the mismatch vector is norm bounded, the worst-case-based beamforming [7] is proposed as an optimization-based technique. An approach utilizing orthogonal interference-plus-noise (IN) subspace projection matrix is introduced in[8] and the dimension of estimated IN subspace is determined by an artificial energy percentage parameter. Sparse reconstruction[9] has been recently introduced as a new beamforming design principle. The algorithm performs perfectly only for ideal array geometry and presence of array perturbations would lead to severe performance degradation. Besides, methods in [7-9] are optimization-based and the computation complexity is high.

In this paper, a low complexity robust adaptive beamforming method with orthogonality constraint is proposed. By analyzing the advantage of orthogonality constraint, we construct a new optimization problem. Low complexity Lagrange multiplier methodology is employed to obtain closed-form solution. Proposed low complexity method is shown to outperform existing beamforming methods by simulation results. Furthermore, proposed method reduces computation complexity significantly.

2. The Signal model

We consider a uniform linear array (ULA) with M omni-directional sensors. The received signal $\mathbf{x}(k)$ is given by:

$$\mathbf{x}(k)=\mathbf{s}(k)+\mathbf{i}(k)+\mathbf{n}(k)=s_0(k)\mathbf{a}_0+\mathbf{i}(k)+\mathbf{n}(k)$$

where $\mathbf{a}_0$ is desired steering vector, k is time index , $s_0(k)$ is waveform of desired signal, $\mathbf{s}(k)$, $\mathbf{i}(k)$ and $\mathbf{n}(k)$ represent desired signal, interference, and noise, respectively. $\mathbf{s}(k)$ and $\mathbf{i}(k)$ are assumed to be statistically independent to each other. The output of beamforming method can be written as

$$y(k)=\mathbf{w}^H\mathbf{x}(k)$$

where $\mathbf{w}$ is the weight vector of the array and $(\cdot)^H$ stands for the Hermitian transpose. The optimal weight vector can be formulated as:

$$\mathbf{w}_{\text{opt}}=\frac{\mathbf{R}_{i+n}^{-1}\mathbf{a}_0}{\mathbf{a}_0^H\mathbf{R}_{i+n}^{-1}\mathbf{a}_0} \quad \text{* MERGEFORMAT (1)}$$

where $\mathbf{R}_{i+n} \quad E\left\{(\mathbf{i}(k)+\mathbf{n}(k))(\mathbf{i}(k)+\mathbf{n}(k))^H\right\}$ is IN covariance matrix. In practical situations, $\mathbf{R}_{i+n}$ and $\mathbf{a}_0$ may be unknown, therefore, sample covariance matrix $\hat{\mathbf{R}}=1/N\sum_{k=1}^{N}\mathbf{x}(k)\mathbf{x}^H(k)$ with N snapshots and presumed steering vector $\bar{\mathbf{a}}$ are employed to replace them. is transformed to sample matrix inversion (SMI) adaptive beamforming:

$$\mathbf{w}_{\text{SMI}}=\frac{\hat{\mathbf{R}}^{-1}\bar{\mathbf{a}}}{\bar{\mathbf{a}}^H\hat{\mathbf{R}}^{-1}\bar{\mathbf{a}}}. \quad \text{* MERGEFORMAT (2)}$$

If gap exists between aprior $\bar{\mathbf{a}}$ and actual $\mathbf{a}_0$, the beampattern of $\mathbf{w}_{\text{SMI}}$ won't point to the direction where desired signal is arriving. Furthermore, with limited snapshots, sample covariance matrix cannot precisely represent $\mathbf{R}_{i+n}$, then desired signal may be regarded as interference and be suppressed. Influenced by above two drawbacks, SINR of degrades severely compared to optimal weight vector. The objective of this paper is to obtain better estimated desired signal vector $\mathbf{a}$ to increase output SINR.

3. Proposed Method

3.1. Beamforming Based on Orthogonality

In [10], A. Khabbazibasmenj et al. proposed an optimization problem to estimate desired signal vector based on norm constraint:

$$\begin{aligned} &\min_{\mathbf{a}} \quad \mathbf{a}^H\hat{\mathbf{R}}^{-1}\mathbf{a} \\ &\text{s.t.} \quad \|\mathbf{a}\|^2=M \\ &\qquad \mathbf{a}^H\tilde{\mathbf{C}}\mathbf{a}\le\Delta_0=\max_{\theta\in\Theta}\mathbf{d}^H(\theta)\tilde{\mathbf{C}}\mathbf{d}(\theta) \end{aligned} \quad \text{* MERGEFORMAT (3)}$$

where $\tilde{\mathbf{C}}=\int_{\tilde{\Theta}}\mathbf{d}(\theta)\mathbf{d}^H(\theta)d\theta$, $\mathbf{d}(\theta)$ is the steering vector corresponding to direction θ under the assumption of ideal array geometry, Θ is angular sector where desired signal is located, $\tilde{\Theta}$ is the complement of Θ. Unfortunately, optimization problem is non-convex due to non-convex equality constraint $\|\mathbf{a}\|^2=M$. The solution can be obtained by imposing SDP relaxation and (4) is transformed into:

$$\begin{aligned} \min_{\mathbf{A}} \quad & \mathrm{Tr}\left(\hat{\mathbf{R}}^{-1}\mathbf{A}\right) \\ \text{s.t.} \quad & \mathrm{Tr}\left(\mathbf{A}\right) = M \\ & \mathrm{Tr}\left(\tilde{\mathbf{C}}\mathbf{A}\right) \le \Delta_0 \\ & \mathbf{A} \ge \mathbf{0} \end{aligned}$$ * MERGEFORMAT (4)

where $\mathbf{A}$ $\mathbf{a}\mathbf{a}^H$. The optimization problem is not equivalent to because $\mathbf{A}$ in may be not rank-one. Furthermore, is usually solved with matlab toolbox such as Sedumi and CVX, resulting in high computation complexity.

The non-convex property of results from non-convex equality constraint, hence the problem without non-convex equality constraint is convex and can be expressed as:

$$\min_{\mathbf{a}} \quad \mathbf{a}^H\hat{\mathbf{R}}^{-1}\mathbf{a} \quad \text{s.t. } \mathbf{a}^H\tilde{\mathbf{C}}\mathbf{a} \le \Delta_0 .$$ * MERGEFORMAT (5)

Obviously, the solution to is $\mathbf{a} = \mathbf{0}$. To avoid this trivial solution, another constraint is needed. Let mismatch vector $\mathbf{e} = \mathbf{a}_0 - \bar{\mathbf{a}}$, $\mathbf{e}$ can be decomposed as $\mathbf{e} = \mathbf{e}_\perp + \mathbf{e}$. $\mathbf{e}_\perp$ represents the part orthogonal to $\bar{\mathbf{a}}$ and $\mathbf{e}$ denotes the part parallel to $\bar{\mathbf{a}}$. Since norm of $\bar{\mathbf{a}}$ does not influence accuracy of weight vector $\bar{\mathbf{a}}$, only the estimation of $\mathbf{e}_\perp$ is essential. Based on the physical meaning of $\mathbf{e}_\perp$, we have:

$$\bar{\mathbf{a}}^H\mathbf{e}_\perp = 0 .$$ * MERGEFORMAT (6)

can be rewritten as:

$$\bar{\mathbf{a}}^H\left(\mathbf{a} - \bar{\mathbf{a}}\right) = 0 .$$ * MERGEFORMAT (7)

Combined with and , the new optimization problem is given by:

$$\begin{aligned} \min_{\mathbf{a}} \quad & \mathbf{a}^H\hat{\mathbf{R}}^{-1}\mathbf{a} \\ \text{s.t.} \quad & \bar{\mathbf{a}}^H\left(\mathbf{a} - \bar{\mathbf{a}}\right) = 0, \quad \mathbf{a}^H\tilde{\mathbf{C}}\mathbf{a} \le \Delta_0 \end{aligned}$$ * MERGEFORMAT (8)

Compared to , is a convex optimization problem and SDP relaxation is not needed. Consider
$\bar{\mathbf{a}}^H\left(\mathbf{a} - \bar{\mathbf{a}}\right) = \bar{\mathbf{a}}^H\mathbf{a} - \bar{\mathbf{a}}^H\bar{\mathbf{a}} = \bar{\mathbf{a}}^H\mathbf{a} - M$.

is transformed as:

$$\begin{aligned} \min_{\mathbf{a}} \quad & \mathbf{a}^H\hat{\mathbf{R}}^{-1}\mathbf{a} \\ \text{s.t.} \quad & \bar{\mathbf{a}}^H\mathbf{a} = M, \quad \mathbf{a}^H\tilde{\mathbf{C}}\mathbf{a} \le \Delta_0 . \end{aligned}$$ * MERGEFORMAT (9)

Mention that both and are quadratically constrained quadratic program (QCQP) problem.

3.2 Closed-form Solution

Consider the function:

$$f_1\left(\mathbf{a}, \lambda, \mu\right) = \mathbf{a}^H\hat{\mathbf{R}}^{-1}\mathbf{a} + \lambda\left(\mathbf{a}^H\tilde{\mathbf{C}}\mathbf{a} - \Delta_0\right) + \mu\left(2M - \bar{\mathbf{a}}^H\mathbf{a} - \mathbf{a}^H\bar{\mathbf{a}}\right)$$ * MERGEFORMAT (10)

where $\lambda \ge 0$, $\mu \ge 0$. Assuming

$$\mathbf{a}^H\tilde{\mathbf{C}}\mathbf{a} \le \Delta_0 .$$ * MERGEFORMAT (11)

The second constraint in is satisfied and the solution to is given by:

$$\mathbf{a} = \frac{M\hat{\mathbf{R}}\bar{\mathbf{a}}}{\bar{\mathbf{a}}^H\hat{\mathbf{R}}\bar{\mathbf{a}}} .$$ * MERGEFORMAT (12)

By inserting into , the condition in is equivalent to:

$$\frac{M^2\overline{\mathbf{a}}^H\hat{\mathbf{R}}\tilde{\mathbf{C}}\hat{\mathbf{R}}\overline{\mathbf{a}}}{\overline{\mathbf{a}}^H\hat{\mathbf{R}}\overline{\mathbf{a}}\overline{\mathbf{a}}^H\hat{\mathbf{R}}\overline{\mathbf{a}}}\le\Delta_0 \quad .$$ * MERGEFORMAT (13)

Therefore, if holds, is the solution to optimization problem . However, if doesn't hold, second constraint in is active. For fixed λ and μ, the minimization of results in:

$$\hat{\mathbf{a}}_{\lambda,\mu}=\mu\left(\hat{\mathbf{R}}^{-1}+\lambda\tilde{\mathbf{C}}\right)^{-1}\overline{\mathbf{a}} \quad .$$ * MERGEFORMAT (14)

Through the differential of subject to λ and μ, solution to the maximization of $f_1(\mathbf{a},\lambda,\mu)$ with respect to λ and μ is given by:

$$\hat{\mu}=\frac{M}{\overline{\mathbf{a}}^H\left(\hat{\mathbf{R}}^{-1}+\lambda\tilde{\mathbf{C}}\right)^{-1}\overline{\mathbf{a}}} \quad .$$ * MERGEFORMAT (15)

$$g_1(\lambda)\quad\frac{\overline{\mathbf{a}}^H\left(\hat{\mathbf{R}}^{-1}+\lambda\tilde{\mathbf{C}}\right)^{-1}\tilde{\mathbf{C}}\left(\hat{\mathbf{R}}^{-1}+\lambda\tilde{\mathbf{C}}\right)^{-1}\overline{\mathbf{a}}}{\overline{\mathbf{a}}^H\left(\hat{\mathbf{R}}^{-1}+\lambda\tilde{\mathbf{C}}\right)^{-1}\overline{\mathbf{a}}\overline{\mathbf{a}}^H\left(\hat{\mathbf{R}}^{-1}+\lambda\tilde{\mathbf{C}}\right)^{-1}\overline{\mathbf{a}}}=\frac{\Delta_0}{M^2} \quad .$$ * MERGEFORMAT (16)

Optimal Lagrange multiplier $\hat{\lambda}$ can be solved with the Newton's method. In the following part, we try to obtain $\hat{\lambda}$. Obviously, $g_1(\lambda)$ is a monotonically decreasing function with respect to λ when $\lambda\ge 0$. The Newton's method can be adopted upon until the upper bound and lower bound of $\hat{\lambda}$ is obtained. When $\lambda=0$, based on ,

$$g_1(0)=\frac{\overline{\mathbf{a}}^H\hat{\mathbf{R}}\tilde{\mathbf{C}}\hat{\mathbf{R}}\overline{\mathbf{a}}}{\left(\overline{\mathbf{a}}^H\hat{\mathbf{R}}\overline{\mathbf{a}}\right)^2}>\frac{\Delta_0}{M^2}$$ * MERGEFORMAT (17)

hence $\lambda=0$ can be regarded as the lower bound. As $\lambda\to\infty$,

$$g_1(\lambda)\to g_1(\infty)=\frac{\overline{\mathbf{a}}^H\tilde{\mathbf{C}}^{-1}\overline{\mathbf{a}}}{\overline{\mathbf{a}}^H\tilde{\mathbf{C}}^{-1}\overline{\mathbf{a}}\overline{\mathbf{a}}^H\tilde{\mathbf{C}}^{-1}\overline{\mathbf{a}}}<\rho\quad\frac{\overline{\mathbf{a}}^H\tilde{\mathbf{C}}\overline{\mathbf{a}}}{\overline{\mathbf{a}}^H\overline{\mathbf{a}}\overline{\mathbf{a}}^H\overline{\mathbf{a}}}<\frac{\Delta_0}{M^2} \quad .$$ * MERGEFORMAT (18)

Based on and , we learn that optimal Lagrange multiplier $\hat{\lambda}$ always exists and the solution to $g_1(\lambda)=\rho$ is large than $\hat{\lambda}$, hence the upper bound of solution to $g_1(\lambda)=\rho$ is also the upper bound of $\hat{\lambda}$. We perform following eigenvalue composition:

$$\hat{\mathbf{R}}^{1/2}\tilde{\mathbf{C}}\left(\hat{\mathbf{R}}^{1/2}\right)^H=\mathbf{U}\boldsymbol{\Gamma}\mathbf{U}^H \quad ,$$ * MERGEFORMAT (19)

where $\mathbf{U}$ denotes the set of eigenvectors, $\boldsymbol{\Gamma}=\mathrm{diag}(\gamma_1,\cdots,\gamma_M)$ and γ_i are eigenvalues sorted in descending order. can be rewritten as:

$$\tilde{\mathbf{C}}=\hat{\mathbf{R}}^{-1/2}\mathbf{U}\boldsymbol{\Gamma}\mathbf{U}^H\left(\hat{\mathbf{R}}^{-1/2}\right)^H \quad .$$ * MERGEFORMAT (20)

Insert into and let $\mathbf{h}=\mathbf{U}\hat{\mathbf{R}}^{1/2}\overline{\mathbf{a}}$, we have

$$g_1(\lambda)=\frac{\sum_{i=1}^{M}\frac{|h_i|^2\phi_i}{(\phi_i+\lambda)^2}}{\left(\sum_{i=1}^{M}\frac{|h_i|^2\phi_i}{\phi_i+\lambda}\right)^2}$$ * MERGEFORMAT (21)

where h_i is the ith variable of $\mathbf{h}$, $\phi_i = \frac{1}{\gamma_i}$ for $i = 1, \cdots, M$ and are sorted in ascending order. Then

$$\rho \le \frac{\frac{1}{(\phi_1+\lambda)^2}\sum_{i=1}^{M}|h_i|^2 \phi_i}{\frac{1}{(\phi_M+\lambda)^2}\left(\sum_{i=1}^{M}|h_i|^2 \phi_i\right)^2} = \frac{(\phi_M+\lambda)^2}{(\phi_1+\lambda)^2 \sum_{i=1}^{M}|h_i|^2 \phi_i}.$$

Upper bound of $\hat{\lambda}$ is given by:

$$\lambda_{up} = \frac{c_2\phi_1 - \phi_M}{1-c_2}$$

where $c_2 = \sqrt{\rho\sum_{i=1}^{M}|h_i|^2 \phi_i}$. With lower bound and upper bound of optimal Lagrange multiplier $\hat{\lambda}$, the Newton's method is employed in $\left[0, \lambda_{up}\right]$ to obtain $\hat{\lambda}$ satisfying $g_1\left(\hat{\lambda}\right) = \Delta_0 / M^2$.
Inserting into , we have:

$$\mathbf{a} = \frac{M\left(\hat{\mathbf{R}}^{-1} + \hat{\lambda}\tilde{\mathbf{C}}\right)^{-1}\bar{\mathbf{a}}}{\bar{\mathbf{a}}^H\left(\hat{\mathbf{R}}^{-1} + \hat{\lambda}\tilde{\mathbf{C}}\right)^{-1}\bar{\mathbf{a}}}. \quad \text{* MERGEFORMAT (22)}$$

The weight vector is given by:

$$\hat{\mathbf{w}} = \frac{\hat{\mathbf{R}}^{-1}\mathbf{a}}{\mathbf{a}^H\hat{\mathbf{R}}^{-1}\mathbf{a}}. \quad \text{* MERGEFORMAT (23)}$$

For proposed method, if holds, the computation complexity of estimating actual steering vector in can be neglected, hence the total computation complexity originates from the matrix inversion operation when computing weight vector in , which is $O(M^3)$. However, if doesn't hold, another matrix inversion operation is performed in . Since the binary search and Newton's method only involves several multiplications and derivative calculations, the total computation complexity is still $O(M^3)$. By solving QCQP problems and with optimization software, the computation complexity is $O(M^{3.5})$. Overall, the proposed method has a remarkable advantage to QCQP methods in the view of the computation complexity and is competent in real-time application scenarios.

4. Experimental results and analysis

Basic simulation conditions are the same as Example 1. All simulations are performed using matlab 2016b running on an Intel Core i7 Duo, 2.5 GHz processor with 4GB of memory, under Windows 7 Service Pack 1. 100 Monte-Carlo runs are performed.

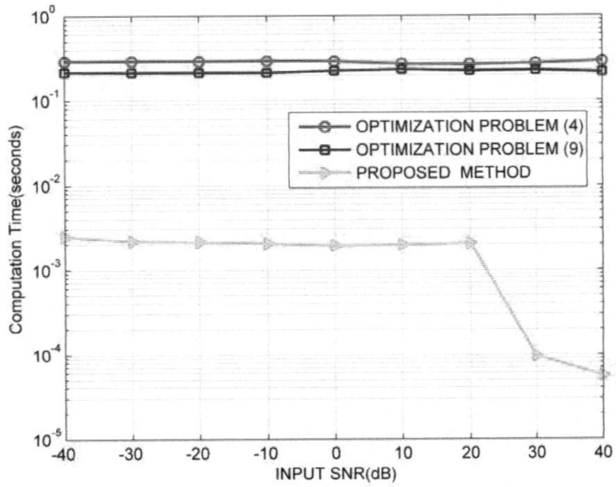

Figure 1. The computation time comparison.

Simulation Example 1: Computation Time. In this example, the computation time of optimization problem , optimization problem solved by CVX matlab Toolbox and proposed low complexity method is displayed in Figure 1. For optimization problem , the number of unknown variables is M times of optimization problem solved by CVX matlab Toolbox, hence more iterations are needed for the convergence of . Proposed low complexity method significantly reduces the computation complexity compared with optimization-based methods. When input SNR is higher than 30dB, the computation complexity is reduced further since is satisfied. However, the computation complexity has no matter with input SNR for optimization-based methods.

The proposed method is compared to the eigenspace-based method[6], the worst-case-based method[7], IN subspace projection matrix method[8], the sparse reconstruction method[9] and optimization problem in terms of the output SINR.

Simulation Example 2: Exactly known desired steering vector. In this example, the desired steering vector $\mathbf{a}_0$ is assumed to be equivalent with presumed steering vector $\bar{\mathbf{a}}$. It can be seen from Figure 2 that proposed low complexity method outperforms other beamforming techniques except for sparse reconstruction method. Under the assumption that desired steering vector is known exactly, the ideal IN can be reconstructed and the output SINR is almost equal to optimal value. At low SNR, eigenspace-based method suffers from severe performance degradation because of a high probability of subspace swap. Proposed method performs better than optimization problem , verifying that SDP relaxation results in performance loss.

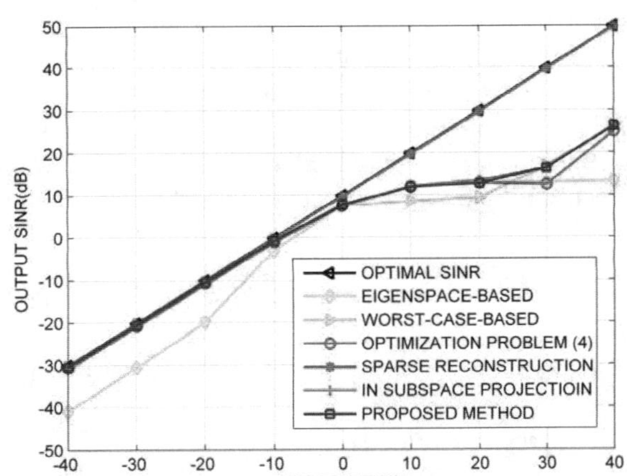

Figure 2. Output SINR versus SNR with Exactly known desired steering vector.

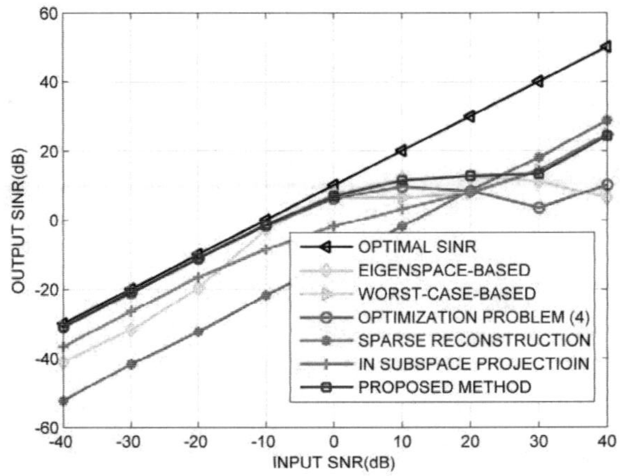

Figure 3. Output SINR versus SNR with pointing error, gain error and phase error.

Simulation Example 3: Pointing error, gain error and phase error. In this example, the presumed DOA of desired signal is not known exactly. Also, array geometry is distorted by gain and phase error. The DOA of desired signal is uniformly distributed in $[\overline{\theta}_s - 3°, \overline{\theta}_s + 3°]$. Array gain and phase error are uniformly distributed in [-3dB, 3dB] and [-20°, 20°], respectively. As can be observed from Figure 3, the performance of [8] is terrible. Since the construction of IN matrix is based on ideal array geometry, the occurrence of array geometry, i.e., gain and phase error, results in severe performance degradation. Besides, compared with proposed method, the disadvantage of SDP relaxation is evident.

5. Conclusion

In this paper, a new beamforming method is proposed. We construct a convex optimization problem by imposing orthogonality constraint. Lagrange multiplier methodology is employed to obtain closed-form solution. As a low complexity method without SDP relaxation, the proposed method is demonstrated to outperform existing beamforming methods and takes less computation time by simulation results.

References

[1] Qian J, He Z, Zhang W, et al. “Robust adaptive beamforming for multiple-input multiple-output radar with spatial filtering techniques,” Signal Processing, Vol. 143, 152-160, 2018.

[2] Huang, Yongwei, Mingkang Zhou, and Sergiy A. Vorobyov. "New designs on MVDR robust adaptive beamforming based on optimal steering vector estimation," IEEE Transactions on Signal Processing, 2019.

[3] Liao B, Guo C, Huang L, et al. “Robust adaptive beamforming with precise main beam control,” IEEE Transactions on Aerospace and Electronic Systems, Vol. 53, 345-356, 2017.

[4] Xiao Z, Wang J, Geng L, et al. “On the Robustness of Covariance Matrix Shrinkage-Based Robust Adaptive Beamforming,” Proceedings of the 2018 International Conference on Electronics and Electrical Engineering Technology, 196-201, 2018.

[5] Feng Y, Liao G, Xu J, et al. “Robust adaptive beamforming against large steering vector mismatch using multiple uncertainty sets,” Signal Processing, Vol. 152, 320-330, 2018.

[6] Feldman D. D. and Griffiths L. J., “A projection approach for robust adaptive beamforming,” IEEE Transactions on Signal Processing, Vol. 93, No.4, 867-876, 1994.

[7] Vorobyov S. A., Gershman A. B., and Luo Z.-Q., “Robust adaptive beamforming using worst-case performance optimization: A solution to the signal mismatch problem,” IEEE Transactions Signal Processing, Vol. 51, No.2, 313-324, 2003.

[8] Jia W., Jin W., Zhou S., et al., “Robust adaptive beamforming based on a new steering vector estimation algorithm,” Signal Processing, Vol. 93, No. 9, 2539-2542, 2013.

[9] Xie J., Li H., He Z., et al., “A robust adaptive beamforming method based on the matrix

reconstruction against a large DOA mismatch," EURASIP Journal on Advances in Signal Processing, Vol. 2014, No. 1, 91, 2014.

[10] Khabbazibasmenj A., Vorobyov S. A., and Hassanien A., "Robust adaptive beamforming based on steering vector estimation with as little as possible prior information," IEEE Transactions Signal Processing, Vol. 60, No. 6, 2974-2987, 2012.

FFT based unambiguous acquisition method for BOC modulation signal

Bin Li[1,2], Dan Qi[1], Shaozhen Zhang[1], Dan Cheng[1], Jun Mu[1] and Zheng Yan[1]

[1]Space Star Technology Co, Ltd, Beijing, China

[2]E-mail: libin@spacestar.com.cn

Abstract. This paper presents an unambiguous acquisition algorithm for Binary Offset Carrier (BOC) modulation signals, which is adopted by modernized Global Navigation Satellite System (GNSS). By utilizing FFT, the spectrum of BOC signals is derived. This paper proposes an unambiguous acquisition method based on FFT to get rid of the BOC modulation drawback. By utilizing FFT, the spectrum of BOC signal is derived. The up and down sideband are separated in frequency domain by rectangular function, which means sideband filtering. Due to sparsity of filtered BOC spectrum, frequency compression method is proposed to reduce the number of IFFT points. Theoretical and simulation results show that this technique completely removes the ambiguity threat in acquisition process.

1. Introduction

With rapid development of Global Navigation Satellite System (GNSS), several systems have to coexist in a limited number of frequency bands. In order to reduce the interference with Binary Phase Shift Keying (BPSK) modulated signal, John Betz developed Binary Offset Carrier modulation (BOC modulation) [1]. BOC modulation method enhances interoperability and achieves high accuracy code estimation in multi-path condition. Currently, this modulation method has been adopted by almost all GNSSs. However, the ambiguous problem challenge signal acquisition, which caused by notches and peaks within $\pm$ one chip in the correlation envelope.

Some unambiguous acquisition techniques are developed getting rid of these drawbacks. Sideband filtering in time domain can reduce the effect of side peaks [3,4]. However, filtering and the dual sideband processing increase the implementation complexity and cause a loss of power. Autocorrelation Side-Peak Cancellation Technique (ASPeCT) algorithm is proposed to realize unambiguous acquisition [5]. However, it is only dedicated to sin-BOC (n, n) signals. Similarly, Gamma-Ray Ablation Sensing System (GRASS) method is only suitable for sin-BOC [6]. In addition, both ASPeCT and GRASS method use non-coherent integral value to reconstruct correlation function, which will cause extra noise and reduce acquisition sensitivity.

In this paper, an unambiguous acquisition based on fast Fourier transform (FFT) is proposed. This method eliminates side peaks by employing a frequency domain filter and is suitable for all BOC modulated signals. The remainder of this paper is organized as follows. Firstly, BOC signal model and spectral property is discussed. Next the unambiguous acquisition is explained in detail. At last, simulation results is given and analyzed. It is shown that proposed method can eliminate side peak effectively and achieve signal acquisition.

2. BOC Signal Model

As shown in Fig.1, the subcarrier modulation step is introduced into BOC modulation, which is different from BPSK modulation. Signal energy will be separate into 2 parts, and shift away from the carrier frequency. Therefore, BOC modulation is also known as a split-spectrum modulation.

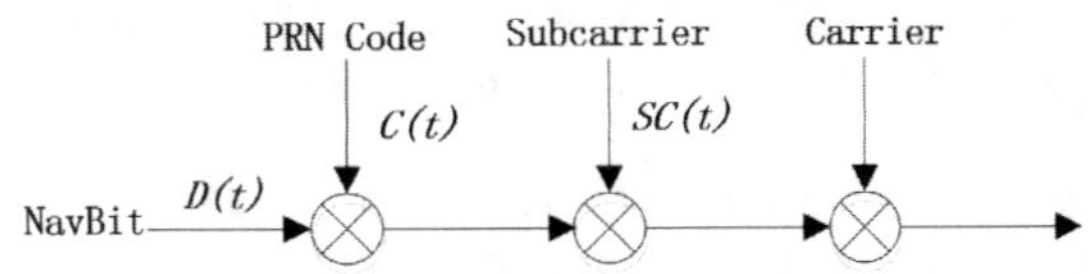

Figure 1. BOC Signal Modulation Block Diagram.

There are two type of subcarriers, sine subcarrier $SC_{sin}(t)$ and cosine subcarrier $SC_{cos}(t)$, which are expressed as:

$$SC_{\sin}(t) = sign(\sin(2\pi f_s t)) \tag{1}$$

$$SC_{\cos}(t) = sign(\cos(2\pi f_s t)) \tag{2}$$

where f_s is subcarrier frequency. BOC modulated signal is denoted as BOC(m, n), where m means the ratio of the subcarrier frequency f_{sc} to 1.023MHz, and n represents the ratio of the spreading chip rate f_c to 1.023MHz. Moreover, the ratio $N = 2m/n$ is referred to as BOC-modulation order.

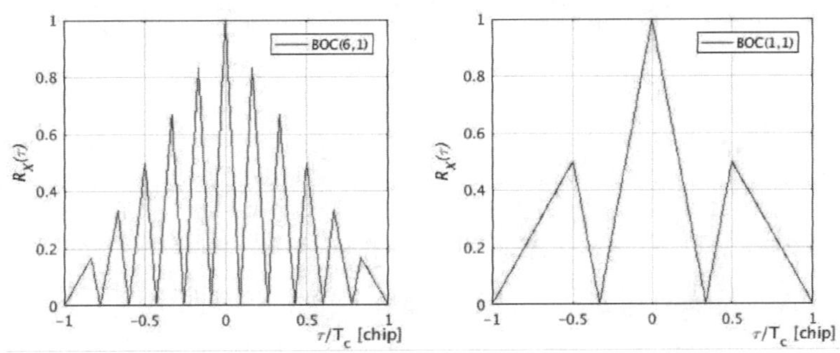

Figure 2. Correlation function of BOC modulated signals.

Fig.2 illustrates the correlation function of BOC modulated signals. For both BOC (1,1) and BOC (6,1) modulation, subcarriers of BOC signals create side peaks on both sides of the center peak of the sharp autocorrelation function (ACF), which leads to ambiguities or notches. These are very challenging in the acquisition process. Moreover, according to the simulation results and theoretical analysis, the number of correlation peaks is 2N-1, which depends on the BOC modulation order.

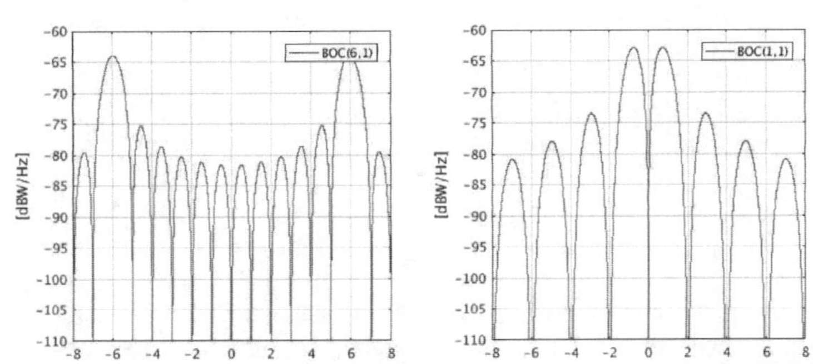

Figure 3. Spectrum of BOC modulated signals.

The spectrum of BOC modulation signals is shown in Fig.3. The main spectral energy of BOC signal concentrates two main lobes which locate at the side of the carrier frequency. The band width between main lobes depends on modulation order. Especially, two main lobes of BOC (1,1) signals locate next to each other. The signal characteristic raises challenges to filter design. It is impossible to implement a filter in time domain, which can remove one of main lobes without energy loss. It leads to acquisition performance reduction.

3. FFT Based Unambiguous Acquisition Method

The flow chart of FFT based unambiguous acquisition method is shown in Fig.4. Proposed acquisition method has 6 parts, carrier cancellation, down sampling, frequency-domain filtering, parallel correlation, frequency compression and non-coherent accumulation.

The received signal is sampling data of intermediate frequency, which is expressed as:

$$s_i = c(t-\tau)sc(t-\tau)\cos[2\pi(f_{IF}+f_D)+\theta] \tag{3}$$

where $c(t-\tau)$ is modulated Pseudo-Random Noise (PRN) code, $sc(t-\tau)$ denotes subcarrier, f_{IF} is the intermediate frequency, f_D is the doppler frequency, θ is carrier phase. The carrier modulated in received signal is cancelled by multiplying the in-phase and quad-phase local carrier.

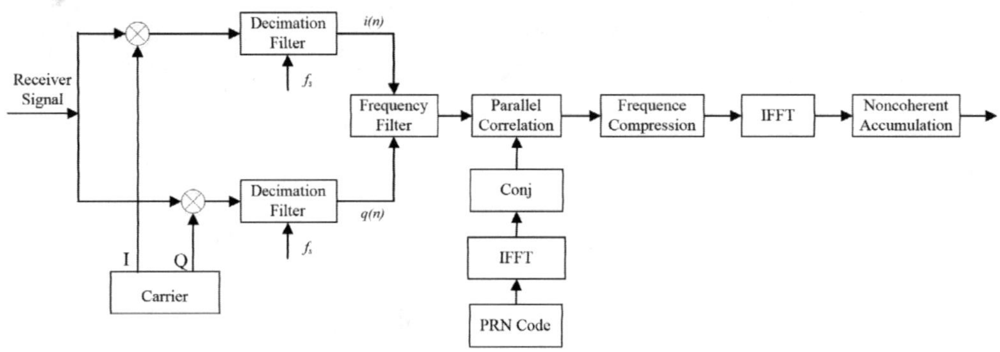

Figure 4. FFT based unambiguous acquisition structure.

In order to reduce the burden of calculating correlation function, a decimation filter is introduced to down sampling. After carrier cancellation, intermediate signals transfer to complex signals, and down sampling in complex field is adopted. According to the sampling theorem, sample frequency f_s should be more than $2(f_c + f_{sc})$, which is BOC signal band width. For example, the sample frequency of BOC (1,1) signals should be over 4.096MHz.

3.1. Frequency-domain filtering

The rectangular function is used as frequency-domain filter, which can separate up and down sideband signals. As shown below, the amplitude-frequency response of filter is expressed as:

$$G_u(f)=\begin{cases}1, & (f_{sc}-f_c)\le f\le(f_{sc}+f_c)\\ 0, & others\end{cases} \tag{4}$$

$$G_d(f)=\begin{cases}1, & -(f_{sc}+f_c)\le f\le-(f_{sc}-f_c)\\ 0, & others\end{cases} \tag{5}$$

where, f_{sc} is subcarrier frequency, and f_c is chip frequency.

As shown in Fig.5, the down sampling data of in-phase and quad-phase signals is handled as complex number at first. And then, FFT is used to transform signals from time to frequency domain. The spectrum of signal multiply up-sideband filtering function $G_u(f)$ and down-sideband filtering function $G_d(f)$, separately. The filtered signals are expressed as:

$$SF_u(f)=S(f)G_u(f)=\begin{cases}S(f), & (f_{sc}-f_c)\le f\le(f_{sc}+f_c)\\ 0, & others\end{cases} \tag{6}$$

$$SF_d(f) = S(f)G_d(f) = \begin{cases} S(f), -(f_{sc}+f_c) \le f \le -(f_{sc}-f_c) \\ 0, \quad\quad others \end{cases} \tag{7}$$

Obviously, there is only one main lobe in the band. And the filtered signals can be handled as BPSK signals.

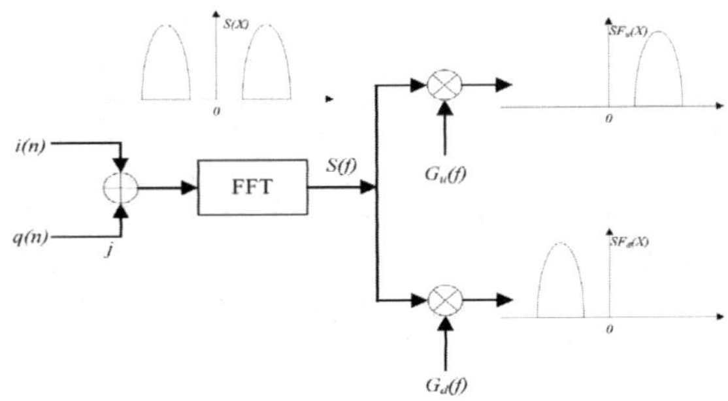

Figure 5. Block diagram of frequency-domain filter.

3.2. Parallel Code Correlation

In order to reuse filtered spectral data, proposed acquisition technique adopts a parallel code correlation. According to the properties of convolution, the correlation function is calculated by frequency multiplying. However, cyclic convolution method is not suitable for calculation correlation function, since GNSS PN code period is not a power of 2.

To avoid code phase mismatch of cyclic convolution, zero padding method is introduced in correlation step. Zero padding parallel code correlation method is shown in Fig.6. For example, 2ms received signal have 4092 points after down sampling. It extends to 4096 points by padding 4 zero elements. Similarly, 2046 local chips will pad 2050 zeros to form 4096 points sequence. Calculating 4096-point FFT for both 4096-point sequences and then computing the complex conjugate of the follow one. By taking multiplying and inverse fast Fourier transform (IFFT), the correlation values are obtained.

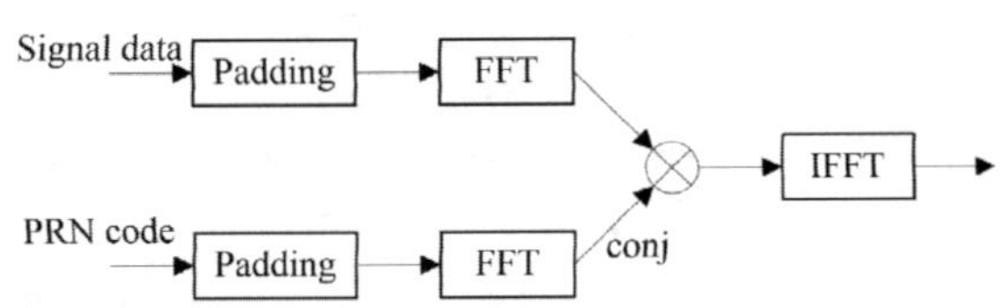

Figure 6. Block diagram of zero padding parallel code correlation.

3.3. Frequency Compression

It is noted that the multiplying results have lots of zeros elements due to the rectangular function. By utilizing the sparse feature of multiplying results, frequency compression method is developed to reduce the number of IFFT points. As shown in Fig.5, only up-band or down-band energy is retained in filtered spectrum. And there is no energy in other frequency band. It seems that the signal energy is compressed. Thus, only nonzero frequency is transformed to time domain, and the number of IFFT is reduced. It seems that the frequency band is compressed. Take BOC (1,1) signal as example. The down sampling frequency f_s is 8.184MHz, which is 8 times of chip frequency. After filtering, up or down band is only 2 times of chip frequency, 2.046MHz. Assuming 2ms signal data is used. The FFT points N_{FFT} is 8192 and the IFFT points N_{IFFT} is 2048. The ratio of compression is given as $R_{fc} = N_{FFT}/N_{IFFT}$.

The sample frequency in time domain is reduced after frequency compression since only pass band is transformed by taking IFFT. According to FFT/IFFT definition, fs decreases to B/2, where B is the pass bandwidth.

3.4. Non-coherent Accumulation

By transforming the results of frequency compression to time domain, we can correlation results of received signal and replica code which are complex numbers. The correlation values are denoted as $i(n)+jq(n)$, $i(n)$ and $q(n)$ are the in-phase and quad-phase correlation value. The expressions are:

$$i(n) = aR(\tau)\text{sinc}(f_e T_{coh})\cos(\phi_e) \tag{8}$$

$$q(n) = aR(\tau)\text{sinc}(f_e T_{coh})\sin(\phi_e) \tag{9}$$

where a is the signal amplitude, τ is the delay between received code phase and local replica code phase, f_e is the frequency error between received signal and local carrier and ϕ_e is the phase error. It is noticed that subcarriers are eliminated after filtering.

Because the phase-frequency response of proposed filter is nonlinear, the value of ϕ_e is different for up-band and down-band signal. Therefore, the up-band and down-band is expressed as:

$$s_{up} = i_{up} + jq_{up} = aR(\tau)\text{sinc}(f_e T_{coh})\mathrm{e}^{-\mathrm{j}\phi_{up}} \tag{10}$$

$$s_{dn} = i_{dn} + jq_{dn} = aR(\tau)\text{sinc}(f_e T_{coh})\mathrm{e}^{-\mathrm{j}\phi_{dn}} \tag{11}$$

Here s_{up} and s_{down} are complex expression. Generally, ϕ_{up} is not equal to ϕ_{dn}, so the signals can not be merged directly. Non-coherent integral is introduced to eliminate the effect of phase error. And the test criterion used in this technique is given by:

$$V = \sqrt{i_{up}^2 + q_{up}^2} + \sqrt{i_{dn}^2 + q_{dn}^2} \tag{12}$$

The criterion V has no relation with phase error, so the nonlinear of phase-frequency response will not affect signal detection.

4. Simulation Results and Discussion

Proposed method is simulated on MATLAB platform. The signal power is -130dBm, intermediate frequency is 46.42MHz, doppler frequency is 500Hz, and the sampling frequency is 62MHz. In addition, the PRN code design refers to BDS B1C signal, the code period is 10230 and chip frequency is 1.023MHz, and the initial code phase is 6270.

As shown in Fig.6, the peak of correlation function is obvious, and there are no side peaks. Therefore, proposed algorithm can eliminate the side peaks effectively for both high and low order BOC signals. Moreover, sin-BOC signal and cos-BOC modulation signal are also simulated. The simulation shows similar results.

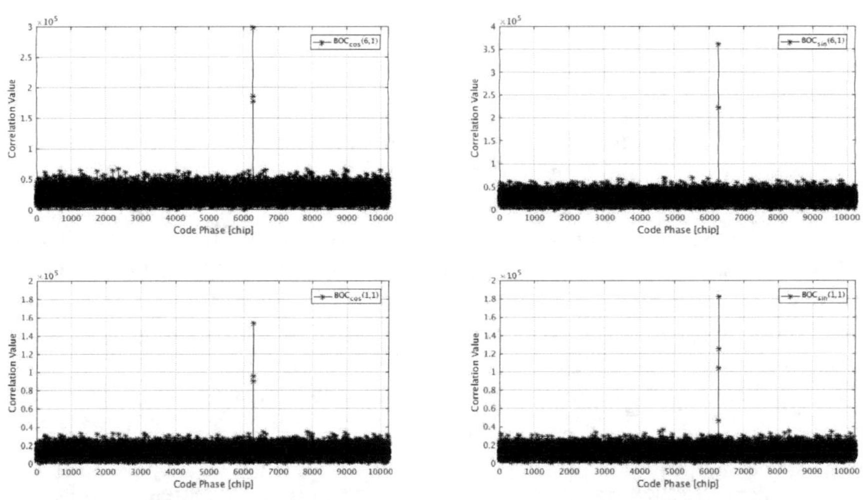

Figure 7. Block diagram of zero padding parallel code correlation.

The simulation results in details with and without frequency compression are shown in Fig.7. BOC (1,1) signal is used to simulate frequency compression. The sample frequency depends on the frequency bandwidth. There are 8 sampling points within one chip. After adopting frequency compression, sampling points number decreases to 2. Both with and without frequency compression, correct code phase is obtained by proposed filtering. Therefore, it is possible to adjust the ratio of frequency compression according to code phase searching accuracy.

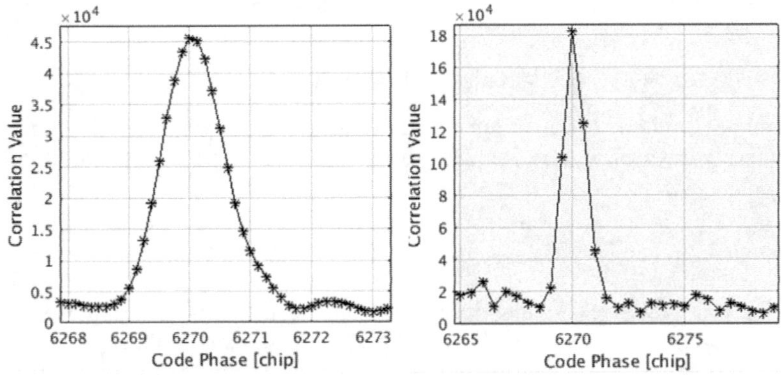

(a) without frequency compression (b) with frequency compression

Figure 8. Acquisition results at correlation peak.

Conclusion

In order to eliminate side peak of BOC signals correlation function, an unambiguous acquisition method is proposed, which introduce a frequency domain filter based on FFT. To reduce calculation burden, parallel code correlation based on data reuse and frequency compression are developed. Simulation results show proposed acquisition method can eliminate side peak effectively and exact code phase can be identified. The number of frequency compression points is determined according to search accuracy. The experiment results show proposed algorithm can be applied to all kinds of BOC modulated signal.

Acknowledgments

The authors express their appreciation to many members of ASIC Design Center whose innovative suggestions and hard work contributed to the implementation of proposed unambiguous algorithm. The ASIC Design Center also provides experimental facilities and simulation platform. This work was supported by BeiDou Navigation Receiver Special Program of Space Star Technology Co. Ltd.

References

[1] Betz J W 1999 The offset carrier modulation for GPS modernization *Proc. ION GPS* P 639-648

[2] Lohan E S, Burian A and Renfors M 2008 Low-complexity unambiguous acquisition methods for BOC-modulated CDMA signals *Int. J. Commun. Syst. Network* **26** p 503-522

[3] Fishman P, Betz J W 2000 Predicting performance of direct acquisition for the M code signal *Proc. ION NTM* p 574-582

[4] Martin N, Leblond V, Guillotel G and Heiries V 2003 BOC (*x*, *y*) signal acquisition techniques and performances *Proc. ION GNSS* p 188-198

[5] Julien O, Macabiau C, Cannon E and Lachapelle G 2007 ASPeCT: unambiguous sine-BOC(n, n) acquisition/tracking technique for navigation applications *IEEE Trans. Aero. Electron. Syst.* **43** p 150-162

[6] Zheng Yao, Mingquan Lu and Zhenming Feng 2010 Unambiguous Sine-Phased Binary Offset Carrier Modulated Signal Acquisition Technique *IEEE Trans on Wireless Communicatins* **9** (2) 577-580

Automated model generation for electronic devices using neural network approaches

Q J Zhang[1,3], W Na[2], M Li[1], Q Ding[1] and G Wu[1]

[1]China Communication Microelectronics Technology Co Ltd, Shenzhen, China
[2]Faculty of Information Technology, Beijing University of Technology, Beijing, China

[3]E-mail: qjzhang64@163.com

Abstract. This paper describes artificial neural network approaches to convert the conventionally expensive process of electronic modelling into an automated model generation process. Artificial neural networks are trained using machine learning algorithm to learn the electronic behaviour, and the trained neural network becomes a model to help predict the electronic device behaviour. The automated model generation algorithm performs adaptive data sampling to determine the amount and the distribution of training data needed to train neural networks. The algorithm also determines the number of hidden neurons needed to achieve a compact and accurate model. Also incorporated into the automated model generation method is an efficient interpolation approach to make the process much faster. The objective of the described method is to generate a compact neural-network based model with better accuracy and in less time than conventional approach. Examples of automated modeling of radio-frequency and microwave filters used in wireless electronic systems are described showing the advantage of this technique.

1. Introduction

Artificial neural networks are important vehicles in data processing and machine learning. In addition to popular applications such as image processing, speech recognition and control, artificial neural networks have also found interesting applications in engineering fields such as electronic modeling and design. Artificial neural networks are trained to learn the behavior of high-frequency electronic devices through machine learning algorithms. The trained neural networks can then represent the model of electronic devices [1,2]. Because neural networks can extract underline relationships from electronics data, the neural network is a much more efficient way to evaluate electronic design solutions than direct look-up table approach using measurement data. Various approaches to apply neural networks to electronic modeling and data processing have been presented in the past, such as radio-frequency and microwave modeling and design [1,2], high-power high-frequency transistor modeling [3], electromagnetic modeling/optimization [4,5] and waveguide filter inverse modeling [6]. The combination of electronic knowledge and neural networks led to knowledge-based neural networks for electronic modeling [7] and space mapping concepts [8] has been applied to map empirical models to accurate device behavior through neural networks. Another important direction is automation of neural network model development which was introduced to integrate all subtasks like data generation, neural-network selection, training and test into one unified framework [9]. In addition to making neural network training more efficient, the automated model generation also addresses two critical problems of neural network training, i.e., determining the suitable amount of training data and the suitable number of hidden neurons. The neural network training is performed stage by stage, where in each stage the amount of training data or the neural network size is adjusted, and neural network is re-trained. In this way, the suitable amount of training data and the suitable number of hidden neurons are automatically determined as part of the training process. However, process in [9] involves repetitive training of neural networks while detecting if the training data is sufficient in different regions of the nonlinear electronic parameter space. This reduces the model development efficiency.

This paper describes an advance on automated model generation [10]. We develop systematic algorithm to incorporate data sampling, data generation, neural network training and testing in a more efficient manner. We also incorporate interpolation techniques to avoid training the intermediate neural networks during the automated model generation process. Training is performed only after data sampling process has finished, making the automated modeling generation process faster than the previously published approaches. In addition, we add a procedure to reduce the number of hidden neurons so that the guess of initial neural-network structure can be more flexible and the final size of the neural-network model is more compact. Demonstration of automated model generation for a radio-frequency and microwave filters are also described.

2. Methodology

2.1. Training data and data generation for high-frequency electronics modeling

Let $\boldsymbol{x}$ represent a n-vector containing physical/geometrical parameters of an electronic device, e.g., gate length and gate width of a high-frequency transistor, or width and spacing of a transmission line. Let $\boldsymbol{y}$ represent a m-vector containing the responses of the device/circuit under consideration, e.g., drain current of a transistor, or scattering parameters of a transmission line. Let the physics/ electromagnetics relationship between $\boldsymbol{x}$ and $\boldsymbol{y}$ be represented as $\boldsymbol{y}=\boldsymbol{y}(\boldsymbol{x})$. This relation can be highly nonlinear and multi-dimensional. The theoretical model for this relationship may not be available (e.g., a new semiconductor device), or theory may be too complicated to implement, or the theoretical model may be computationally too intensive for online microwave design and repetitive optimization (e.g., 3D full-wave electromagnetic analysis inside a Monte Carlo statistical design loop).

To overcome these problems, we aim to develop a fast and accurate artificial neural network model for the relation $\boldsymbol{f}$ by teaching/training the artificial neural network to learn the electronic behaviour from a set of data. Let the training data be denoted as input-output sample pairs, $\{(\boldsymbol{x}^{(i)}, \boldsymbol{d}^{(i)}), i=1,2, \ldots, N_d\}$, where $\boldsymbol{d}^{(i)}$ represents the data for output $\boldsymbol{y}$ when the inputs are $\boldsymbol{x}^{(i)}$, and N_d represents the number of training data. The measured/simulated data of the outputs and the inputs are related by,

$$\boldsymbol{d}^{(i)}=\boldsymbol{f}(\boldsymbol{x}^{(i)}) \quad (1)$$

The data generation can be done either by measurement of the device, or by simulation using electromagnetic simulators. Typically, data for nonlinear active devices such as Gallium Nitride transistors are obtained by measurement. The scheme to gather measurement data is that we first apply different voltages to the transistor gate and drain, and then measure the drain current to obtain current voltage data. Secondly we generate small-signal data by measuring the scattering parameter data (called the S-parameter data) over a frequency range, and re-perform such measurement again and again for different operating bias voltages. Thirdly, we perform large-signal measurement by exciting the device with large signal at the gate, and measure the distorted large-signal at the drain. In this way, we obtain the measurement data in the form of harmonic data including signal at fundamental frequency, second-harmonic frequency, and all the high-order harmonic frequencies. This harmonic data measurement is repeated for different operating bias points, different values of fundamental frequencies and different levels of input signal power. Consequently, a large amount of data is collected, which contain valuable information of the device internal dynamic relationships, for which no explicit formula is available.

Data generation for passive electronic components are typically achieved through simulation. This data can be supplemented by measurement data. The reason is that the typical design variables for passive components are geometrical variables such as height of an iris window in a waveguide resonant cavity, or length and width of a microstrip in a filter. It is expensive to manufacture a large number of passive components corresponding to different geometries just for purpose of data collection. Therefore, electromagnetic simulation become typical choice for collecting data for passive high-frequency electronic components. The amount of data becomes large if more geometrical variables (e.g., length 1, length 2, …, width 1, width 2, … for multiple coupled microstrip lines) are needed. In this case, we need to obtain data for various combinations of values from many geometrical parameters.

2.2. Training of neural networks with automated model generation algorithm

To efficiently use data to train the neural networks, our automated model generation process includes initial data generation, neural network training, neural network size adjustment, additional data generation, and neural network testing. Incremental data generation is performed whenever and wherever data are needed as decided by the model generation algorithm during neural network training process.

The overall automated model generation process proceeds in stage-wise process [10]. Each stage includes data generation, neural-network structure adaptation, training and test. Different stages may have different data and different neural-network size. Our process uses three sets of data, namely training, interpolation and test data. Training data are used to adjust neural-network weights during training. Interpolation data are used to create local models in subregions of the model input space R^0 to assess the adequacy of training data during training process. Additional training data are added upon detecting data inadequacy by the interpolation criteria. Test data are used to check the quality of the final neural model.

Let $(\boldsymbol{x}^{(i)}, \boldsymbol{y}^{(i)})$ represent the ith data. Let L^k, I^k, and T^k denote training, interpolation and test data respectively in the kth stage. Let H^k denote the number of hidden neurons of neural network S^k in the kth stage. The algorithm regards the original input space R^0 as one region at first. If the test error in the region is large, the region would be divided subregions and new data are generated. In each stage, the algorithm compares the test error in each subregion. The subregion with the maximum test error is chosen as the worst region R^*. Then R^* would be further divided into 2^n new subregions in the next stage. In order to avoid training intermediate neural networks in each stage, we evaluate the test error of various subregions using interpolation approaches during each stage of the automated model generation process. Different subregions have different interpolations. When calculating the interpolation in one region, the region is called the interpolation region. Even-though the electronic behavior over the entire region of design space is complex and nonlinear therefore requiring a neural networks to learn, the interpolation space is usually in a small local subregion. Therefore, a simpler interpolation function $g(\boldsymbol{x})$ is sufficient for interpolation space. In this way we speedup the overall model generation process. In our work, we use polynomial interpolation. Let $g(\boldsymbol{x})$ be the polynomial interpolation function with N terms. Suppose $\boldsymbol{p}$ is a vector of coefficients in the interpolation function, and $\boldsymbol{h}(\boldsymbol{x})$ is a vector of polynomial terms in the multi-dimensional space of $\boldsymbol{x}$. For example, for a second order polynomial in two variables, $n = 2$, $N = 6$, $\boldsymbol{p} = [p_1, p_2, \_, p_6]$ and $\boldsymbol{h} = [1, x_1, x_2, x_1x_2, x_1, x_2]$.

In general for neural networks with multiple outputs, say m outputs, $\boldsymbol{p}$ would be an N x m matrix. To compute the test error for an interpolation region, we select M training data in and around the interpolation region to form matrices **A** and **B**, such that

$$\mathbf{A}\boldsymbol{p} = \mathbf{B}. \tag{2}$$

where **A** has N columns and M rows, and **B** has m columns and M rows, where N, M and m are the number of elements in $\boldsymbol{h}$, the number of training data selected, and the number of model outputs, respectively. The elements in **B** are output values $\boldsymbol{y}^{(i)}$ from M training data. To determine **A**, the locations of these M training points $\boldsymbol{x}^{(i)}$ are related to $\boldsymbol{x}^*$, which is at the center of the worst region R^*. $\boldsymbol{x}^{(i)}$ is obtained as

$$\boldsymbol{x}^{(i)} = \boldsymbol{x}^* + \boldsymbol{Q}_i\,(\boldsymbol{x}_{max} - \boldsymbol{x}_{min})/2^r \tag{3}$$

where r is the number of times the original region was divided to reach the present worst subregion R^*, $\boldsymbol{x}_{min}$ and $\boldsymbol{x}_{max}$ are minimum and maximum values of the variables in the original $\boldsymbol{x}$ space, $\boldsymbol{x}^*$ is the test point in R^*, and $\boldsymbol{Q}_i$ is a $n \cdot n$ diagonal matrix with the values 0, +1, or -1 on its diagonal. Different combinations of 0, +1, and -1 in the diagonal of $\boldsymbol{Q}_i$ allow us to obtain different training points. After normalizing the $\boldsymbol{x}$ and $\boldsymbol{y}$ values, the $\mathbf{A}$ matrix can be derived as a constant matrix, containing 0, +1, and -1. Additional scaling factors can be applied to particular rows in $\boldsymbol{A}$ in order to emphasis the importance of fitting particular data points. The least squares solution of normalized version of (2) is

$$\boldsymbol{p} = (\mathbf{A}^T\mathbf{A})^{-1}\mathbf{A}^T\mathbf{B} \quad (4)$$

Subsequently the interpolation function is obtained for the subregion R^*. Then, the test error E_{In} of the interpolation regions is calculated by comparing the $g(\boldsymbol{x})$ and test data. The subregion with the largest test error will be designated as the new worst region R^*. Automated model generation process will then enter the next stage where the new worst region is further divided, forming a new set of subregions, and the process described in (2)-(4) is repeated. In this way, we take advantage of availability of training data to produce localized interpolation function, avoiding the training of neural network during intermediate stages of automated model generation process, and speeding up model development. Neural network training is performed only during the last few stages of automated model generation process. The final model is a neural network valid in the entire region. We also add a procedure reducing the number of hidden neurons after good-learning is detected, as shown in figure 1. In the figure, E^d represents user-defined error threshold for training the model. S^k denotes the neural network in the kth stage of automated model generation process, and H^k represents the number of hidden neurons of S^k. Symbols E_{In} and E_l denote interpolation error w.r.t. test data, and training error w.r.t. training data, respectively.

The algorithm will recursively reduce the number of neurons until the error of the reduced neural network starts to increase. This procedure will allow the guess of the initial number of hidden neurons to be more flexible, and make the final neural model more compact. The overall algorithm can start from an initial set of data and initial neural network, and ultimately find the most compact neural network to learn the given electronic problem using data collected at the most critical regions of the input parameter space.

3. Experimental / Theoretical Results and Discussions

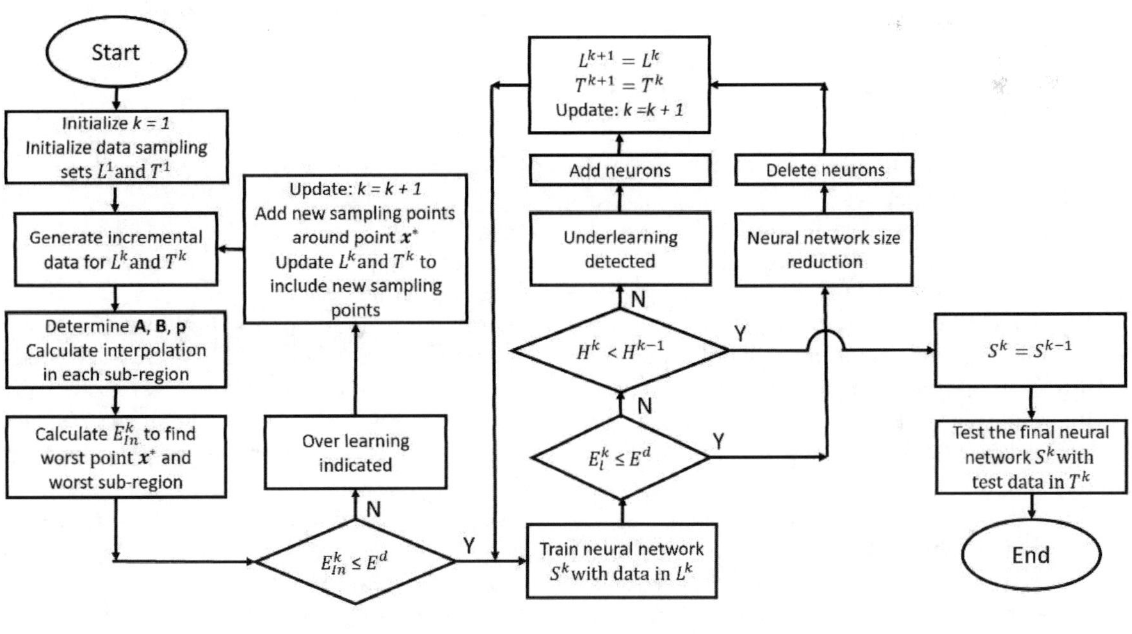

Figure 1. Flow chart of the proposed automated model generation algorithm, including data generation, neural network training, neural network size adjustment and neural network testing.

In this section, we describe results of applying the automated model generation method to an example of neural network modeling of a bandpass filter in wireless communication systems [10]. More specifically, the application of the method is demonstrated by the development of a parametric model of a High-Temperature Superconductor (HTS) quarter-wave parallel coupled-line microstrip filter [10,11] used in radio-frequency and microwave wireless electronic systems [8]. The HTS filter structure is illustrated in figure 2. In order to generate training data for neural networks to learn, the filter geometrical formulation is setup in a 3-dimension electromagnetic simulator. The electromagnetic simulator is executed upon demand to generate training data at the worst sub-regions of the modeling space in each stage as determined by the automated model generation algorithm.

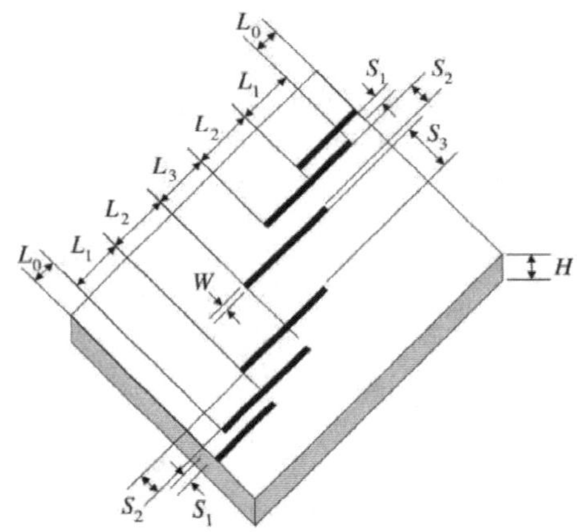

Figure 2. The structure of the HTS coupled line microstrip bandpass filter example.

The response behavior of the radio-frequency/microwave filter is best represented by the transfer function and the return loss of the filter. Since this filter operates in 4 GHz frequency range, conventional resistor/capacitor/inductor representation is not applicable. Electromagnetic oriented models are required to capture the subtle relationship between geometrical parameters and the filter responses. However, standard electromagnetic simulation is time-consuming costing the designer's expensive online design time. We use neural networks to learn from electromagnetic data of the filter. Using the automated model generation process, neural network models can be developed offline, including automated data generation offline and automated neural network training offline. The trained neural network is a fast and accurate model for designer's online use, providing fast estimation of electromagnetic behavior of the filter during filter design. In this way, filter design can be substantially faster.

The input parameters for the neural network model are $\boldsymbol{x} = [L_1, L_2, L_3, S_1, S_2, S_3]^T$, where L_1, L_2 and L_3 are the lengths of the parallel coupled-line sections and S_1 , S_2 and S_3 are the gaps between the sections. The outputs for the neural network are $\boldsymbol{y}=[S_{11}, S_{21}]^T$ where S_{11} represent the return loss of the filter, and S_{21} represent the transfer function of the filter in terms of the input-output wave signals.

Training data generation is done by an electromagnetic simulator, called CST Microwave Studio. In this example, there are six geometrical variables, i.e., $n = 6$. Matrix **A** has 70 columns meaning that the interpolation function has 70 coefficients or 70 polynomial terms (i.e., N=70). The number of rows of matrix A is 729, meaning that the number of training samples around the neighborhood of an interpolation space is 729 (i.e., M=729). Matrix **B** has 2 column and 729 rows. The worst region R^* is divided into 64 subregions, so there are 64 **A** matrices used during interpolation process. The automated modeling methods used 1394 electromagnetic training geometries. After 6 stages of the process in Figure 1, the final neural network training is finished. The test error of the final neural network is 1.59% representing a good accuracy for engineering use. Figure 3 shows the neural network response compared with original electromagnetic response, confirming that the neural network has good accuracy for predicting the electromagnetic behavior of the filter. Notice that the neural network responses in figure 3 are calculated with two sets of testing geometry which are never used in training.

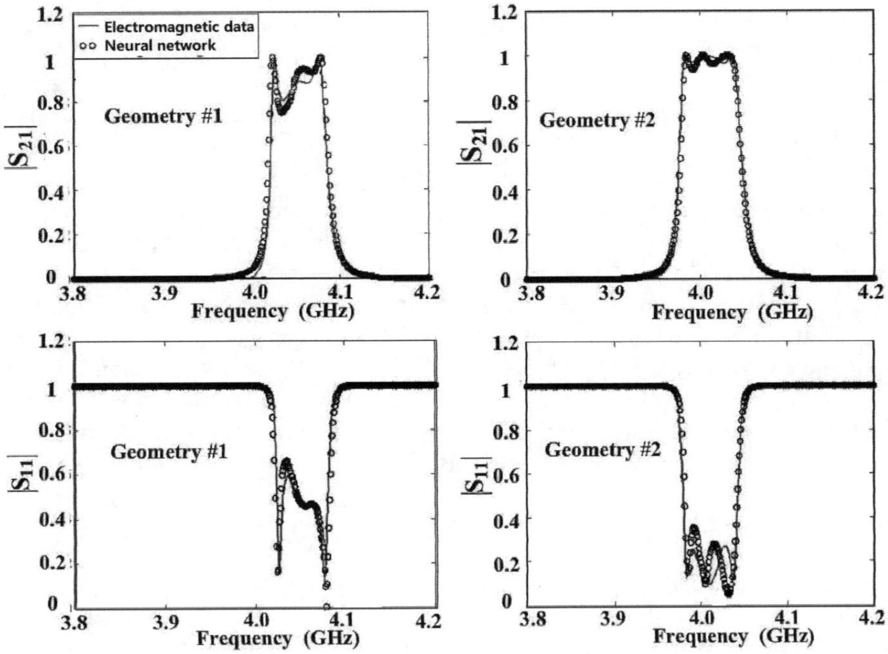

Figure 3. The comparison of neural network result versus original electromagnetic test data of the HTS bandpass filter at two different sets of geometrical values. For geometry #1, $\boldsymbol{x}$=[4.45, 4.81, 4.60, 0.52, 2.32, 2.0]^T (mm). For geometry #2, $\boldsymbol{x}$=[4.58, 4.96, 4.46, 0.58, 2.10, 2.11]^T (mm).

For comparison purpose, we also attempted conventional manual data generation and training of neural networks. To achieve similar test accuracy (1.59%), we endup with much more training data (4096 electromagnetic training geometries), and more training time (85 hours including data generation and neural network training). Using the automated model generation algorithm, the training data is judicially selected such that highly nonlinear sub-regions of the modeling space have more training samples, and smooth subregions have less training samples. In this way, only 1394 electromagnetic training geometries are sufficient to achieve same accuracy as manual training with 4096 training geometries. Subsequently the automated model generation process is much faster than manual data generation and training.

Further comparison is performed by comparing the method in this paper with the automated training algorithm of [9]. The method described in this paper achieved same accuracy of trained neural network as that of [9]. However, the model trained by the method in this paper used less stages (6 stages by the method in this paper versus 9 stages by the existing automated model generation of [9]). The total amount of time for developing the model including electromagnetic data generation and neural network training by the method in this paper is less than the existing automated model generation of [9] (22 hours by automated algorithm in this paper versus 26 hours by previous automated algorithm), and produced a more compact neural network. Other aspects for the comparison are equal between the automated model generation algorithm of this paper and the automated model generation algorithm of [9], including same optimization method for training, same electromagnetic simulator, the same geometrical configurations for the bandpass filters. In training the neural networks, the quasi-Newton algorithm, which is a second-order gradient based optimization algorithm, was used for training. The models trained by the method in this paper and that in [9] are both validated by comparing with original electromagnetic data simulated at 10001 frequency points across the entire frequency range of interests. The reason for a net speedup of model development by the method described in this paper is due to the use of interpolation techniques in two stages of the automated model generation, bypassing intermediate training of neural networks and thus shortening the overall model development time. In this way, the model generation method described in this paper is faster than the previously published automation methods [9].

4. Conclusion

An artificial neural network approach has been described to learn and predict the electromagnetic behaviour of electronic components. The neural network training has been described through an automated model generation process which includes data sampling to determine the amount and the distribution of training data needed to train the neural network, and to determine the number of hidden neurons to obtain a compact and accurate model. We have also described an interpolation approach to improve the speed of the model development process. An example has been described for neural network modeling of radio-frequency and microwave filters to illustrate that the training can be carried out efficiently and the trained neural network model achieves good accuracy in predicting the electromagnetic behaviour of the filter. The potential impact of this work is to employ artificial neural networks to allow accurate electronic models to be developed in shorter time and by automation, subsequently speeding up the design cycle of electronic circuits. This work is useful for electronic designers in their electronic data processing, fast behaviour modelling and design of electronic components and systems.

References

[1] Rayas-Sanchez J E 2004 EM-based optimization of microwave circuits using artificial neural networks: the state-of-the-art *IEEE Trans. Microwave Theory Tech.* **52** p 420-35

[2] Kabir H, Zhang L, Yu M, Aaen P H, Wood J and Zhang Q J 2010 Smart modeling of microwave devices *IEEE Microwave Mag.* **11** p 105-18

[3] Prasad A, Thorsell M, Zirath H and Fager C 2018 Accurate modeling of GaN HEMT RF behavior using an effective trapping potential *IEEE Trans. Microwave Theory Tech.* **66** p 845-57

[4] Na W C, Liu W Y, Zhu L, Feng F, Ma J G and Zhang Q 2018 Advanced extrapolation technique for neural-based microwave modeling and design *IEEE Trans. Microwave Theory Tech.* **66** p 4397-418

[5] Feng F, Gongal-Reddy V, Zhang C, Ma J G and Zhang Q J 2017 Parametric modelling of microwave components using adjoint neural networks and pole-residue transfer functions with EM sensitivity analysis *IEEE Trans. Microwave Theory Tech.* **65** p 1955-75

[6] Zhang C, Jin J, Na W C, Zhang Q J and Yu M 2018 Multivalued neural network inverse modeling and applications to microwave filters *IEEE Trans. Microwave Theory Tech.* **66** p 3781-97

[7] Na W, Zhang Q J 2015 Unified automated knowledge-based neural network modeling for microwave devices *IEEE MTT-S Int. Conf. Numerical Electromagnetic and Multiphysics Modeling and Optimization* (Ottawa, Canada) August 2015

[8] Bandler J W, Cheng Q S, Dakroury S A, Mohamed A S, Bakr M H, Madsen Kaj and Sondergaard Jacob 2004 Space mapping: the state of the art *IEEE Trans. Microwave Theory Tech.* **5** p 337-61

[9] Devabhaktuni V K, Yagoub M C E and Zhang Q J 2001 A robust algorithm for automatic development of neural-network models for microwave applications *IEEE Trans. Microwave Theory Tech.* **49** p 2282-91

[10] Na W C, Zhang Q J 2013 Automated parametric modeling of microwave components using combined neural network and interpolation techniques *IEEE MTT-S Int. Microwave Symp.* (Seattle, Washington) June 2013

[11] Na W C, Zhang Q J 2014 Automated knowledge-based neural network modeling for microwave applications *IEEE Microwave and Wireless Components Letters* **24** p 499-501

Evolution of urban construction land structure in Henan Province based on information entropy and equilibrium model

Chunyang Lu[1,2], Feng Wen[1], Kaidong Yang[1] and Linyun Duan[1]

[1]Henan University of Urban Construction, Pingdingshan, Henan 467036, China

[2]E-mail: luchunyang@hncj.edu.cn

Abstract. The paper aims to accurately identify the evolution characteristics of urban land use structure in Henan Province, clarify the current situation and problems of urban land use structure, and provide reference and basis for the optimization of urban land use structure and urban land allocation. Statistical analysis, comparative analysis, information entropy, equilibrium and dominance were used. From 2007 to 2016, the urban construction land area in Henan Province showed an overall upward trend; from 2007 to 2016, the per capita construction land area in Henan Province showed an overall upward - downward - upward trend. From 2007 to 2016, the information entropy, equilibrium and dominance curves of urban land use structure in Henan Province are in a stable state, which shows that the change of urban land use structure tends to be gentle, the overall fluctuation range is not large, and all types of land are relatively uniform. To a certain extent, it can show that urban land development in Henan Province is relatively mature. In the structure of urban construction land in Henan Province, the proportion of residential land is the highest, and its area and proportion are increasing year by year. This is closely related to the increase of urban population. The proportion of industrial land is higher, but it decreases year by year. The land for urban green space, square and road traffic facilities is increasing year by year, and others. Although the land use area of different types is increasing year by year, the proportion of structure is decreasing year by year, and the urban land use structure of Henan Province is in the stage of continuous adjustment.

1. Introduction

The change of urban land use structure will lead to the change of urban function, which has an important impact on the economic development of a region. The problem between urban functional structure and land use structure has always been one of the important contents of urban planning and socio-economic development, and it is also an important guarantee to deal with the contradiction between urban development and land development and utilization. Scholars' research on urban land use structure mainly focuses on the evolution of urban land use structure [1-3], driving force of urban land use structure change [4-6], optimal regulation of urban land use structure [7-8], and simulation [9-10]. From 2007 to 2016, the area of urban construction sites in Henan Province increased by 649.3 square kilometers, with a growth rate of 36.57%, while the urban population increased by 368,030,000, with a population growth rate of 22.37%, and the growth of urban construction land was relatively fast. The purpose of this study is to accurately identify the evolution characteristics of urban land use structure in Henan Province and clarify the current situation and problems of urban land use structure.

2. Regional overview

Henan Province is located in the coastal open zone and an important junction between the central and western regions of China, and is also the middle zone for China to promote the economic and social development from east to west. It is located in the middle and lower reaches of the Yellow River in the middle and east of China, with Anhui and Shandong in the east, Hubei in the south, Shaanxi in the west, Hebei and Shanxi in the north. The total area of the province is about 167,000 square kilometers, which is between 110 -21'-116 -39', and 31 -23'-36 -22'. By the end of 2016, there were 17 provincial municipalities, one provincial directly administered city Jiyuan, 52 municipal districts, 21 county-level cities, 85 counties, 1120 towns, 682 townships, 633 sub-district offices, 4743 residents'committees and 46831 villagers' committees.

3. Data sources and research methods

3.1. data sources

The data are from Henan Statistical Yearbook from 2007 to 2016 and Henan National Economic and Social Development Statistical Bulletin from 2007 to 2016. In this paper, the urban land use data of Henan Province originated from the Statistical Yearbook of Urban Construction in China, and the social and economic data of Henan Province originated from the Statistical Yearbook of Henan Province.

3.2. research method

3.2.1. Information Entropy. The theory of information entropy is an information theory put forward by American mathematician Shannon, which was later applied to the field of geography. In terms of land information, it can reflect the comprehensive dynamic change and transformation degree of land use types in a region or region at a certain stage, and can also be used to describe the multiple land use types. Sample is used to describe the complexity of land use types in a region or region. This research method is of great practical significance to the adjustment of land use structure in a certain region. We assume that there are n types of land use in a city, the total area of construction land is S, and the area of each type of structure is Si. According to the definition, the information entropy of land use structure is H, and its basic formula is as follows:

$$\mathrm{H} = \sum_{i=1}^{n}(P_i)\ln(P_i)(\mathrm{i} = 1,2,...) \quad (1)$$

In the above formulas, Pi represents the proportion of the type I land use, and the information entropy can reflect the difference between different types of urban land use structure. The higher the entropy value is, the more types of land used for different functions are. If the proportion of different types of land is small, the more balanced the land distribution is. On the contrary, it is not balanced. When the functions of a city are relatively single, the entropy value is 0. When the functions of different types of cities are constantly enriched and improved, the land distribution is more balanced. The richness of the utilization type increases, if the maximum entropy is satisfied, that is, S1 = S2 = ... = Sn, then the maximum value of entropy H is obtained at this time, and the type of urban land use is the largest. If the various types of land in a city tend to be stable and uniform, then the development of the city tends to mature, and the diversity index reaches the maximum at this time.

3.2.2. Equilibrium and Dominance. The concept of equilibrium degree is introduced to reflect the equilibrium distribution degree of various types of urban construction land, which makes the structure of urban land use more clear. The equilibrium degree is the ratio between information entropy and maximum entropy. Its basic formula is evolved from the formula of information entropy calculation. The formula is as follows:

$$F=\sum_{i=1}^{n}(P_i)\ln(P_i)/\ln(n) \tag{2}$$

F ranges from 0 to 1. When F = 0, it means that urban land use is in the most uneven state. When F = 1, the homogeneity of land use types is the strongest. The significance of dominance degree is opposite to that of equilibrium degree, and its calculation formula is as follows:

$$K = 1- F \tag{3}$$

K means dominance, has opposite meanings of equilibrium.

4. Result analysis

4.1. Urban Land Use in Henan Province

4.1.1. Present Situation of Urban Land Use Structure in Henan Province. According to the Statistical Yearbook of Urban Construction of China in 2016, the classification standard of urban construction land has changed partially since 2012, from 9 types of urban land use before to 8 types now, as shown in the following figure 1. Until the end of 2016, the area of urban construction land in Henan Province was 2424.64 square kilometers, and the per capita area of urban construction land was 120.43 square meters. According to the planning of new cities, the per capita index of urban construction land should be 85.1-105.0 per person, far exceeding the upper limit of planning standards. Among them, residential land is 726.83 square kilometers, accounting for the highest proportion, accounting for 30%, occupying the leading position; public management and public service land is 271.26 square kilometers, accounting for 11%, commercial services facilities land 130.34 square kilometers, accounting for 5%, industrial land 378.44 square kilometers, accounting for 16%, logistics storage land 78.02 square kilometers, accounting for 3%. The land for road traffic facilities is 388.32 square kilometers, accounting for 16%, for public facilities 99.1 square kilometers, accounting for 4%, for green space and square land 352.33 square kilometers, accounting for 15%, as shown in Figure 3-2.

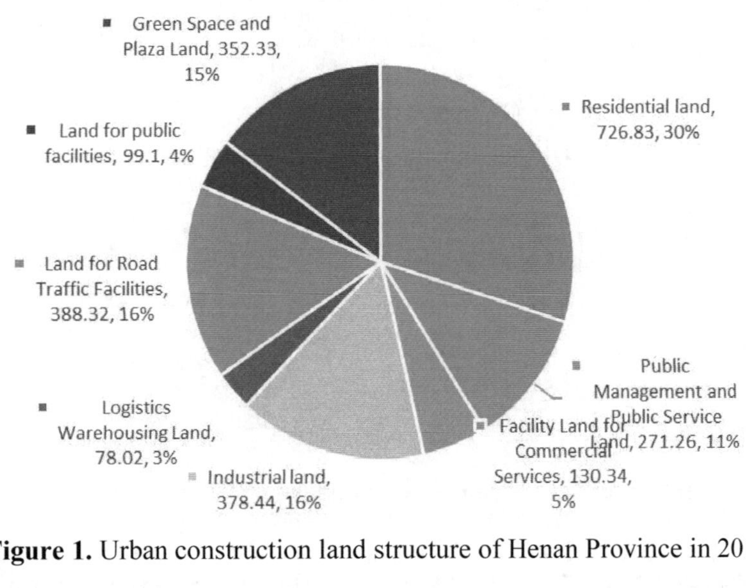

Figure 1. Urban construction land structure of Henan Province in 2016.

4.1.2. Change of Urban Land Use Quantity. From 2007 to 2016, the statistical tables of relevant data of urban construction land structure in Henan Province in the past five years and the statistical tables of urban construction land structure in 18 cities in Henan Province in 2016 are separately counted, as the classification standards of urban land before and after 2012 have changed, as shown in tables 1, 2 ,3 and 4.

Table 1. Urban land use structure of Henan Province, 2007-2011 (unit: square kilometers).

Particular year	Residential land	Land for public facilities	Industrial land	Warehousing land	Land for external transportation	Land for Road Square	Land for Municipal Public Facilities	Greenland	Special land use
2007	465.82	226.72	331.53	57.82	84.73	195.14	61.30	191.83	35.38
2008	499.22	243.60	364.80	59.46	84.87	205.91	65.05	205.01	32.40
2009	525.81	261.74	357.04	61.09	89.02	219.67	67.51	213.69	32.83
2010	578.68	287.38	338.30	61.54	81.05	249.34	68.42	250.70	31.77
2011	594.03	294.60	319.73	72.82	84.00	263.91	82.09	276.76	31.32

Table 2. Urban land use structure of Henan Province, 2012-2016 (unit: square kilometers).

Particular year	Residential land	Public management-service land	Commercial service facility land	Industrial land	Logistics warehousing land	Road traffic facilities land	Utility land	Green land and square land
2012	612.01	242.86	130.96	305.91	72.66	323.04	98.11	297.87
2013	633.82	250.56	110.22	343.22	71.80	334.18	90.25	309.56
2014	660.49	258.28	117.09	347.80	75.46	352.69	93.66	327.44
2015	699.13	272.03	128.02	370.70	79.78	372.29	97.05	344.14
2016	726.83	271.26	130.34	378.44	78.02	388.32	99.10	352.33

Table 3. Urban construction land area of Henan Province, 2007-2016.

Particular year	Urban population (10,000 people)	Urban construction land area (square kilometers)	Per capita urban construction land area (m2 / person)
2007	1645.36	1775.34	107.90
2008	1689.42	1857.24	109.93
2009	1724.48	1913.26	110.95
2010	1798.62	2014.40	112.00
2011	1834.06	2098.07	114.39
2012	1863.60	2083.42	111.80
2013	1906.39	2143.61	112.44
2014	1970.66	2232.91	113.31
2015	2038.81	2363.14	115.91
2016	2013.39	2424.64	120.43

As can be seen from Figure 2, the urban construction land area of Henan Province shows an overall upward trend from 2007 to 2016, which can be roughly divided into two stages. The urban construction land area in 2007-2011 keeps increasing at a constant speed; the urban construction land area in 2011-2012 suddenly decreases by 14.65 square kilometers; and the urban construction land area in 2012-2016 shows a uniform growth trend.

By comparison, Figure 3 shows that the average per capita construction land area in Henan Province shows a trend of increasing, decreasing and rising from 2007 to 2016. The per capita construction land area increased from 107.90 square meters in 2007 to 114.39 square meters in 2011, and dropped to 111.80 square meters in 2012, and has been increasing since then.

Journal of Physics: Conference Series **1419** (2019) 012015 doi:10.1088/1742-6596/1419/1/012015

Table 4. Urban land use structure of 18 cities in Henan Province in 2016 (unit: square kilometers).

Area	Urban construction land area	Residential land	Public management -service land	Commercial service facility land	Industrial land	Logistics ware housing land	Road traffic facilities land	Utility land	Green land and square land
Zhengzhou	452.49	122.38	60.27	19.50	49.30	16.64	76.19	23.86	84.35
Kaifeng	97.95	31.92	12.57	7.04	21.88	2.30	12.68	2.57	6.99
Luoyang	200.90	67.53	24.29	39.05	5.67	10.65	28.67	4.48	20.56
Pingdingshan	114.05	38.50	7.38	5.10	23.01	4.38	16.16	3.26	16.26
Anyang	98.75	33.19	11.76	2.50	18.09	3.52	16.14	2.90	10.65
Hebi	60.82	13.44	5.98	3.97	15.60	1.03	9.61	1.65	9.54
Xinxiang	143.29	46.28	17.74	7.41	31.77	4.11	16.26	4.01	15.71
Jiaozuo	135.49	46.55	22.20	2.86	22.00	2.60	19.99	4.77	14.52
Puyang	45.80	14.40	3.00	5.00	5.50	1.50	6.20	2.00	8.20
Xuchang	139.04	35.12	13.47	6.80	23.73	6.73	22.61	7.77	22.81
Luohe	59.15	16.01	4.50	2.00	5.59	2.11	7.90	7.50	13.54
Sanmenxia	67.14	17.95	9.04	3.17	10.01	1.35	10.21	2.49	12.92
Nanyang	137.55	39.95	17.95	7.00	27.65	4.61	23.77	5.52	11.10
Shangqiu	92.09	20.39	10.48	5.41	5.53	0.99	13.82	4.17	31.30
Xinyang	71.50	20.51	1.91	1.65	15.70	4.05	13.00	10.67	4.01
Zhoukou	74.53	22.26	8.80	5.50	6.08	4.18	13.31	4.80	9.60
Zhumadian	56.00	12.60	8.60	5.00	12.00	0.90	10.22	4.48	2.20
Jiyuan	36.88	13.03	2.92	2.00	6.80	1.01	6.30	1.21	3.61

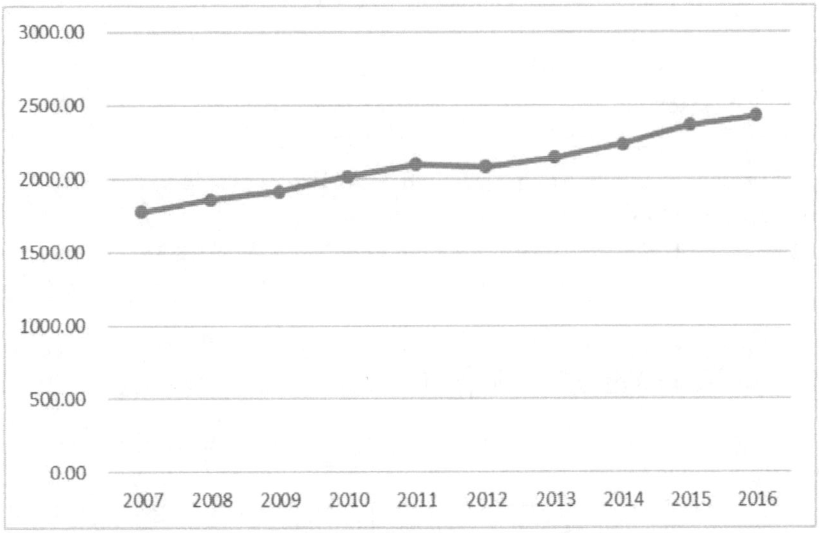

Figure 2. Urban Construction Land Area Change Map of Henan Province, 2007-2016.

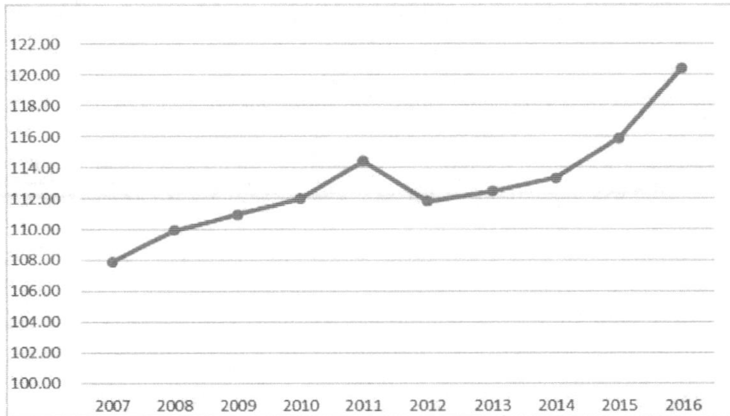

Figure 3. Change of per capita urban construction land area in Henan Province from 2007 to 2016.

4.2. Analysis on Evolution Characteristics of Urban Land Use Structure in Henan Province

4.2.1. Evolution Characteristics of Urban Land Use Structure in the Province. From Table 5, it can be seen that the urban land use structure in Henan Province is developing at a uniform speed in general from 2007 to 2011. The proportion of residential land is the largest, rising from 26.24% in 2007 to 28.31% in 2011, with an average annual increase of 0.52%, and that of public facilities from 12.77% in 2007 to 14.04% in 2011, with an average annual increase of 0.32%. The proportion of industrial land decreased from 18.67% in 2007 to 15.24% and decreased by 3.43%, which was the largest decline in all land use structures; the proportion of storage land decreased gradually from 2007 to 2010, and increased by 0.41% in 2011; the proportion of external transportation land decreased gradually, and the proportion of road square land decreased by 0.77%. The proportion of land for municipal public facilities increased first and then decreased; the proportion of green space increased year by year, and the proportion of special land decreased year by year.

Table 5. Urban land use structure in Henan Province, 2007-2011 (unit: %).

Particular year	Residential land	Land for public facilities	Industrial land	Warehousing land	Land for external transportation	Land for Road Square	Land for Municipal Public Facilities	Greenland	Special land use
2007	26.24	12.77	18.67	3.26	4.77	10.99	3.45	10.81	1.99
2008	26.88	13.12	19.64	3.20	4.57	11.09	3.50	11.04	1.74
2009	27.48	13.68	18.66	3.19	4.65	11.48	3.53	11.17	1.72
2010	28.73	14.27	16.79	3.06	4.02	12.38	3.40	12.45	1.58
2011	28.31	14.04	15.24	3.47	4.00	12.58	3.91	13.19	1.49

From Table 6, it can be seen that the urban land use structure of Henan Province in 2012-2016 is developing steadily and uniformly. The proportion of residential land increased from 29.38% in 2012 to 29.98% in 2016, with an average annual increase of 0.15%. Land for public administration and public service facilities decreased from 11.66% in 2012 to 11.19% in 2016, with an average annual decrease of 0.12%. Land for commercial service facilities decreased from 6.29% in 2012 to 5.38% in 2016, of which, from 2012 to 2013. The proportion of industrial land increased from 14.68% in 2012 to 15.61% in 2016, and increased by 0.93%. The proportion of logistics storage land decreased from 3.49% in 2012 to 3.22% in 2016. The proportion of road traffic facilities land increased year by year, from 15.51% in 2012 to 16.02% in 2016. Facility land use is gradually decreasing, from 4.71% in 2012 to 4.09% in 2016, with an average annual decrease of 0.16%. The proportion of green land and square land increased from 14.3% to 14.66% from 2012 to 2014, and from 14.66% to 14.53% from 2014 to 2016, but overall it has increased.

Table 6. Urban land use structure in Henan Province, 2012-2016 (unit: %).

Particular year	Residential land	Public management-service land	Commercial service facility land	Industrial land	Logistics warehousing land	Road traffic facilities land	Utility land	Green land and square land
2012	29.38	11.66	6.29	14.68	3.49	15.51	4.71	14.30
2013	29.57	11.69	5.14	16.01	3.35	15.59	4.21	14.44
2014	29.58	11.57	5.24	15.58	3.38	15.80	4.19	14.66
2015	29.58	11.51	5.42	15.69	3.38	15.75	4.11	14.56
2016	29.98	11.19	5.38	15.61	3.22	16.02	4.09	14.53

From 2007 to 2016, the classification of urban land, residential land and industrial land, is unchanged, and its structure changes are shown in Figure 4. The proportion of residential land is increasing year by year. There is an obvious uniform upward trend from 2007 to 2010, and it declined in 2011. It began to grow slowly and uniformly in 2012. From 2008 to 2012, the proportion of industrial land continued to decline, and began to stabilize in 2013, with little fluctuation.

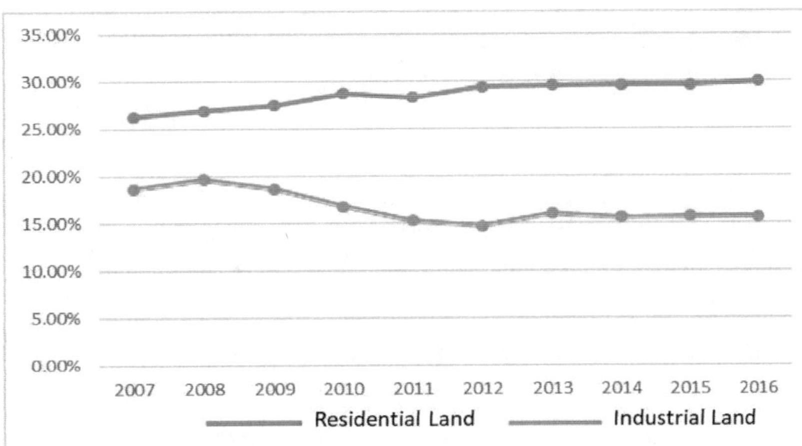

Figure 4. Residential and Industrial Land Change Map 2007-2016.

Table 7. Information Entropy, Equilibrium and Dominance of Urban Land Use in Henan Province, 2007-2016.

Particular year	Information entropy	Equilibrium degree	Dominance
2007	1.861	0.847	0.153
2008	1.866	0.849	0.151
2009	1.874	0.853	0.147
2010	1.870	0.851	0.149
2011	1.883	0.857	0.143
2012	1.894	0.911	0.089
2013	1.873	0.901	0.099
2014	1.875	0.901	0.099
2015	1.875	0.902	0.098
2016	1.868	0.898	0.102

From Table 7 and Figure 5, we can see the change of information entropy, equilibrium degree and dominance degree of urban land use structure in Henan Province. The overall curve is stable. From

Table 7, we can see that the information entropy increased from 1.861 in 2007 to 1.868 in 2016. Among them, the information entropy increased by 0.013 from 2007 to 2009, and decreased by 0.004 and 201 from 2009 to 2010. From 2010 to 2012, it increased by 0.024, and from 2012 to 2016, it decreased by 0.026. The total information entropy experienced a process of decline - rise - gentle - decline. Compared with information entropy, the equilibrium degree experienced two stages around 2012, rising first and then declining, while the dominance degree was the opposite. The overall change is not obvious, the overall trend is gentle, indicating that the change of urban land use structure in Henan Province in the past five years tends to be gentle, the overall fluctuation is not large, all types of land are relatively uniform, to a certain extent, it can explain that the urban land development in Henan Province is more mature.

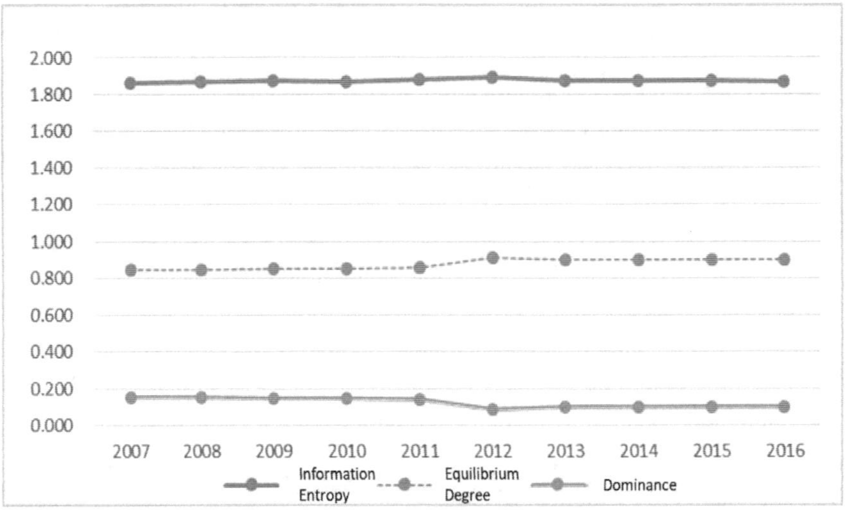

Figure 5. Change Map of Information Entropy, Equilibrium and Dominance of Urban Land Use in Henan Province, 2007-2016.

4.2.2. Characteristics of Urban Land Use Structure in Henan Province. Because of the difference of function orientation and industrial structure of each city, the proportion of land for each functional type in Henan Province is different. From the table 8, it can be seen that among the 18 cities, Jiyuan City has the highest proportion of residential land, which is 35.33%, while Hebi City has the lowest proportion of residential land, which is 22.10%; Jiaozuo City has the highest proportion of residential land. The proportion of management and public service land is the highest, 16.38%, Xinyang is the lowest, 2.67%; the proportion of commercial service facilities land is the highest, 19.44% and 2.11% of Luoyang, the lowest is Jiaozuo; the proportion of industrial land is the highest, 25.65% and 2.82% of Hebi, and the lowest is Luoyang logistics warehouse. The proportion of storage land is the highest 5.66%, the lowest 1.08% of Shangqiu City; the largest proportion of road traffic facilities land is 18.25% in Zhumadian City, and the lowest is 11.35% in Xinxiang City; the highest proportion of public facilities land is 14.92% in Xinyang City, and the lowest is 2.23% in Luoyang City; the largest proportion of green space and square land is Luohe City. The height was 22.89%.

The proportion of residential land accounted for 25.0-40.0% of urban construction land structure, 5.0-8.0% of public management and public service land, 15.0-30.0% of industrial land, 10.0-30.0% of road traffic facilities land, and 10.0-15.0% of green land. Comparing with this land use standard, the residential land in Shangqiu and Zhumadian is less than 25.0% of the lower limit of the national standard. Only the proportion of public management and public service land structure in Xinyang city does not reach 5.0% of the lower limit of the national standard. 13 cities exceed 8.0% of the upper limit of the national standard. The proportion of industrial land structure in seven cities such as Zhengzhou and Luoyang is not. It is 15.0% of the lower limit of national standard. The proportion of land structure of road traffic facilities in 18 cities is within the scope of national standard. The proportion of green space in Kaifeng, Nanyang, Xinyang, Zhumadian and Jiyuan cities does not reach the lower limit of 10% of national standard. Zhengzhou, Hebi, Puyang, Xuchang, Luohe, Sanmenxia and Shangqiu are less than 10%. The proportion of green space and square land structure in these seven cities exceeded the national standard limit of 15.0%.

Table 8. Urban land use structure in Henan Province in 2016 (unit: %)

Area	Residential land	Public management-service land	Commercial service facility land	Industrial land	Logistics warehousing land	Road traffic facilities land	Utility land	Green land and square land
Zhengzhou	27.05	13.32	4.31	10.90	3.68	16.84	5.27	18.64
Kaifeng	32.59	12.83	7.19	22.34	2.35	12.95	2.62	7.14
Luoyang	33.61	12.09	19.44	2.82	5.30	14.27	2.23	10.23
Pingdingshan	33.76	6.47	4.47	20.18	3.84	14.17	2.86	14.26
Anyang	33.61	11.91	2.53	18.32	3.56	16.34	2.94	10.78
Hebi	22.10	9.83	6.53	25.65	1.69	15.80	2.71	15.69
Xinxiang	32.30	12.38	5.17	22.17	2.87	11.35	2.80	10.96
Jiaozuo	34.36	16.38	2.11	16.24	1.92	14.75	3.52	10.72
Puyang	31.44	6.55	10.92	12.01	3.28	13.54	4.37	17.90
Xuchang	25.26	9.69	4.89	17.07	4.84	16.26	5.59	16.41
Luohe	27.07	7.61	3.38	9.45	3.57	13.36	12.68	22.89
Sanmenxia	26.74	13.46	4.72	14.91	2.01	15.21	3.71	19.24
Nanyang	29.04	13.05	5.09	20.10	3.35	17.28	4.01	8.07
Shangqiu	22.1	11.38	5.87	6.00	1.08	15.01	4.53	33.99
Xinyang	28.69	2.67	2.31	21.96	5.66	18.18	14.92	5.61
Zhoukou	29.87	11.81	7.38	8.16	5.61	17.86	6.44	12.88
Zhumadian	22.50	15.36	8.93	21.43	1.61	18.25	8.00	3.93
Jiyuan	35.33	7.92	5.42	18.44	2.74	17.08	3.28	9.79

Table 9. Information entropy, equilibrium and dominance of urban land use in Henan Province in 2016.

Area	Information entropy	Equilibrium degree	Dominance
Zhengzhou city	1.889	0.908	0.092
Kaifeng city	1.790	0.861	0.139
Luoyang city	1.793	0.862	0.138
Pingdingshan	1.787	0.859	0.141
Anyang city	1.783	0.857	0.143
Hebi city	1.838	0.884	0.116
Xinxiang city	1.802	0.867	0.133
Jiaozuo city	1.755	0.844	0.156
Puyang city	1.866	0.897	0.103
Xuchang city	1.923	0.925	0.075
Luohe city	1.874	0.901	0.099
Sanmenxia	1.855	0.892	0.108
Nanyang city	1.848	0.889	0.111
Shangqiu city	1.757	0.845	0.155
Xinyang city	1.793	0.862	0.138
Zhoukou city	1.920	0.923	0.077
Zhumadian	1.875	0.902	0.098
Jiyuan city	1.778	0.855	0.145

From Table 9 and Figure 6, we can see that the information entropy and equilibrium degree of Xuchang and Zhoukou are the highest, and the information entropy of Xuchang and Zhoukou exceeds 1.9. The equilibrium degree is higher, which indicates that the urban land development of Xuchang and Zhoukou is more mature, but the dominance degree is the lowest. There are Zhengzhou, Puyang, Luohe, Sanmenxia and Zhumadian whose information entropy is above 1.85, which belong to the medium level of development and need to be improved. Jiaozuo, Shangqiu, Xinyang and Jiyuan which are below 1.8 have low information entropy, which indicates that the land development is not mature enough, the dominance degree is relatively high and the urban soil is relatively high. The land still needs to be developed and further developed.

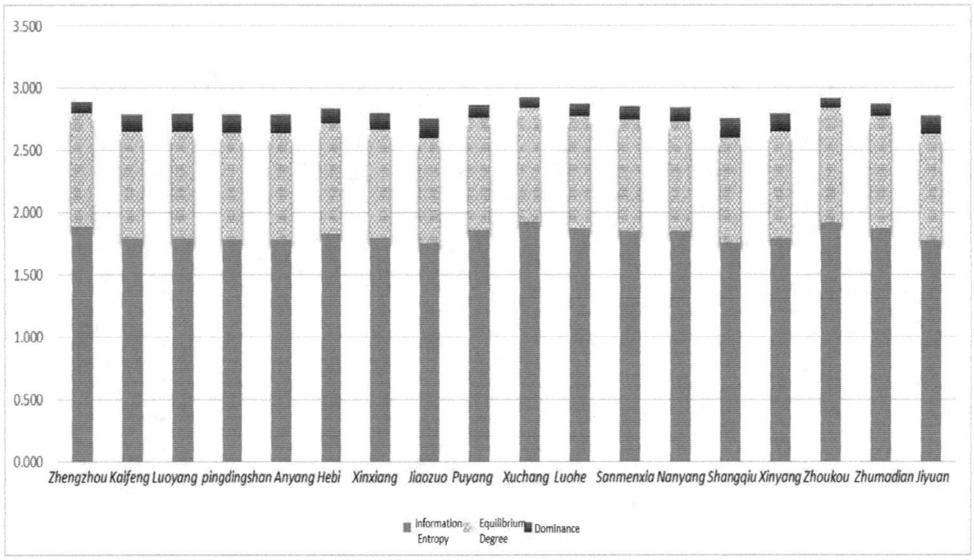

Figure 6. Change Map of Information Entropy, Equilibrium and Dominance of Urban Land Use in Henan Province in 2016.

4. Conclusions

From 2007 to 2016, the urban construction land area in Henan Province showed an overall upward trend; from 2007 to 2016, the per capita construction land area in Henan Province showed an overall upward-downward-upward trend.

From 2007 to 2016, the information entropy, equilibrium and dominance curves of urban land use structure in Henan Province are in a stable state, which shows that the change of urban land use structure tends to be gentle, the overall fluctuation range is not large, and all types of land are relatively uniform. To a certain extent, it can show that urban land development in Henan Province is relatively mature.

In the structure of urban construction land in Henan Province, the proportion of residential land is the highest, and its area and proportion are increasing year by year. This is closely related to the increase of urban population. The proportion of industrial land is higher, but it decreases year by year. The land for urban green space, square and road traffic facilities is increasing year by year, and others. Although the land use area of different types is increasing year by year, the proportion of structure is decreasing year by year, and the urban land use structure of Henan Province is in the stage of continuous adjustment.

Acknowledgments

The project of Pingdingshan Science and Technology Bureau: Research on Key Technologies of Urban Land Use Performance Evaluation under the Background of Improving Urban Construction 2017008(8.6)); the project of Henan Science and Technology Department: Research on Risk Identification and Prevention of New Urbanization in Henan Province Based on Problem Orientation (172400410290).

References

[1] Lu Chunyang 2011 *Study on the relationship between urban land use structure evolution and industrial structure evolution* [D] Southwest University
[2] Li Jiang, Guo Qingsheng 2002 Dynamic evolution analysis of urban land use structure based on information entropy *Resources and environment in the Yangtze River Basin* (05): 393-397
[3] Chen Zhe 2007 *Study on spatial change of urban land use in Nanjing* [D] Nanjing Normal University
[4] Ji Yu, Wang Chengxin 2017 Analysis of the Evolution and Driving Force of Urban Land Use Structure in Shandong Province *Journal of Natural Science* Hunan Normal University **40** (3): 1-6
[5] Kang Ruohe 2015 *Evolution of Yinchuan Urban Spatial Structure since 1949* [D] Xi'an University of Architectural Science and Technology
[6] El GHORAB H K, SHALABY H A 2016 Eco and Green cities as new approaches for planning and developing cities in Egypt *Al-exandria Eng J.* **55** (1): 495-503
[7] Baty M, Xie Y C 1994 Modeling in GIS: Part 1. Modelstructures, exploratory spatial data analysis and aggregation *Intermational Jourmal of Geographical Information Systems* (8): 291-307
[8] Danicl Z Sui 1997 Reconstruction Urban Reality: From GIS to Electropolis *Urban Geography* **18** (1): 74-89
[9] Li Tongsheng 2012 *Study on the structure and layout of urban land use in Hunan Province* [D] Hunan Agricultural University
[10] Sun Li 2016 *Study on the Impact of Urban Land Use Structure Change on Land Use Efficiency* [D] Nanjing University
[11] Li Shan 2016 *Evaluation and optimization of urban construction land structure* [D] Chongqing University
[12] Hu Jieju 2011 *Study on land use structure and layout in county-level urban areas* [D] Jiangxi University of Technology

Journal of Physics: Conference Series **1419** (2019) 012016 doi:10.1088/1742-6596/1419/1/012016

Evaluation on intensive use of urban land based on PSR model taking Henan Province as an example

Hongmin Zhang[1,2], Chunyang Lu[1], Kaidong Yang[1] and Linyun Duan[1]

[1]Henan University of Urban Construction, Pingdingshan, Henan 467036, China

[2]E-mail: 987712672@qq.com

Abstract. This paper evaluates the level of urban land intensive use in Henan Province and analyzes the influence of different indexes on urban land intensive use in Henan Province. PSR model was adopted. The conclusions are as follows: (1) the intensive use level of urban land in Henan Province gradually increased from 2008 to 2016, but it was still in the stage of overuse, only in 2010 reached a moderate level of utilization; (2) the intensive use level of urban land in Henan Province is not high. Population is one of the main factors restricting the intensive use level of urban land in Henan Province; (3) the government of Henan Province should comprehensively consider the actual situation of Henan Province, increase the intensity of intensive use of urban land, and formulate relevant policies to improve the level of intensive use of urban land in Henan Province.

1. Introduction

As the core area of regional social and economic development, the city's land use change plays an important role in improving the overall land use efficiency of the region. Therefore, carrying out the analysis of the characteristics of urban land use change and exploring the general law of land use evolution are of great significance to the internal motivation of the urban land use structure evolution. At present, some scholars have studied the impact of urban land intensive use, such as He Fang believed that to improve the economic benefits and output efficiency of the land, it is necessary to increase the investment in land assets or improve land management [1]. Zhao Xiaofeng and Huang Xianjin believed that urban land intensive use and excessive pursuit of economic benefits would bring traffic congestion, disorderly land occupation, land environmental pollution etc [2]. Some scholars have also carried out research on urban land intensive use evaluation. For example, scholar Yang Donglang selected indicators such as capital input, GDP per capita, average fiscal revenue, population density, and average investment amount to comprehensively evaluate the status and potential of land intensive use in Xi'an [3]. Scholar Zhao Dandan used the Yangtze River Delta City as the evaluation object to establish an evaluation system from the input-output-sustainability aspect [4]. Based on the bearing capacity of land function, scholar Liu Yiying constructed an evaluation index system based on land use efficiency and land use intensity and land use benefit [5]. However, the existing evaluation criteria for urban land intensive use are not uniform [6-7], and the evaluation system and specific indicators are mostly static [8-10], and the dynamics are insufficient. This study is to make up for the existing research deficiencies. Taking Henan Province as an example, based on the Pressure-State-Response (PSR) model, the urban land intensive use evaluation framework is constructed to provide reference for the improvement of urban land intensive use level and urban land use in Henan Province.

2. Regional overview

The research data in this paper comes from "China Urban Construction Statistical Yearbook" (2008-2016), "China Urban and Rural Construction Statistical Yearbook" (2008-2016), "China Statistical Yearbook" (2008-2016), "Henan Province Statistical Yearbook" (2008-2016), "Statistical Bulletin of Henan Province" (2008-2017).

The total area of Henan Province is about 167,000 square kilometers, 17 provinces under the jurisdiction of the province, one province directly under the jurisdiction of the city, 52 municipal districts, 21 county-level cities, 85 counties, 1120 towns, 682 townships, 633 Sub-district offices, 4743 residential committees, and 46,831 village committees. At the end of 2017, the resident population of Henan Province reached 95.591 million. In the same period, the total GDP was 4498.816 billion yuan, of which the output value of the first, second and third industries increased by 433.149 billion yuan, 2144.99 billion yuan and 191.868 billion yuan respectively. The area of urban construction land in Henan Province has increased year by year. From 2008 to 2011, the area of residential and industrial land has increased year by year. After 2012, the area of industrial land has decreased. The proportion of road traffic land was 16.0%. Logistics warehousing has the smallest proportion, with a proportion of 3.2%. Among the top three urban construction land in 2016, residential land area, road traffic land area and industrial land area were 726.83 square kilometers, 388.32 square kilometers and 378.44 square kilometers respectively. The logistics and storage land area with the lowest proportion was 78.02 square kilometers.

3. Research methods

3.1. PSR model

The PSR model can well show the comprehensive benefits and influencing factors of urban land intensive use. The PSR framework comprehensively considers the evaluation of land use benefits in land use. The purpose is to analyze the motivations, internal mechanisms of urban land use change and the impact of various factors on land change and various response measures made by society. To better understand the interaction and mutual constraints of human activities, social stability, and sustainable land use in land use. Therefore, the PSR model "pressure-state-response" was selected in the specific research to construct the urban land intensive use evaluation system.

3.2. Entropy method

The entropy method is used to calculate the index weights, which reflects the comprehensive situation of land intensive use level in the study area.

In this paper, the entropy method is used to determine the degree of influence of selected indicators on the intensive use of urban land in Henan Province, that is, the weight of indicators. The calculation steps are as follows:

Entropy calculation formula (1): e_i is the entropy value of the *i*-th index, and p_{ij} is the normalized index value.

$$e_j =- k\sum_{i=1}^{n} p_{ij}\ln\left(p_{ij}\right), among\ them, k > 0, k = \frac{1}{\ln(n)}, e_j > 0 \tag{1}$$

Redundancy calculation formula (2):

$$d_j = 1 - e_j \tag{2}$$

Weight calculation of indicators as shown in formula (3): w_j is proportional to the importance of the indicator:

$$w_j = \frac{d_j}{\sum_{i=1}^{n} d_j} \tag{3}$$

Determine the weight of each subsystem as shown in formula (4): the pressure system redundancy is recorded as D_P, the state system redundancy is recorded as D_S, the response system redundancy D_R, D_P, D_S and D_R are combined, denoted as D, then the weight of each subsystem for:

$$W_P = \frac{D_P}{D}; W_S = \frac{D_S}{D}; W_R = \frac{D_R}{D} \tag{4}$$

4. Construction of Urban Land Intensive Use Evaluation System

4.1. Index system selection basis

(1) The essence of urban land intensive use. The article summarizes the connotation of urban land intensive use from the aspects of output efficiency, layout structure and land use intensity of land input. We select economic indicators such as the average fixed asset investment, land GDP, and average fiscal revenue representing land input and output; population density reflecting land use intensity and urban population land growth elastic coefficient; land use structure layout optional built-up area green coverage rate, etc.

(2) Comprehensive benefits of urban land use. In practice, the comprehensive benefits of urban land pay more attention to the rational and appropriate use of urban land. The evaluation index should consider more ecological factors. The ultimate goal is to control the expansion of the urban extension of the occupied land, and promote the three-way unification of urban land resources to the land economy, society and environment.

(3) Influencing factors of urban land intensive use. The article analyzes the specific factors affecting the intensive use of urban land from the perspective of economic, social and ecological environment. The details can be divided into population density, economic development level and land policy. However, because the evaluation indicators are to be quantified, and the land policy indicators have strong subjective concepts, they do not meet the principle of objective quantification. Therefore, the land policy indicators cannot be used as evaluation indicators. The economic factors can be selected from per unit area GDP, per unit area and average financial revenue, and the retail sales of the average social consumer goods. The social factors can be selected from the urban population density and the number of employees per capita.

(4) Land resource carrying capacity. As a kind of natural resource, land provides various functional services for human beings. Because of the difference in land quality and location, each land has different functions, which can be divided into land production, land space, land ecology and land security. With the rapid development of urbanization, the contradiction between people and land has become increasingly acute. Rational use of land resources, coordinated economic development and the environment have an important impact on the carrying capacity of land resources. The land resource carrying capacity is used as the guarantee of land function to evaluate the carrying capacity by per unit area GDP and population density.

(5) PSR model determination index. The PSR model, the pressure subsystem is the activity of people cause the pressure to natural resources, under the pressure of natural resources quantity and quality of restricting human production status as the state of subsystems, react to these changes become response subsystems, is composed of three subsystems such a circulation system, to reflect the causality between human activities and natural environment. The pressure index can select indicators such as population density and per capita road area, indicating the pressure caused by production activities to land resources; under the influence of pressure, land use degree, land input and output, etc. can be used as status layer to present pressure, and corresponding indicators can be selected. Under the influence of pressure, the degree of land use, land input and output can be used as the state layer to present the pressure. The corresponding indicators can be expressed by the fiscal revenue, the GDP per capita, and the number of employees in the land.

4.2. Index system construction

Table 1. Pressure system evaluation index.

Target layer	Criteria layer	Indicator layer	Indicator description
Urban land intensive use evaluation E	Pressure system P	C1 Population density (person/square kilometer)	Urban population/urban construction land area (+)
		C2 Urban population and land use growth elastic coefficient	Urban population growth rate / construction land growth rate (+)
		C3 Green coverage rate of built-up area (%)	Urban greening area / urban built-up area (+)
		C4 Per capita park green area (m2 / person)	Park Green Area / Total City Population (+)
		C5 Per capita road area (m2 / person)	Urban road area / total urban population (+)

Table 2. State system evaluation index.

Target layer	Criteria layer	Indicator layer	Indicator description
Urban land intensive use evaluation E	State system S	C6 GDP per capita (100 million yuan / square kilometer)	Gross domestic product / land area for urban construction land (+)
		C7 Average fiscal revenue (100 million square kilometers)	City government fiscal revenue / urban construction land area (+)
		C8 City per capita disposable income (yuan)	City government fiscal revenue / total urban population (+)
		C9 Second and third industry output value and land use growth elasticity coefficient	Growth rate of output value of the second and third industries / growth rate of urban construction land (+)
		C10 Industrial land proportion (%)	Industrial land area / urban construction land area (+)
		C11 Employment by the average number of people (10,000 people per square kilometer)	Urban employment / urban construction land area (+)

The PSR model divides all influencing factors into three categories: pressure factors, state factors and response factors; and constructs an evaluation system for urban land intensive use in Henan Province based on the PSR model. The evaluation system of urban land intensive use in Henan Province is divided into target layer, criterion layer and index layer, including three subsystems: pressure index, state index and response index. A total of 15 indicators are selected. Detailed indicators are shown in Table 1, Table 2, and Table 3.

Table 3. Response system evaluation index.

Target layer	Criteria layer	Indicator layer	Indicator description
Urban land intensive use evaluation E	Response system R	C12 The output value of the secondary industry as a percentage of GDP (%)	Secondary industry output / GDP (+)
		C13 Average fixed assets investment (10,000 yuan / square kilometer)	Total urban fixed assets investment / urban construction land area (+)
		C14 Retail sales of average social consumer goods (100 million square kilometers)	Retail sales of social consumer goods / urban construction land area (+)
		C15 Sewage treatment rate (%)	Sewage Disposal Capacity / Sewage Discharge (+)

5. Result analysis

5.1. Data source and standardization
The research data in this paper is shown in Table 4.

Table 4. Evaluation indicator raw data.

Year	2008	2009	2010	2011	2012	2013	2014	2015	2016
C1	20297.45	20553.49	20809.58	21072.08	21469.51	21659.72	21581.70	21255.62	21578.46
C2	0.07	0.13	0.73	0.13	0.16	0.19	0.14	0.10	0.24
C3	35.38	36.29	36.56	36.64	36.90	37.60	38.32	37.69	39.33
C4	8.20	8.72	8.65	8.90	9.23	9.58	9.93	10.16	10.43
C5	8.14	8.21	8.15	8.18	7.22	7.20	7.32	7.41	7.42
C6	10.28	10.71	11.93	13.42	14.30	15.13	15.76	15.77	16.69
C7	1.01	1.05	1.18	1.41	1.58	1.72	1.83	1.87	1.94
C8	13231.11	14371.56	15930.26	18194.80	20442.62	22398.03	24391.45	25575.61	27232.92
C9	3.00	2.32	2.87	4.89	3.25	3.11	2.18	1.12	3.62
C10	20.72	19.53	17.37	15.83	14.68	16.01	15.58	15.69	15.61
C11	5544.45	5835.70	5787.86	6373.62	6638.12	7160.82	7671.60	7782.02	7935.20
C12	55.99	55.21	55.68	55.28	53.91	52.25	51.26	48.70	47.63
C13	724.06	986.63	1151.58	1214.95	1658.09	1711.47	1677.62	1960.12	2089.68
C14	3.30	3.69	4.11	4.68	5.24	5.80	6.27	6.66	7.27
C15	77.59	83.90	87.60	89.04	87.83	90.84	92.52	93.57	95.91

Since the statistical units and magnitudes of the obtained index values are different, the original values of the obtained indicators should be standardized. This paper uses the specific gravity method to perform dimensionless processing. The algorithm is as follows:

$$\mathrm{p}_{ij} = \frac{x_{ij}}{\sum_{i=1}^{n} x_{ij}} (i = 1,2...n; j = 1,2...m) \tag{5}$$

$$\mathrm{p}_{ij} = 1 - \frac{x_{ij}}{\sum_{i=1}^{n} x_{ij}}, (i = 1,2...n; j = 1,2...m) \tag{6}$$

p_{ij} indicates the standard value of the *i*-year *j* index in Henan Province; x_{ij} indicates the actual value of the i-year j index in Henan Province. According to the positive and negative effects of the index on the system, the standardized value of urban land intensive use in Henan Province is shown in Table 5.

Table 5. Evaluation index raw data normalization value.

Year	2008	2009	2010	2011	2012	2013	2014	2015	2016
C1	0.1067	0.1080	0.1094	0.1107	0.1128	0.1138	0.1134	0.1117	0.1134
C2	0.0395	0.0677	0.3848	0.0714	0.0859	0.1010	0.0732	0.0512	0.1254
C3	0.1057	0.1084	0.1092	0.1095	0.1102	0.1123	0.1145	0.1126	0.1175
C4	0.0979	0.1041	0.1032	0.1062	0.1101	0.1143	0.1185	0.1212	0.1245
C5	0.1175	0.1186	0.1177	0.1181	0.1043	0.1039	0.1057	0.1070	0.1072
C6	0.0829	0.0864	0.0962	0.1082	0.1153	0.1220	0.1271	0.1272	0.1346
C7	0.0744	0.0773	0.0866	0.1039	0.1159	0.1265	0.1349	0.1378	0.1428
C8	0.0728	0.0791	0.0876	0.1001	0.1125	0.1232	0.1342	0.1407	0.1498
C9	0.1138	0.0879	0.1088	0.1855	0.1234	0.1179	0.0828	0.0425	0.1373
C10	0.1372	0.1293	0.1150	0.1048	0.0972	0.1060	0.1031	0.1039	0.1033
C11	0.0913	0.0961	0.0953	0.1050	0.1093	0.1179	0.1263	0.1281	0.1307
C12	0.1176	0.1160	0.1170	0.1161	0.1133	0.1098	0.1077	0.1023	0.1001
C13	0.0550	0.0749	0.0874	0.0922	0.1259	0.1299	0.1273	0.1488	0.1586
C14	0.0703	0.0785	0.0874	0.0996	0.1114	0.1233	0.1334	0.1417	0.1545
C15	0.0971	0.1050	0.1097	0.1115	0.1100	0.1137	0.1158	0.1171	0.1201

5.2. Index weight determination

The index weights calculated by the entropy method are shown in Table 6.

Table 6. Evaluation of the weight value of the indicator system.

Target layer	Criteria layer	Indicator layer	Entropy value	Redundancy	Weights	Subsystem weight
Urban land intensive use	Pressure indicator	Population density (person/square kilometer)	0.8113	0.1887	0.0625	0.3479
		Urban population and land use growth elasticity coefficient	0.706	0.294	0.0974	
		Green coverage rate in built-up areas (%)	0.8112	0.1888	0.0625	
		Per capita park green area (m2)	0.8103	0.1897	0.0628	
		Per capita road area (m2 / person)	0.8108	0.1892	0.0627	
	Status indicator	GDP per capita (100 million square kilometers)	0.8065	0.1935	0.0641	0.3922
		Average fiscal revenue (100 million square kilometers)	0.8016	0.1984	0.0657	
		Urban per capita disposable income (yuan)	0.8008	0.1992	0.066	
		The second and third industry output value and land use growth elasticity coefficient	0.7897	0.2103	0.0697	
		Industrial land ratio (%)	0.809	0.191	0.0633	
		Employment per capita (10,000 people per square kilometer)	0.8083	0.1917	0.0635	
	Response indicator	The output value of the secondary industry as a percentage of GDP (%)	0.8108	0.1892	0.0627	0.2599
		Average fixed assets investment (10,000 yuan / square kilometer)	0.7941	0.2059	0.0682	
		Average retail sales of social consumer goods (100 million square kilometers)	0.7997	0.2003	0.0663	

Sewage treatment rate (%)	0.8107	0.1893	0.0627

5.3. Comprehensive evaluation results

The urban land intensive use system will change with the passage of time. Therefore, when evaluating the level of urban land intensive use, it is necessary to analyze the current urban land intensive use in terms of comprehensive pressure, status and response, and use the following formula to calculate the level of urban land intensive use:

$$E = W_P \sum_{i=1}^{n} p_i w_i + W_S \sum_{i=1}^{n} p_j w_j + W_R \sum_{i=1}^{n} p_k w_k \tag{7}$$

Among them, E represents the comprehensive evaluation index of urban land intensive use, Wp is the weight of the pressure index, W_S is the weight of the state index, W_R is the weight of the response index, W is the weight of the system occupied by the indicator, and P is the standardized value of the indicator. Through calculation, the comprehensive evaluation results of urban land intensive use in Henan Province from 2008 to 2016 are obtained, as shown in Table 7.

Table 7. Comprehensive Evaluation Index of Urban Land Intensive Use in Henan Province from 2008 to 2016

Year	2008	2009	2010	2011	2012	2013	2014	2015	2016
Pressure system	0.0306	0.0341	0.0650	0.0348	0.0358	0.0377	0.0354	0.0333	0.0412
State system	0.0374	0.0362	0.0385	0.0466	0.0441	0.0467	0.0462	0.0442	0.0523
Response system	0.0219	0.0242	0.0260	0.0272	0.0300	0.0310	0.0315	0.0333	0.0349
Comprehensive evaluation	0.0310	0.0324	0.0445	0.0374	0.0375	0.0395	0.0386	0.0376	0.0439

The pressure index P of urban land intensive use evaluation in Henan Province fluctuated greatly. It rose sharply from 2009 to 2010, and rapidly declined from 2010 to 2011. However, the overall process is rising, but the increase is not large. The pressure index P is consistent with the fluctuation of the comprehensive evaluation index E, indicating that the pressure index has a greater impact on the level of urban land intensive use. The state index S and the response index R have no significant changes during the period from 2009 to 2011.

In terms of economic benefits, the per capita GDP, per capita fiscal income, per capita disposable income and per capita fixed asset investment of Henan Province have all been steadily increasing from 2008 to 2016, which greatly promotes the intensive use of urban land in Henan Province. From the social benefits, the number of employees in the land, the retail sales of the average amount of social consumer goods and so on, are also increasing year by year. In terms of ecological environment, the government has invested heavily in the per capita park green area, the green coverage area of the built-up area, and the sewage treatment rate, and has also achieved great results. In 2016, the green coverage rate of the built-up area was 39.33%, which was 11.2% higher than that of 35.38% in 2008. In 2016, the per capita park green area was 1043 square meters/person, compared with 2008, the per capita park area of 8.20 square meters/person increased by 27.2%. Henan's investment in urban ecological environment not only improved the urban ecological environment of Henan Province, but also promoted the improvement of urban land intensive use in Henan Province.

The comprehensive evaluation index E of intensive land use in Henan Province showed a slow rising trend of fluctuation. The highest value is in 2010, when the comprehensive evaluation index E is 0.0445. The minimum value is in 2008, when the comprehensive evaluation index E is 0.0310. However, only in 2010 did the intensive use of urban land in Henan reach a moderate level, all other years were overused. Although the overall trend of the comprehensive evaluation index is rising, the overall result is not ideal. From this result, although the level of urban land intensive use in Henan

Province is rising, the level of urban land intensive use in Henan Province is still very low, and the overall level is still in the stage of overuse.

6. Conclusions and recommendations

6.1. Conclusions

The level of urban land intensive use in Henan Province has gradually increased from 2008 to 2016, but it is still in the stage of excessive utilization. Only in 2010, it has reached a moderate level of utilization.

Henan Province has a low level of intensive use of urban land, and population is one of the main factors restricting the intensive use of urban land in Henan Province.

The government of Henan Province should comprehensively consider the actual situation of Henan, increase the intensity of urban land intensive use, and formulate relevant policies to improve the level of urban land intensive use in Henan Province.

6.2. Recommendations

(1) Reasonable control of the scale of urban land use. By controlling the size of the town and suppressing the excessive growth of urban areas, it leaves room for urban construction and meets its current development needs. According to the development of smart growth methods, we can combine current and long-term interests and better promote land intensive use.

(2) Implementing differentiated land policies. For the influencing factors of urban land intensive use, even if the same influencing factors, the impact may be very different for different places. Therefore, when formulating policies on intensive use of urban land, the government must make relevant policies based on the actual situation of the region and comprehensively consider various factors.

Acknowledgments

The project of Pingdingshan Science and Technology Bureau: Research on Key Technologies of Urban Land Use Performance Evaluation under the Background of Improving Urban Construction 2017008(8.6); the project of Henan Science and Technology Department: Research on Risk Identification and Prevention of New Urbanization in Henan Province Based on Problem Orientation (172400410290).

References

[1] He Fang 2004 *Economic and urban land use* Tongji University Press
[2] Zhao Xiaofeng, Huang Xianjin and Chen Yi 2011 Research Progress in Urban Land Intensive Use *J. of Natural Resources* **25** (11): 1980-1996
[3] Yang Donglang, An Xiaoli 2007 Colligation Evaluation of Xi'an Urban Land Intensive Use *Economic Geography* (3): 102-107
[4] Zhao Dandan, Hu Yecui 2016 Spatial-temporal Characteristics of Intensive Land Use in the Yangtze River Delta Region *Chinese Agricultural Science Bulletin* **20** (32): 77-84
[5] Liu Yiying 2015 Study on Land Carrying Capacity in South of Jiangsu Area Based on Land Function *J. of Shanxi Agricultural Sciences* **43** (4): 481-485
[6] Trusilova K, Jung M and Churkina G 2009 On Climate Impacts of a Potential Expansion of Urban Land in Europe *J. of Applied Meteorology and Climatology* **48** (9): 1971-1980
[7] Wang Jiating, Zhao Liang 2009 The Research of Resource Intensive Efficiency and Influencing Factors of Cities in China *China Population, Resources and Environment* **19** (5): 45-52
[8] Yang Shuhai 2012 China Low-cost *Intensive Urbanization Model and Its Land Use Research* Nankai University Press
[9] Zhang Nian, Xiao Rong-ge 2013 Review of Current Studies and Trend Outlook of Urban Intensive Use *China Mining Magazine* **22** (S1): 110-112

[10] Li Guoping, Li Zhi and Zhang Zuo 2009 Urban Land Intensive Use Based on the PSR Model—A Case Study of Shanxi Province *East China Economic Management* **23** (10): 98-105

Predict post spreading in online social networks based on independent cascade model

O V Yakovleva[1,2], I V Rudakov[1] and Y V Stroganov[1]

[1]Bauman MSTU, ul. Baumanskaya 2-ya, 5/1, Moscow, 105005, Russian Federation

[2]E-mail: olg-199774@mail.ru

Abstract. Today, online social networks play an important role in the information spreading. Based on the social influence effect, posts can be spread across the network by repost mechanism. This paper addresses the problem of using independent cascade model for repost dynamics prediction in online social networks. A spreading post probability between two connected users is presented as function of preassigned features and coefficients fitted by the EM algorithm. This approach combines ideas of previously proposed methods. The list of considered features is composed and it includes users, their relationships and posts features. The iteratively solved by the Newton's method system of equation is derived for independent cascade model to obtain coefficients of a probability function. The experiments with the Sina Weibo dataset show that the EM algorithm more accurately estimates unknown coefficients of probability function that the logistic regression method. Since actual values of diffusion probabilities are unknown for a real network, a problem of their estimation is considered as a binary classification problem to determine accuracy of various methods. The independent cascade model tuned by proposed method of diffusion probability estimation was verified. The experimental results show than the model can commendably predict the reposting dynamics for a given post.

1. Introduction

An online social network (OSN) is a web-service that allows its users to create a personal account, explicitly connect to other users thus creating social relationships, and publish messages, called posts, to share various kinds of information, such as review of any product or service, political views, ideas, opinions, etc [1].

Based on the social influence effect, posts can be spread across the network through the principle of informational cascade [1] by repost mechanism. If a large amount of users has adopted a post, i.e. have made a repost of this post, it is considered viral. Modeling the post propagation process is useful for estimation whether the post will become viral. Also, it allows defining a group of users who will be interested in one or another content. In addition, users in OSN form communities [2] and such modeling can be helpful to detect information transmission paths from one community to another. Finally, information diffusion models are used to tackle the influence maximization problem, which asks to find a set of the most influential users in the network [3]. This analysis of OSN can be carried out for marketing, advertising and business purposes [4] or in the case of destructive information diffusion (promotion of radical ideas, suicidal content, drug distribution, incitement to ethnic and religious hatred, etc.) when a government needs to assess the situation [5,6].

2. Background and related work

An OSN can be formally represented by a directed graph $G(V, E)$, where a set of nodes $V =$

$\{v_1, ..., v_N\}$ are users and a set of edges E are relationships. If $(v_i, v_j) \in E$, the user v_j is called the follower of user v_i and the user v_i is called the friend of user v_j.

There are explanatory and predictive information diffusion models. Predictive models can be graph and non-graph based and proposed to predict spreading process in OSN by learning from past diffusion traces. The spreading process is characterized by two aspects: its structure, i.e. the spreading cascade that transcribes who reposted whom, and its temporal dynamics, i.e. the evolution of the diffusion rate which is defined as the amount of users that adopts the post over time [1].

This paper deals with the graph based Independent Cascade (IC) model [7]. In the IC model nodes can have two states: active (it means that the node has already adopted the post), and inactive (the node is unaware of the post or it is aware, but it did not adopt it). The process runs in discrete steps and starts from a set of initially activated nodes. In each step, an active node tries to influence each of its inactive followers. Regardless of its success, the same node will never get another chance to activate the same inactive follower. The success of node u in activating the node v depends on the propagation probability associated to the edge $e = (u, v)$. The process terminates when no further node gets activated.

There is the asynchronous extension of IC model named AsIC [8]. The difference between AsIC and IC models is that the first one unfolds a diffusion process along continuous time axis, and the second one is discrete. The AsIC model also requires a time-delay parameter on each edge of the graph.

Estimation of information diffusion probabilities for a real OSN is quite a challenge. In addition, verification of such methods also causes difficulties, since actual values of probabilities are unknown.

Saito *et al.* [7] proposed a discrete diffusion probability function, the values of which were learned by the Expectation Maximization (EM) algorithm. This method requires only a several past spreading cascades and does not need to calculate any node, edge or post features. It does not scale for large graphs since each edge must be present in training sample for accurate results. Therefore, this method can't be used in practice since real online social networks contain millions of users and billions of connections.

In the paper "Learning diffusion probability based on node attributes in social networks" [9], Saito *et al.* addressed the problem of learning parameters for AsIC model. A diffusion probability and a time-delay are considered as continuous functions of the *J*-dimensional vector of node attributes. This approach allows training the model in one small subgraph $G' \subset G$ and then predicting the spreading of new cascades in any other subgraph. According to the research of Saito *et al.* [7], objective function is introduced and maximized by the EM algorithm. However, authors only experimented with synthetic data and didn't provide a practical solution.

Guille *et al.* [10] also dealt with AsIC model. They solved the problem of estimating a diffusion probability as a classification problem for which the input data were semantic, social, and temporal features. That is, the diffusion probability is defined as the probability that the edge $e = (v,w)$ belongs to the "successful activation" class. To solve the classification problem, the logistic regression method is used. The advantages of Guille's approach include the simplicity and the high model training speed, since there is no need to consider all attempts of successful and unsuccessful activations in training cascades, as well as high prediction accuracy, confirmed by practical experiments. However, learning by logistic regression method contravenes the IC model assumption that any of the active friends could have influence on the newly activated node.

3. Notations

Let $G = (V, E)$ be a directed graph without self-links, where V and E ($\subset V \times V$) stand for the sets of all the nodes and edges, respectively. For each node $v \in V$, let $F(v)$ be the set of all the nodes that have edges from v, i.e., $F(v) = \{u \in V; (v,u) \in E\}$, and let $B(v)$ be the set of all the nodes that have edges to v, i.e., $B(v) = \{u \in V; (u,v) \in E\}$.

Now let's look at the diffusion process of some post. Let $D(t)$ be a set of nodes newly made active at time-step t, where $D(0)$ is a set of initially activated nodes. The diffusion episode is denoted by $D = D(0) \cup D(1) \cup ... \cup D(T)$.

Let $C(t)$ be a set of nodes made active by time-step t, i.e. $C(t) = \cup_{\tau \leq t} D(\tau)$.

Next, let there be a training sample of S independent diffusion episodes. Then $D_s(t)$ equals $D(t)$ and $C_s(t)$ equals $C(t)$ for an episode with number $s = 1, ..., S$.

Let M be a set of all posts in OSN. A set of all posts written by user u is denoted by M_u. A set of posts written by user u and contained the mentions of other users is denoted by $\mathcal{D}_u \subset M_u$. Then, let M_u be a set of users who are mentioned in the posts of a user u and let $t\mathrm{M}^u$ be a set of all the posts which have mentioned a user u. Each post m is associated with the set of keywords K_m.

4. Proposed method

According to Guille's method [10], a diffusion probability of post m from newly activated node v to its still inactive follower w is denoted as a simple and smooth function of the feature vector and the coefficient vector as in equation (1). However, unlike Guille's method [10], the coefficient vector is learned by EM algorithm, according to Saito [7,9], not by logistic regression method. In addition, the list of features is changed relative to Guille's method [10].

$$p_{v,w,m} = p(\boldsymbol{x}_{v,w,m}, \boldsymbol{\theta}) = \frac{1}{1+\exp(-\boldsymbol{\theta}^T \boldsymbol{x}_{v,w,m})} \quad (1)$$

where $\boldsymbol{x}_{v,w,m}$ is the $(J+1)$-dimensional vector, first element of which is set to 1, i.e. $x_{v,w,m,0} = 1$, and others elements are composed of the J normalized features of user v, user w, relationship $e = (v,w) \in E$ and post m; $\boldsymbol{\theta}^T = (\theta_0, ..., \theta_J)$ is transposed coefficient vector, where $\theta_0 = const$.

4.1. Features

So it is proposed to consider the features of users (nodes), relationships (edges) and posts. The list of features described below was obtained by the following way. All features from Guille's method [10] were taken without changes, each of boolean type features was replaced on similar a real type one and several features were added by the authors. Then the best features were selected by the recursive feature elimination algorithm.

For user, five features are captured.

- User's activity as in equation (2). This feature characterizes user's ability to post and is computed as the average amount of posts, including reposts, emitted per hour.

$$I(u) = \begin{cases} \frac{|M_u|}{\Delta t}, & \text{if } |M_u| > 1 \\ 0, & \text{Otherwise} \end{cases} \quad (2)$$

where Δt is the difference between the earliest and latest posts' times measured in hours.

- The ratio of posts with mentions as in equation (3). This feature is taken without changes from the paper "A predictive model for the temporal dynamics of information diffusion in online social networks" [10] and provides an idea about the ability of a given user to distribute a content to other users.

$$dTR(u) = \begin{cases} \frac{|\mathcal{D}_u|}{|M_u|}, & \text{if } |M_u| > 0 \\ 0, & \text{Otherwise} \end{cases} \quad (3)$$

Last three user features characterize his/her popularity and socialization. They are determined by the capacities of appropriate sets.

- Number of user's followers.
- Number of user's friends.
- Number of posts which have mentioned a user.

For a relationship, i.e. for a pair of connected users, six features are considered:

- Social homogeneity by mentions as in equation (4). This feature is also taken without changes from the paper "A predictive model for the temporal dynamics of information diffusion in online social networks" [10] and for users u_1 and u_2 it reflects the similarity of the sets of users they talk to. It is computed with the Jaccard's coefficient.

$$HM(u_1,u_2) = \frac{|\mathrm{M}_{u_1} \cap \mathrm{M}_{u_2}|}{|\mathrm{M}_{u_1} \cup \mathrm{M}_{u_2}|} \quad (4)$$

- Homogeneity by posts. This feature reflects the similarity of users' interests. It is computed with the Jaccard's coefficient. There are three cases when post m_1 and post m_2 are equals: (i) m_1 is repost of m_2, (ii) m_2 is repost of m_1, (iii) m_1 and m_2 both are reposts of another post m_3.
- Social homogeneity by followers. This index describes homogeneity of users' social circles. It is computed with the Jaccard's coefficient.
- Social homogeneity by friends. This index is similar to the previous one, but it describes another aspect of users' social circles homogeneity.
- The mentioning rate of the second user by the first one as in equation (5). This feature is introduced instead of the similar boolean type one in "A predictive model for the temporal dynamics of information diffusion in online social networks" [10] and it reflects a measure of social relation between users.

$$rM(u_1,u_2) = \begin{cases} \frac{|\mathrm{M}_{u_1} \cap t\mathrm{M}^{u_2}|}{|\mathrm{M}_{u_1}|}, & if\ |\mathrm{M}_{u_1}| > 0 \\ 0, & \text{Otherwise} \end{cases} \quad (5)$$

- Correlation of temporal activity vectors as in equation (6). According to Guille's method [10] a day is considered as a partition of 6 blocks of 4 hours. So the fraction of tweets the user emitted during each block is computed and a 6-dimensional vector noted V is filling: $\sum_{i=0}^{5} V^i$. For two users coefficient of such vectors correlation shows whether users are online at the same time.

$$C = \sum_{i=0}^{5} V_{u_1}^i * V_{u_2}^i \quad (6)$$

Finally, one post feature is presented:

- User's interest in the topic of post as in equation (7). It is computed as the ratio of the number of user posts with at least one keyword of given post to the total number of user posts. This feature replaces the similar boolean type one described by Guille in "A predictive model for the temporal dynamics of information diffusion in online social networks" [10].

$$hK(u,m) = \begin{cases} \frac{|\{m: K_m \cap K_u \neq \varnothing\}|}{|M_u|}, & \text{if } |M_u| > 0 \\ 0, & \text{Otherwise} \end{cases} \quad (7)$$

where $K_u = \cup_{m \epsilon M_u} K_m$

Before fitting coefficients, the all feature values are normalized by maximum value in the training dataset. Normalization coefficients will be saved to divide on them subsequently computed when modeling features values.

4.2. Coefficient fitting method

According to Saito [7,9], the system of equation (8) can be derived for IC model. The method of learning coefficients is described below. Firstly, initial approximation of vector $\boldsymbol{\theta}$ is chosen. Then the next two steps repeat until convergence:

1. The values of probabilities $\hat{p}_{v,w,m_s}$ for each of posts m_s is computed by formula (1) for current approximation of vector $\boldsymbol{\theta} = \hat{\boldsymbol{\theta}}$
2. Computed values $\hat{p}_{v,w,m_s}$ are substituted into equation (8) and the system equation (8) is solved by Newton's method for obtaining the next approximation of vector $\boldsymbol{\theta}$ as a result.

$$\sum_{s=1}^{S}\sum_{t=0}^{T_s-1}\sum_{v\in D_s(t)}\left(\sum_{w\in F(v)\cap D_s(t+1)}\left(\frac{\hat{p}_{v,w,m_s}}{\hat{P}_w^{(s)}}-p_{v,w,m_s}\right)\boldsymbol{x}_{v,w,m_s}-\sum_{w\in F(v)\setminus C_s(t+1)}p_{v,w,m_s}\boldsymbol{x}_{v,w,m_s}\right)=0 \tag{8}$$

where $p_{v,w,m_s} = p(\boldsymbol{x}_{v,w,m_s}, \boldsymbol{\theta})$ and $\hat{P}_w^{(s)}$ is calculated by the formula (9).

$$\hat{P}_w^{(s)} = 1 - \prod_{v\in B(w)\cap D_s(t)} (1-\hat{p}_{v,w,m_s}) \tag{9}$$

5. Experimental Evaluation

5.1. Dataset

For the experimental evaluation, the Weibo-Net-Tweet dataset gathered by Jing Zhang *et al.* [11] was used. The 65 diffusion episodes with sizes ranging from 50 to 600 reposts were selected from the total dataset. For each selected episode the spreading cascade was restored relies on the assumption that a user is influenced by a last friend who reposted the post before him. Then, the obtained cascades were split into training and testing ones in proportion 75:25, respectively.

The examples of unsuccessful activation attempts are also needed to train and test the model, therefore, for each spreadable post, author's followers who did not adopt this post were randomly selected. Their number was equal to the number of post adopters to equalize probabilities of making type I and type II errors (the ratio of successful and unsuccessful attempts was obtained experimentally). It is considered that a type I error ("false positive" finding) was made on time step t, if the probability computed by the formula (10) is more than 0.5, and the user w did not make a repost. Reciprocally, a type II error ("false negative" finding) is to see the probability less than 0.5 when the repost is made by the user w. The condition of equality of "false positive" and "false negative" finding probabilities must be observed to avoid underestimating or overestimating a diffusion episode size when modeling.

$$P_w = 1 - \prod_{v\in B(w)\cap D(t)} (1-p(\boldsymbol{x}_{v,w,m}, \boldsymbol{\theta})) \tag{10}$$

5.2. Accuracy estimate

Since actual values of diffusion probabilities in real social networks are unknown, to assess accuracy of various methods of learning coefficient vector $\boldsymbol{\theta}$ in equation (1), we will assume that a probability is determined correctly if neither a type I error nor a type II error was made. The results of testing the EM algorithm and the logistic regression method are illustrated in Table 1.

Table 1. Learning method accuracy with the training: testing ratio 75:25.

Learning method	Accuracy
EM-algorithm	0.75
Logistic regression	0.71

So model trained by the EM algorithm provides better accuracy than model trained by the logistic regression method. The accuracy of the logistic regression method is lower than obtained in "A predictive model for the temporal dynamics of information diffusion in online social networks" [10] which can be explained by incompleteness of the dataset used to compute the features.

5.3. Modeling diffusion process

For each of the 15 test posts, a spreading cascade was obtained by IC model. Results for two various posts are shown on figure 1. In the case of first post, the number of reposts maximally increases in step 2 that means author's followers are mainly adopters. In the case of second post, only a minor proportion of adopters are influenced by author, but in the step 2 there are one or more influential users who made their followers adopt the post. In both cases predicted number of reposts is close to real one in each model time steps.

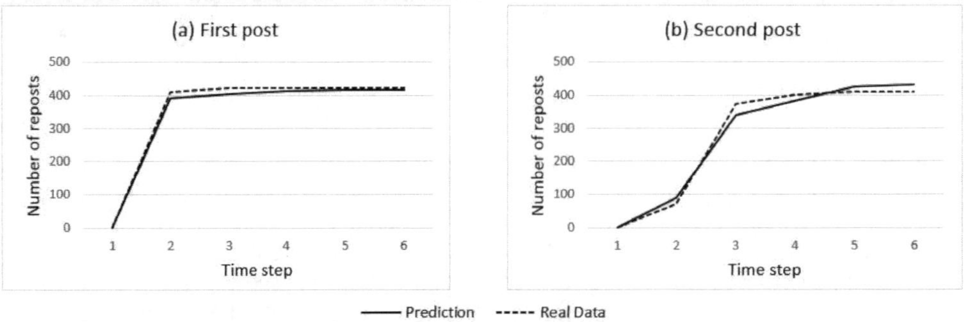

Figure 1. The number of reposts as a function of time step for two test posts.

6. Conclusion and future work

This paper addresses the problem of setting up and using IC model for obtaining a spreading cascade [1] of post in OSN. A method of learning diffusion probability is proposed. It combines the ideas of Guille [10] and Saito [9] methods. A diffusion probability is considered as a function of known features and unknown coefficients. The list of features of users, their relationships and post was composed. According to Saito [7,9], unknown coefficients are estimated with the EM algorithm. The experimental results validate the ability of IC model tuned by proposed method to correctly predict the reposting dynamics for a given post.

In future we plan to analyze post text [12] to increase an accuracy of estimation diffusion probabilities. In addition, we plan to develop a method of time-delay diffusion estimation and research the AsIC model. Finally, we need to compare an accuracy of the EM algorithm and the gradient boosting algorithm.

References

[1] Guille A, Hacid H, Favre C and Zighed D 2013 Information diffusion in online social networks: a survey *ACM SIGMOD Record* **42** pp 17-28

[2] Chesnokov V 2017 Overlapping community detection in social networks with node attributes by neighborhood influence *Models, Algorithms, and Technologies for Network Analysis* ed V Kalyagin, A Nikolaev, P Pardalos and O Prokopyev (Cham: Springer) vol 197 pp 187-203

[3] Yang W, Brenner L and Giua A 2019 Influence maximization in independent cascade networks based on activation probability computation *IEEE Access* **7** pp 13745-13757

[4] Chernova V, Tretyakova O and Vlasov A 2018 Brand marketing trends in russian social media *Media Watch* **9** pp 397-410

[5] Okhapkina E, Okhapkin V and Kazarin O 2017 The cardinality estimation of destructive information influence types in social networks *Proc. of 16th European Conf. on Cyber Warfare and Security* (Ireland: University Colledge Dublin) pp 282-287

[6] Okhapkina E, Tarasov A and Kazarin O 2017 Adaptation of information retrieval methods for identifying of destructive informational influence in social networks *Proc. of 31st Int. Conf. on Advanced Information Networking and Applications Workshops* (Taipei: Tamkang University) vol 1 pp 87-92

[7] Saito K, Nakano R and Kimura M 2008 Prediction of information diffusion probabilities for independent cascade model. *Knowledge-Based Intelligent Information and Engineering Systems* ed I Lovrek I, R J Howlett and L C Jain (Berlin, Heidelberg: Springer) pp 67-75

[8] Saito K, Ohara K, Kimura M and Motoda H 2013 Learning asynchronous-time information

diffusion models and its application to behavioral data analysis over social networks *J. Comput. Eng. and Inf.* **1** pp 30-57

[9] Saito K, Ohara K, Yamagishi Y, Kimura M and Motoda H 2011 Learning diffusion probability based on node attributes in social networks *Proc. of 19th Int. Symp. On Methodologies for Intelligent Systems* (Warsaw) pp 153-162

[10] Guille A and Hacid H 2012 A predictive model for the temporal dynamics of information diffusion in online social networks *Proc. Int. Workshop on Mining Social Network Dynamics* (Lyon) pp 1145-1152

[11] Zhang J, Liu B, Tang J, Chen T and Li J 2013 Social influence locality for modeling retweeting behaviors. *Proc. of 23st IJCAI* (Bejing) pp 2761-2767

[12] Kanev A, Cunningham S and Valery T 2017 Application of formal grammar in text mining and construction of an ontology *Proc. of 2017 Internet Technologies and Applications* (Wrexham) pp 53-57

Journal of Physics: Conference Series **1419** (2019) 012018 doi:10.1088/1742-6596/1419/1/012018

Applications of a robust particle tracking velocimetry algorithm in the multiple-parametric measuring system for coal exploration industry

Weifeng Xue[1,2], Changsong Wu[3], Sujian Wang[2], Ruijun Ji[2] and Yang Zhang[3,4]

[1]College of Geology and Environment, Xi'an University of Science and Technology, 710065, Xian, China
[2]Shaanxi Coal and Chemical Technology Institute, 710065, Xian, China
[3]School of Energy and Power Engineering, Xi'an Jiaotong University, 710049, Xi'an, China

[4]E-mail: zhangyang1899@mail.xjtu.edu.cn

Abstract. With the gradual increase in the average exploration depth of domestic coal industry, the conflict between deep coal exploration technology and exploration requirements is becoming increasingly prominent. A series of exploration disaster accidents caused by geological disasters require a high level of survey technology before the deep coal exploration. In this study, a high-resolution single hole measuring system for aquifers in coal seam is proposed. The entire system can be divided into two parts: hardware and software. The hardware probe integrates various sensors for different physical parameters using optical fibre to communicate with the controlling part above the ground. The home-made software is then used to receive and process the signals of temperature, pressure and orientation, as well as the instantaneous image sequence of aquifer with flowing contaminants. A home-made novel particle tracking velocimetry algorithm based on Voronoi diagram is used as the core algorithm to analyse the image sequence and obtain the magnitude and direction of flow at the measuring spot of aquifer. After practical tests, the system is confirmed to be able to effectively monitor the hydrological conditions of the aquifer in coal seam. Therefore, the proposed algorithm can be applied to the evaluation of hydrogeological conditions before well establishment for coal-exploration and to the prevention of potential dangers during the exploration work.

The precise exploration of hydrogeological conditions in coal explorations plays an important role in determining the safe coal production [1,2]. Effective monitoring methods are the fundamental solutions to avoid exploration disasters and accidents. Traditional methods for measuring the velocity and flow direction of aquifer are indirect; empirical formulas are used to calculate the aquifer velocity, which cannot ensure the accuracy of the measurement [3]. With the development of electronic technology, the direct measurement of seepage in aquifers through integrated single hole equipment has been widely studied and applied [4,5]. The velocity and flow direction of aquifers and other hydrogeological parameters are tested by the following methods. Firstly, aquifers are marked with radioisotope tracers, and the aquifer flow velocity and direction are determined through the tracer method. Secondly, the natural potential and charging methods in hydro-geophysical prospecting are adopted. Thirdly, the flow velocity and direction are measured using the thermo sensitivity method. However, the abovementioned methods pose disadvantages in the long-term field application: (1) the methods have low sensitivity in terms of detection of aquifer flow velocity and direction; (2) there is a gap (or a tuning coefficient that is local and thusly difficult to determine) between the indirect observation result and the real flow field. Hence, this study proposes a high-resolution system for directly measuring the flow velocity and direction and other hydrogeological parameters of aquifers in coal seam. This system is dedicated to a full reflection of the hydrogeological information of aquifers in coal seam, and to improve the reliability of the exploration of hydrogeological conditions and reduce the prediction error of water inflow in explorations. This system is expected to provide technical support for the improvement of water resource monitoring and water environment protection in mining areas.

1. Core algorithm

The traditional single hole measuring system for flow velocity and direction of aquifer is equipped with isotope tracer. Radioisotopes are used to mark the aquifer: the concentration in the tracer continuously decreases with the dilution of water flow, so the velocity of the borehole can be obtained by recording the change of isotope concentration [6,7]. However, if radioisotopes are used as tracer, protective measures should be applied to prevent any possible contamination during the operation. This process is generally difficult for technicians. Aquifer contains micron-sized sand and gravel particles; if the flow viscosity is higher than that of air, the carrying capacity of fluid is large. Therefore, sand and gravel particles can be used as natural tracer particles for water flow. High-speed digital photography and particle tracking velocimetry (PTV) are used to perform the cross-frame matching (tracking) of suspended particles in aquifer to obtain the velocity in the "forward difference format" on the feature plane. The operation is simple, has no interference and exhibits high measurement accuracy. Moreover, the measurement errors are mainly concentrated in the recognition process of the particle centre (about 0.1 pixels). Therefore, a self-made particle tracking algorithm is developed to process the image sequence of the flow field to obtain the flow characteristics of aquifer.

PTV algorithms have been continuously developing as a means of reconstructing particle trajectories in a flow field. Amongst the algorithms, the particle cluster algorithm is the easiest to operate because it delivers the quickest updates by considering two- and three-dimensional versions with few pre-setting parameters [8,9]. Classical particle cluster PTV algorithms include spring model based PTV and velocity gradient tensor PTV [10,11]. However, both algorithms demonstrate difficulty in defining the size of the particle cluster. In particle matching, the size of a particle cluster will change with the size of particle displacement. This problem can be avoided by using appropriate meshing techniques. In addition, the result obtained for a single image using the same partitioning algorithm is unique, thus overcoming the uncertainty in the size of the particle cluster. Song et al. connected the discrete particle points in an image to nonoverlapping triangles using the Delaunay Tessellation (DT) technique. Then, they conducted double-layer matching for the triangle and the vertex of the triangles in the two consecutive frames to obtain the matching results of particles [12]. This algorithm is referred to as the DT-PTV algorithm. However, triangles are simple and have few characteristics; thus, triangles become intrinsically "the same" when the particle density is high. As a result, the accuracy of the algorithm is limited [13,14]. To overcome the high self-similarity of triangles, the present authors developed a cytochemical treatment of discrete points using the Voronoi Diagram (VD). In this study, this new PTV algorithm is adopted to postprocess the images and obtain the flow velocity and direction of contaminants in aquifer.

The aquifer image collected by the camera is initially background-erased using matrix subtraction in a "central difference format". Then, the Gaussian filtering function is adopted to denoise the images. The reconstructed image is shown in Figure 1(a). The coordinates of the identified particle domain centres are used as the input parameters for PTV, as shown in Figure 1(b). After obtaining the coordinates of the particle centre in image, the image is divided into polygonal cells with different sizes and shapes using the VD grid generation technology. In other words, after VD division, the planar regions with several discrete points are divided into nonoverlapping polygons, which are called Voronoi cells, as shown in Figure 1(c) and Figure 1(d). The above VD division is applied to the two consecutive images.

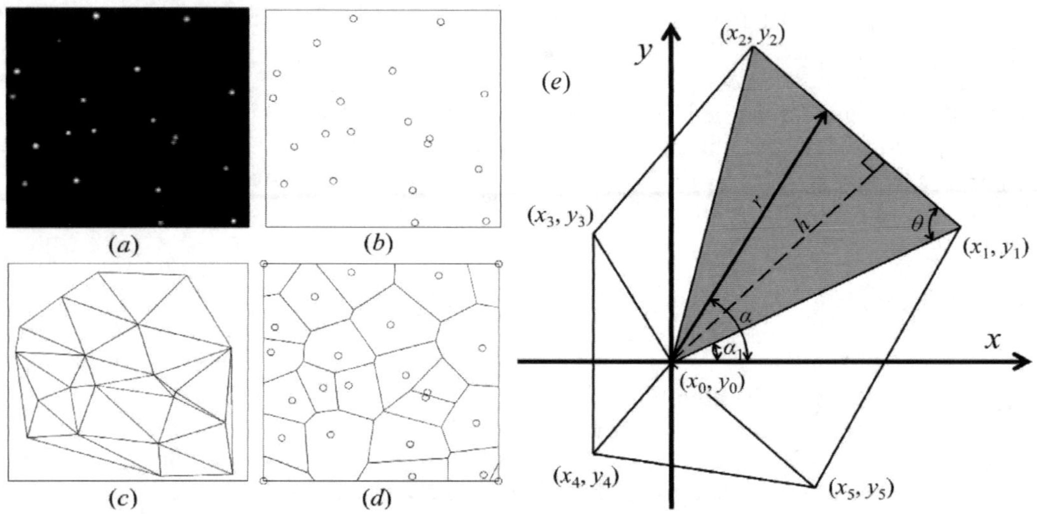

Figure 1. PTV processing: (a) Particle image generated after background removal and denoising; (b) Identification of particles from the images; (c) Triangular grid from DT; (d) Cell arrays obtained by DT transposing VD. (e) Calculation of the characteristic curve of a cell comprising vertices (x_i, y_i) (i = 1~5).

Based on the hypothesis of limited deformation in continuous flow, the corresponding cells of the same particle will not undergo large deformation across the frames. Therefore, the new PTV can be utilised to obtain the matching result of the two frames of particles by determining that of the two frames of cells. Firstly, the characteristic curve of a single cell is extracted. The radius from the particle point to the outer edge of the cell is used as the characteristic curve of the cell. The calculation formula for the characteristic curve $\boldsymbol{r}$ is shown in equation (1), where the parameters are shown in Figure 1(e).

$$\boldsymbol{r}(\boldsymbol{\alpha}) = \frac{h}{|\sin(\boldsymbol{\alpha}+\theta_1-\alpha_1)+1-\text{logical}(\theta_1-90^\circ)|} \qquad \text{logical}(x) = \begin{cases} 1 & (x \neq 0) \\ 0 & (x = 0) \end{cases},$$

$$h = \frac{|Ax_0+By_0+C|}{\left(A^2+B^2\right)^{1/2}} \qquad \begin{cases} A = y_2 - y_1 \\ B = x_1 - x_2 \\ C = x_2 y_1 - x_1 y_2 \end{cases} \tag{1}$$

Secondly, to compare a pair of characteristic curves ($\boldsymbol{r_1}$ and $\boldsymbol{r_2}$) from two respective frames, both are nondimensionalised by dividing each by $(|\boldsymbol{r_1}|^2+|\boldsymbol{r_2}|^2)^{1/2}$ to avoid the effect of cell size on matching results. The covariance of the two curves is calculated using equation (2). The particles with the smallest cell-to-cell difference are considered as matching particles.

$$C_r = \frac{\text{cov}(\boldsymbol{r}_1,\boldsymbol{r}_2)}{\left[\text{cov}(\boldsymbol{r}_1,\boldsymbol{r}_2)\right]^{1/2}} \tag{2}$$

The matching results can only be considered effective if the corresponding cells are sufficiently similar. Therefore, a judgment must be added to the matching result. If the covariances between the cell corresponding to a particle and any of the cells of all candidate particles are less than 0.5, the particle does not match in the second frame. In applications, 0.5 corresponds to the median value between two cases in which the cell is completely different and completely identical, respectively. For the flow with strong deformation, the judgement is suggested to between 0.25 and 0.5.

Based on the new PTV, the cells obtained from the Voronoi partitions are used to perform particle matching. The individual difference of cells is much larger than that of triangles. The triangular convergence like DT-PTV will not occur when the density is high. Therefore, the new PTV performs well when the particle concentration and local distortion of the flow field are large, which is consistent with the wide operating conditions of aquifer flow. In addition, this algorithm has simple structure and high computational efficiency. With only one pre-setting parameter, the robustness is high after realising the soft– and hardware connection. This new algorithm is integrated into the Aquifer Monitoring Program V1.0 to achieve the real-time calculation of aquifer flow information.

2. Hardware

The proposed high-resolution single hole measuring system for aquifers in coal seam consists of a probe, a signal processing module and a computer terminal. The probe will go deep underground to collect the hydrological characteristics of aquifer. The signal processing module and personal computer terminal are used for data processing. The probe and the ground are connected by optical fibres for signal transmission shown in Figure 2. The probe is used to collect the temperature and pressure of aquifer and flow field images. The new PTV algorithm is adopted to process the image sequence of the flow field. Then, based on the absolute geographic direction collected by the compass, the flow velocity and direction of aquifer is finally obtained.

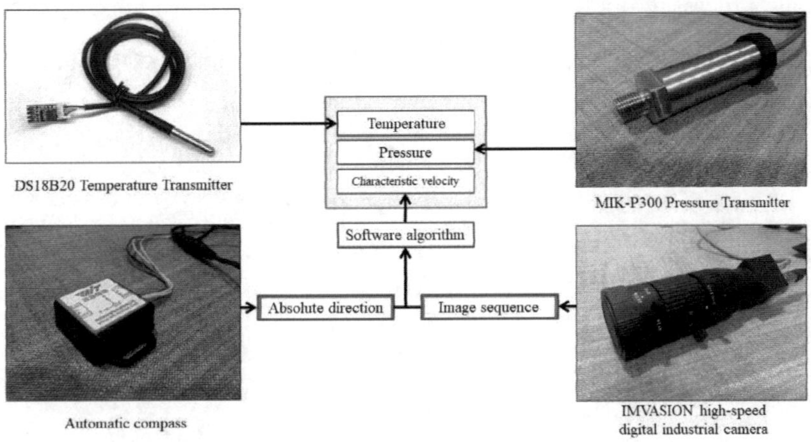

Figure 2. Constitution of the single hole measuring system for aquifer flow.

The sensor, camera and wiring are protected by encapsulating the probe part in a stainless and steel cylinder with a length of 700 mm and a diameter of 65 mm. The cylinder consists of two parts: top and bottom. The two parts are connected by welding three steel tubes arranged in a regular triangle to save framing space. The lower cylinder is responsible for collecting the temperature and pressure information of aquifer. A DS18B20 temperature sensor module and an MIK-P300 pressure transmitter are encapsulated, which is installed behind the water hole. DS18B20 is a commonly used digital temperature sensor that outputs digital signals. The temperature measurement range is from −55°C to 125°C. The inherent temperature measurement error is ±1°C between −10°C and 85°C. The measurement error is ±0.5°C. The range of MIK-P300 pressure transmitter is 0–2 MPa, and the measurement accuracy is 0.1% FS. A silicon piezoresistive pressure oil-filling core served as the pressure sensitive core. The integrated circuit converts the millivolt signal of the transducer into a standard voltage signal. All signals are transmitted along their respective lines to the integrated chip inside the upper cylinder. The holes on top of the lower cylinder are used to install lighting components for framing. The longer cylinder on the upper side collects aquifer images and monitors the status of the probe. An IMVASION high-speed industrial digital camera and lens are installed at the lower hole, whereas the digital posture sensor compass and camera-sensor integrated chip for data reception and transcoding are installed on top. The maximum framing frequency can be set as high as 50 Hz. The measuring range of the electronic compass is ±180° and the measuring error in static rate is ±0.05°. In all, one can tell that the entire system possesses good temporal resolution in acquiring data.

The collected data are transmitted by optical fibre to ensure the quality and efficiency of communication. The image data collected by the camera and the signal collected by the sensor are transmitted by two independent optical fibres to avoid mutual interference. Optical fibers are integrated with the power lines of each component, and the signal processing module reaches the surface through the wiring holes. The signal processing module is composed of corresponding signal converters, which convert optical signals into digital signals that can be recognised by a computer. Data are eventually transmitted to the personal computer terminal for further processing. The whole system is powered by a 220V AC power.

3. System calibration

All parts should be calibrated before using the overall measuring system. The temperature and pressure sensors and electronic compass adopt the integrated sensor module; thus, additional calibration is not required. However, the flow information collected by the camera only depends on the reference system where the camera is located; hence, the velocity and direction must be calibrated before use. The particle velocity obtained by the algorithm should be multiplied by the object-to-distance ratio of the camera to obtain the actual size. The object-to-distance ratio can be determined by capturing a photo of a reference object with known length. The actual direction can be obtained by jointly processing the velocity direction with the direction information collected by the electronic compass.

As for the calibration of the flow direction angle, considering that the camera and compass are installed on transition plates parallel to each other by threads in the probe, the photographic plane of the camera is parallel to the horizontal plane of the compass. Therefore, only the correction of horizontal angle should be considered. For the velocity direction angle, when the fluid is required to move to the east, the velocity direction angle is 0°. The anti-clockwise direction is set as the positive orientation. For example, when the velocity direction angle is 90°, the fluid moves northward.

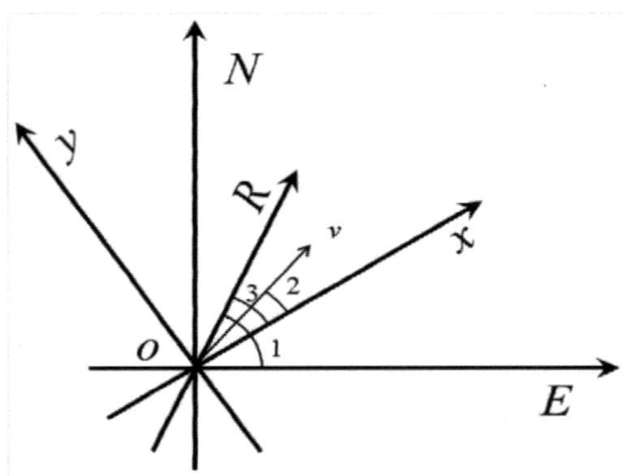

Figure 3. Angle calibration diagram for flow direction.

In Figure 3, R refers to the compass direction axle, the EoN coordinate system represents a geographic coordinate system, N refers to the due north, E refers to the due east, and the xoy coordinate system represents the photograph coordinate taken by a camera. The geographic coordinate system moves based on the earth's movement. No rotation is observed. The camera and electronic compass are installed in the probe, and no relative movement with the probe is set. The xoy coordinate system and R can freely rotate around point o. However, the angle $\angle 3$ between R and x does not change in the rotation. $\angle 3$ is also the target angle in the angle calibration. In addition, due to the problems on computer image data storage, the picture coordinate system is usually a left-handed coordinate system. Backward processing for the y-axis is therefore necessary. The angle between the movement velocity v and the x-axis is $\angle 2$, which is obtained through PTV. $\angle 1$ refers to the angle between the geographic coordinate system and R. The output of the compass in the horizontal level is recorded. Figure 3 shows that the actual direction of the particle velocity $\angle v$ is shown in equation (3):

$$\angle v = \angle 1 - \angle 3 + \angle 2 \quad (3)$$

Where ∠1 refers to the read value of the angular sensor in the horizontal level, ∠2 can be calculated through particle tracking algorithm, and ∠3 is the angle waiting for calibration. Ensuring that the geographic coordinate system overlaps with the image coordinate system is necessary. The horizontal angle reading output from the angle sensor is ∠3 after inputting this angle in the calibration parameters of the software. This phenomenon indicates the completion of the systems angle calibration. After encapsulation and calibration, the whole probe is lowered to a predrilled hole for 3 min before the official test.

4. Software interface

A self-developed software called Aquifer Monitoring Program V1.0 is used to control the self-designed measurement system. The self-developed software adopts the Qt application development framework. Qt supports numerous operating systems and has strong cross-platform capabilities. This framework can run on mainstream platforms, such as Windows, Linux, Android and IOS. The developed software has good performance due to the close combination with C++, which is close to the performance of that developed by Native. Moreover, the Qt development framework contains various mature class libraries with abundant functions. The Qt framework supports open graphic library (OpenCV) and provides great convenience for graphic processing in this research. OpenCV is an open-source cross-platform computer vision library that contains numerous general algorithms for image processing and computer vision [15]. Meanwhile, these platforms are written in C++ language, which reduces the difficulty of any further maintenance.

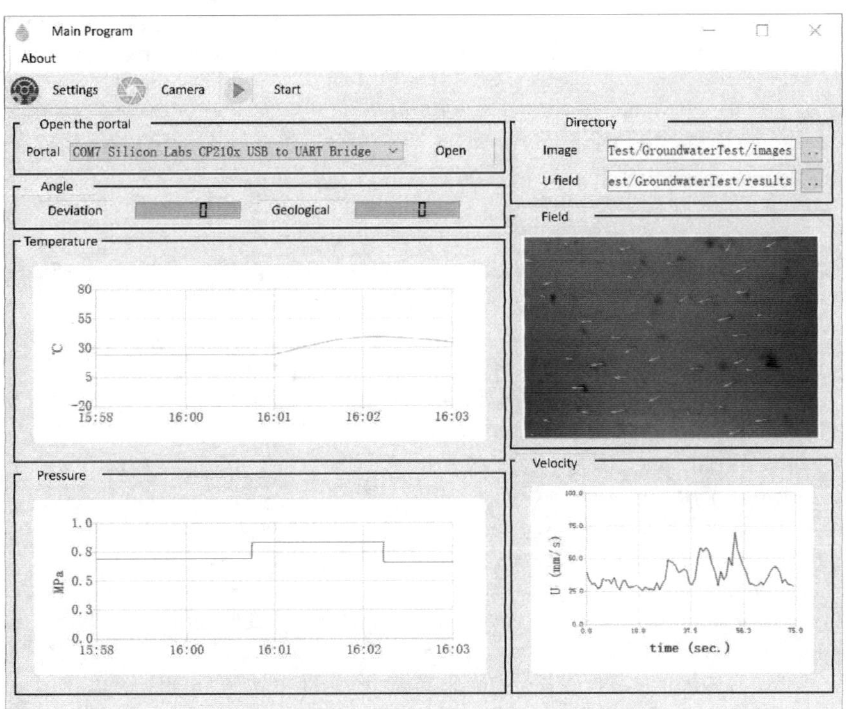

Figure 4. Software interface of Aquifer Monitoring Program V1.0.

As shown in Figure 4, camera-related parameters can be controlled by the parameter configuration algorithm in the upper left corner of the software interface, such as the radius of PTV interrogation domain and the camera framing frequency. On the left side of the main interface, the software reads the real-time temperature and pressure data, as well as the data from the posture sensors through the question-and-answer mode for the combined serial port. The default framing frequency is set as 10 Hz. Data related to temperature and pressure are displayed on the screen in real time. Data from the posture sensor will be combined with particle velocity to display the real-time mainstream geographic orientation and the elevation of the probe relative to the horizontal plane. When the camera is switched on after configuring all the parameters, sampling can be initiated by choosing the directories of image backup and PTV result storage on the right side of the main interface. The image sequence will be saved in the local hard drive for later viewing. The PTV algorithm module is then used to process the image sequence and obtain real-time aquifer flow velocity and direction. In addition, the variation of characteristic velocity (spatial-averaged) with time in the flow field is shown in the lower right corner. After the calculation is completed at the end of sampling, the software will generate hydrological reports based on the collected data and compress the flow field images to save storage space. The flow field of aquifer obtained from PTV matching is shown in the upper right corner. The large and stationary stains in the background are successfully removed. The cross marker is the coordinate of the identified particle points, and the arrows refer to the velocity vector of particles. After placing the probe in the position, the measured characteristic velocity appeared continuous, which is consistent with the actual flow condition of aquifers. In other words, we found that the flow fields across the two consecutive frames are relevant to each other, and no velocity abnormality is observed during the monitoring. Therefore, the present system works effectively in measuring a classical aquifer in coal seam.

5. Conclusion

Around a homemade PTV algorithm, a high-resolution single hole measuring system for the flow of aquifers in coal seam has been established. Owing to the simple structure (good efficiency) and reliable principles (satisfactory accuracy) of PTV algorithm, the system has been demonstrated to be effectively measuring the hydrological condition of the aquifers in coal seam. The collected data can serve as a reliable reference in evaluating the influence of aquifer on the production of coal industry.

Acknowledgement

This work is funded by National Basic Research Program of China (2014CB047100) and National Natural Science Foundation of China (11402190).

References

[1] Zhang X D, Liu X R 2011 A study on the feature and formation mechanism of super karst water burst in deep karst tunnel *Industrial Construction* **41** (4): 79

[2] Zhang L H, Wang S D, Tang Z G, Wu H J, Duan X and Peng S Y 2008 The harm of water to high way and influence of high way on water environment *J. Catastrophology* **23** (1): 76-77

[3] Wang Y J 2006 Hydrogeological exploration method for measuring aquifer flow velocity and its effect *West-china Exploration Engineering* (3): 124-125

[4] Mu Y Q 2007 Technical analysis of single hole steady flow pumping test *Heilongjiang Science and Technology of Water Conservancy* (6): 50

[5] Gu S Q and Song Jisu 2010 Discussion on the test method and quality requirements of deep hole field pumping *Energy Technology and Management* (2): 73-74

[6] Zhao X H, Wang Y M, Pan G F and LI Y J 2011 Development of aquifer flow rate measurement system based on isotope logging technology *Science & Technology Information* (10): 783-784

[7] Lu C P, Shu L C, Yuan L B, Zhang R R, Huang B J and Wang B B 2009 Determination of hydrogeologic parameters of karst auifer based on tracer test *J. Jilin University (Earth Science Edition)* **39** (4): 717-719

[8] Cui Y T, Zhang Y, Jia P, Wang Y, Huang J C, Cui J L and Lai W T 2018 Three-dimensional particle tracking velocimetry algorithm based on tetrahedron vote *Experiments in Fluids* **59** (2): 31

[9] Zhang Y, Wang Y and Jia P 2014 Improving the Delaunay tessellation particle tracking algorithm in the three-dimensional field *Measurement* (49): 1-14

[10] Okamoto K, Hassan Y A and Schmidl W D 1995 New tracking algorithm for particle image velocimetry *Experiments in Fluids* **19** (5): 342-347

[11] Ishikawa M, Murai Y, Wada A, Iguchi M, Okamoto K and Yamamoto F 2000 A novel algorithm for particle tracking velocimetry using the velocity gradient tensor *Experiments in Fluids* (29): 519-531
[12] Song X, Yamamoto F, Iguchi M and Murai Y 1999 New tracking algorithm of PIV and removal of spurious vectors using Delaunay tessellation *Experiments in Fluids* (26): 371-380
[13] Zhang Y, Wang Y and Jia P 2015 A particle tracking velocimetry algorithm based on the Voronoi diagram *Measurement Science and Technology* (7): 075302
[14] Jia P, Wang Y and Zhang Y 2013 Improvement in the independence of relaxation method-based particle tracking velocimetry *Measurement Science and Technology* (24): 055301
[15] Xi Y D, Zhang Y S, Yu Y and Tang Z H 2018 Design of Low Slow Small Target Tracking System Software Based on OpenCV *Industrial Control Computer* **31** (07): 103-104

The method to annotate buildings with images from CCTV cameras with an intersecting field of view

N V Nazarenko[1,2], D E Bekasov[1]

[1]Bauman Moscow State Technical University, 105005 Baumanskaya 2th st., 5, Moscow, Russian Federation

[2]E-mail: nazarenkonv@student.bmstu.ru

Abstract. Video surveillance and tracking systems help to increase the safety of human activity in a given area and improve detecting the geo-location of an observer (a swarm of drones, autopilot) in modern cities. This article proposes the method for determining the full or partial address (annotation) of buildings by using images from several CCTV cameras with an intersecting field of view. As input parameters, the method takes several images from surveillance cameras, their geographical coordinates and directions of the optical axes and the vector map of the area where the cameras are located. The outputs are annotations of buildings for the initial images. The proposed method allows solving the problem of georeferencing the video analytics system and improving the work of the surveillance and object tracking systems. The resulting annotations of buildings provide an opportunity to determine more accurately the situations with overlapping of the tracked objects with buildings and predict the probable paths of the objects. A lost object on one camera can be found by using annotations from other cameras, which field of view is not overlapped by this building. Besides, the proposed ap0070roach is sustainable to full or partial overlapping of buildings and allows solving to solve the problem in the absence of a detailed height map of the terrain and buildings. Also, the article researches the restrictions of the method. It is necessary to find an angle between two cameras, at which the best result can be achieved. That is, the method should return the correct addresses of all buildings visible in the image.

1. Introduction

Video analytics is used in such spheres of modern life activity as retail and service, logistics, and security. In the field of security, video analytics can be used to analyze the video image of a limited area automatically [1]. The main functions are:

- Counting objects number in the control zone;
- Objects identification located in the control zone, according to specific criteria (personnel detection by uniform, etc.);
- Events detection in the specified perimeter (motion detection, objects absence in the perimeter, etc.);
- Tracking the movement of the object;
- Objects geolocation in the given perimeter.

The modern video analytics systems are required to automate the geolocation function. For example, most systems use a manual setup of camera geolocation and its optical axes in the task of georeferencing. Geolocation is required in the tracking task to improve the quality of object tracking. This task can be divided as follows:

- Object paths construction;

- Classification of situations with object loss (overlapping by an obstacle, algorithm error, etc.).

This article proposes a method to annotate buildings using CCTV cameras with an intersecting field of view and researches the restrictions of the method. This method allows classifying situations with objects loss and, in some cases, avoiding it. Besides, it can be used to detect which buildings are in the camera field of view.

2. Related Work

The existing systems available in open sources allow solving the problem only partially:

- VisualSFM [2], Bundler [3], and other similar systems based on the SfM (Structure from Motion) method are systems for 3D reconstruction [4] as a point cloud. There are two types of reconstruction: sparse and dense. Such systems take a set of images, their elements or matches as input data and create a 3D reconstruction of the camera and scene geometry as a result. Such systems are successfully used to create the first version of a scene point cloud further texturing it.
- Mask R-CNN is a neural network based on a convolutional neural network. Mask R-CNN segments objects, even if there are several instances, they have different sizes and partially overlap [5].

As already mentioned above, the existing solutions solve the indicated problem only partially. They either detect and segment the building in the image or determine the relative depth of a particular image point (pixel). Thus, none of the systems determines which buildings are being observed in the image.

3. The Method to Annotate Buildings with Images from CCTV Cameras with an Intersecting Field of View

Two cameras with the intersecting field of view can be used to solve the task of annotating buildings. It does not matter which camera to use (CCTV, mobile, etc.) for the method to work. The main limitation imposed on the camera is the high resolution of frames obtained from the video stream of the camera. The higher the resolution, the more characteristic points can be highlighted. Therefore, the whole method will work more accurately. Then images taken at any time from these two cameras are processed by this method and annotations of the building are returned.

The following scheme can be proposed for the buildings annotation method based on the above-mentioned partial solutions, which consists of the next steps:

1. Building a point cloud of the area by images from CCTV cameras to obtain the relative depth of each image point;
2. Filtering a point cloud to clean the cloud from noise and points do not belong to buildings. These points may belong to cars or people;
3. Clustering of individual buildings from a point cloud to determine cloud points association to specific real buildings;
4. Comparison of clusters and area vector map where the cameras are located to determine the annotations of buildings.

The method takes area images from the video surveillance cameras, these cameras coordinate and the directions of their optical axes, and a vector map as input data. The method result is the annotations of buildings.

3.1. Point cloud

Initially, it is necessary to determine the relative depth of each building seen by the cameras. To do this, it needs to find the relative depth of each point in the image, or other words, to build a point cloud [6] of the area observed by cameras.

The set of photogrammetric methods SfM and MVS (Multi-view Stereo [7]) was chosen to build a point cloud. These methods do not require camera mobility or any additional sensors (lidars, IR sensors). Moreover, they are not affected by the external factors interfering with the method (scattered light, reflective surfaces, etc.). At the same time, there is a requirement for the presence of two or more cameras with an intersecting field of view [8]. The result of this step is a dense point cloud, built on the input images.

Figure 1. An example of a dense area reconstruction from two CCTV cameras with an intersecting field of view.

The SfM method in this set is used to obtain such camera parameters as its location in the point cloud and its projection matrix. This matrix with 3x4 dimension contains a camera rotation angle relative to the origin, focal length, and distortion. Then the MVS method uses the obtained parameters to reconstruct a dense point cloud (figure 1).

3.2. Point cloud filtration

The resulting point cloud must be corrected by removing the points formed by the error of cloud reconstruction methods ("outliers") and points belonging not to buildings, but any other objects from this cloud. For example, to people or cars.

In order to keep in the cloud the points belonging only to buildings, it was initially proposed to use the filtering cluster method. The idea is based on the fact, that clusters, describing buildings, are much larger than clusters, belonging to cars or people. However, it is impossible to determine the exact cluster size that can be considered large enough for a building. It is also quite possible when reconstructing a point cloud that one of the buildings consists of much smaller points number than the others. In this case, the points that need to be saved will still be deleted.

Another option of the filtering method is the edge detectors usage. However, since the outlines of buildings are very different, it is difficult to determine whether this edge belongs to any building.

The most suitable variant for filtering a point cloud is the neural networks usage. Neural networks can find buildings on the images and segment them. So it is possible to save the points in the cloud belonging to the buildings. In addition, this method can also filter out "outliers" resulting from errors in the cloud reconstruction methods.

Mask R-CNN was chosen for the filtering algorithm implementation. It is a neural network architecture based on a convolutional neural network [9]. It can segment objects on images (determines whether an image pixel belongs to this object or not). Mask R-CNN takes any RGB image as input.

The proposed filtering algorithm consists of two main parts. Firstly, it segments objects in the input images from surveillance cameras using the Mask R-CNN. Secondly, it removes points not belonging to buildings from the dense cloud obtained in the first step of the annotation method. As a result, the filtering algorithm returns a dense cloud containing points that belong only to buildings.

To remove extra points from a dense cloud, we can use the projection matrix P of the camera from the SfM algorithm output. This matrix can be used to accurately determine the point location on the image caught by this camera using its coordinates in the cloud (equation (1)):

$$d\begin{pmatrix} u \\ v \\ 1 \end{pmatrix} = (P) * \begin{pmatrix} x \\ y \\ z \\ 1 \end{pmatrix}, \quad (1)$$

where *u, v* denotes coordinates of the pixel in the image; *d* is the point depth with respect to the current camera; *P* denotes a 3x4 projection matrix of the camera; *x, y, z* denotes coordinates of the point in the cloud.

3.3. Buildings clusterisation from point cloud

Filtered point cloud needs to be clustered into separate buildings. This step is necessary to determine the affiliation of points to real buildings.

The k-means [10] algorithm was chosen for the clustering realization. Moreover, the neural network removes noises from the cloud that could affect the quality of the k-means algorithm. As a metric, we use the Euclidean distance [11] in three-dimensional space (equation (2)):

$$d_{ij} = \left[\sum_{i=1}^{3} (p_{il} - p_{jl})^2 \right]^{1/2}, \quad (2)$$

where p_i is the i-th point in the cloud, and *l* is the component (coordinate) x, y or z of this point.

3.4. Matching of clusters and map

The last step of the method to annotate buildings is necessary for matching clusters from the previous step with the original buildings on the map. No algorithm was found in open sources that solves the similar problem. Therefore the proprietary solution was proposed.

Initially, we know the real geographical coordinates of the cameras and their optical axes direction. From the SfM output, we also know the coordinates of the cameras in the point cloud space. So it is possible to obtain the transition matrix *MT* from the geographical coordinates to the point cloud coordinates and the scaling factor of the coordinate system. It suffices with the matrix *MT* to take the central or middle point k_i of each cluster and apply the transformation into the geographic coordinates to this point, given the calculated scaling factor. The resulting coordinates using the reverse geocoding by the vector map determines the current building address. The pixel coordinates of the annotations are then calculated using the projection matrix obtained earlier from the SfM method.

As a geographic coordinate system, this algorithm can use the ECEF (Earth-Centered, Earth-Fixed [12]). The coordinates of the point cloud can be represented by the rotated ENU system (East, North, Up [12]). Then, using the available input data, it will not be difficult to calculate a transition matrix from one system to another.

4. Restrictions of the Method

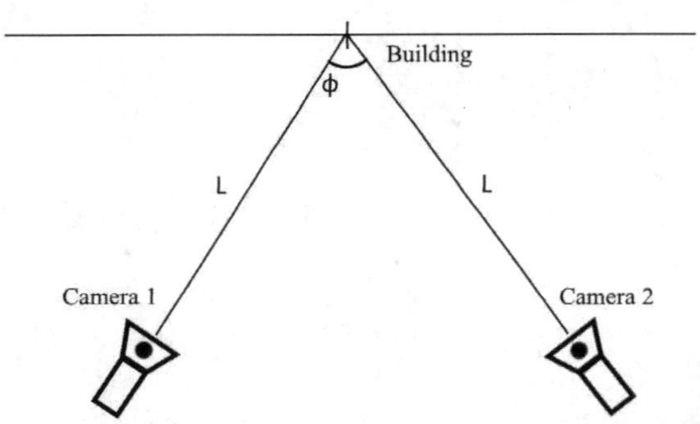

Figure 2. The position of the cameras during the test environment creating.

One of the most critical parameters of the system configuration, which affects the method applicability, is the angle φ between the cameras (figure 2). Therefore, it is necessary to find such an angle, so the result obtained by the method to annotate buildings is the best. The sample for the experiment consists of 120 images of different scenes and buildings. The following characteristics are used as criteria for evaluating the performance of the method:

- The number of points in the constructed cloud. The more points cloud will contain, the more buildings we can identify and annotate. Therefore, the result will be better and more accurate;
- The correctness of the received addresses of buildings.

A generalized graph of the experiment results is presented in figure 3. It can be seen that the highest number of points is reached at an angle of 30 degrees. The angles of 0 and 90 degrees show the worst results. At these angles, the number of constructed points is zero. Therefore, the address of the building in the image cannot be determined at all. In all other cases, the addresses of buildings were detected correctly. Hence the method to annotate buildings worked correctly. Thus, the proposed method can work correctly at angles between two cameras from 15 to 60 degrees. However, the best result achieved at the angle of 30 degrees. These restrictions make it possible to apply this method in real-life tasks when cameras with an intersecting field of view are located on different sides of the same building or on neighbouring buildings placed close to each other.

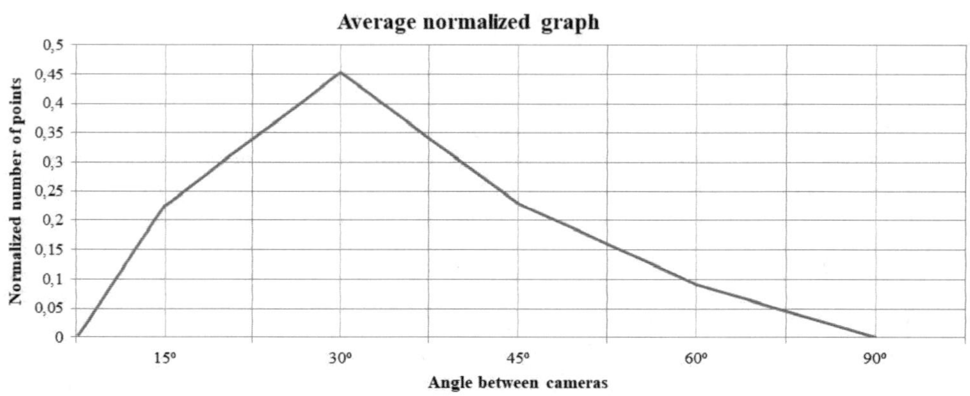

Figure 3. The average normalized graph of the number of points versus the rotation angle.

The ratio of correctly determined addresses to incorrect (accuracy of the method) on the images used above was 0.83. As already noted, mostly incorrect addresses are determined in images made at angles of 0 and 90 degrees.

Figure 4. Example results of the method to annotate buildings.

Examples of the proposed method to annotate buildings are shown in figure 4. Annotation attaches to each building on each initial image and consists of the number, street, city, and country where the building is located. It can be accurately determined when and which building overlaps the tracked object by using these annotations. This information can be further used to detect an overlapped object on the camera with the field of view that is not blocked by this building. Thus, cameras quickly locate the object, and moreover, it is always possible to predict its probable path.

5. Conclusion

This article has proposed the method to annotate buildings with images from CCTV cameras with an intersecting field of view and reviewed all its steps. Also, this article has researched the restrictions of the proposed method at different angles between two cameras to establish the range of applicability. During this study, the accuracy of the method was found.

As a further development of this method, it is possible to modify the neural network so that it would work directly with point clouds, and not with input images, as it has been done in the current implementation. This modification will increase the overall speed of the method.

References

[1] Taranyan A R, Devyatkov V V, Alfimtsev A N 2018 Selective Covariance-based Human Localization, Classification and Tracking in Video Streams from Multiple Cameras *Proc. of the 11th Int. Joint Conf. on Biomedical Engineering Systems and Technol. (BIOINFORMATICS* vol 4*)* (Funchal: SciTePress) pp 81–8 http://dx.doi.org/10.5220/0006538100810088

[2] Wu C 2019 *VisualSFM: A Visual Structure from Motion System* http://ccwu.me/vsfm

[3] Snavely N 2019 *Bundler: Structure from Motion (SfM) for Unordered Image Collections* http://www.cs.cornell.edu/~snavely/bundler

[4] Ostroukh A, Vakhrushev O, Maikov K, Kolbasin A 2019 Combined identification method by reconstruction and analysis of face 3D structure *ARPN J. of Engineering and Appl. Sci.* **14** pp 1385-1388

[5] He K, Gkioxari G, Dollar P, Girshick R 2017 Mask R-CNN *IEEE Int. Conf. on Computer Vision (ICCV)* (Venice: IEEE) pp 2980-2988 http://dx.doi.org/10.1109/iccv.2017.322

[6] Kozov A V, Volosatova T M, Vukolov A Y 2018 Structural Obstacle Recognition Method and Its Application in Elevated Terrain Objects Search *Int. Russian Automation Conf. (RusAuto Con)* (Sochi: IEEE) pp 1-5 http://dx.doi.org/10.1109/ rusautocon.2018.8501765

[7] Furukawa Y, Hernández C 2015 *Multi-View Stereo: A Tutorial (*Now Publishers) p 166

[8] Kuznetsov A O, Gorevoy A V, Machikhin A S 2019 Image rectification for prism-based stereoscopic optical systems *Computer Vision and Image Understanding* **182** pp 30-7

[9] Anishchenko L 2018 Machine learning in video surveillance for fall detection *Ural Symp. on Biomedical Engineering, Radioelectronics and Information Technol. (USBEREIT)* (Yekaterinburg: IEEE) pp 99-102

[10] Piech C 2019 *K Means* https://stanford.edu/~cpiech/cs221/handouts/kmeans.html

[11] Barret P 2019 *Euclidian Distance* https://www.pbarrett.net/techpapers/euclid.pdf

[12] Leick A, Rapoport L, Tatarnikov D 2015 *GPS Satellite Surveying* (John Wiley & Sons, Inc) p 807

Journal of Physics: Conference Series **1419** (2019) 012020 doi:10.1088/1742-6596/1419/1/012020

Method of determining vehicle speed according to video stream data

I Murashov[1,2], Y Stroganov[1]

[1]Bauman Moscow State Technical University, 105005 Baumanskaya 2th st., 5, Moscow, Russian Federation

[2]E-mail: murashovid@student.bmstu.ru

Abstract. This article presents a method of determining the speed of a dynamic object from the data of a video stream. The determination of speed occurs based on the analysis of repeated objects of the environment, the distance between which varies according to a previously known law, such as lamp posts, milestones or bump stops on roads, intermittent road marking lines, emergency exits in tunnels. The method of determining vehicle speed includes three steps: pre-processing of the frame, searching for a pattern in the image, calculating the average vehicle speed obtained on the basis of the number of repeated objects, taking into account the law of their repetition and the distance between them for some fixed time. The advantage of the proposed method is that there is no need to install any additional sensors on the dynamic object except video camera. The development of an additional method for calculating speed has a significant application value for the complexes of control of autonomous traffic, as the existing methods of determining speed have their disadvantages (not high enough accuracy, the requirement for the presence of additional equipment). This algorithm was studied on regularly recurring objects (lampposts) under different lighting conditions (day, night), the distance between repeated objects and different weather conditions. It was also compared with other methods of determining the speed of the data from the video stream, the results of which the proposed method has a significant advantage in accuracy as compared with analogs in the absence of interference with video.

1. Introduction

The creation of autonomous vehicles is one of the leading areas of the automotive industry. These vehicles have many significant advantages over conventional vehicles, one of which is road safety.

An autonomous vehicle [1] is a vehicle equipped with an automatic control system that can be driven without human intervention. Autopilot [2] is a device or hardware and software system, leading a vehicle along a specific, given to its trajectory. Autopilots are most often used to control aircraft (due to the fact that the flight often takes place in a space that does not contain a large number of obstacles), as well as to drive vehicles. Modern autopilot allows automating all stages of vehicle movement. As main tasks are considered such tasks as object detection, determination of the distance to objects [3], navigation, speed measurement, creation of 3D map.

An autonomous car has different types of sensors. The main ones are [4]:

- Cameras. They "look" in different directions. This "picture" is analyzed by a computer program to identify obstacles, people, cars, and other objects on the roadway, its borders, markings, and even signs.
- Stereo cameras. They determine the distance to objects.

- Radar. It determines the distance to objects with the help of radio waves, as on warships or airplanes.
- Lidar. Laser-based, all-around laser sensor that provides a three-dimensional map of the space around the car.
- Sensors. GPS/GLONASS receivers, inertial meters, and sensors that show the speed at which the wheels rotate, where they turn.

The more sensors are installed on the vehicle, and the more data processing methods these sensors provide, the more reliable the on-board computer can drive an autonomous vehicle. One of the essential characteristics of vehicle movement is speed. Today, the vehicle's speed is measured in several ways:

- Direct acquisition of the instantaneous speed of the car from its speedometer
- Receiving car speed values from GPS receiver data
- Speed measurements with sonar and radar mounted on the vehicle are also possible

As each of these methods has some error, and devices that can fail are used to determine speed, redundant systems should be used to obtain a more reliable result. Therefore, the development of an additional method for calculating speed will have a significant application value for autonomous traffic control systems.

In the article proposed the method of determining the speed of a dynamic object using video stream, and studied the accuracy of the method under different conditions.

2. Related works

At present, the following main methods of speed determination based on video stream data can be identified:

- Speed Challenge [5] is a system based on optical flow [6] analysis and convolution neural networks [7]. This method is based on the preliminary preparation of data by which the system will then determine the speed in real-time. Initially, created pairs of images from the video stream, which follow each other. For a pair of frames, each of which knows the time between them, as well as the actual speed of the car, measured by a calibrated speedometer, is determined by the optical flow. Then the data (optical flow, time difference, speed difference) is fed into the neural network for further training. After training, the neural network is used for further determination of the speed on the real video stream.
- The algorithm developed by Nicolo Valigi [8] is an algorithm that uses only optical flux to determine the speed. The input comes from a video recorder, as well as parameters that include the focal length of the camera and the height of the camera above ground.
- The method [9] based on the determination of the distance from the images obtained with the help of stereo vision. It is one of the variants of determining the distance to the required object and further calculating the speed relative to this object. The method involves the use of two identical cameras with a certain fixed distance between them, which is called the base. The formula for calculating the distance to the object has the following form:

$$r_n = \frac{dH}{tg\alpha(x_1 - x_2)} \qquad (1)$$

where d – the distance between cameras; H – horizontal image resolution; α – camera angle; x_1 and x_2 – coordinates of the point to which the distance is determined in the coordinate system of the first and second cameras, respectively.

- A method for measuring the distance to an object from the known width (height) of the object in the image and further calculating the speed relative to the object. If the actual dimensions of the object (its height or width) are known, the distance to it can be determined:

$$r_n = \frac{WF}{P} \qquad (2)$$

where W – the actual height or width of the object, F – focal length, P – the height or width of the object in pixels.

3. Method for determining vehicle speed according to video stream data

To determine the speed of a dynamic object, a method of processing data from the video stream, consisting of three stages, was proposed. At the first stage, there is preliminary processing of the video. At the second stage, for each successive incoming frame, an attempt is made to search for an object, which is used to measure the average speed. The third step is to calculate the speed of the vehicle over the distance between the detection of the previous object and the detection of the next one.

Input data are: images from video cameras located on dynamic objects, an image of the object on which the measurement will be made, the distance between objects, as well as the number of frames per second. Objects must be at a fixed distance from each other. The output data is a video file with marked speed.

3.1. Pre-processing of the frame

Before recognizing an object, the frame and template should be processed first to optimize the operating time. First of all, it is necessary to define the sector on the image, in which the template will be searched. In this case, was chosen a sector that occupies the right side of the frame by 15 percent, as there is no need to consider the entire image.

To reduce the number of calculations, there is a need to convert the color images of the target and reference images to grayscale. The following formula is used to calculate the brightness of each pixel in shades of grey [10]:

$$Y = 0.299R + 0.587G + 0.114B \quad (3)$$

where R - is the intensity of red, G - is the intensity of green, B - the intensity of blue.

After the image and template are pre-processed, a template is searched for in the specified sector.

3.2. Object detection

The process of recognizing an object by its template, in relation to which the speed measurement will be performed, is to search for a template in the source image [11].

At each iteration step, one of the formulas is applied to the selected area in the target image to obtain the matching coefficient each time. If you get a larger coefficient value than in the previous calculation, this value is saved and considered maximal at this stage. The result of the loop will be the maximum matching rate of the pattern and the target image. If the obtained coincidence rate is above a specific threshold value, it can be considered that a reference image is present on this target image. The disadvantage of this search method in the image is the impossibility of finding the object if its size is very different from the image of the template. As a result, it is necessary to improve this algorithm by applying scaling to the original image. The difference is that the target image is scaled while searching for a template.

At the beginning of the template search method on the image, the scale of the target image is changed depending on the specified parameters. Once the scale has been changed, it should be made a check to ensure that the size of the scaled target image is not smaller than the size of the reference image (template).

3.3. Speed determination

After the object is found, the average speed that the car has kept at over the distance between the current and previous objects should be calculated.

If the distance between objects is known, the speed can be calculated by the formula of average speed [12]:

$$\overline{v} = \frac{S}{t} \quad (4)$$

where $\bar{v}$ is average speed, S is the distance between the objects relative to which measurements are made, t – time.

4. Accuracy of the method

The use of this method may lead to errors in speed measurement. The following characteristic was selected as the accuracy estimation criterion:

ΔD – speed calculation error – the difference between the speed value calculated by the method implemented in this work and the actual speedometer values.

This characteristic may depend on the following parameters:

- driving speeds
- weather conditions
- the distance between the pillars
- time of day

As part of the work, a virtual environment was created, reproducing the road and simulating weather and time conditions with different characteristics of the distance between the objects relative to which the measurement was made (25, 35, 65 meters), weather conditions (clear weather, rain), time of day (day, night). Below is example of the results.

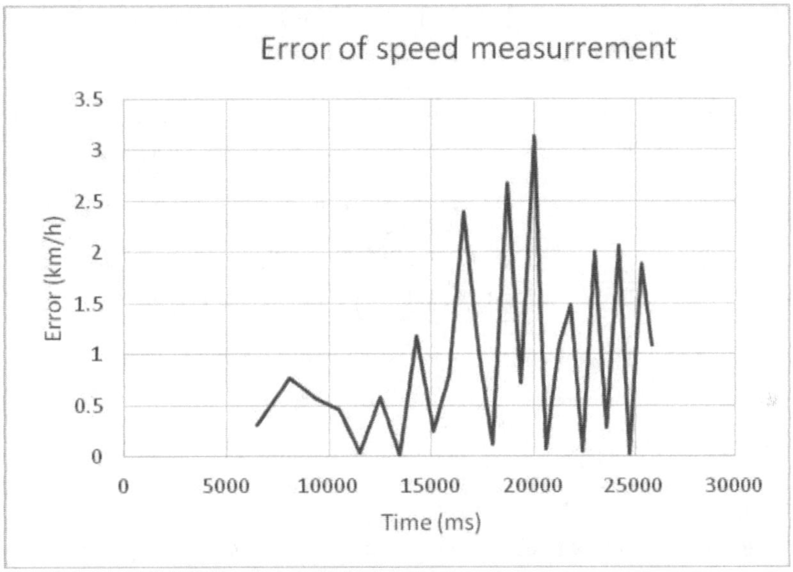

Figure 1. Error of measurements in daytime conditions without precipitation at a distance of 25 meters between lamp posts

It was also recorded several videos from car DVRs to study the application of the method in real conditions.

According to the results of the study, the following conclusions can be drawn:

- The error does not depend on the speed of the vehicle and the distance between the objects (lamp posts)
- The error does not depend on the time of day in conditions without unnecessary objects on the video
- Excess objects significantly impair the accuracy of the method

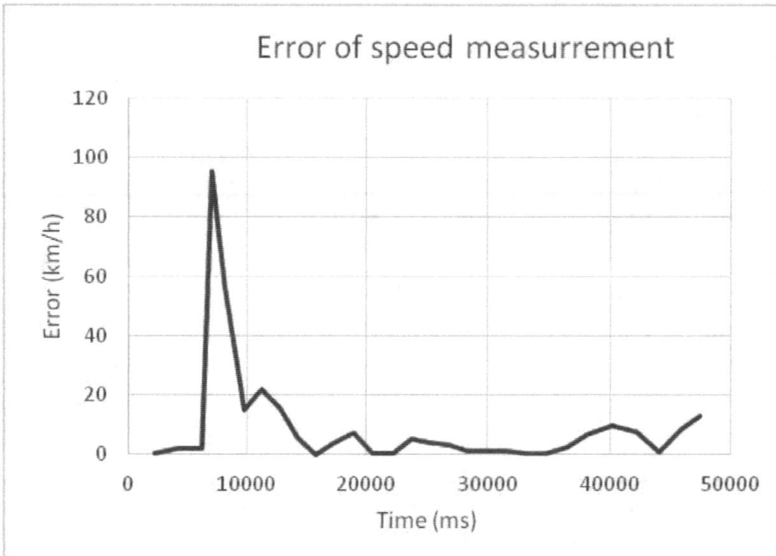

Figure 2. Error of speed measurement in the section of the Northeast chord of Moscow at a distance of 35 meters between lamp posts

5. Comparison of the proposed method and analogs

A comparison was made of the implemented method with existing analogues. As the evaluation criterion, the mean square error was chosen, which is calculated by the following formula:

$$MSE = \frac{1}{n}\sum_{1}^{n}(y_i - \breve{y}_i)^2 , \tag{5}$$

where $\breve{y}_i$ is the actual speed, y_i is the speed calculated by one of the methods.

Table 1. Table with comparative results of methods error

Method	MSE
SpeedChallenge	20.16
Nikolo Valigi's algorithm	27.35
Implemented method, results obtained in the simulated system in the absence of precipitation	2.32
Implemented method, results obtained in the simulated system in rain conditions	35766.93
Implemented method, results in the simulated system in rain conditions without taking into account the peak values	2.63
Implemented method, real data	475.68
Implemented method, real data without taking into account peak values	4.63

Peak values in the accuracy of the proposed method arise because of objects similar in their parameters to the objects, in relation to which measurements are made, entering the frame. This can be avoided by improving the method of object detection that would use neural networks.

Methods based on optical flow give less accurate results (20.16, 27.35), in contrast, the proposed method due to the work with objects along the roads has higher accuracy. However, methods based on optical flow are more universal, as they do not require the presence of objects on the road for which measurements are made.

Journal of Physics: Conference Series **1419** (2019) 012020 doi:10.1088/1742-6596/1419/1/012020

According to the results of the comparison of the proposed method and the existing analogues, it can be concluded that the proposed method has higher accuracy than SpeedChallenge and the Nicolo Valigi's algorithm in the absence of unnecessary volumes.

6. Conclusion

In this article the method of determining the speed of the car by the data of video stream was proposed, and the variant of its realization was described. The work of the proposed method under different weather conditions was studied. Based on the results of the accuracy study, it can be concluded that the method does not depend on the speed of the vehicle, time of day and distance between the objects to be measured. But excess objects significantly impair the accuracy of the method. Also comparison with existing analogues was made. Based on the results of comparison methods, it can be concluded that the proposed method has higher accuracy than existing analogues, if the video stream will not contain objects similar to those for which speed measurements are made.

As a further development of this method, it is possible to improve the algorithm of recognition of objects, in relation to which the speed measurement takes place, which will improve the accuracy of the method.

References

[1] Avadon P 2018 Autonomous vehicle for begginers Available from: https://habr.com/ru/post/431758/

[2] Sychev I 2016 History of autopilots Available from: https://habr.com/ru/post/371603/

[3] Loktev D A, Loktev A A 2016 Estimation of measurement of distance to the object by analyzing the blur of its image series *2016 Int. Siberian Conf. on Control and Communications (SIBCON)* (Moscow: IEEE) pp 1–6 Available from: http://dx.doi.org/10.1109/sibcon.2016.7491683

[4] Vdovin D S, Khrenov I O 2019 Systems of the self-driving vehicle *IOP Conf. Series: Materials Science and Engineering* **534** 012016 Available from: http://dx.doi.org/10.1088/1757-899x/534/1/012016

[5] Mitchell J 2017 Autonomous Vehicle Speed Estimation from dashboard cam https://chatbotslife.com/autonomous-vehicle-speed-estimation-from-dashboard-cam-ca96c24120e4

[6] Lychkov I I, Alfimtsev A N, Sakulin S A 2018 Tracking of Moving Objects With Regeneration of Object Feature Points 2018 *Global Smart Industry Conf. (GloSIC)* (Chelyabinsk: IEEE) pp 1–6 Available from: http://dx.doi.org/10.1109/glosic.2018.8570061

[7] Anishchenko L 2018 Machine learning in video surveillance for fall detection *2018 Ural Symp. on Biomedical Engineering, Radioelectronics and Information Technology (USBEREIT)* (Yekaterinburg: IEEE) pp 99–102 Available from: http://dx.doi.org/10.1109/usbereit.2018.8384560

[8] Valigi N 2016 Car speed estimation from a windshield camera Available from: https://nicolovaligi.com/car-speed-estimation-windshield-camera.html

[9] Loktev D 2013 Estimation of measurement of distance to the object using complex system of videomonitoring Available from: http://engjournal.ru/articles/996/996.pdf

[10] Gomes J, Velho L 1997 *Image Processing for Computer Graphics* (New York: Springer) pp 94–95 Available from: http://dx.doi.org/10.1007/978-1-4757-2745-6

[11] Devyatkov V V, Alfimtsev A N, Taranyan A R. 2018 Multicamera Human Re-Identification based on Covariance Descriptor *Pattern Recognition and Image Analysis* **28** p 232–242 Available from: http://dx.doi.org/10.1134/s1054661818020025

[12] Naumov VN, Mashkov KY, Byakov KE 2019 Autonomous tracked vehicles effectiveness estimation *IOP Conf. Series: Materials Science and Engineering* **534** 012006

Design of component cooling water system for Comprehensive Research Facilities in Support of CFETR

Weibao Li[1,2], Jiansheng Hu[1,2], Lei Yang[1,2], Bin Guo[1,4], Lili Zhu[1], Jinxuan Zhou[1,2], Zhe Liu[1,2], Lidong Yao[1,2] and Qianglin Xu[3]

[1]Institute of Plasma Physics, Chinese Academy of Sciences, Hefei, China
[2]University of Science and Technology of China, Hefei, China
[3]Anhui Jianzhu University, Hefei, China

[4]E-mail: guobin@ipp.ac.cn

Abstract. Comprehensive Research Facilities in Support of CFETR is aimed to demonstrate a comprehensive research platform with the highest parameters and the most complete functions in the field of magnetic confinement. The cooling water system is an integral part of a comprehensive research platform. The cooling water system includes component cooling water system (CCWS), chilled water system (CHWS) and heat removal system (HRS). The component cooling water system is mainly responsible for transferring the heat load generated by each subsystem during the operation, preventing the system from overheating. This paper presents the thermal hydraulic design of CCWS for Comprehensive Research Facilities in Support of CFETR. Through the development of the thermal hydraulic model of the CCWS, the AFT Fathom code is used to calculate the key thermal hydraulic parameters and predict the thermal hydraulic behaviour. Based on the results of thermal hydraulic analysis, the key thermal hydraulic parameters have been plotted to evaluate the capability of CCWS, such as the required pump head, inlet pressure, flow rate, inlet and outlet temperature and pressure drop of the customer. This paper will provide design guidance for the construction of component cooling water system (CCWS) for Comprehensive Research Facilities in Support of CFETR.

1. Introduction

Comprehensive Research Facilities in Support of CFETR is a comprehensive research platform which serves for Chinese fusion reactor science and technology research. Multiple research systems are contained, such as include power supply system (PWS), cryogenic system (CGS), lower hybrid current driven (LHCD), Electron-cyclotron-resonance heating (ECRH), host vacuum vessel system (HVVS), vacuum vessel system (VVS) and integration of several cooling system (ISCS). These research systems are also called subsystem of Comprehensive Research Facilities in Support of CFETR in the following. During the subsystems operation, more than 56.6 MW of head load will be generated which need to be removed. Therefore, the cooling water system is necessary to remove the thermal power generated by the subsystems, the component cooling water system (CCWS), the chilled water system (CHWS) and the heat removal system (HRS) are include in the cooling system. As the most important and complex part of the cooling water system, the CCWS is discussed in this paper.

The component cooling water system (CCWS) has the function to transfer the thermal power to the heat rejection system (HRS) with required flow rate and temperature, which can ensure the stable operation of the subsystems. Hence, it is necessary to design the component cooling water system (CCWS) to supply cooling water with certain pressure, temperature and flow for subsystems.

Presently, many researchers have been made on the thermal hydraulic analysis of the component cooling performance itself. In References [1-4], water-cooled divertor which are focus on the thermal performance of the divertor itself have been analyzed. However the external cooling water system is usually be neglected, such as CCWS. So, this paper presents a method for the CCWS process design and developing the thermal-hydraulic model to predict the behavior of the system.

In this paper, the CCWS description of the requirement and process design are presented firstly. Secondly, thermal-hydraulic analysis theory and pressure loss model is discussed. Thirdly, thermal hydraulic model analysis is introduced to determine pump-head and other thermal-hydraulic parameters. The discussion and conclusion are showed in the end.

2. CCWS description of the requirement and process design

Numerous clients are included in each subsystem thereby the large number of clients are labeled by the uniform rule. Letter S, L, and N are applied to description of the clients, respectively represent the system serial number, cooling loop serial number, and clients serial number. For example, S1-LHCD-L1-N1 represents the first client is located in the cooing loop 1 of LHCD system which is also named first system [5].

According to requirement of each subsystem, as shown in the Table 1, it is obvious that the overall heat load is 56.6MW, and the total cooling water mass flow is about 2176.43 kg/s. In addition, the requirement of inlet pressure, temperature for some clients is different, as well as the cooling water conductivity. Also, the 5 independent cooling loops are requested on the GCS system. Taking these factors into account, the clients with similar requirement are designed in a same cooling loop, thereby 32 clients are designed to be 17 cooling loops. However, the each cooling loop has the same function so that the schematic of all cooling loops can be illustrated in Fig. 1.

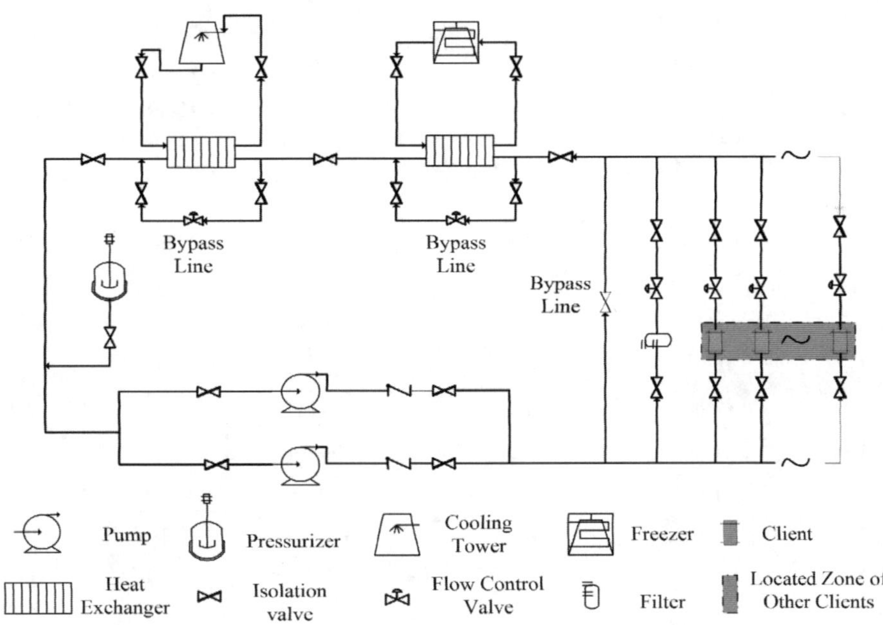

Figure 1. Schematic of cooling loop.

To satisfy the requirement of different clients, some branches in parallel are designed to feed each client. The clients put in the green zone can be replaced to other clients. Flow control valve are installed in the downstream of the clients so that the mass flow and inlet pressure of the clients can be controlled at the specific level. Isolation valve is located at the inlet and outlet of each client to isolate the clients when the clients are in maintenance. Bypass line of the clients is designed to protect the

Journal of Physics: Conference Series **1419** (2019) 012021 doi:10.1088/1742-6596/1419/1/012021

clients form exceeding the requirement. Similarly, the bypass line of the heat exchanger has the same function to satisfy the temperature. To avoid the conductivity is lower than the requirement of clients, filter is placed in parallel of branches. Pressurizer is at the main returned head, is also at the upstream of the pump to ensure that the coolant does not vaporize, which will guarantee the system safely. The pumps are located in parallel to decrease chance of system breakdown [6].

Table 1. Equipment water cooling system client.

Named	Flow rate (kg/s)	Max pressure loss (Mpa)	Max heat load (kW)	Conductivity (μs/cm)	Inlet temperature (°C)
S1-LHCD-L1-N1	66.67	0.6	44500	< 0.3	≤ 28
S1-LHCD-L1-N2	16.67	0.8	195	< 0.3	≤ 28
S1-LHCD-L1-N3	5.56	0.8	100	< 0.3	≤ 28
S1-LHCD-L2-N4	27.78	1	210	< 0.3	≤ 28
S1-LHCD-L2-N5	22.22	1	500	< 0.3	≤ 28
S1-LHCD-L3-N6	13.89	1.8	220	< 0.3	≤ 28
S2-ECRH-L1-N1	41.67	0.8	150	< 0.1	≤ 30
S2-ECRH-L1-N2	250	0.6	12000	< 0.3	≤ 30
S2-ECRH-L1-N3	5.56	0.6	120	< 0.3	≤ 15
S3-VVS-L1-N1	8.33	0.5	400	≤ 1	≤ 20
S4-CGS-L1-N1	277.78	0.55	3	< 20	≤ 33
S4-CGS-L2-N2	333.33	0.55	0.25	< 20	≤ 33
S4-CGS-L3-N3	111.11	0.55	1	< 20	≤ 33
S4-CGS-L4-N4	55.56	0.55	0.5	< 20	≤ 33
S4-CGS-L5-N5	27.78	0.55	0.2	< 20	≤ 33
S5-PWS-L1-N1	27.78	0.6	500	< 1	≤ 32
S5-PWS-L1-N2	111.11	0.6	2500	< 1	≤ 32
S5-PWS-L2-N1	1.39	1	14	< 1	≤ 32
S5-PWS-L2-N2	41.67	0.6	1400	< 1	≤ 32
S5-PWS-L2-N3	77.78	0.6	4300	< 1	≤ 32
S5-PWS-L2-N4	88.89	0.5	7200	< 1	≤ 32
S5-PWS-L2-N5	133.33	0.5	6000	< 1	≤ 32
S5-PWS-L3-N1	138.89	0.5	4450	< 10	≤ 32
S5-PWS-L3-N2	55.56	0.6	1500	< 10	≤ 32
S5-PWS-L3-N3	8.33	1	100	< 10	≤ 32
S6-ISCS-L1-N1	5.56	0.5	300	< 1	≤ 35
S6-ISCS-L1-N2	27.78	0.7	2000	< 1	≤ 35
S6-ISCS-L1-N3	27.78	0.5	2000	< 1	≤ 35
S6-ISCS-L1-N4	27.78	0.5	1000	< 1	≤ 35
S6-ISCS-L1-N5	27.78	0.6	1000	< 1	≤ 35
S6-ISCS-L1-N6	83.33	0.5	3000	< 1	≤ 35
S6-HVVS-L1-N1	27.78	0.5	1000	< 1	≤ 32

3. Thermal hydraulic analysis theory

During the progress of flow with incompressible fluid, the fluid is required to lose mechanical energy to compensate for the energy is consumed by the friction loss between the fluid and the solid wall, and the fluid when the local flow boundary changes abruptly. In the engineering design calculation, the energy loss is divided into two categories: the resistance loss ΔP along the path and the local resistance loss ΔP_s. Through the calculation of the pressure loss, system sizing parameters for the pump, pipe, control valve and other fittings can be determined [7,8].

The pressure loss of the pipeline can be calculated by the Darcy formula, referring to (1):

$$\Delta P = \rho g \left(f \frac{L}{D} \frac{v^2}{2g} \right) \tag{1}$$

Where f is the pressure loss coefficient, ρ is the fluid density, g is the gravitational acceleration, L is the pipe length, D is the pipe diameter, and v is the fluid velocity. If the tube is laminar flow (Re < 2000), f can be calculated by equation (2).

$$f = \frac{64}{\mathrm{Re}} \tag{2}$$

If it is turbulent (Re > 4000), it can be calculated by Colebrook equation (3), where ε is the absolute surface roughness of the pipe:

$$\frac{1}{\sqrt{f}} = -2\log\left[\frac{\varepsilon}{3.7D} + \frac{2.51}{\mathrm{Re}\sqrt{f}}\right] \tag{3}$$

Local resistance loss of components in the model is calculated by local pressure loss formula (4), ξ is the local drag coefficient:

$$\Delta P_s = \xi \frac{\rho v^2}{2} \tag{4}$$

The conservation of pressure between the two components in the cooling water loop is described by the Bernoulli equation (5):

$$P_1 + \frac{\rho {v_1}^2}{2} + \rho g Z_1 = P_2 + \frac{\rho {v_2}^2}{2} + \rho g Z_2 + \Delta P_s \tag{5}$$

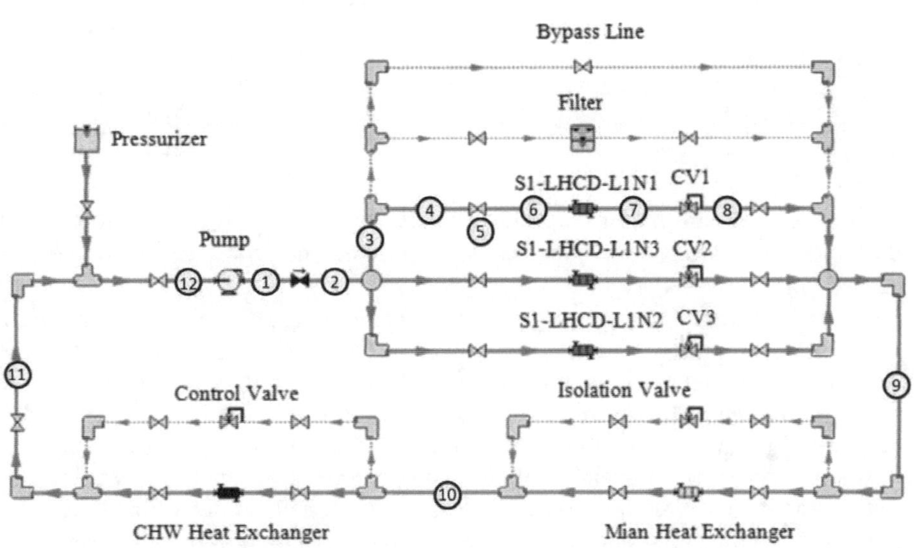

Figure 2. Thermal hydraulic analysis model of the LHCD cooling loop 1 (circle with yellow present pressure reference point).

4. Thermal hydraulic simulation analysis

For steady-state operating conditions, the AFT Fathom code is used to determine the fitting parameters for each cooling loop, including pressure distribution throughout the piping system, flow rate and cooling water temperature changes [9]. Due to all cooling loop model being too large to shows here, only the cooling loop 1 is given in Fig. 2 as an example. In the model, clients are modeled as heat exchanger in red to reject heat power to the coolant flowing through it. Valve in green is control valve at the downstream of the client to keep the flow and pressure. Once the component is stay in maintenance, valve in gray is isolation valve to isolate the component. Heat exchanger in Cambridge blue represents main heat exchanger located at the main returned head Heat exchanger in dark blue represents CHW heat exchanger which can make the coolant temperature lower [10].

4.1. Simulation results

The output file calculated by Fathom code is mainly showing pump head, the inlet pressure of the client, the inlet and outlet temperature.

a) Pump. Pump sizing is determined by the head result of the system provided by the Fathom code.

Table 2. Pump minimum requirement.

Subsystem	name	Mass Flow (kg/s)	dH (meter)	Total power (KW)	NPSHa (meter)
LHCD	L1	88.9	99.99	87.1	11.74
	L2	50.00	119.95	58.78	11.74
	L3	13.90	193.10	26.28	11.18
ECRH	L1	41.50	96.02	39.21	11.05
	L2	249.90	75.95	186.09	10.95
	L3	5.55	70.30	3.83	11.62
VVS	L1	8.30	55.00	4.48	11.94
CGS	L1	276.77	67.51	183.11	11.95
	L2	332.12	70.76	230.31	11.68
	L3	110.71	72.48	78.63	11.67
	L4	55.35	69.49	37.70	11.68
	L5	27.68	71.48	19.39	11.67
PWS	L1	138.9	75	102.1	11.73
	L2	343.1	130	437.1	11.68
	L3	202.8	120	238.5	11.72
ISCS	L4	199.30	90	175.8	11.72
HVVS	L5	27.68	60	16.27	11.77

Table 2 is mainly showing pump minimum requirement, 20% margin on the minimum required pump head should be included when specifying these pumps. Therefore, the required pump parameters are provided as Table 3.

b) Pressure. Fig. 3 shows the scatter plot of client inlet pressure and client demand pressure. In this figure, the vertical axis represents the pressure value, and the horizontal axis represents the number of clients. The blue points represent simulation result and the red points represent the minimum requirement of the clients. Clearly, inlet pressure of the clients simulated is much larger than the minimum requirement. And the maximum pressure is 19.85 bar, the minimum pressure is 6.49 bar.

Through the simulation, the branch in parallel with pressure drop has been calculated as the critical path to decide system total pressure drop. Once the critical path is determined, 12 points taken as shown in the Fig. 2 are the pressure reference points. Fig. 4, Fig. 5, and Fig. 6 present the critical path pressure curves for the seven cooling subsystems. In these figures, the vertical axis represents the pressure value, and the horizontal axis represents reference points. Obviously, maximum pressure drop happens between the inlet and outlet of clients, and minimum pressure is larger than the NHPs of the pump shown as Table 3.

Table 3. Pump requirement.

Subsystem	name	Mass Flow (kg/s)	dH (meter)	Total power (KW)	NPSHa (meter)
LHCD	L1	88.9	124.99	108.88	11.74
	L2	50.00	149.94	73.48	11.74
	L3	13.90	241.38	32.85	11.18
ECRH	L1	41.50	120.03	49.01	11.05
	L2	249.90	94.94	232.62	10.95
	L3	5.55	87.88	4.79	11.62
VVS	L1	8.30	68.75	5.6	11.94
CGS	L1	276.77	84.39	228.89	11.95
	L2	332.12	88.45	287.89	11.68
	L3	110.71	90.6	98.29	11.67
	L4	55.35	86.86	47.13	11.68
	L5	27.68	89.35	24.24	11.67
PWS	L1	276.80	93.75	102.1	11.73
	L2	311.10	162.5	437.1	11.68
	L3	257.4	150	238.5	11.72
ISCS	L1	199.30	112.5	175.8	11.72
HVVS	L1	27.68	75	20.3375	11.77

c) Temperature. Figure 7 shows the scatter plot of client inlet temperature, outlet temperature, and client requirement temperature. In this figure, the vertical axis represents the temperature value, and the horizontal axis represents the number of clients. The blue points and red points respectively represent client inlet and outlet calculation temperature, and the pink points represent the requirement temperature of the clients. Clearly, inlet temperature of the client is lower than the required maximum temperature, which has satisfied the requirement of the clients. Also, the maximum outlet temperature simulated is 51.46 °C, the minimum outlet temperature simulated is 28.64 °C.

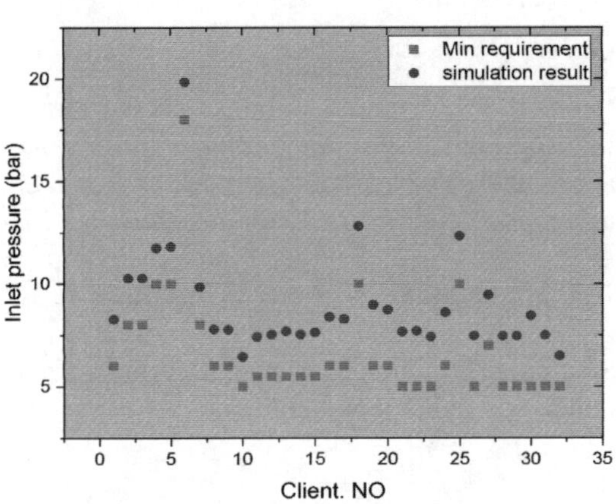

Figure 3. Inlet pressure of clients.

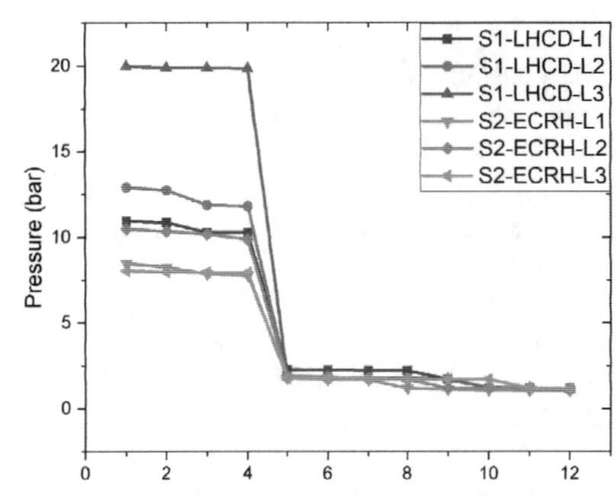

Figure 4. The critical path pressure distribution of S1 and S2.

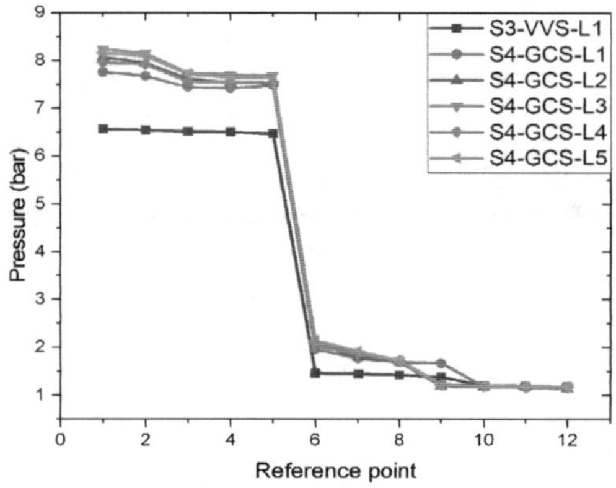

Figure 5. The critical path pressure distribution of S3 and S4.

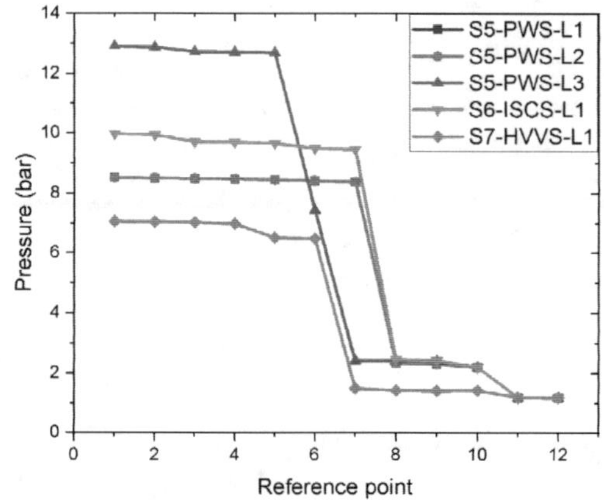

Figure 6. The critical path pressure distribution of S5, S6 and S7.

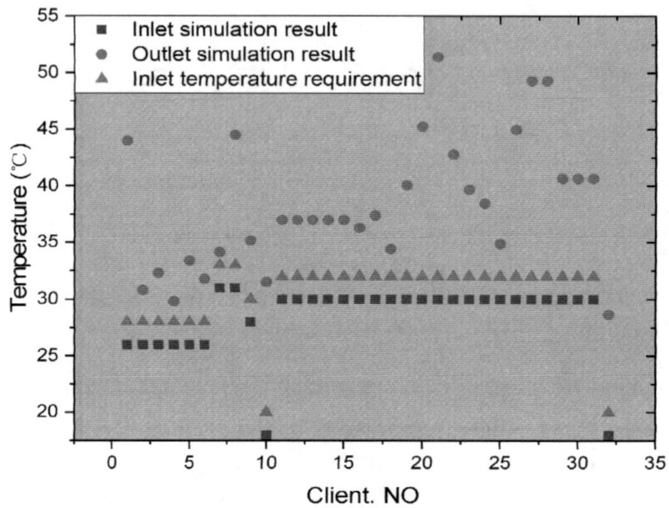

Figure 7. Temperature distribution of inlet and outlet of different clients.

5. Discussion

Comparing Fig. 4, Fig. 5, Fig. 6, it is clear that different clients required different pressure loss will cause the different pressure distribution through the flow path, therefore multiply cooling loop is necessary to be designed to satisfy different client requirement. The size of pump selected in the table 3 can meet all system head requirement. Also, the maximum pressure loss of the client S1-LHCD-L3N6 is 18 bar.

The Fig. 3 indicated that the inlet pressure can meet the requirement of the clients, and the largest pressure is 19.85 bar at the inlet of the S1-LHCD- L3N6. Meanwhile, the Fig. 7 indicated that the temperature distribution of the client inlet can also satisfy the requirement of the clients, and the highest temperature 51.46 °C is distributed at the branch pipe of S5-PWS-L2-N2.

6. Conclusion

The AFT fathom code is used to design the cooling water system for Comprehensive Research Facilities in Support of CFETR. The thermal-hydraulic analysis results indicate that cooling water system designed can satisfy the thermal-hydraulic parameter of all clients, such as pressure, temperature, which can provide guarantee for safety operation. The design methodology introduced in this paper can be a good reference for design the cooling water system in the future, thereby providing a solid guarantee for the stable operation of the fusion device.

Acknowledgments
The work has been supported by the National Key Research and Development Program of China No. 2017YFE0300500, 2017YFE0300504.

References

[1] Shijun Q, Yang Z, Tiejun X, Lei C, Kun L, Zhuang X, Damao Y, Shengming W, Yuntao S 2017 RAMI analysis for EAST upper full tungsten divertor cooling system, *Fusion Engineering and Design* **124** p 1181-86
[2] G.Dell O and Ancona A 2017 Steady state and transient thermal-hydraulic analyses on ITER divertor mudule, *Fusion Engineering and Design* **75** p 457-461
[3] Rene A and Merola M 2012 SOverview of the design and R & D of the ITER blanket system, *Fusion Engineering and Design* **87** p 769-776
[4] Zanino R, Bonifetto R, Cau F, Portone A and Savoldi R L 2014 CFD analysis of the ITER first wall 06 panel. Part II: Thermal-Hydraulics, *Fusion Engineering and Design* **89** p 431-441
[5] Bin G, Giovanni D, O, Teodoros L, Ployhar S, Jun T, Peng F, Lei Y Ajith K, Dinesh G, Nirav P and Mahesh J 2016 Thermal-Hydraulic Analysis of ITER Component Cooling Water System Loop 2B *Journal of Fusion Energy* **35** p 335-340
[6] Bin G, Zhiquan S, Peng Fu, Xuesong X, Chuan L, Min W and Lin D 2016 Design of coolant distribution system (CDS) for ITER PF AC/DC convertor, *Fusion Engineering and Design* **108** p 21-27
[7] Chuan H and Lijun G 2008 *Pump and Fans* (China Electric Power Press) p 94-104
[8] Qingyuan M and Jiping G 2010 *Fluid mechanics and distribution network* (Metallurgical Industry Press) p 67-75
[9] Applied Flow Technology 2008 *AFT Fathom User's Guide*
[10] Bin G, Peng Fu, Giovanni D, O, Teodoros L, Jun T and Lei Y 2015 Thermal behaviour analysis on ITER component cooling water system loop 2B, *Fusion Engineering and Design* **100** p 243-253

Journal of Physics: Conference Series **1419** (2019) 012022 doi:10.1088/1742-6596/1419/1/012022

Research on pedestrian detection system based on intelligent terminal with limited computing ability

Ning Chen[1,2], Menglu Li[1] and Zhijian Liu[1]

[1]School of Electronics Information, Xi'an Polytechnic University, Xi'an, China

[2]E-mail: chennvictor@gmail.com

Abstract. Pedestrian detection technology is the use of computer vision technology to determine whether there are pedestrians in static images or video images, and accurately mark pedestrians. Intelligent vehicles, intelligent monitoring, human behaviour analysis and other aspects are related to pedestrian detection technology, so pedestrian detection technology has been widely concerned by all walks of life. However, in some application environments, the network setting is flexible and the location of the device can be changed at any time, so it is difficult to establish a pedestrian detection system by wired way, and the pedestrian detection system can only be established through WSN. It is necessary to consider how to complete pedestrian detection through intelligent terminal under the condition of limited computing power. Many researchers have put forward a lot of algorithms, which are constantly optimized, but there are still many problems to be solved in practical application. In this paper, aiming at the intelligent terminal in WSN environment with limited computing power, a pedestrian detection system based on HOG+SVM is designed, which can detect moving pedestrians in video in real time. The detection system is mainly composed of detection algorithm and video acquisition module. After the test of INRIA and CVC data set, the accuracy of the detection algorithm in this paper is 89.06%. This method is tested under the intelligent terminal with limited computing power, with a good detection effect.

1. Introduction

With the development of society and economy, more and more temporary outdoor activities, such as marathons and concerts, have attracted more and more attention. At present, UAV monitoring is mainly used to monitor security. However, there is a lack of complete and comprehensive coverage of the monitoring site. Therefore, the pedestrian detection system based on intelligent terminal combined with Wireless Sensor Network (WSN) network appears.

The main problem of pedestrian detection is to detect pedestrians in pictures or videos, and then the position and size of pedestrians are represented by rectangular boxes. Face detection is also a kind of target detection and analysis. Pedestrian detection technology is very practical, which is involved in autopilot system, intelligent robot, intelligent video surveillance, human behaviour analysis, pedestrian tracking, pedestrian identification, passenger flow statistics, transportation and other fields. However, in some application environments, the network structure is flexible and the location of the device can be changed at any time, so it is difficult to establish a pedestrian detection system by wired way, and the pedestrian detection system can only be established through WSN. In this way, it is necessary to consider how to complete pedestrian detection through intelligent terminal under the condition of limited computing power. All these factors make pedestrian detection a very challenging topic in the field of computer vision [1,2].

At the beginning of the development of pedestrian detection, the detection method based on sliding window is the most widely used. This method was proposed by Papageorgiou and Poggio [3]. They also proposed Haar wavelet features and trained pedestrians with SVM to test samples including the front and back of pedestrians. After that, Viola and Jones [4,5] used cascade AdaBoost and Haar-like feature learning algorithm to detect pedestrians, and applied this technology to identify pedestrians in video surveillance, which was considered to be the basis of modern video detection. In 2005, Dalal and Triggs and others proposed that the gradient direction histogram feature (HOG), HOG feature depends on the gradient information detection of the image, and allows the image information to overlap, which makes it insensitive to the change of light and transition. And it can effectively characterize the edge characteristics of the human body. In recent years, pedestrian detection algorithm based on deep learning has achieved a great success. Deep learning model training requires a large number of pedestrian sample data, and the more complex the network layer, the greater the amount of computation required, so it is urgent to improve the optimization technology [6,7].

The detection method based on sliding window and the pedestrian detection algorithm based on deep learning require the intelligent terminal to have strong computing power, but this is precisely what the intelligent terminal lacks [8-10]. Therefore, aiming at the intelligent terminal in WSN environment with limited computing power, this paper designs a pedestrian detection system based on HOG + SVM, which can detect moving pedestrians in video in real time. The detection system is mainly composed of detection algorithm and video acquisition module. After the test of INRIA and CVC data set, the accuracy of the detection algorithm in this paper is 89.06%. This method is tested under the intelligent terminal with limited computing power, with a good detection effect.

2. Related algorithm

2.1. HOG feature

When calculating HOG features, the entire image is divided into small junction regions called cells, creating a directional gradient histogram in each unit. When describing the detected object, the combination of these histograms can be used.

2.2. Support vector machine

Support vector machine [11,12] is a binary classification model to find the hyperplane used to separate samples. The principle is to maximize the interval and turn the problem into a convex quadratic programming problem to find the best answer.

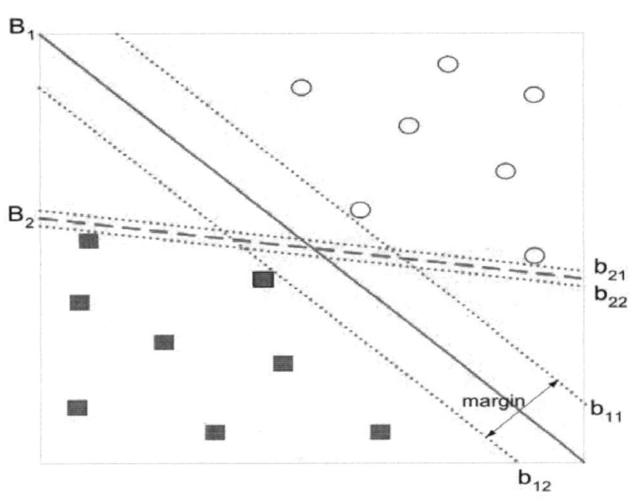

Figure 1. Hyperplane.

In the binary classification problem, there is data set T = {(x_1, y_1), (x_2, y_2), ... (xn, yn)}, $y_i \in$(0, 1). We can find a hyperplane w * x + b = 0. The corresponding decision function f(x) = sign (w * x +

b) separates the data as shown in Figure 1.

From the above figure, it is obvious that B1 has a better classification effect. The distance between the two points closest to the optimal hyperplane is called margin. From the graph, we can see that the larger the range of margin, the higher the accuracy of classification. The calculation of margin is formula (1):

$$margin = \frac{2}{\|\omega\|} \tag{1}$$

3. Pedestrian detection algorithm based on intelligent terminal with limited computing power

This is the core part, using HOG+SVM to achieve pedestrian detection algorithm. The implementation is divided into the following processes: (1) Dataset reprocessing includes negative sample cutting and positive sample scaling; (2) HOG feature extraction; (3) Classifier training (Figure 2); (4) Algorithm detection, including detecting pedestrians in still images and image sequences.

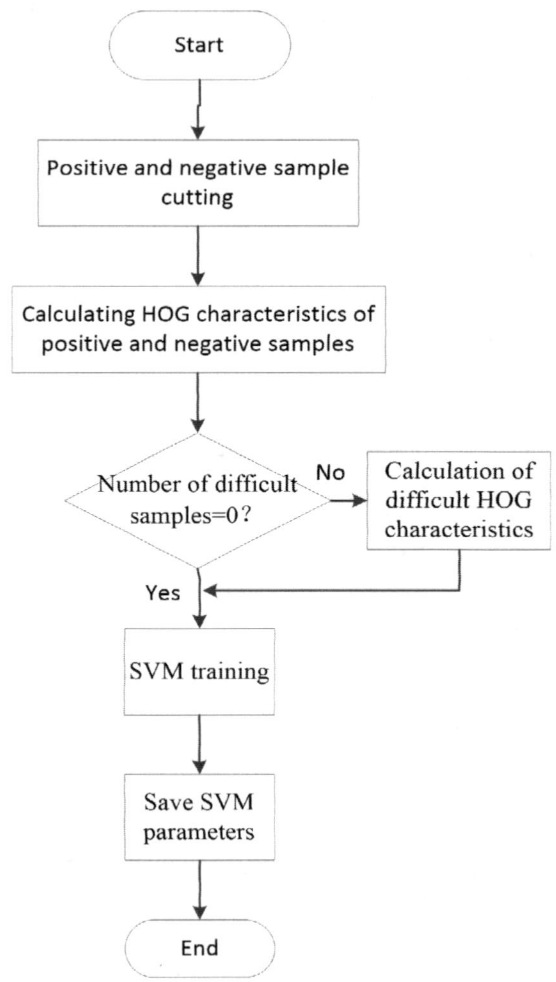

Figure 2. Training process.

4. Experiment

As a mature and convenient classifier, SVM is used by many researchers. There are many software packaging libraries implemented by SVM, among which LIBSVM is familiar to everyone. OpenCV collates the functions of LIBSVM, on the basis of which the SVM algorithm is implemented.

Therefore, we use OpenCV to develop a HOG+SVM pedestrian detection program for intelligent terminals with limited computing power, and use INRIA and CVC data sets to test. INRIA is a static image database and CVC is an image sequence database. The training set in INRIA database is used to train SVM classifiers, and the test set is used to test the accuracy of classifiers. The database is the most commonly used static pedestrian database, providing the image and the corresponding tag file. Positive and negative samples were included, of which 2416 pedestrians were included in the positive sample. 2441 positive samples were randomly cut from 2416 images for training, and some difficult images were selected for repeated training. The program interface is shown in Figure 3:

Figure 3. Our pedestrian detection program.

The experimental results are as follows in Table 1 and Table 2:

Table 1. Experimental results.

Sample set			Total
Positive sample (1082)	TP = 825	FP = 10	835
Negative sample (1359)	FN = 257	TN = 1349	1606

Table 2. Statistics of experimental results.

	Result
TPR (True Positive Rate)	76.26%
FPR (False Positive Rate)	0.74%
ACC (ACCuracy)	89.06%

From the experimental data, it can be concluded that the SVM classifier trained in this paper is better than the positive samples in dealing with negative samples.

5. Summary and prospect

In this paper, OpenCV, is used to complete the HOG+SVM pedestrian detection algorithm on the intelligent terminal with limited computing power. After testing, a good detection effect has been achieved. But there are still some problems:

- The training set in the prototype system is small, and the selection of parameters is not further optimized. In the future, it will be gradually improved in the product stage.
- In the case of complex background conditions and a large number of people, TPR is not high; there will be error detection, missed detection and so on. With the development of intelligent

terminal technology and the continuous improvement of computing power, higher complexity algorithms can be used to improve the accuracy [13].

- Combining HOG operator with other operators, an improved directional gradient histogram algorithm has been proposed by researches. The characteristics dimension of the deep model extracted by Res Net is reduced by principal component analysis algorithm (PCA), which greatly improves the accuracy.

This paper first introduces the development of pedestrian detection technology at home and abroad. Secondly, it describes the HOG+SVM pedestrian detection scheme used by intelligent terminals with limited computing power, as well as the implementation of the algorithm and the test of the detection system, which verifies the feasibility and effectiveness of the system. Intelligent terminal technology and pedestrian detection technology are still developing. The arrival of big data and artificial intelligence era has put forward higher requirements for pedestrian detection algorithm, in order to make the detection technology better applied in various fields, further research is needed.

Acknowledgment
This work was supported by Project of China Textile Industry Federation (2018097).

References

[1] Yang Y, Chen N and Jiang S 2018 Collaborative strategy for visual object tracking *Multimedia Tools and Applications* **77** (6) 7283-7303
[2] Yang Y, Wu Y and Chen N 2019 Explorations on visual localization from active to passive *Multimedia Tools and Applications* **78** (2) 2269-2309
[3] Papageorgiou C, Poggio T 2000 A trainable system for object detection *Int. J of computer vision* **38** (1) 15-33
[4] Viola P, Jones M J and Snow D 2005 Detecting pedestrians using patterns of motion and appearance *Int. J of Computer Vision* **63** (2) 153-161
[5] Jones M J, Snow D 2008 Pedestrian detection using boosted features over many frames *2008 19th Int. Conf. on Pattern Recognition* pp. 1-4 IEEE
[6] Levi K, Weiss Y 2004 Learning object detection from a small number of examples: the importance of good features *Proc. of the 2004 IEEE Computer Society Conf. on Computer Vision and Pattern Recognition* CVPR 2004 Vol. 2 pp II-II IEEE
[7] Danal N 2005 Histogram of Oriented Gradients for Human Detection *Proc. of IEEE Conf. on Computer Vision and Pattern Recognition (CVPR)* pp 886-893
[8] Tuzel O, Porikli F and Meer P 2008 Pedestrian detection via classification on Riemannian manifolds *IEEE transactions on pattern analysis and machine intelligence* **30** (10) 1713-1727
[9] Kieritz H, Becker S, Hübner W and Arens M 2016 Online multi-person tracking using integral channel features *2016 13th IEEE Int. Conf. on Advanced Video and Signal Based Surveillance (AVSS)* pp 122-130 IEEE
[10] Benenson R, Omran M, Hosang J and Schiele B 2014 Ten years of pedestrian detection, what have we learned? *European Conf. on Computer Vision* pp 613-627 Springer, Cham
[11] Cheng H, Zheng N and Qin J 2005 Pedestrian detection using sparse Gabor filter and support vector machine *IEEE Proc. Int. Conf. on Intelligent Vehicles Symposium* pp 583-587 IEEE
[12] Vapnik V 2013 The nature of statistical learning theory *Springer science & business media*
[13] LeCun Y 1989 Generalization and network design strategies *Connectionism in perspective* Vol. 19 Amsterdam: Elsevier

Decision support system for an early-stage keratoconus diagnosis

B R Salem[1,2,3], V I Solodovnikov[2]

[1]Bauman Moscow State Technical University, 2nd Baumanskaya str., 5-1, Moscow, 105005, Russian Federation
[2]Center of Information Technologies in Engineering RAS, Marshal Biryuzov Str. 7a, Odintsovo, Moscow region, 143000, Russian Federation

[3]E-mail: isub97@gmail.com

Abstract. Currently, there is a wide variety of different diseases that exist, a lot of which can be hardly prompt diagnosed even by medical specialists. This paper presents a method for early-stage diagnosis of ophthalmologic disorder keratoconus. Working with medical imagery that was captured by a rotating Scheimpflug camera system for anterior segment analysis, the goal was to create a decision support system to aid ophthalmologists in prompt detection of the disorder to eliminate the chance of further surgical intervention. Given approach uses several steps to achieve that goal, such as find the region of interest on the medical imagery using a Single Shot MultiBox Detector, filter and binarize the image, locate the cornea and approximate the curve with least squares fitting, classify the stage according to the previously labeled dataset by medical specialists with a random forest method. The suggested approach achieves 76% precision on a dataset containing 500 images of the patients with the first stage of keratoconus and healthy patients. The final result was compared with modern existing medical methods that are usually used in ophthalmologic clinics by medical specialists.

1. Introduction

Information technologies are currently used in almost all possible areas of human activity. Medicine is especially actively developing, offering modern specialized means for the detection and subsequent diagnosis of various diseases. Often, as a result of the examination, the doctor receives a set of images that need to be studied, interpreted, draw conclusions about the patient's condition and outline a further course of treatment. Given the vast range of possible diseases, some of which are characterized by mild symptoms, even qualified doctors are not always able to notice the initial stages of the pathological process. This problem is particularly relevant to diseases of unknown etiology or less studied. In particular, ophthalmologic disease keratoconus [1], which is a degenerative non-inflammatory ophthalmopathy can be attributed to a change in the optical properties of the pupil due to a change in the thickness of the cornea and its deviation from the ideal spherical healthy form. Moreover, there are cases when the symptoms of the disorder in the early stages are confused for myopia or astigmatism due to similar visual sensations, while its progression can lead to severe visual impairment. Keratoconus has 4 different stages [2], numbered from 1 to 4 which are stated in Table 1.

In this regard, issues related to the development of processing tools obtained in the process of examining the cornea of images, identifying abnormalities and decision-making support in making the diagnosis as early as possible are of particular relevance.

Table 1. Keratoconus classification based on disease evolution. VA, visual acuity; D, dioptres.

Stage	*Description*
1	Frustre or subclinical form; diagnosed by corneal topography; ~6/6 VA achievable with spectacle correction
2	Early form; mild corneal thinning; corneal scarring absent
3	Moderate form; corneal scarring and opacities absent; Vogt's striae; Fleischer's ring; <6/6 VA with spectacle correction, but ~6/6 VA with contact lens correction; irregular astigmatism between 2.00-8.00 D; significant corneal thinning
4	Severe form; corneal steepening > 55.00 D; corneal scarring, <6/7.5 VA with contact lens correction; severe corneal thinning and Munson's sign

In this paper, the proposed mathematical apparatus, algorithmic and software tools can be used in the construction of an automated decision support system [3,4] for diagnosing and determining the stage of keratoconus based on the analysis of sets of medical imagery.

2. Image pre-processing

Before starting to work on the curve of the cornea, it is necessary to select a region of interest on the image, which includes only the corneal arc. This will considerably facilitate further image processing since the output image will not include many default white marks of the capturing device.

This pre-processing will also help to get rid of possible components of the images that can lead to an inaccurate finding of the contours of the cornea. Such areas include tears, part of the eyelid, as well as other random details in the picture. Besides, the selection of the working area will reduce the number of processed pixels. Considering the nature of cornea medical imagery, that is, there are not many objects on it, and the required arcs usually have similar appearance on different images, it is not important to determine the high accuracy of the working area in the image, the Single Shot MultiBox Detector [5] was chosen based on a single pass through the image. This pre-processing is carried out using the Tensorflow [6] machine learning library for building and training a neural network, which will automatically find the working area. It is necessary to provide that the cornea in the image can be presented in different sizes. In the given case, training and test datasets were prepared, containing 30% and 70% of the total number of images respectively. On the training set, the necessary areas of the image, which include the corneal arc was manually pre-marked. For the subsequent finding of contours, it is necessary to process the obtained region of interest. The main idea is to eliminate possible distortions on the image and bring it into a binary form.

For the most accurate results, the original image is pre-converted to halftone using the formula:

$$y(i,j) = 0.299 * R(i,j) + 0.587 * G(i,j) + 0.114 * B(i,j) \quad (1)$$

Where R, G, B are for red, green and blue components of the image respectively. It should be noted that medical imagery is sometimes quite noisy, which can negatively affect the results. For this, it was decided to use a Gaussian Blur [7], which would significantly reduce image noise, and prevent the possible case where the cornea would represent a somewhat rough arc.

After filtering, it is necessary to binarize the image, which will uniquely divide the pixels into foreground and background. The most suitable for this is binarization with setting the global intensity threshold. This method is the most primitive. Since the original images have a similar set of colours and pixel intensity, as part of the development of the software, the images after binarization were inspected to determine the most suitable threshold value.

3. Finding the arc of the cornea

After binarization of the image, the contours of the cornea can already be distinguished, since it is possible to determine the boundaries between the background of the image and its foreground, which are the cornea and the adjacent parts of the eye. Boundaries were detected by using the algorithm of

Suzuki and Abe [8], which allows determining the external boundaries on the binary image. The resulting contours are presented in the form of an array of points that clearly define the arc of the cornea.

Images are often not centred so to find the arc it necessary to find a point as far as possible from the pupil of the eye. It can be accomplished by ignoring the edges of the image since they can be distorted which negatively affects the final result. The resulting contour is divided into three equal sections. In the middle section, the chosen point is the farthest from the pupil.

Considering the standards of capturing the eye of a human and the characteristics of the human eye, the maximum distant point from the pupil will be the middle of the cornea, given this fact this point will be chosen as the starting point for analysing its upper surface of the cornea and highlighting the corresponding arc. Since the cornea is bounded by the upper and lower arcs, it is necessary to choose only the nearest points corresponding to the upper part of the cornea. The resulting point is further considered as a reference point. Using the Manhattan distance, the coordinates of neighbouring points nearest to the reference point are found. Considering the angle of inclination between the points, the left and right parts of the arc can be distinguished. Due to the nature of the human eye, if the point is not the border of the image, then an abrupt change in the angle of inclination between the points of the cornea will be a clear indication that the end of the arc has been reached. As a result of the operations performed, an array of points describing the arc of the cornea will be obtained.

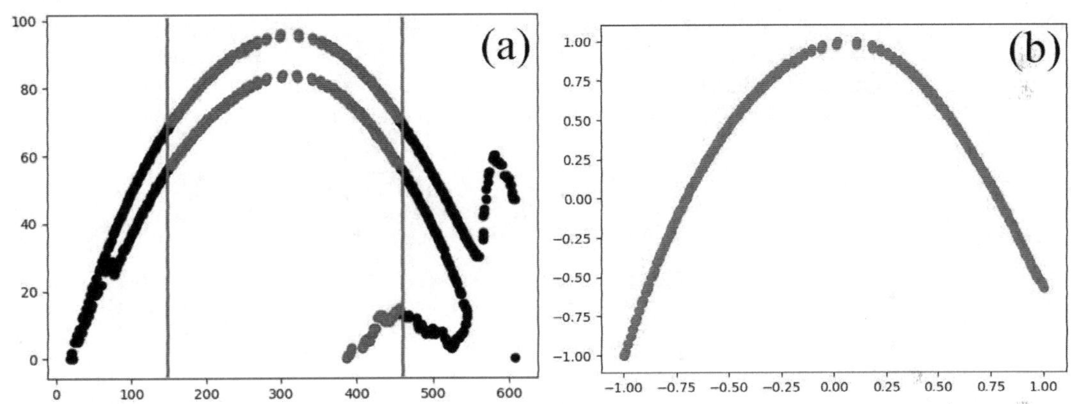

Figure 1. (a) The Initial array of points describing all contours of the image. The section inside the blue bounding lines highlights the points from which reference point is found. (b) The array of points describing the outer corneal arc, which was obtained after processing the initial array of points.

4. Curve fitting

To have a correct approximating function, it is necessary to provide for the possibility to consider the left and right parts of the cornea arc separately.

The slope will take a negative value if the points represent the left half of the arc and a positive value otherwise.

To eliminate the possibility that the obtained halves could be having different lengths additional checking was implemented which compares both parts of the arc to only include equidistant points from the reference point, as a result, an array of points is obtained that uniquely describes the cornea. Based on the given points, it is possible to construct a mathematical function, which in general will help in further determining the keratoconus stage from the initial images. For the subsequent classifications of the disorder stages, it is worthwhile to single out certain parameters reflecting the properties of the cornea. To compare each point of the arc is time-consuming and not an accurate way. One of the possible ways of representing this array in another form is an approximation. Let there be N values of the argument x_i and corresponding values of the function F_i. The main task is to find a function from the class of algebraic polynomials of the Nth degree. For a set of N, the maximum degree of a polynomial is $k = N - 1$. The maximum degree of a polynomial is determined by the number of data points used to generate it. However, it is generally recommended that you use the lowest degree as possible to accurately represent your data set, since higher-order polynomials passing directly through each data point may exhibit erratic behaviour between these points. A general polynomial regression model can be developed using the least squares fitting which can be defined as follows:

$$F(x) = a_0 + a_1 x + a_2 x^2 + \dots + a_n x^n \quad (2)$$

The least squares fitting [9] is aimed to minimize the differences between the values estimated by a polynomial and the expected values from the data set. Given the characteristics of the polynomial coefficients, the general form of which is presented below, the value of the first coefficient can be ignored, since it only affects the position of the curve along the y-axis. In the given dataset 8th degree polynomial was sufficient.

5. Stage classification

Based on the polynomial coefficients obtained and similar works [10,11], a random forest method [12], which is using a collection of decision trees, was chosen to classify stages of the disorder. The number of trees was selected using an experiment that took into account the classification accuracy based on a prepared sample of 200 images: 100 with the parameters of the first stage of keratoconus and 100 without deviations from the norm. The training sample included 70% of the original sample, and the remaining 30% entered the test sample. Taking into account possible deviations in the accuracy values, it was decided to average the classification values, therefore as shown in figure 2, each point on the graph represents the average value obtained by the classifier in the five previous experiments.

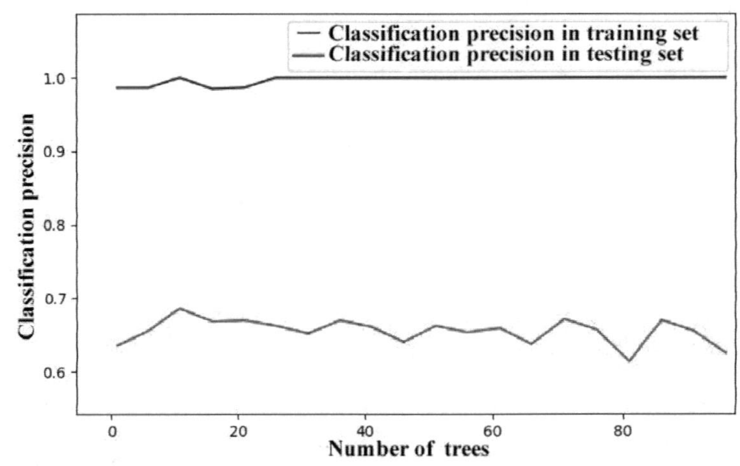

Figure 2. Classification accuracy of random forest method with different number of tress used.

The importance of certain coefficients of the polynomial was also examined. For evaluation, a Gini impurity index [13] was calculated, which is a measure of how often a random element is incorrectly classified. In general, the index is as follows:

$$G = \sum_{i=1}^{n_c} p_i(1 - p_i) \quad (3)$$

Where n_c is the number of classes in target variable and p_i is the ratio of this class.

As can be seen from figure 3, the most important coefficients of 8th degree polynomial are a_2, a_8, a_3, a_6 . But the remaining coefficients are also significant since their Gini impurity index > 0.050 $ and the value of the coefficient a_2 exceeds a_5 only 1.7 times, which cannot be considered a significant difference.

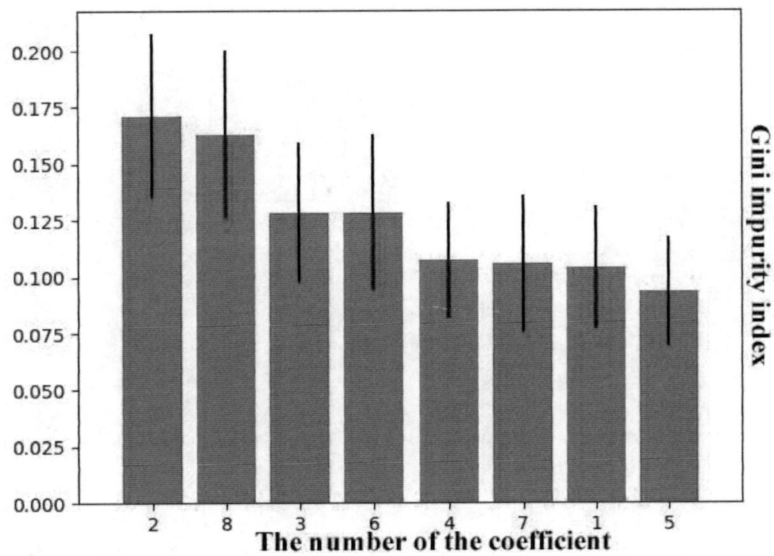

Figure 3. Coefficient importance of 8th degree polynomial. Black line is a standard deviation of the given coefficient.

6. Experimental results

For the experimental part, it was decided to test the results of the decision support system based on previously marked images by ophthalmologists.

In mathematical statistics to test statistical hypotheses concepts called type I and type II errors [14] were used. Type I error is the rejection of a true null hypothesis and type II error is the non-rejection of a false null hypothesis. In a given case, the medical specialist needs to confirm or deny the diagnosis of the disorder.

Results of the classification were based on the following measurements:

- Accuracy
- Recall
- F-score

The result of medical testing is often the doctor's verdict, which confirms the presence of the disease or denies it. Diagnosis of keratoconus is not an exception and the most important is the detection of corneal changes that occur in the first stage, which will prevent surgical intervention promptly. Given these characteristics of the disease, you can take the first stage of keratoconus and the absence of keratoconus as statistical hypotheses. Results of classifying 500 images of the eye with the first stage of keratoconus and healthy eye, which were equally divided evenly are given in table 2. The training set contained 70% of images, and the testing set contained 30% of the images.

Table 2. Precision, recall and F-score obtained for patients with the first stage of keratoconus and healthy patients.

	Precision	*Recall*	F-score
First stage	0.76	0.83	0.79
Healthy	0.73	0.62	0.67

Results of the suggested system were compared with modern medical approaches [15] and are given in figure 4.

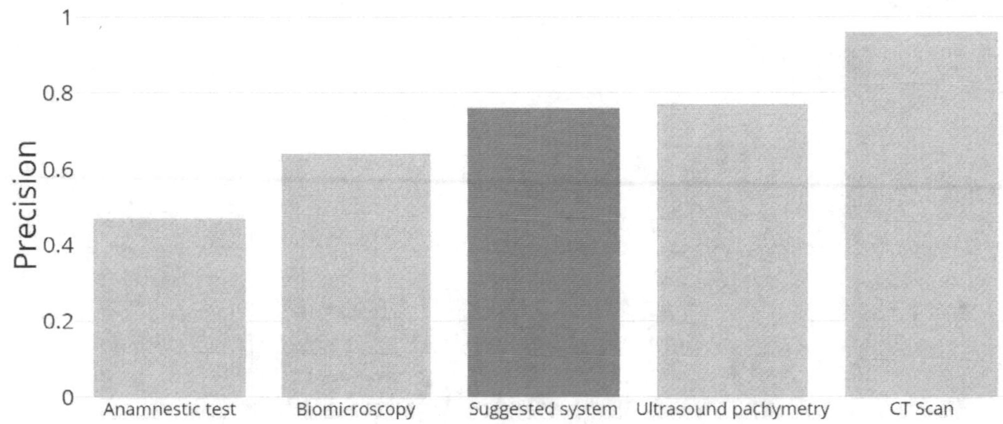

Figure 4. Comparisons of the different medical approaches used for diagnosing keratoconus.

7. Conclusion

As a result of the current research, a suggested decision support system was created for the diagnosis of keratoconus disease in its early stages. The developed system offers medical staff to import a set of medical images, based on which the program will give the textual and visual results with the information about the possible current stage of the disease the patient. Despite 76% precision, it is projected to take into account the inner part of the cornea, as well as the thickness between the inner and outer arcs. Increased precision could be achieved with further thorough research of the keratoconus disorder in its earliest form, especially the causes of occurrence. The designed software and the article will be useful for ophthalmologists not only to use but also to have a feedback for future discussions to produce a multifunctional decision support system, which can be used not only to diagnose ophthalmologic disorders but a wide range of human diseases by analysing medical imagery.

References

[1] Rabinovitz Y S 1998 Keratoconus *Surv Ophthalmol* **42** 297-319

[2] Romero-Jiménez M, Santodomingo-Rubido J, Wolffsohn J S 2010 Keratoconus: a review *Contact Lens & Anterior Eye.* **33** (4): 157-66

[3] Konnova N S, Basarab M A, Basarab D A 2018 Image processing using artificial intelligence methods in cardiovascular decision support systems *2018 Int, Conf. on Image and Video Processing, and Artificial Intelligence (IVPAI 2018)* vol. 10836

[4] Skvortsova M, Grout V 2018 through 1 February 2018 Basic approaches to assessing risks and threats in decision support systems *2018 IEEE Conf. of Russian Young Researchers in Electrical and Electronic Engineering* St. Petersburg and Moscow, Russian Federation pp 1563-1566

[5] Liu W, Anguelov D, Erhan D, Szegedy C, Reed S, Fu C Y, et al. 2016 *SSD: Single shot multibox detector Proc.* ECCV 9905 pp 21-37

[6] Martín Abadi, Paul Barham, Jianmin Chen, Zhifeng Chen, Andy Davis, Jeffrey Dean, Matthieu Devin, et al. 2016 TensorFlow: A System for Large-Scale. *Proc. of the 12th USENIX Symposium on Operating Systems Design and Implementation (OSDI'16)* pp 265-283

[7] Davies E 1990 *Machine Vision: Theory, Algorithms and Practicalities* Academic Press p 42-44

[8] Satoshi Suzuki and Keiichi Abe 1985 Topological structural analysis of digitized binary images by border following *Computer Vision, Graphics, and Image Processing* **30** p 32-46

[9] Orlov A I 2018 Nonparametric Method of Least Squares: Accounting for Seasonality *Journal of Mathematical Sciences (United States)* **228** p 501-509

[10] Sergeevna K N, Alekseevich B M 2019 Identification of Functional States of the Cardiovascular System According to Flowmetry Data Using Machine Learning Methods *11th Int. Symp. on*

Computational Intelligence and Design, ISCID 2018 Hangzhou, China vol 1 p 221-224

[11] Mokrova N V, Mokrov A M, Safonova A V, Vishnyakov I E 2017 Machine Learning Methods for Production Cases *Analysis 4th Int. Conf. on Advanced Engineering and Technology* Incheon, South Korea

[12] Breiman, Leo 2001 *Random Forests* Machine Learning **45** p 5-32

[13] Louppe G, Wehenkel L, Sutera A and Geurts P *Advances in Neural Information Processing Systems* C. J. C. Burges, L. Bottou, et al. Red Hook, NY, USA 26 p431-439

[14] Banerjee A, Chitnis U B, Jadhav S L, Bhawalkar J S, Chaudhury S 2009 Hypothesis testing, type I and type II errors *Ind Psychiatry J.* **18** (2): 127-131

[15] Nerpina M E, Pozharitskaya Ye M and Pozharitskiy M D 2013 *Modern algorithm of keratoconus early diagnostics* Vestnik OGU **4** p 189-192

Journal of Physics: Conference Series **1419** (2019) 012024 doi:10.1088/1742-6596/1419/1/012024

Faulty node modeling and diagnosis in interconnection networks (extended abstract)

E Cheng[1], Y P Mao[2], K Qiu[3] and Z Shen[4,5]

[1]Dept. of Mathematics and Statistics, Oakland University, Rochester, MI 48309, USA
[2]School of Mathematics and Statistics, Qinghai Normal University, Xining, Qinghai, 810001, China
[3]Dept. of Computer Science, Brock University, St. Catharines, Ontario, L2S 3A1, Canada
[4]Dept. of Computer Science and Technology, Plymouth State University, Plymouth, NH 03264, USA

[5]E-mail: zshen@plymouth.edu

Abstract. Faulty processing node analysis, in particular, self-diagnostic paradigm, is an important topic in the area of interconnection network studies. Several diagnostic models have been introduced, with the PMC and MM* models being two of the most popular ones. Researchers have also proposed various extensions and enhancements of fault-tolerance models to better capture the distribution, and identification, of faulty nodes in realistic scenarios. One of them is the g-extra fault-tolerance model, where each cluster in a network with faulty nodes contains at least g+1 fault-free nodes. This paper, following an analytical and constructive approach based on applied graph theory, suggests a general process to identify the maximum number of faulty nodes in a network in terms of the g-extra fault-tolerance model, and as a demonstrative example, provides a specific result for the (n, k)-star graphs.

1. Introduction

Nowadays, thanks to the rapid progress in computer engineering technology, multi-processor systems have become a common reality in both our work and lives, where an interconnection network plays a crucial role. It is unavoidable that some of the processing nodes in such a network will become faulty, sometimes disabling the normal function of the computer system. Thus, there is often a need to identify, and then correct/replace, these faulty nodes in such a system to restore its connectivity, and the normal operation of the computer system. For obvious reasons, one would want to have a self-diagnosable system where the computing nodes are able to detect faulty ones themselves. It turns out that we could simulate such a diagnostic task of an engineering nature via a mathematical process, with three components: the associated topology of such a system, the intended fault-tolerance model, and the diagnostic model.

The topology of a network system is usually modeled with a connected graph $G(V, E)$, where V represents a collection of processing nodes, and E the connections in between pairs of nodes. Many network topologies have been suggested and studied in the literature, including the hypercube [7], the star graph [1], and the (n, k)-star graph [5], denoted by $S_{n,k}$ in the rest of this paper.

Let $G(V, E)$ represent an interconnection network, we use a *faulty set* to identify a collection of faulty processing nodes, which is just a collection of vertices $F \subseteq V$, effectively removed from G. Since such an unrestrictive fault-tolerance model could imply the highly unlikely situation that all the neighbors of a vertex would be faulty, several more sophisticated fault-tolerance models have been suggested in the literature. With the *conditional faulty set* model [8], every vertex, faulty or not, has at least one *fault-free* neighbor in the *survival graph G-F*. With a *g-good neighbor faulty set* [12] every fault-free vertex has at least g fault-free neighbors in the survival graph; and, with a *g-extra faulty set* [18], every component in the survival graph contains at least $g+1$ vertices. In this paper, we will focus on this latter fault-tolerance model associated with a g-extra faulty set.

By a *neighbor* of a vertex, v, in G, we mean a vertex u such that (u, v) is an edge in G. The MM* model [14], which extends the *comparison diagnostic model,* i.e., the MM model [10,11] is a rather popular diagnostic model, where each processing node, called a *testing node*, sends a test message to each and every pair of its distinct neighbors, called a *tested node*, and then compares their responses. Based on all such comparison results, the fault status of the system, as well as the faulty nodes, can be determined. With another diagnostic model, the PMC model [13], a node sends a test message to each of its neighbors and judges its fault status based on the received responses. Various efficient algorithms to identify such faulty sets have been proposed in, e.g. [6,14]. We notice that, with both the MM* and PMC models, when a testing node is faulty, responses from those tested nodes will be unreliable.

A collection of all the aforementioned test results obtained with a diagnostic model is called a *syndrome* of the diagnosis. A subset F ($\subseteq V$) is *compatible with a syndrome* [8] if the latter can be generated when all the vertices in F are faulty and all those in $V \setminus F$, i.e., the difference between V and F, are fault-free. Since faulty testing nodes lead to unreliable results, as observed in [10, 11], two faulty sets may be compatible with the same syndrome, thus making such a faulty set unidentifiable. This observation leads to the notion of a graph being *t-diagnosable* [14] when up to t faulty vertices in G can be identified. And the *diagnosability* of a graph G, denoted by $t(G)$, is defined to be the maximum number of faulty vertices that G can guarantee to identify.

Many diagnosability results have appeared in literature, including [8,9,12,15,18], where we notice that much of the derivation details as reported in those papers devoted to different structures are essentially shared among themselves, and even with those used to derive results on the g-good-neighbor diagnosability. We strongly believe this practice is unnecessary. In this paper, we will continue our work in [4], and outline a general process to derive diagnosability results under various fault-tolerance, and diagnostic, models.

The rest of this paper proceeds as follows: We will start with some basic definitions regarding the self-diagnostic paradigm at a system level in the following section, and describe a general process of seeking both upper bounds, and lower bounds, of general diagnosability results in Section 3, and demonstrate the value of such a general process in Section 4. We conclude this paper with some final remarks in Section 5.

2. Basic notions of general diagnosability

Let $G(V, E)$ represent an interconnection network, and let M stand for a certain fault-tolerance model. *An M-faulty set* is a faulty set, $F \subseteq V$, consistent with M. For example, F ($\subseteq V$) is a g-extra faulty set if every component in G-F contains at least $g+1$ vertices. G is said to be *M t-diagnosable in terms of a diagnostic model D*, if G is diagnosable for any M-faulty set of size at most t in D, where D refers to either the PMC model or the MM* model in the rest of this paper.

Let F_1 and F_2 be two distinct M-faulty sets, $F_1 \subseteq V$, $F_2 \subseteq V$, (F_1, F_2) is *distinguishable* in G if and only if they do not lead to the same syndrome. They are *indistinguishable* otherwise. Then, $t_M(G, D)$, the *M-diagnosability of G, in terms of a diagnostic model D*, equals the maximum number t such that G is M t-diagnosable, i.e., for all distinct M-faulty set pairs (F_1, F_2), such that $F_1 \subseteq V$, $F_2 \subseteq V$, and $|F_1| \leq t$, $|F_2| \leq t$, (F_1, F_2) is distinguishable in terms of D.

Thus, the diagnosability problem really comes down to a decision problem of whether two faulty sets are distinguishable within a diagnostic model. In this regard, the following result specifies a necessary and sufficient condition of two faulty sets being distinguishable in the MM* model, where $F_1 \Delta F_2$ stands for $(F_1 \setminus F_2) \cup (F_2 \setminus F_1)$, i.e., the symmetric difference of F_1 and F_2.

Theorem 2.1. (following as [14]) Let $G(V, E)$ be a connected graph. For any two sets F_1, F_2 such that $F_1 \subseteq V$, $F_2 \subseteq V$ and $F_1 \neq F_2$, F_1 and F_2 are distinguishable under the MM* model if and only if at least one of the following three conditions holds: 1) there are two distinct vertices v and w, $\{v, w\} \subseteq V \setminus (F_1 \cup F_2)$ and there is a vertex x, $x \in F_1 \Delta F_2$ such that (v, w, x) is a path in G; 2) there are two distinct vertices v and x, $\{v, x\} \subseteq F_1 \setminus F_2$ and there is a vertex w, $w \in V \setminus (F_1 \cup F_2)$ such that (v, w, x) is a path in G; and 3) there are two distinct vertices v and x, $\{v, x\} \subseteq F_2 \setminus F_1$ and there is a vertex w, $w \in V \setminus (F_1 \cup F_2)$ such that (v, w, x) is a path in G.

The conditions associated with the PMC model are somewhat simpler, as expected, since the PMC model is a special case of the MM* model when the two vertices being tested are the same.

Theorem 2.2. (following as [6]) Let $G(V, E)$ be a graph. For any two sets F_1, F_2 such that $F_1 \subseteq V$, $F_2 \subseteq V$ and $F_1 \neq F_2$, F_1 and F_2 are distinguishable under the PMC model if and only if there exist a vertex u, $u \in V \setminus (F_1 \cup F_2)$, and another vertex v in $F_1 \Delta F_2$, such that (u, v) is an edge of G.

3. A general process of deriving diagnosability results

Recall that an *M*-faulty set is just a faulty vertex set F in a graph $G(V, E)$, related to a certain fault-tolerance model M. It is also an *M-cut* if G-F is disconnected. Although an *M*-faulty set does not need to be an *M*-faulty cut, the construction of such an *M*-faulty cut in G turns out to be a crucial step to derive the upper bound of the M t-diagnosability of G.

The size of a minimum *M*-faulty cut of a graph G is referred to as its *M-connectivity*, denoted by $\kappa_M(G)$. The derivation of the *M*-connectivity of a graph depends on its topology, and is often tedious and challenging. On the other hand, this quantity plays a critical role in deriving the lower bound of the related M t-diagnosability of a graph. Many results to this regard have appeared in literature. Readers are referred to [4] for the connection between the g-good neighbor connectivity of the arrangement graph and its g-good-neighbor diagnosability, $g \in [1, 2]$. The 2-extra connectivity of the bubble-sort graph is derived in [16], where its connection to its g-extra diagnosability is also given. The g-good-neighbor connectivity, $g \in [0, n\text{-}k]$, of the (n, k)-star graph is given in [17], and its connection to its g-good-neighbor diagnosability is explored in [4]. Recently, the g-extra connectivity of the (n, k)-star graph, and its connection to its g-extra diagnosability, are discussed in [9]. The major contribution of this paper is to provide an alternative and largely structure independent, approach to deriving a diagnosability result for a graph, once its related connectivity result is available.

Given a graph $G(V, E)$, we use $N(v)$ to denote the neighbors of v in V, i.e., $N_G(v) = \{w: (v, w) \in E\}$. Let $S \subseteq V$, we use $N_G(S)$ to denote all the neighbors of vertices in S, excluding those in S; and use $N^c_G(S)$ to denote $N_G(S) \cup S$. Notice that $N_G(S)$ and $N^c_G(S)$ are also referred to as *open* and *closed neighborhood* of S, respectively, in literature. We will drop the subscript G when the context is clear.

To derive an upper bound of $t_M(G, D)$, we carefully choose a set $Y(\subsetneq V)$ such that $V \neq N^c(Y)$ and both $N(Y)$ and $N^c(Y)$ are *M*-faulty sets.

The following result follows from Theorems 2.1 and 2.2. (Due to page restrictions, we have to omit technical proofs of all the results as contained in this paper, which will be included in its full version.)

Proposition 3.1. Let $G(V, E)$ be a connected graph, M stand for a certain fault-tolerance model, and let $Y \subsetneq V$. If both $N(Y)$ and $N^c(Y)$ are *M*-faulty sets, and $V \setminus N^c(Y) \neq \emptyset$, then $t_M(G, D) \leq |N^c(Y)|$-1.

It turns out that we can derive the following result for two specific fault-tolerance models.

Corollary 3.1. Let $G(V, E)$ be a connected graph, and let Y be a subset of V. If $N(Y)$ is an *M*-faulty set, where M stands for either the g-good-neighbor or the g-extra fault-tolerance model, and $V \setminus N^c(Y) \neq \emptyset$. Then $t_M(G, D) \leq |N^c(Y)|$-1.

Recall that $S_{n,k}$ stands for the (n, k)-star graph, the theorem below gives one specific result, obtained via such a construction, where we use $t_g(G, D)$ to denote the g-good-neighbor diagnosability of G in terms of a *diagnostic model D*.

Theorem 3.1. (following as [4]) For $n \geq 4$, $k \in [2, n)$, $g \in [0, n\text{-}k]$, $t_g(S_{n,k}, D) \leq n+g(k\text{-}1)\text{-}1$.

To show that t is a lower bound of $t_M(G, D)$, i.e., $t_M(G, D) \geq t$, we need to show that, for any two distinct M-faulty sets, F_1, F_2, such that $|F_1| \leq t$, $|F_2| \leq t$, and $V \setminus (F_1 \cup F_2) \neq \emptyset$, (F_1, F_2) is distinguishable in G in terms of the diagnostic model D. We start with a useful, and indeed a critical, result.

Proposition 3.2. Let $G(V, E)$ be a connected graph, M stand for a fault-tolerance model, and let $F_1, F_2 \subsetneq V$. If both F_1 and F_2 are M-faulty sets in G, so is $F_1 \cap F_2$.

The structure independent part of the general process of deriving a lower bound of M-diagnosability of a graph G in terms of the PMC model can then be summarized as follows.

Proposition 3.3. Let $G(V, E)$ be a connected graph, $\kappa_M(G)$ be the M-connectivity of G, and F_1 and F_2 be a pair of distinct M-faulty sets, such that $|F_1| \leq t$, $|F_2| \leq t$, and $V \setminus (F_1 \cup F_2) \neq \emptyset$. Assuming that $F_1 \setminus F_2 \neq \emptyset$ (Since $F_1 \neq F_2$, either $F_1 \setminus F_2 \neq \emptyset$ or $F_2 \setminus F_1 \neq \emptyset$). If $t \geq |F_1 \setminus F_2| + \kappa_M(G)$ leads to a contradiction with the PMC model assumption, then $t_M(G, \text{PMC}) \geq t$.

In particular, let $\tau_g(G, D)$ stand for the g-extra diagnosability of G in terms of D, we have the following result.

Corollary 3.2. Let $G(V, E)$ be a connected graph, and let $\kappa_g(G)$ be its g-extra connectivity, $g \geq 1$. If $|V| > 2\,(\kappa_g(G) + g)$, then $\tau_g(G, \text{PMC}) \geq \kappa_g(G) + g$.

We give a sketch of a proof of the above result as follows: Let (F_1, F_2) be a pair of distinct g-extra faulty sets in G such that $|F_1| \leq t = \kappa_g(G) + g$ and $|F_2| \leq t = \kappa_g(G) + g$. We want to show that such a pair is always distinguishable in terms of the PMC model, thus G is t-diagnosable. Just assume such a pair (F_1, F_2) is indistinguishable in terms of the PMC model.

Assume $F_1 \setminus F_2 \neq \emptyset$, and let C be a component that shares vertices with $F_1 \setminus F_2$. Since, (F_1, F_2) is assumed indistinguishable in terms of the PMC model, by Theorem 2.2, no vertex outside $F_1 \cup F_2$ is adjacent to any vertex in $F_1 \setminus F_2$. Hence, $C \subseteq F_1 \setminus F_2$, and, as a result, $|F_1 \setminus F_2| \geq |C| \geq g+1$. Since G is connected, we would have to conclude that any vertex u of $V \setminus (F_1 \cup F_2)$ has to go through a vertex in $F_1 \cap F_2$ to be connected to another vertex in $F_1 \setminus F_2$. Thus, $F_1 \cap F_2$ is a cut. By Proposition 3.2, since both F_1 and F_2 are g-extra faulty sets, $F_1 \cap F_2$ is also a g-extra faulty set, thus a g-extra cut. In other words, $|F_1 \cap F_2| \geq \kappa_g(G)$. Finally, the assumption $\kappa_g(G) + g = t \geq |F_1 \setminus F_2| + \kappa_g(G) \geq (g+1) + \kappa_g(G)$ leads to a contradiction, hence (F_1, F_2) has to be distinguishable in terms of the PMC model, and the result follows.

The process of deriving a lower bound result of the M-diagnosability in terms of the MM* diagnostic model is essentially the same as that for the PMC model, except that we also need to show that no isolated vertex exists in $V \setminus (F_1 \cup F_2)$, where (F_1, F_2) is the pair of indistinguishable M-faulty sets that we would use to construct the desired contradiction. The reason for this additional requirement is that, for this MM* case, if a vertex u is isolated in $V \setminus (F_1 \cup F_2)$, it can be adjacent to some vertex in $F_1 \Delta F_2$. Then, $F_1 \cap F_2$ would not be a cut, and we could not use the argument that we just used in proving Corollary 3.2 to reach the desired conclusion. We instead have the following parallel result.

Proposition 3.4. Let $G(V, E)$ be a connected graph, $\kappa_M(G)$ be the M-connectivity of G, and let F_1, F_2 be two M-faulty sets in terms of the MM* diagnostic model, such that $|F_1| \leq$ t, $|F_2| \leq$ t, and the non-empty set of $V \setminus (F_1 \cup F_2)$ contains no isolated vertex. Assume that $F_1 \setminus F_2 \neq \emptyset$, if $t \geq |F_1 \setminus F_2| + \kappa_M(G)$ leads to a contradiction, then $t_M(G, \text{MM*}) \geq t$.

It turns out that no vertex in $V \setminus (F_1 \cup F_2)$ could be isolated when $g \geq 2$ for both the g-good-neighbor and the g-extra fault-tolerance models, as the following result shows.

Corollary 3.3. Let $G(V, E)$ be a connected graph, and let $\kappa_g(G)$ be the g-extra connectivity of G, $g \geq 2$. If $|V| > 2(\kappa_g(G)+g)$, then $\tau_g(G, \text{MM*}) \geq \kappa_g(G) + g$.

4. The g-extra diagnosability of the (n, k)-star graph

The star graph, denoted by S_n, was proposed in [1] as an attractive alternative to the well-known hypercube structure when used as an interconnection network. However, the requirement that the number of vertices in the star graph be $n!$ results in a large size gap between S_n and S_{n+1}.To address this scalability issue, the (n, k)-star graph was suggested in [5], which brings in a flexibility in choosing its size, while preserving many attractive properties of the star graph, including vertex symmetry. This class of graphs has been well studied in the literature, together with its fault-tolerance properties, e.g., [3,4,9,17].

Vertices of $S_{n, k}$, $n \geq 3$, $k \in [2, n)$, is simply the collection of all the k-permutations taken out of $\{1, 2, \ldots, n\}$. Thus, $S_{n, k}$ contains $n!/(n-k)!$ vertices. Let $u = [p_1, \ldots, p_k]$, $v = [q_1, \ldots,, q_k]$, (p, q) is an edge of $S_{n, k}$ either, for some $i \in [2, k]$, v can be obtained from u by swapping p_1 and p_i (called an *i-edge*}); or, for some symbol $e \in \{1, 2, \ldots, n\} \setminus \{p_1, \ldots, p_k\}$, v can be obtained from u by replacing p_1 with e (called a*1-edge*). Thus, a vertex is incident to k-1 i-edges, $i \in [2, k]$; and n-k 1 edges, a total of n-1 edges. As a result, $S_{n, k}$ contains a total of $((n-1)n!/[2(n-k)!]$ edges.

As shown in Figure 1, $S_{4, 2}$ contains 4!/2! (= 12) vertices, and 18 edges. For example, [1, 2] (=12) is a vertex in $S_{4, 2}$, and ([1, 2], [2, 1]), ([1, 2], [3, 2]), and ([1, 2], [4, 2]) are all edges of this graph. In particular, [1, 2] is incident to three edges, including two 1-edges: ([1, 2],[3, 2]) and ([1, 2], [4, 2]) , and [1, 2] is also incident to exactly one 2-edge ([1, 2], [2, 1]).

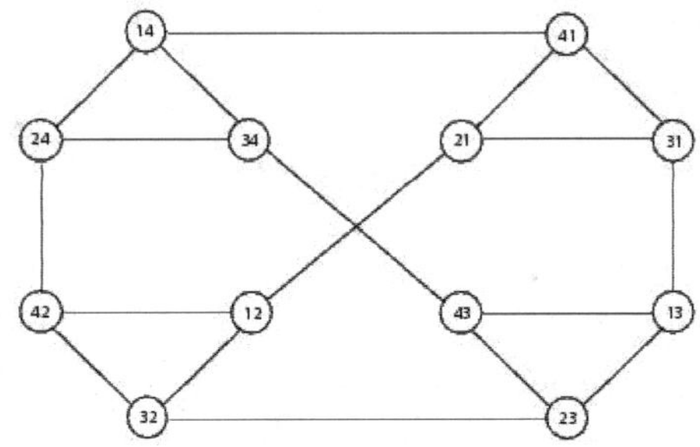

Figure 1. $S_{4, 2}$

Issues related to the g-good-neighbor diagnosability of the (n, k)-star graphs have been addressed in [4,15]. Its g-extra diagnosability has also been derived recently in [9] by following a structure dependent approach. It turns out that these results can all be derived following the general approach that we have described in the previous section.

Recall that $t_g(G, D)$ stands for the g-good-neighbor diagnosability of a graph G in terms of a diagnostic model D, and $\tau_g(G, D)$ its g-extra diagnosability. We start with the following result, since a vertex u has at least one neighbor if and only if the component containing u has at least two vertices.

Lemma 4.1. (following as [16]) Let $G(V, E)$ be a connected graph, then $\tau_1(G, D) = t_1(G, D)$.

Since it has already been proved that $t_1(S_{n, k}, D) = n+k-2$ (Theorem 5.3 in [4] for the PMC case and [15] for the MM* case), by Lemma 4.1, we immediately have the following result.

Corollary 4.1. Let $n \geq 4$, $k \in [2, n)$, $\tau_1(S_{n, k}, D) = n+k-2$.

For the cases of $g \geq 2$, the following result provides an upper bound of $\tau_g(S_{n, k}, D)$ since if a fault-free node has at least g neighbors, thus the component contains at least g+1 vertices.

Lemma 4.2. (following as [16]) Let G be a connected graph, let $g \geq 0$, then $\tau_g(G, D) \leq t_g(G, D)$.

We then obtain an upper bound result for the g-extra diagnosability of the (n, k)-star graph, by Theorem 3.1 as follows.

Lemma 4.3. For $n \geq 4$, $k \in [2, n)$, $g \in [0, n\text{-}k]$, $\tau_g(S_{n, k}, D) \leq n+g(k\text{-}1)\text{-}1$.

As pointed out, to obtain a hopefully tight lower bound of $\tau_g(G, D)$, we need a result of its M-connectivity, which could be derived by making use of its super-connectedness property. More specifically, a graph is *super m-vertex connected of order q* if, with at most m vertices being deleted, the survival graph is either connected or it consists of a large component and the small components containing at most q vertices altogether [2,17].

For a detailed discussion about this inspiring and important structural property of a graph, and its close relationship to various fault-tolerance properties, including g-good-neighbor connectivity, g-extra connectivity, component connectivity, cyclic connectivity, as well as the Menger connectedness, readers are referred to [2]. To address our current issue, we make the following observation.

Theorem 4.1. (following as Theorem 8 of [17]) Let n, k, and r be positive integers such that $k \in [2, n)$ and $r \in [1, n\text{-}k\text{+}1]$. If F is a set of vertices of $S_{n,k}$ such that $|F| \le n+(r-1)k-2r$, then $S_{n,k} - F$ is either connected or has a large component and small components with at most r-1 vertices in total.

Taking $g = r-1$, if we want to have a component, besides the larger component, in a survival graph $S_{n,k} - F$, which contains at least g+1 (=r) vertices, $g \in [0, n\text{-}k]$, we have to take at least $n+g(k-2)-1$ vertices. As a result, any g-extra cut must contain at least $n+g(k-2)-1$ vertices. In particular, $\kappa_g(S_{n,k}) \ge n+g(k-2)-1$.

It is easy to show that $|V(S_{n,k})| > 2(\kappa_g(S_{n,k})+g) = 2(n+g(k-1)-1)$. Thus, by Corollaries 3.2, 3.3, 4.1, and Lemma 4.3, we have the following result.

Theorem 4.2. Let $n \ge 4$, $k \in [3, n)$, $g \in [0, n\text{-}k]$, $\tau_g(S_{n,k}, D) = n+g(k-1)-1$.

Incidentally, since $S_{n,n-1}$ is isomorphic to the star graph (Lemma 4 in [5]), and $S_{n,n-2}$ is isomorphic to the alternating group network [3], denoted by AN_n, the g-extra diagnosability results of these latter two graphs immediately follow.

Corollary 4.1. Let $n \ge 4$, $\tau_g(S_n, D) = \tau_g(S_{n,n-1}, D) = 2n-3$.

Corollary 4.2. Let $n \ge 4$, $g \in [1, 2]$, $\tau_g(AN_n, D) = \tau_g(S_{n,n-2}, D) = n+g(n-2)-1$.

5. Concluding remarks

In this paper, we summarized and unified a process that we can effectively apply to derive diagnosability results for multiple fault-tolerance models in terms of such mainstream diagnostic models as the PMC and the MM* models. Simply put, to derive such a diagnosability result for a graph G in terms of a fault-tolerance model M, and a diagnostic model D, we need to obtain $\kappa_M(G)$, the M-connectivity of G, and the details of the rest of the process in terms of model D are mostly structure independent, thus can be spared.

We also demonstrated the value of such a general process by obtaining the g-extra diagnosability of the (n, k)-star graphs. Although this result has already been reported in the literature, this alternative derivation is straightforward, and even mechanical, thus has its clear advantage.

As future research topics, besides attempting to achieve the diagnosability results of other interconnection structures under the existing fault-tolerance models in light of this general process, we will also look into other appropriate fault-tolerance models, and the feasibility of applying this general process to derive various fault-tolerance property under such alternative models.

References

[1] Akers S B, Krishnamurthy B 1989 A group theoretic model for symmetric interconnection networks *IEEE Trans. Comput.* **38** (4): 555-566

[2] Cheng E, Hao R X, Qiu K and Shen Z 2019 Structural properties and fault resiliency of interconnection networks, in: Adamatzky A, Selim A and Sirakoulis G ed. *From Parallel To Emergent Computing* CRC Press p 77-101

[3] Cheng E, Qiu K and Shen Z 2012 A note on the alternating group network *J Supercomput.* **59** (1): 246-248

[4] Cheng E, Qiu K and Shen Z 2019 A general approach to deriving the g-good-neighbor conditional diagnosability of interconnection networks *Theor. Comput. Sci.* **757** (24): 56-67

[5] Chiang W K, Chen R J 1995 The (n, k)-star graph: a generalized star graph *Inf. Process. Lett.* **56** p 259-264

[6] Dahbura A T, Masson G M 1984 An n2.5 faulty identification algorithm for diagnosable systems *IEEE Trans. Comput.* **33** (6): 486-492

[7] Harary F, Hayes J P and Wu H J 1988 A survey of the theory of hypercube graphs *Comput. Math. Appl.* **15** (4): 277-289
[8] Lai P L, Tan J J M, Chang C P and Hsu L H 2005 Conditional diagnosability measures for large multiprocessor systems *IEEE Trans. Comput.* **54** p 165-175
[9] Lv M J, Zhou S M, Sun X L, Lian G G and Liu J F 2019 Reliability of (n, k)-star network based on g -extra conditional fault *Theor. Comput. Sci.* **757** p 44-55
[10] Maeng J, Malek M 1981 A comparison connection assignment for self-diagnosis of multiprocessor systems *Proc. 11th Int. Symp. Fault-tolerance Comput.* p 173-175
[11] Malek M 1980 A comparison connection assignment for diagnosis of multiprocessor systems *Proc. 7th Int. Symp. Comput. Archit.* p 31-35
[12] Peng S L, Lin C K, Tan J J M and Hsu L H 2012 The g-good-neighbor conditional diagnosability of hypercube under the PMC model *Appl. Math. Comput.* **218** (21): 10406-10412
[13] Preparata F P, Metze G and Chien R T 1967 On the Connection assignment problem of diagnosable systems *IEEE Trans. Electron. Comput.* EC-**16** (6): 848-854
[14] Sengupta A, Dahbura A T 1992 On self-diagnosable multiprocessor systems: diagnosis by the comparison approach *IEEE Trans. Comput.* **41** p 1386-1396
[15] Xu X, Li W W, Zhou S M, Hao R-X and Gu M M 2017 The g-good-neighbor diagnosability of (n, k)-star graph *Theor. Comput. Sci.* **659** p 53-63
[16] Wang S, Wang Z and Wang M 2016 The 2-extra connectivity and 2-extra diagnosability of bubble-sort star graph networks *The Compt. Journal* **59** (12): 1839-1856
[17] Yuan A, Cheng E and Liptak L 2011 Linearly many faults in (n, k)-star graphs *Int. J. of Found. Comput. Sci.* **22** (7): 1729-1745
[18] Zhang S, Yang W 2016 The g-extra conditional diagnosability and sequential t/k-diagnosability of hypercubes *Int. J. Comput. Math.* **93** p 482-497

Adaptive emergency control system with an algorithm for rapid assessment of sustainability united energy system

Maxim G Popov[1,3], Zhiyu Lu[1,2]

[1]Power Plants & Power Automation Department, Peter the Great St. Petersburg Polytechnic University, Saint-Petersburg, Polytechnicheskaya St., 29, 195251, Russia
[2]SPbPU Representative Office in Shanghai, Shanghai, Pudong New Area, Zhangjiang Hi-tech Park, China

[3]E-mail: PopovMG@eef.spbstu.ru

Abstract. In the present study described the methodology for determination of sustainable modes of long equivalent of the interconnected power systems of various structures with simplified representation of excitation controllers of generators. According to the results of numerous studies, the criterion of oscillatory disturbance of static stability based on the control of the partial derivative $\partial A_N / \partial\delta$ is formulated. To confirm the formulated criterion of violation of the static vibrational stability performed quantitative mapping of polynomial coefficients of haracteristics equations and additional applications of the matrix equations is redundant points of the equivalent circuit of the Electrical Power Systems (EPS) obtained under the criterion Hurwitz. Quantitative comparison and qualitative analysis of the nature of the change in the roots of the characteristic equation and the Hurwitz determinants confirm that the transition across the boundary of the stability region, found by the formulated criterion on the basis of the control of the partial derivative $\partial A_N / \partial\delta$ taking into account the proposed simplified model of generators at the constant excitation voltage U_f = const, is accompanied by the appearance of reciprocal values of the Hurwitz determinants. Taking into account the generalizations and conclusions obtained, a modified method of studying the static oscillatory stability of the EPS is proposed. The efficiency of the developed algorithm for determining the boundary of the region of potentially stable regimes is estimated. It is shown that the proposed method has the best performance indicators and can reduce the computational cost.

1. Introduction

Maintaining the stability of power systems is presently being done to ensure standardized inventory to specific design image of the limit of the transmit power regimes. Determination of these modes in the course of research can be performed using both static and dynamic models of electric power systems implemented in modern specialized software systems. The desire to improve the accuracy of modeling in the verification of the used models leads to an increase in the order of differential equations, which greatly complicates the use of existing methods of quantitative evaluation of reserves for stability. Subsequent qualitative analysis of electromechanical transients and improvement of damping parameters using a complete system of equations leads to an increase in computational cost and increase the complexity of the calculations performed. Despite this, a quick search for the region of stable modes can be performed using simplified models that reliably reflect the electromechanical motion of the generators. Made in [1-9, 14, 17-26] functional analysis of the vibrational characteristics of violation of static stability of simplified models of the power system in the representation of the excitation regulators of the generators in the constancy of the excitation voltage U_f = const [6-17] allowed to improve the methodology to study the static oscillatory stability of power grids. The use of this model of automatic excitation regulator allows to determining the area of potentially stable modes of the investigated (real) power system with high-speed excitation regulators at the preliminary stage.

In this article the modified technique of research of static oscillatory stability which allows to carrying out primary, qualitative assessment of the modes of power systems of various structure limiting on the transmitted power is stated. The tool is formulated and proved a criterion of oscillatory violations of static stability $\partial A_N / \partial \delta$.

2. Mathematical model and method research description

Numerical experiments were carried out in relation to the design schemes containing a different number of equivalent power systems (from 2 to 6). For ring and chain schemes of power connections, containing N generators with a simplified model of automatic excitation regulators (U_f = const), without taking into account the equations of damper circuits and when replacing loads with constant resistances, the characteristic equation of free movement of the Electromechanical system takes the form:

$$p^{3N-2} + A_1 \cdot p^{3N-3} + ... + A_{k-2} \cdot p^{3N-k} + ... + A_{3N-3} \cdot p^1 + A_{3N-2} = 0 \qquad (1)$$

Computer studies were carried out in relation to typical structural schemes of chain and ring power transmission with variations in stiffness and installed capacities of equivalent power systems. Equivalent ring schemes are typical for system-forming networks of 330 kV and 500 kV of large megacities and regional energy associations (for example, the North-West Power Pool). Intersystem transmission lines completed 500 kV and 750 kV, chain structure characteristic of the major of power grids of the Power Pools of East and Siberian. When allocating groups of in-phase relative motion of equivalent generators, the Russian Electricals Power Systems can also be represented by an equivalent power system of a chain structure.

It is established that the region of potentially stable regimes, which corresponds to the unregulated (U_f = const) representation of equivalent generators, regardless of the structure and design conditions, is near the maximum angular characteristic (margin from 0.01% to 10%) and characterizes the greatest use of power. The performed computational studies found that the stability region determined by the simplified model U_f = const is always to the left of the minimum of the free term A_N. In all the studied circuit-regime conditions, regardless of the structure (chain or ring) and the stiffness of power transmission, the boundary of the static vibrational stability region is characterized by a minimum of dependence $A_N = f(\delta)$ without changing the sign of the free term. This is explained by the fact [9-17] that the transition across the boundary of root decay with further weighting of the regime is accompanied by a relatively rapid increase in the absolute value of the real parts of two pairs of complex conjugated roots ($+\alpha \pm j\,\omega$ and $-\alpha \pm j\,\omega$).

The required normative reserves of 8 % and 20 % correspond to the regimes located within the area of potentially stable regimes determined using the proposed structural criteria and characterized by the introduction of stabilization by the damping decrement α = -0,3...-0,4 s^{-1} [2-14].

The analysis of the change in the roots of the characteristic equation (column 1 of the table 1) with the variation of its coefficients A_1 ... A_{3N-2} found that near the boundary of the stability region there is a significant increase in the increment of electromechanical oscillations and a decrease in the damping time constant of aperiodic processes in the circuits of automatic excitation regulators.

Comparison of polynomial coefficients of the characteristic equation and additional minors of the matrix of equations of excess moments of the equivalent two-machine circuit of the Electrical Power Systems, obtained using the algebraic Hurwitz criterion, confirms that the transition across the boundary of the stability region (δ_{g21} = 87.36°) of the model at U_f = const is accompanied by the appearance of negative values of Hurwitz determinants. In the course of qualitative studies of the product of the roots of the characteristic equation and the free term in the function of the relative angle, some important results were obtained – in General, with an odd number of equivalent generators, A_{3N-2} is located in the region of negative values of the free term.

Table 1. Coefficients of the characteristic polynomial unregulated model (U_f = const) and Hurwitz determinants.

The angle between the EMF, degree	Roots characteristic equation, s^{-1}	Coefficients of the characteristic polynomial					Hurwitz determinants			
		A_0	A_1	A_2	A_3	A_4	Δ_1	Δ_2	Δ_3	Δ_4
δ_{g21}=58,75°	–0,314; –0,209; –0,074±j0,640	1,0	+0,670	+16,195	+8,409	+1,055	+0,670	+2,447	+20,108	+21,212
δ_{g21}=87,36°	–0,333; –0,333; –0,001±j0,425	1,0	+0,668	+7,260	+4,753	+0,814	+0,668	+0,095	+0,089	+0,073
δ_{g21}=106,50°	–1,743; –0,357; **+0,714±j0,083**	1,0	+0,672	**–1,595**	+0,754	+0,487	+0,672	**–1,826**	**–1,570**	**–0,777**

In the domain of stable modes of the model U_f = const, the product of the dominant components of the relative motion is proportional to the change of the product of all roots of the characteristic equation. The obtained results confirm the competence of the proposed model of excitation regulators for a rapid and reliable analysis of the stable regimes.

3. Results

The results presented in this publication are a generalization of studies [9-26] using the previously proposed criterion of vibrational disturbance of static stability, based on the control of the change of the free term of the characteristic equation.

The computation time of the coefficients of the characteristic polynomial and eigenvalues of the matrix of coefficients of the equivalent system of linearized equations of transients in the unregulated model U_f = const is two orders of magnitude less than the same calculation time in the complete system of differential-algebraic equations, which allows us to talk about the efficiency of the developed algorithm (figure 1, 2).

On the basis of the obtained results was c method reliable determination of the region of stable regimes, based on the control of the partial derivative $\partial A_N / \partial\delta$, which consists of the following sequence of operations:

– formation of a system of algebraic equations of nodal voltages of the electrical network;

– reduction of the order (equivalent) of the subsystem of algebraic equations by means of elimination of variables (node voltages of the electric network) by the Gauss method;

– formation of expressions for electrical power in the nodes;

– the formation of a subsystem of the linearized differential equations of the relative motion of the Electromechanical synchronous machines;

– numerical analysis of the free term, roots of the characteristic equation and determination of the domain of stable regimes;

– calculation of the eigenvalues of the coefficient matrix of the complete system of linearized differential equations of transients in synchronous generators, excitation systems and their automatic regulators;

– calculation of eigenvalues of the coefficient matrix of a system of linearized differential equations in simplified models.

Proposed an improved method for the rapid assessment of steady-state modes of the controlled power systems can be used in the control algorithms of the interconnection with the structure of devices of automatics prevent loss of stability. On the basis of rapid assessment using structural stability criteria can be determined (calculated) and formed the corresponding control effects of automatic unloading equipment (unloading stations and etc.). The problem of rapid assessment can be solved by implementing several options:

1) rigidly defined (known in advance) the structure and parameters of the sections controlled in it, for example, two-(n)-machine circuit of intersystem power transmission;

2) only the structure of the equivalent power system is known, and its parameters and "weak" (controlled) sections must be determined;

3) it is necessary to identify the structure and parameters of the equivalent circuit of the Electrical Power Systems, as well as to determine the "weak" interconnection.

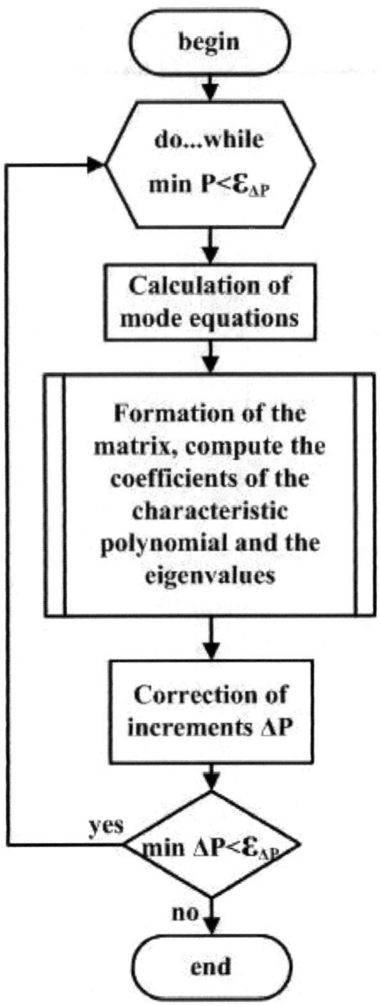

Figure 1. Algorithm of the procedure for determining the boundary of the domain of stable regimes.

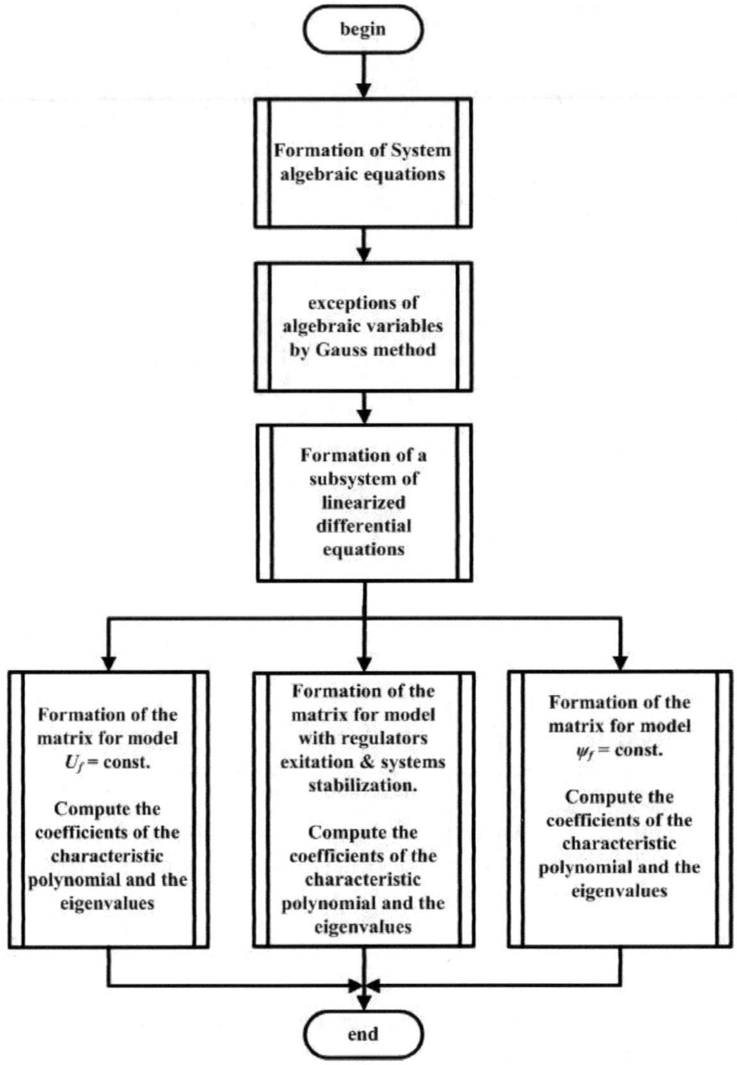

Figure 2. The algorithm of the procedure of formation of the matrix, compute the coefficients of the characteristic polynomial and the eigenvalues of the linearized system of differential equations.

In all three cases, it is necessary to know the equivalent electromechanical parameters of the combined power systems, while the third method is the most time-consuming, in terms of computational costs.

4. Conclusions

The identified characteristic features of the disturbance of stability and associated changes in polynomial coefficients allowed to develop methods for studying the dynamic properties of combined power systems of arbitrary structure and to formulate on this basis new practical criteria for assessing stability.

The proposed improved method of evaluation of static oscillatory stability can be used by design organizations, joint and regional dispatching offices for rapid analysis of the static stability in modern complex power systems. Computational studies using the developed software allow us to formulate requirements for the speed and selectivity of the main and backup protection and emergency automation.

As the results of numerous studies have shown, the application of the developed method for determining the boundary of the region of stable regimes can significantly reduce the computational cost (processor time) even for the Electrical Power Systems schemes characterized by a small order of the system of equations of transients.

The most effective application of the developed algorithm as part of modern microprocessor means of emergency control of power line equipment overload limitation. The implementation of this algorithm in the means of emergency control can improve the efficiency and efficiency of AC power transmission due to a reasonable and safe increase in the load of overhead lines.

References

[1] Jdanov P 1979 *Voprosy ustojchivosti elektricheskih sistem* [*Electrical system stability issues*]
[2] Ragozin A A, Popov M G 2002 *Power of Russia* **2** p 26 - 28
[3] Ragozin A A, Popov M G 2002 *Power of Russia* **10** p 18-24
[4] Methodical instructions on stability of power systems 2003 CO 153-34.20.576-2003
[5] Popov M G, Bogdanov A V 2010 News of higher educational institutions *Electromechanics* **2** (special issue) p 25-28
[6] Popov M G, Kozyrev A V 2010 News of higher educational institutions *Electromechanics* **2** (special issue) p 30-33
[7] Ilyshin P, Chusovitin P 2014 *Relay Protection and Automation* **4** p 16-22
[8] Mokeev A V, Bovykin V N 2015 *The New in Russian Electric Power Industry* (4): 40-48
[9] Popov M G, Zakharova E V and etc. 2015 Study of static stability of complex interconnected power systems of arbitrary structure *Electric Power Plants* (11) p 61 (9)
[10] Antonov V I, Naumov V A, Soldatov A V and Ivanov N G 2015 A Method of Extracting an Electrical Component Value *Byul. Otkryt. Izobret* 28
[11] Bulychev A V and etc. 2015 *Releinaya Zashch. Avtom* **4** (21)
[12] Vanin V K, Ambrosovskaya T D and etc. 2016 *Power Technology and Engineering* **50** (1): 87-92
[13] *Wide Area Protection & Control Technologies* 2016 *CIGRE*, Working Group B5.14 Paris
[14] Zakharova E 2016 (PhD thesis) *Criterion of rapid assessment of static stability of integrated power systems* Saint-Petersburg
[15] Soldatov A V, Antonov V I and etc. 2016 *Releinaya Zashch. Avtom* **3** (24): 12-20
[16] Bulychev A V, Dementii Y A and etc. 2017 Measurement of currents in systems of power-system protection from single-phase earth faults and in automated control by arc-suppression reactors *Russian Electrical Engineering* **88** (7): 430-436
[17] Popov M G 2018 (ScD thesis) *Modern means of emergency control of integrated power systems* Saint-Petersburg
[18] Vanin V, Bulychov A and etc. 2018 About influence of non-sinusoidal currents and voltages on the amount of the electric energy *MATEC Web of Conferences* **245** Article No. 06009
[19] Vanin V, Bulychov A and etc. 2018 Measurement of currents and voltages non-sinusoidal parameters in power supply systems with rectifier load. *MATEC Web of Conferences*, Vol. **245** Article No. 06007
[20] Popov M G 2019 Adaptive correction of current longitudinal differential protection of power equipment *3rd International Scientific and Technical Conference Energy Systems (ICES 2018)* Vol. **552** Issue 1 Article No. 012017 DOI: 10.1088/1757-899X/552/1/012017
[21] Abounekirah S S and etc. 2019 Dynamic identification of electromagnetic parameters for adaptive protection devices *Proceedings of the 2019 IEEE Conference of Russian Young Researchers in Electrical and Electronic Engineering (ElConRus 2019)* Article No. 8656689 p 639-643
[22] Sinianskii I V and etc. 2019 Development of an algorithm for the cascade of vuoksin HPPS automated transfer to parallel operation with nordel *Proceedings of the 2019 IEEE Con. of*

Russian Young Researchers in Electrical and Electronic Engineering (ElConRus 2019) Article No. 8657262 p 681-685

[23] Zelenin A S and etc. 2019 Development and implementation experience of microprocessor emulators for cyber-physical AVR testing complexes *Proceedings of the 2019 IEEE Conference of Russian Young Researchers in Electrical and Electronic Engineering, (ElConRus 2019)* Article No. 8657237 p 752-756

[24] Petrushin D E, Efimov N S and etc. 2019 Increasing the emergency control systems efficiency in Kola's and Karelia's power systems *Proceedings of the 2019 IEEE Conference of Russian Young Researchers in Electrical and Electronic Engineering (ElConRus 2019)* Article No. 8657093 p 1040-1043

[25] Abounekirah S S, Refaat A 2019 A modified restraint method to enhance the performance of differential protection for overhead transmission lines *Proceedings of the 2019 IEEE Conference of Russian Young Researchers in Electrical and Electronic Engineering, (ElConRus 2019)* Article No. 8656690 p 634-638

[26] Popov M G, Bocharov Y N and etc. 2019 Dynamic methods of identification of electromagnetic parameters of power transformers in non-stationary mode *E3S Web of Conferences* Vol. **91** Article No. 01002

Design and implementation of flight test parameter calibration data management system

Xue Bai[1,2], Ning Tang[1] and Wei Li[1]

[1]Chinese Flight Test Establishment, Xi'an 710089, China

[2]E-mail: 446234448@qq.com

Abstract. Given the shortcomings of current flight test calibration data management mode, in order to improve the calibration data management technology, based on user requirements, a test flight parameter calibration data management system with multiple functional modules was built under Visual C# and SQL Server platform. The C919 and HO300-903 flight test parameter calibration data was used for system function test. The efficiency of the system was tested with another sample of calibration data. The results show that the system is scientific and effective compared to the traditional file-based mode calibration data management method. The efficiency is increased by at least 2 times. The design of the system enhances the ability to calibrate data management techniques and reserve for calibration data mining for flight test sensors.

1. Introduction

The flight test parameter calibration data is used as the value transfer standard for post-flight data processing, which is the medium for converting recorded digital quantities into physical quantities. However, the number of calibration data is increasing because of the initial calibration, periodic calibration and visual calibration. It is urgent to carry out reasonable and efficient data management. At present, the flight test parameter calibration data management technology is still in the file stage. For one aircraft, each calibration data is named according to an independent name and stored in a computer in the form of an excel file [1]. Despite its long contribution, File storage method has a number of problems in use, such as poor sharing, large redundancy, and high data maintenance time cost. The current calibration data management mode is the supervisor retention system. Because different supervisors manage the calibration data in different ways, the same calibration data has different management modes according to calibration time, parameter type and so on, thus the management mode is difficult to unify.

With the development of computer application technology, database-based data management technology has emerged, and various industries at home and abroad are conducting data management technology research [2,3]. Test calibration is a special work link in the field of flight test. The storage situation of calibration data is unique. At present, advanced data management technology research has not been carried out. In order to improve the calibration data management technology, unify the data management mode, strengthen the standardization of test flight quality and standardize the operation process system, The .NET framework and SQL database technology is used in this paper to develop the flight test parameters calibration data management system based on the database model. The main aim of this paper is to design and implement a unified standardized management platform for all types of calibration data [4].

2. System requirements analysis

2.1. Functional design
The flight test parameter calibration data management system designed in this paper needs to have professional counterpart, convenience and good interactivity [5]. The system uses a top-down design, setting up three different user levels for common users, calibration supervisors, and system administrators. Through the construction of multiple management modules, all the functions of uploading, storing, querying, updating, deleting, exporting, parameter expiration management and data sharing of all flight test calibration data are realized. According to the calibration data supervisor responsibility system, set the data operation authority of different users to ensure the safety of calibration data operation. Set up system administrators to implement user management and resource configuration.

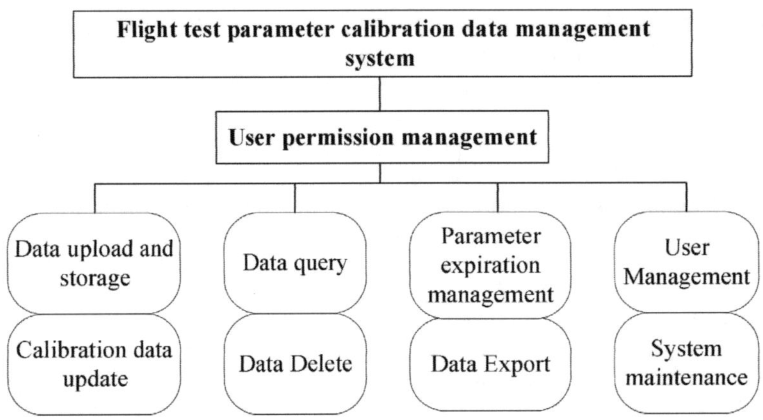

Figure 1. Function of the flight test parameter calibration data management system.

The overall function of the system is shown in Figure 1. The common user can use the calibration data query and export function. On the basis of this, the calibration supervisor has the authority to upload, store, update, delete and expire the calibration data; the system administrator is mainly responsible for the user rights management and database security maintenance.

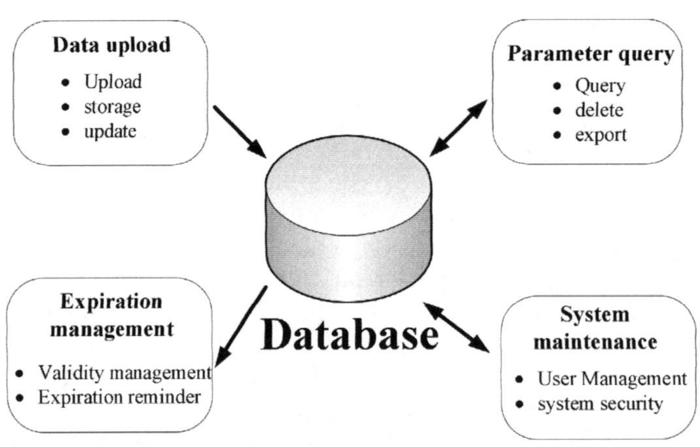

Figure 2. Structure of the flight test parameter calibration data management system.

2.2. Key technologies
The flight test parameter calibration data management system is oriented to the parameter calibration data of the full test flight cycle. The client/server mode is adopted as its technical architecture in order to facilitate the interface with the automated calibration system. Select Visual C# and SQL Server

under Visual Studio platform to realize its functions [6]. Data interactive operations are implemented through ADO.NET technology in the .NET framework.

2.3. System structure
According to the top-down modular design concept, design data upload, parameter query, expiration management, system maintenance four parts to achieve all the functions of the calibration data management system as shown in Figure 2.

3. Calibration data management system design

3.1. Database design
The key to the design of the flight test parameter calibration data management system is database construction. The system realizes the storage of the calibration curve and the setting of the system management authority by creating the information table, the data table, and the user table according to the content of the calibration data. The integrity of the calibration data storage is achieved by setting the aircraft model plus the parameter name as the primary key of the information table and the foreign key of the data table. The construction of data tables, information tables and user tables is shown in Tables 1 to 3.

Table 1. Calibration information table.

No.	Field name	Type	No.	Field name	Type
1	Aircraft Type	Varchar	13	Humidity	Varchar
2	Parameter	Varchar	14	Air pressure	Varchar
3	Standard	Varchar	15	Physical unit	Varchar
4	Sensor model	Varchar	16	Correction unit	Varchar
5	Sensor number	Varchar	17	Precision	Varchar
6	Collector model	Varchar	18	Calibration date	Varchar
7	Collector number	Varchar	19	Calibrator	Varchar
8	Regulator model	Varchar	20	Proof reader	Varchar
9	Regulator number	Varchar	21	Remarks	Varchar
10	Board model	Varchar	22	Date of Expiry	Date
11	Board number	Varchar	23	Page number	Varchar
12	Temperature	Varchar			

Table 2. Calibration data table.

No.	Field name	Type	No.	Field name	Type
1	Aircraft Type	Varchar	7	Code value3	Int
2	Parameter	Varchar	8	Code value4	Int
3	Physical quantity	Varchar	9	Voltage value	Small int
4	Correction	Varchar	10	Average code value	Int
5	Code value 1	Int	11	ID	Big int
6	Code value 2	Int			

Table 3. User table

No.	Field name	Type
1	User number	Varchar
2	Name	Varchar
3	Department	Varchar
4	Password	Varchar
5	Permission	Varchar

Journal of Physics: Conference Series **1419** (2019) 012026 doi:10.1088/1742-6596/1419/1/012026

3.2. System module design

The flight test parameter calibration data management system realizes the design and function realization of the interface of the system main interface, user login, data upload, parameter query, expiration management and system maintenance through the WPF-based user interface framework.

3.2.1. User login and system main interface. To maintain the login rights of the operating system, perform user authentication before entering the main interface. First, the user enters the job number and password, connects to the user table in the SQL database for security verification, and then obtains the login qualification after confirming the identity security exists; Secondly, before the user enters the main interface, his rights need to be confirmed. Only the administrator user has the right to access the system maintenance module. The module is hidden from other users, and at the same time, restricts the uploading, deleting, and expiration management functions of the ordinary users. Users are provided with three consecutive password error opportunities. Once the user is proved as unauthorized, the system enters a locked state and exits the login interface.

After the user successfully logs in, the welcome dialog box and the main interface of the system pop up, and the test data calibration system is officially entered. The user personal information maintenance module is provided in the system, which is convenient for the operation user to update and maintain the information.

3.2.2. Data upload. The data uploading module realizes the uploading, storing and updating functions of the calibration data. The data uploading module is designed with a calibration data selection unit, a database connection and upload unit, an upload process control unit, and an upload result display unit. Through the specific class to achieve the selection of multiple calibration data, after confirming, click the upload button. Through the OLEDB interface, the calibration data in the EXCEL template is read into the Dataset, uploaded to the database through the SQL connection class to achieve storage and update. The upload control module displays the parameter information of both waiting for upload and uploaded, and the parameter information of the upload failure is marked red to realize the user reminder. The upload result display module implements the statistics of the number of parameter upload results.

3.2.3. Parameter query module. The parameter query module implements the query, delete and export functions of the calibration data. The system designed two query modes, one is the hierarchical retrieval mode based on aircraft model, parameter type, and parameter name, and the other is a fuzzy query mode based on four keys information: aircraft model, parameter name, sensor model, and sensor number. The list of related items that are queried is displayed in the table of the interface. Double-click the parameter name to pop up the new interface to display the full calibration data. The last column of the table displays the delete button, and the calibration supervisor double-clicks to confirm the deletion of the calibration data. In order to realize data transfer, download and local save function, the calibration data management system opens the download and export port to the user, through the check box, realizes the selection of multiple parameter calibration data, clicks the export to local button, and writes the calibration data to the excel template.

3.2.4. Expiration management. This module mainly implements the expiration management function of the flight test calibration data. The validity period management of calibration data is an important part of the quality and safety management of flight test. The validity period of different types of parameters varies according to the calibration standard of the flight test. For example, the displacement parameter calibration data is valid for 1 year; the electrical parameter and pressure parameter calibration data are valid for 2 years; and the strain and temperature parameters are valid for 3 years. When the calibration data is uploaded, the calibration standard information is used to judge the parameter type, and the expiration time is calculated according to the calibration time, and is set to the Date format and stored in the corresponding field of the calibration information table.

In the expiration management module, we first set the query time period, and then click the query button to achieve the expiration parameter statistics. The system also provides the list local save function, which is convenient for the calibration supervisor's expiration control.

3.2.5. System maintenance. The system maintenance module mainly implements user management and database maintenance functions. System user management mainly implements system user addition, query, deletion, information update and password reset by accessing the user table in the database. The database maintenance function mainly includes backup and recovery of the database.

4. System implementation and testing

4.1 The specific implementation of the system

Using Visual C# under Visual Studio platform, the developer writes the program to realize the function module of the flight test parameter calibration data management system. The system is connected with the SQL Server database, the main interface and login interface of the data management system are shown in Figure 3 and Figure 4.

Figure 3. Login interface of the flight test Parameter calibration data management system.

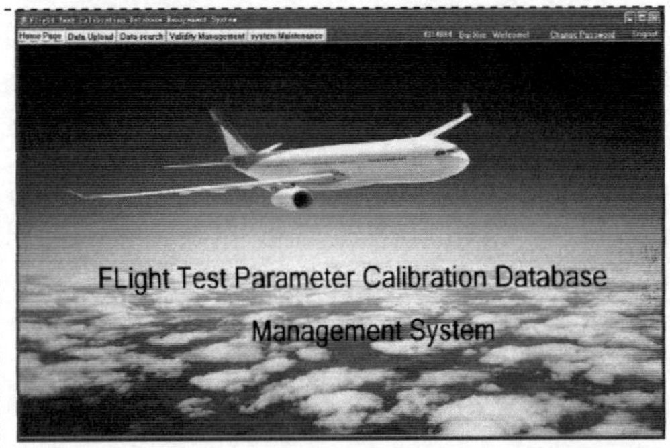

Figure 4. Main interface of the flight test parameter calibration data management system.

4.2. Functional test

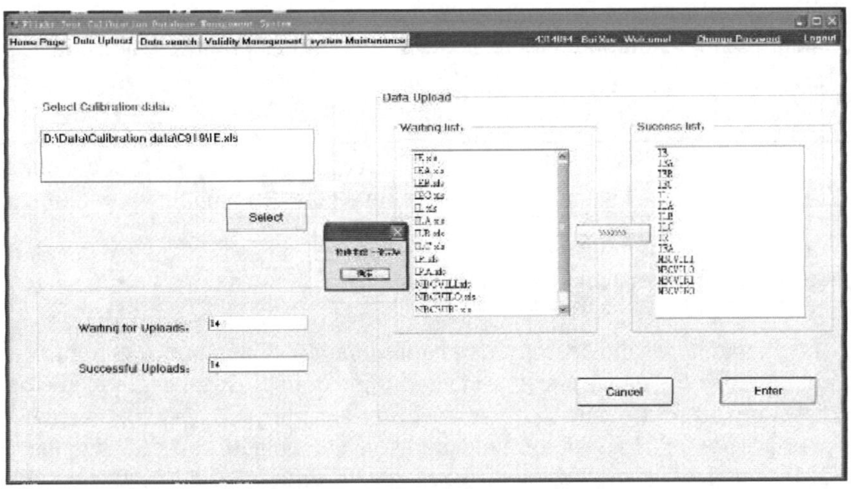

Figure 5. Data upload module test chart.

In order to verify the validity of the data management system, the calibration data of the civil aircraft C919-01 and the seagull HO300-903 are used as test samples to verify the function of the system. Upload nearly 1000 calibration data to the database to test the system's upload, query, delete, download and export, expiration management, and maintenance functions. The test chart of the upload and parameter query function module is shown in Figure 5 and Figure 6. After half a year of practical application, the validity and convenience of the data management system is proved. The test flight parameter calibration data management system developed in this paper solves the shortcomings of the current file management mode and provides a scientific information management platform for calibration data. Efficient management of calibration data is achieved, which improves the standardization of calibration work quality.

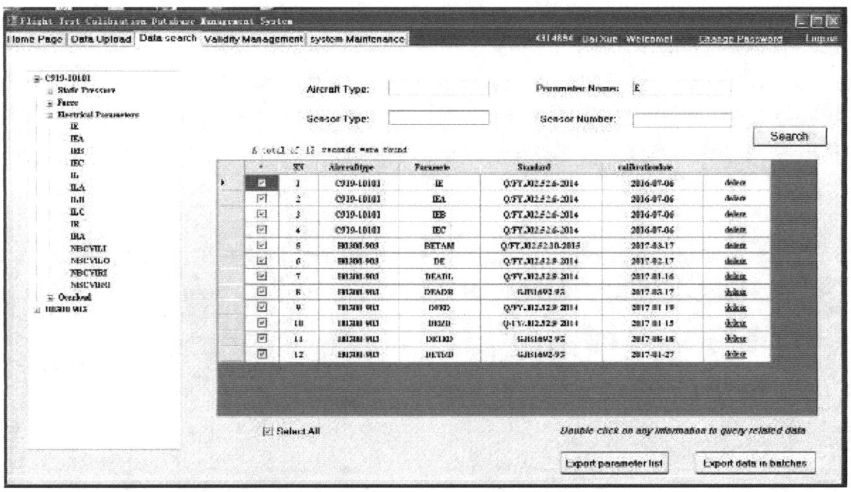

Figure 6. Parameter query module test chart.

4.3. Efficiency test

In order to test the execution efficiency of the flight test parameter calibration data management system, the software tester selects the traditional file-based data management method and the system designed in this paper to carry out the efficiency comparison experiment. A total of 5000 calibration data of 30 aircraft were selected as test samples, and the average query time of the data and the expiration management query time were compared and analyzed. The experiment compares the time taken by the two systems to query the calibration data of 50 specified parameter names, and the time taken to count and list the expiration parameters of the past three months. The comparison results are shown in Table 4.

Table 4. Efficiency comparison.

	Time spent querying 50 calibration data	Time spent counting three-month expiration parameter
Traditional file management mode	10min	30min
The system designed in this paper	3min	1min

After testing, it was found that the efficiency of calibration data operation was improved by at least 3 times. The calibration data management system designed in this paper uses ado.net technology, through the establishment of a database, to achieve unified import, query, list display and export of parameters, to make up for the lack of human-based file operations, showing the advantages of systematic and systematic data management. It has been verified that the test data calibration system using the flight test parameters improves data management efficiency and shortens the data management time of technicians.

5. Conclusion

Test calibration is one of the important technical aspects in flight test. Calibration data is a measure of the value of post-flight data processing, while scientific and efficient management of it is of great significance. The flight test parameter calibration data management system designed and implemented in this paper realizes the functions of uploading, querying, exporting and expiring the calibration data, and at the same time improves the efficiency of various management operations of the calibration data by at least 3 times, makes up for the shortcomings of traditional file system-based data management methods and improves the efficiency of the work. As a standardized flight test calibration data management system, not only can the current relatively backward calibration data management mode be changed, but also the centralized analysis and mining of the calibration data stored in the system can be facilitated. The system has been successfully applied to the data management process of multiple testing machines such as C919 and ARJ, and achieved good results.

References

[1] Dang Huaiyi 2017 Application study of flight test project big data management and mining *Computer Measurement and Control* **25** (11): 299
[2] Liu Jian 2019 Design and implementation of asset management information system *Nanchang University*
[3] Xie Chi 2018 Design of computer-aided dance teaching resource management system *Modern electronics technique* **41** (16): 100
[4] Jiang Hongna and Gong Haibo 2017 Research and Implementation of System Calibration Automation Technology *China Science and Technology Information* **10** p 25
[5] Jabeen N and Ghani A 2015 Comparison of the traditional chalk and board lecture system versus Power Point presentation as a teaching technique for teaching gross anatomy to the first professional medical students *Edinburgh working papers in applied linguistics* **4** (11): 1811
[6] Hammer J V 2015 *Beginning C# 6.0 Programming with Visual Studio 2015* (Birmingham: Wrox Press Ltd)

Journal of Physics: Conference Series **1419** (2019) 012027 doi:10.1088/1742-6596/1419/1/012027

Research on infrastructure of cloud computing for power application

Hao Qin

China Electric Power Research Institute, Nanjing, 211106, China

E-mail: qinhao@epri.sgcc.com.cn

Abstract. The key development directions of state grid corporation of China, such as smart grid and centralized data centre, all put forward a lot of demands for cloud computing. In general, it is necessary to build a more flexible, convenient and secure information support platform through cloud computing virtualization, distributed, parallel computing and other technical means. Finally let hardware, software to the direction of centralized service evolution.

1. Introduction

Cloud computing is an internet-based computing model of mass participation. Its computing resources (including computing power, storage capacity, and delivery capabilities) are dynamic, scalable, virtualized, and delivered as services. This new mode of organization, distribution and use of computing resource is conducive to rational allocation of computing resources and improving their utilization rate, promoting energy conservation and emission reduction, and achieving green computing. Cloud computing includes several key technologies such as virtualization, distributed storage, distributed data management, distributed programming, etc. Its outstanding advantage is to improve the computing power of hardware and software while improving resource utilization, so as to facilitate the implementation of distributed computing and parallel computing. For large-scale, complex, parallel computation, such as massive data processing, information acquisition, search engine, simulation calculation, multi-linear high-precision numerical calculation model has a good application value [1].

The State Grid Corporation informatization experienced the rapid development stage of the 1980s with the focus on dispatching production automation and individual management information systems, and the rapid development stage of enterprise informationization in the 1990s and the Tenth Five-Year Plan period. During the "eleventh five-year plan" period, remarkable achievements have been made in the informatization of SG186 engineering and communication network construction. The information system has become the basic means for the daily grid business of the company's employees, and has played an increasingly important role in power grid production control and company management. However, with the promotion and deepening application of a large number of information systems within the company, the accumulated data and services tend to be massive, user experience requirements continues to grow, the data is massive, the distribution is heterogeneous, the processing is complicated, the use is cumbersome, and the maintenance is difficult. Cloud computing technology is an effective solution to solve the above problems. At the same time, cloud computing can also provide a solid foundation and strong support for power grid, power grid planning and design, unified search platform, collaborative work platform, intelligent electricity huge amounts of information processing and analysis, decision support system, unstructured data storage system, and mass real-time data storage and disaster recovery systems and more Grid business scenarios [2-6]. At present, under the new situation of the State Grid Corporation building a strong smart grid and SG-ERP project, under the general direction of building a "one strong three excellent" modern company, fully utilizing the current advanced cloud computing technology can effectively help the intelligent transformation of power grid operation and the intensive transformation of enterprise operation. Specific needs can be reflected in the following aspects:

1.1. Information collection, storage and processing technology

With the large-scale construction of information systems and the large-scale implementation of the Internet of Things, information collection is developing towards a wider range, more types, higher frequencies, and more precise directions. At the same time, higher requirements are put forward for information processing and storage technology. As the source increases, the type of information becomes more complex. Real-time, non-real-time, structured, semi-structured, unstructured information will be flooded in various business systems, and the amount of information will show a big explosion. According to statistics, after the completion of the SG186 project, the average daily data growth of the eight business applications reached more than 50 million (144G); with the construction of the smart grid and SG-ERP, the company's data growth rate will be repeated several times. Ultra-large-scale composite information storage, retrieval, backup and disaster recovery will become important technical fields. The construction effect of data center and disaster recovery center will directly affect the continuity of the overall business. Cloud computing pools computing resources and storage resources, and can dynamically expand as demand increases; through distributed computing, parallel computing and other technologies can significantly improve the efficiency of information processing; through distributed storage and normal backup, you can achieve high reliability and parallel read of storage [7, 8].

1.2. Unified architecture, unified deployment, unified operation and maintenance, and unified management

The information system of State Grid Corporation has the problems of complex architecture and low operation and maintenance efficiency [9]. The standards and specifications of the existing IT infrastructure are not perfect. The IT architectures within the provinces and the provinces are not uniform. The existing application systems all configure hardware and software resources according to their respective business needs, and various heterogeneous hardware and software flooded in the machine room. The utilization of software and hardware resources of information systems is insufficient, and the complexity of deployment and operation and maintenance is high, and personnel consumption is large. Through hyper-scale virtualization, cloud computing can realize hardware and software resource integration and centralized management, and uniformly allocate and centralize tuning are carried out according to the characteristics of specific applications. The way in which hardware and software equipment are configured according to the needs of a single business application will be improved to the way of unified planning and design of the overall architecture, overall arrangement of equipment procurement and configuration on demand.

1.3. Energy saving and emission reduction, building a green data center

Building a data center is an important measure for the construction of State Grid Corporation's informatization, but the data center is also a major energy consumer that is unanimously recognized by the industry. Traditional data center must purchase large server queues according to the peak demand of the enterprise; in fact, enterprises have their computing needs ebb and flow, and don't always need so many servers. According to statistics from Internet, the average utilization rate of software and hardware for large enterprises is only about 10%. However, traditional servers, even in idle state, power consumption is also 60% of the peak. The power consumption of idle devices and the energy consumption of cooling equipment in data center will increase energy consumption by 50% to 100%. Green IT is an important way to save energy and reduce emissions, and data center energy conservation bears the brunt. Cloud computing abstracts IT resources and services from the underlying architecture. Computational resources, storage resources, and network resources are all integrated into a virtualized resource pool to provide highly scalable on-demand services over IP networks in a multi-user and resilient environment. The traditional resource full operation mode is changed to the on-demand operation mode, and the temporarily unnecessary hardware resources are turned off, thereby effectively reducing energy consumption.

1.4. Business management strategy
In the next 5 to 10 years, it is a key period for the company to deepen the management requirements of "two transformations" and "three episodes and five majors" and realize the "one strong and three excellent" modern company. At present, the main business applications of the State Grid adopt two-level deployment and three-level application, while the future needs to adopt one-level deployment. According to the SG-ERP plan, between 2010 and 2015, more than 50% of the application systems will be deployed in a centralized manner. Realize the first-level deployment of enterprise core applications,and build a set of integrated comprehensive management platform that can not only meet the business needs of the corresponding business departments, but also can stably operate ,improve the management level of business departments, and achieve intensive management objectives, which can strengthen the management and control of the headquarters. Enhance data reliability, save investment costs, and reduce operation and maintenance workload, but at the same time put forward high requirements for the reliability, load capacity and dynamic balance of the operating environment. Cloud computing technology can realize the dynamic expansion of operating resources with capacity load, which can provide strong support for the first-level deployment of applications.

In order to make better use of cloud computing for power system services, based on the overall overview of power cloud computing, this paper analyses the power application cloud platform reference architecture from the perspective of development status, key technologies and reference architecture.

The rest of the paper is organized as follows: Part 2 focuses on the development status; Part 3 introduces the key technologies of the power cloud platform; Section 4 gives its platform reference architecture; Section 5 summarizes and looks forward to it.

2. Development status

2.1. Technological development
In recent years, the cloud computing market has grown rapidly, industry standards have been gradually established, technology levels have been continuously improved, and the industrial environment has gradually matured. Cloud computing has experienced a development process from small to large, from hype to maturity, and is entering the 2.0 era of great change. The cloud computing 1.0 era is mainly used by the Internet or start-ups, but with the continuous development of cloud computing, enterprise customers are gradually realizing the huge advantages of cloud computing and starting to use cloud computing. In the era of cloud computing 2.0, all cloud computing enterprises are building an ecosystem with their own core or an ecosystem with open source as the core.

In the era of cloud computing 2.0, OpenStack and CloudStack are typical open source cloud computing management platforms, which promote the rapid development of cloud computing. Among them, OpenStack is a free software and open source project licensed by NASA (National Aeronautics and Space Administration) and Rack Space. It provides computing services, object storage services, virtual image services, and 7 types of service components, such as dashboard, authentication and authorization service, network service, and data block storage service, covering various aspects such as network, virtualization, operating system, and server. OpenStack supports almost all types of cloud computing environments, and its goal is to provide a cloud computing management platform that is simple to implement, scalable, rich, and standardized. OpenStack provides Infrastructure as a Service (IaaS) solutions through a variety of complementary services, each providing APIs for integration. In addition to the strong support of Rack Space and NASA, OpenStack also has the contributions and support of heavyweight companies such as Dell, Citrix, and Cisco.

CloudStack was originally developed by Cloud.com and is divided into commercial and open source versions. The open source version is licensed under the GPL General (3) license, and after Citrix acquired Cloud.com in 2011, the entire code will be available. Open source, and in 2012, will contribute CloudStack to the Apache Software Foundation as an Apache incubator project, and change the license agreement to a more relaxed and business-friendly Apache license agreement. CloudStack was upgraded to Apache's official project in March 2013. CloudStack is an open source, high-availability and scalable cloud computing platform that provides highly available, highly scalable open cloud platforms for large-scale virtual machine deployment and management, supporting the management of most mainstream hypervisors, such as KVM. XenServer, VMware, Oracle VM, Xen, etc.

In summary, OpenStack and CloudStack are currently typical open source cloud computing management platforms. Technically, OpenStack and CloudStack have their own advantages.

With the development of cloud computing technology and the clear demand, the virtual container engine Docker technology came into being in March 2013, which opened the curtain of the transformation of the product release method based on the cloud computing platform. Docker is a PaaS provider dotCloud open source LXC (Linux Containers)-based advanced container engine, source code is hosted on Github, based on the language and Apache2.0 protocol open source. Currently, Docker is very popular in the industry, including DotCloud, Google Compute Engine, and Baidu Application Engine (BAE), which all use Docker. Docker is a container engine built on top of LXC technology by Docker. Inc. The source code is hosted on GitHub, based on Go and Apache 2.0.

2.2. Standard development

The wide application of cloud computing can affect a large number of products, systems and services, and the standard will lay a solid foundation for it. Since 2008, cloud computing has become one of the hotspots of standardization in the world. There are 33 standardization organizations and associations at home and abroad to carry out cloud computing standardization work from various angles, including: ISO/IEC First Joint Technical Committee (ISO/IEC JTC1), the International Telecommunication Union Telecommunication Standardization Sector (ITU-T), the National Institute of Standards and Technology (NIST), and the European Network and Information Security Agency (referred to as the European Network and Information Security Agency, ENISA), Distributed Management Task Force (DMTF), Storage Networking Industry Association (SNIA), Cloud Security Alliance (CSA), China Communications Standards Association (CCSA) which are among the most important organizations. The release of cloud computing standardization organizations and standards research at home and abroad.

In addition, the International Organization for Standardization (ISO), the International Electro-technical Commission (IEC) and the International Telecommunication Union (ITU) jointly developed the international cloud computing standard ITU-T Y.3500 and ISO for the first time in the field of cloud computing in 2014: ISO/IEC 17788 "Overview and Vocabulary for Information Technology Cloud Computing" and ITU-T Y.3502, ISO/IEC 17789 "Information Technology Cloud Computing Reference Architecture". Among them, "Information Technology Cloud Computing Overview and Vocabulary", including cloud computing overview, terminology and definitions, provides the basic standards for cloud computing; "Information Technology Cloud Computing Reference Architecture", which specifies cloud computing reference architecture specifications, including cloud computing roles and activities, functional components and their relationships.

2.3. Application development

Cloud computing related standards system is gradually established, technology has become increasingly mature, and risks are controllable. Large enterprises at home and abroad have developed cloud computing applications, built internal cloud and achieved obvious social and economic benefits.

Foreign countries, such as United Energy Corporation, France Gas Suez Group, and Korea Power Company have achieved positive results through cloud computing applications. United Energy Corporation manages self-service orders and other and repetitive IT services by implementing a private cloud, reducing the cost of corporate IT services, reducing annual IT services by about 12%, and achieving higher IT service reliability. IT departments have won business. The trust of the department; the French gas Suez Group, through the private cloud to build a "free cooling" data center, and the definition, implementation and maintenance of the group's internal IT services, improve the ability of IT collaboration and sharing, reduce IT Service cost; Korea Electric Power Company, implementing the "Cloud Computer" project to convert office computers into virtualized terminals, is expected to reduce spending by more than 5 billion won, improve the convenience of employees, and reduce the risk of data leakage.

Domestically, cloud computing has been listed as a national strategic emerging industry. Several ministries, provincial and municipal governments have successively introduced relevant policies to promote the development of cloud computing. The Ministry of Industry and Information Technology and the National Development and Reform Commission jointly issued Beijing, Shanghai, Shenzhen, Hangzhou and Wuxi as cloud computing. Pilot demonstration city first, "China Cloud" industrial development plan approved by the State Council. In addition, some large domestic enterprises and institutions, such as State Grid Corporation, China Mobile, China Telecom, PetroChina, and Sinopec, have built private clouds for enterprise applications. State Grid Corporation, built the company's power cloud, cloud computing has become the core operating model of the company's management and service information systems, realizing the innovation of IT infrastructure, the concentration of business application systems and providing innovative IT service capabilities; China Mobile, built IT Supporting cloud, business cloud, and public service cloud three cloud computing platforms, realizing the rational allocation of resources within the network, providing external IT services, reducing the overall cost of construction and operation and maintenance, and improving the utilization and reliability of IT systems and networks. Improve the sustainability of infrastructure; China Telecom, cloud computing main business mainly includes IDC and CDN, and cloud hosting and cloud storage based on these IT infrastructure resources, which reduces the overall cost of construction and operation and maintenance, and improves Infrastructure sustainability.

3. Key technologies for power cloud platforms

The key technologies involved in cloud computing include: virtualization technology, parallel computing, distributed technology and cloud computing security.

3.1. Virtualization technology

Virtualization refers to the abstraction of resources, the logical representation of resources, and is not subject to physical constraints. It is divided into server virtualization, storage virtualization and network virtualization. Virtualization is the foundation of resource pooling, which can effectively improve the level of resource management.

Table 1. Key Lifting Points and Optimization Contents of Virtualization Technology.

Key lifting point	Optimize content
Centralized deployment and refined management of IT resources	The use of virtualization technology to improve the control accuracy of the entire life cycle of information resources, real-time centralized monitoring of resource status, to provide accurate decision-making basis for the overall planning of information resources.
Low coupling between system construction and physical resources	Realize resource pool management of IT infrastructure such as servers, storage, and networks, quickly allocate resources as needed, and realize rapid and automatic installation and configuration of the operating environment, effectively improving operation and maintenance management efficiency and human investment costs.
Application high availability	Realize server fault automation management, applications running on failed servers are automatically migrated to idle servers in the resource pool, reducing application downtime from hours to minutes.
Resource application configuration is not exclusive, high resource utilization	Dynamically allocate resources such as computing, storage, and network available in the resource pool, and supply them to new applications and applications that need to expand resources, which can greatly improve the efficiency of server resource utilization.

The technical advantages of virtualization technology are mainly reflected in the following four aspects: (1) reducing IT hardware investment costs; (2) simplifying server operation and maintenance management, reducing operation and maintenance management costs; (3) improving the flexible adaptability of IT architecture for business applications; (4) supporting the integration of heterogeneous environments and the continuous operation of business applications. After the introduction of virtualization technology, the key lifting points and optimization content are listed in Table 1.

3.2. Parallel computing

Parallel computing is based on serial computing, which refers to the process of using multiple computing resources to solve computational problems. It is the main way to implement high-performance, high-availability computer systems.

The technical advantages of parallel computing are mainly reflected in the following three aspects: (1) improve the computational efficiency for massive data processing; (2) improve the concurrency of data processing, simplify the difficulty of parallel programming; (3) provide efficient, flexible, and scalable computing service capabilities for customized development. After the introduction of parallel computing technology in cloud computing, the key lifting points and optimization contents are listed in Table 2.

Table 2. Key Lifting Points and Optimization Contents of Parallel computing technology.

Key lifting point	Optimize content
high performance	Use a variety of computing resources to quickly solve large complex computing problems and provide high-performance cloud computing services.
High concurrency	Computation of computational task decomposition and integration of computational results through concurrent work mode, while simplifying the difficulty of parallel programming, while having high throughput, low latency computing tasks concurrent processing performance advantages.
High availability	Improved Parallel Computing Task Allocation Mechanism based on Resource Dynamic Scheduling Mechanism, realizing the dynamic joining of resource nodes and the repair and exit of the effective nodes to ensure the high availability of cloud computing services.
Highly scalable	Use PC servers, blade servers, etc. to replace large computers, overcome the limitations of the scale expansion of mainframes, and use the network to connect computing resources to work together, which with flexible expansion capabilities.

3.3. Distributed File System

The distributed file system is the core of the storage system. It provides a unified and object-oriented access interface to the user through the abstraction of the storage space managed by the operating system, and shields the direct operation and resource management of the physical device.

The technical advantages of distributed file system are mainly reflected in the following four aspects: (1) providing efficient information integration, exchange, storage and access services, and realizing centralized management of data services; (2) effectively integrating data to achieve centralized management of data, improving cross-regional and inter-departmental information sharing; (3) forming a reasonable data layout, reducing data interaction costs of service terminals, and concentrating superior resources to improve service performance; (4) forming a physically dispersed and logically centralized IT architecture to effectively improve the efficiency of system operation and maintenance and management. After cloud computing introduced a distributed file system, the key lifting points and optimization contents are listed in Table 3.

Table 3. Key file system lifting points and optimization content.

Key lifting point	Optimize content
Virtuality	Further improve the parallelism of work and the level of collaborative autonomy, and enhance system operational reliability and system transparency.
Availability	Improve the continuous availability of file service capabilities and reliable data service capabilities in a distributed environment.
Scalability	Increased performance scalability and management scalability, the performance growth of the system is linear with the scale expansion, and the expansion of system size will not bring about an excessive increase in management complexity.
safety	Improve data storage security and trust management to protect data in the file system from being stolen, tampered and destroyed.

3.4. Cloud Computing Security

Cloud computing security is a new technology, which accompanying the development of cloud computing. It is the innovation and deep development of traditional information security technology in the cloud computing environment, including the security between cloud computing internal security and cloud computing. With the development of cloud computing and the popularity of applications, the importance of cloud computing security has gradually increased, which is the basic premise and guarantee for the construction and promotion of cloud computing.

4. Platform reference architecture

Combining the functional and non-functional requirements of the cloud platform of the power industry, as well as the cloud computing implementation cases and experiences of the State Grid Corporation, through the organic combination of the traditional IT architecture system and the newly expanded cloud computing model, build a reference architecture for the next-generation power industry application cloud platform (as shown in Figure 1), it effectively supports cloud computing data center to be efficient, flexible, and scalable, and promote innovative development of IT services and business models.

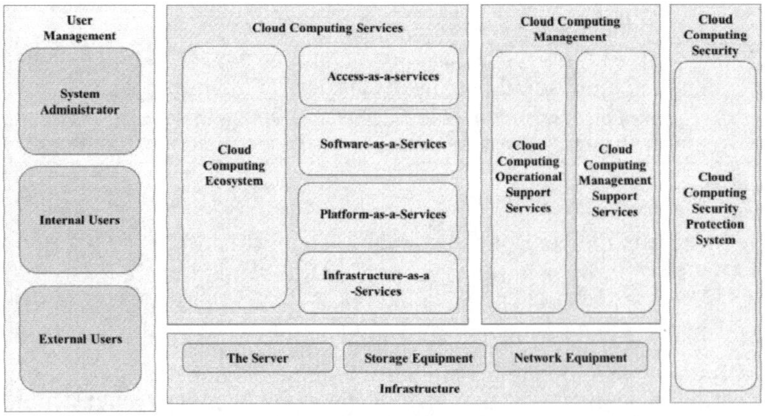

Figure 1. Power application cloud computing reference architecture.

The cloud computing reference architecture provides basic principles, technical routes, application modes, etc. for the construction of the cloud computing platform, and can effectively guide the definition, design, implementation, and application requirements of the application of the cloud computing platform by the Chongqing Electric Power Company. The technical architecture and deployment architecture are described in detail below.

4.1. Technical Architecture

Based on the above cloud computing reference architecture, according to the overall architecture of the State Grid Corporation (SG-ERP), combined with the cloud industry application requirements of the power industry, the hardware resource pool, the data resource pool, and the application resource pool are taken as the core to form a power application cloud platform technical architecture, which is shown in the Figure 2.

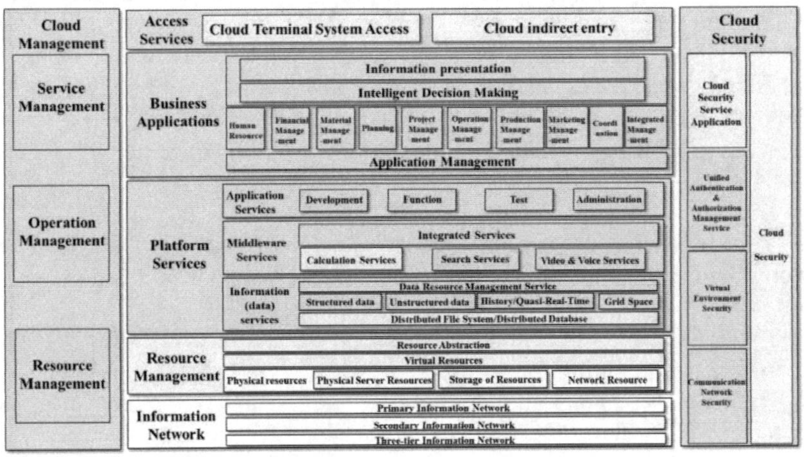

Figure 2. Technical architecture diagram.

4.2. Deployment Architecture

The entire power application cloud platform should support distributed deployment, centralized management mode, and has strong scalability.

4.2.1. Storage domain. The storage domain is the basic data support environment of the cloud computing platform, and it can adopt the SAN storage architecture and the distributed storage architecture hybrid mode. SAN storage architecture. From the perspective of application types, storage is based on high-performance grid storage arrays that support scale-out, while taking into account the need for large concurrent external user access, hardware-based storage virtualization systems are used to store different types of storage in the network. The devices work together to provide data storage and service access functions.

4.2.2. Computing domain. The computing domain is the core environment for the cloud computing platform to support application services. It can be divided into traditional application pools, x86 physical pools, and x86 virtual resource pools.

4.2.3. Network domain. The network domain is the basis for connecting various functional domains to implement intelligent data storage and high-performance computing. It is recommended to adopt dual-system cluster, 10 Gigabit fibre backbone, and 10 Gigabit to PC server configuration.

4.2.4 Management domain. The management domain mainly carries the cloud computing service platform service management, including resource management, service management, portal management, security management, etc. It can flexibly allocate resources dynamically, define data security access, storage, and backup mechanisms, and support unified management to multiple virtualization technologies and heterogeneous storage.

4.2.5 Security domain. According to the information system security protection requirements, information security protection includes four levels: boundary protection, network protection, host protection, and application protection. The power industry application cloud platform security service domain is designed for network security, host security, virtual machine security and application security.

5. Conclusion

Cloud computing marks an important transition of information technology from product to service, and it is the inevitable development direction of integration and intensification of enterprise information construction. The State Grid Corporation's information support platform needs to introduce cloud computing technology to achieve more flexibility, simplicity and security, and realize the gradual evolution of hardware and software to centralized services. Based on the construction of power cloud platform, this paper focuses on the power application cloud platform reference architecture from the three aspects of development status, key technologies and reference architecture, and provides technical support for the cloud computing construction of State Grid Corporation.

References

[1] WANG Ji-Ye, MENG Kun, CAO Jun-Wei and CHENG Zhi-Hua 2015 Information Technology for Energy Inernet: A Survey *Journal of Computer Research and Development* **52** (5): 1109-1126
[2] WANG De-Wen, SONG Ya-Qi and ZHU Yong-Li 2010 Information Platform of Smart Grid Based on Cloud Computing *Automation of Electric Power Systems* **34** (22): 7-12
[3] MU Lian-Shun, CUI LI-Zhong and AN Ning 2011 Research and Practice of Cloud Computing Center for Power System *Power System Technology* **35** (6): 171-175
[4] ZHAO Jun-Hua, WEN Fu-Shuan, XUE Yu-Sheng and LIN Zhen-Zhi 2010 Cloud Computing: Implementing an Essential Computing Platform for Future Power Systems *Automation of Electric Power Systems* **34** (15): 1-8
[5] ZENG Hai-Long, ZHANG Bo-Feng, ZHANG Li-Hua, HE Bing and WU Geng-Feng 2010 Virtual Cluster Constructing Based on Cloud Computing Platform *Microelectronics & Computer* **27** (8): 31-35
[6] WANG Xiao-Yu, CHENG Liang-Lun 2014 Study of multi-source information resources cloud systems and cloud services model on cloud computing *Application Research of Computers* **31** (3): 784-788
[7] ZHANG Shui-ping, LI Ji-zhen, ZHANG Feng-qin, LI Xiao-bo and YU Kan-min 2011 Research and implementation of data center security system based on cloud computing *Computer Engineering and Design* **32** (12): 3965-3968
[8] PAN Rui, LIU Jun-Yong and GUO Xiao-Ming 2009 Cloudy Computing of Power System Oriented to Smart Grid *Sichuan Electric Power Technology* **32** (S1): 6-10
[9] ZHANG Da-jun, LI Yun-fa and ZHENG Zhou 2012 A Secure Sharing Mechanism for Data Resources in Cloud Computing *Netinfo Security* (8): 79-82

Journal of Physics: Conference Series **1419** (2019) 012028 doi:10.1088/1742-6596/1419/1/012028

Comparative analysis for rock slope stability with different constitutive models of rock materials

H Yu[1,3], S J Li[2]

[1]School of Architecture and Civil Engineering, Shenyang University of Technology, Liaoning, Shenyang 110870, China
[2]State Key Laboratory of Structural Analysis for Industrial Equipment, Dalian University of Technology, Liaoning, Dalian 116024, China

[3]E-mail: yhdut9954053@163.com

Abstract. Based on the principle of reducing shear strength of rock materials, a rock slope stability with different constitutive models of rock materials is performed by using finite element method. The safety factor of rock slope stability is computed based on convergence property of elastic-plastic numerical simulations by reducing shear strength of rock materials until the numerical computing is divergent or plastic zone is connected. Three kinds of Drucker-Prager yield criterions with different yield surfaces are utilized to investigate the influence of yield criterions with different yield surfaces to safety factor of rock slope stability. The investigation shows that the computed safety factors of rock slope stability by using Drucker-Prager yield criterions with different yield surfaces are approximately equal. The maximum relative error for safety factors is less 10% by using different Drucker-Prager yield criterions.

1. Introduction

Rock slope stability is commonly affected by shear strength parameters of rock materials. Mohr-Coulomb yield criterion and Drucker-Prager yield criterion are widely used to evaluate rock slope stability. However, yield criterions with different yield surfaces will affect safety factor of rock slope stability. Raghuvanshi comprehensively reviewed on governing parameters and various stability analysis techniques for plane mode of failure in rock slopes [1]. Elmo introduced an improved terminology commonly used by the discrete fracture network community to define rock bridge intensity relative to the sampling region. In rock engineering, the measurement of rock bridges is exacerbated by the fact that rock bridges are not visible unless the rock mass is exposed by human activities or by natural events such as rockfalls [2]. Schlotfeldt proposed an integrated approach of designing overhanging rock slopes where the relative dimensions of the slope exceed the scale of fracturing and the rock mass failure needs to be considered rather than kinematic release of individual blocks[3]. Tang developed a model that combines the FEM and DDA approaches. The main concept of this approach is to first apply FEM to model crack growth behaviour and then automatically switch to the DDA module to model the post-failure process when the slip surface forms [4]. Lv proposed a procedure to determine the shear strength parameters through back analysis based on stability and failure status before and after earthquake and monitoring data after earthquake [5]. Belghali adopted a pseudo-static method to account for the inertial forces induced in the rock mass by seismic events. The strength properties of the rock material are described by a modified HoekeBrown strength criterion, whereas the passive bolts are modelled as bar-like inclusions that exhibit only resistance to tensile-compressive forces [6]. Chen presented a case history of the 205m high intake slope at the Huangjinping (HJP) hydropower station on the Dadu River. Deformation and cracks developed on the cut slope during excavation, and the deformation extended onto the natural slope above the cut, giving rise to serious safety concerns [7]. Stead emphasized the importance of structural geology to slope stability assessments, reviewing how structures control slope failure mechanisms, how engineering geologists measure structures and include them in slope stability analyses, and how numerical simulations of slopes incorporate geological structures and processes [8]. Wong developed the numerical manifold method as a tool to investigate the progressive failure in rock slopes. The entire processes of the progressive slide surface development related to crack initiation, propagation, coalescence and degradation to eventual catastrophic failure are successfully captured [9]. The objective of the paper is to analyse the influence of yield criterions with different yield surfaces to safety factor of rock slope stability and evaluate rock slope stability for Bijia mountain.

2. Determination of parameters of constitutive models for rock materials

Mohr-Coulomb yield criterion is widely used to characterize soil and rock yielding property and is fund to hold well in actual tests, and is expressed as

$$\frac{\sigma_1-\sigma_3}{2}=[\frac{\sigma_1+\sigma_3}{2}+\frac{C}{\tan\varphi}]\sin\varphi \tag{1}$$

Where C and φ denote the cohesion and internal frication angle of rock mass, respectively. σ_1 and σ_3 denote maximum principal stress and minimum principal stress. C and φ can be determined from direct shear tests of rock specimens with different normal pressure. Main mechanical parameters of rock materials of some rock slope are listed in Table 1.

Table 1. Main mechanical parameters of rock materials of some rock slope.

Parameter	ρ/kg/m³	C/MPa	φ/°	E/GPa	μ
Values	2300	0.20	35	15.0	0.30

The yield surface defined in Mohr-Coulomb yield criterion has a serious drawback due to its angular nature in the principal stress space. In order to deal with the problem, Drucker-Prager proposed smooth surface by writing the yield criterion as a continuous relationship between the stress invariants. Drucker-Prager yield criterion with linear yield surface is expressed as

$$\begin{cases} F=t-p\tan\beta-d=0 \\ t=\frac{q}{2}[1+\frac{1}{k}-(1-\frac{1}{k})(\frac{r}{q})^3] \end{cases} \tag{2}$$

Where β, q, d, p, k and r are parameters of Drucker-Prager yield criterion, and can be determined from following equations and Table 2 when parameters (C and φ) in Mohr-Coulomb yield criterion are known. k is strength ratio between tensile and compressive tests, and $0.778< k<1.0$. p is average stress.

$$\begin{aligned} p&=-\frac{1}{3}(\sigma_1+2\sigma_3) \\ q&=\sigma_1-\sigma_3 \\ r&=-q \end{aligned} \tag{3}$$

$$d=\begin{cases} (1-\frac{1}{3}\tan\beta)\sigma_c & \text{uniaxial compressive test} \\ (\frac{1}{k}-\frac{1}{3}\tan\beta)\sigma_t & \text{uniaxial tesile test} \\ \frac{\sqrt{3}}{2}\tau(1+\frac{1}{k}) & \text{direct shear test} \end{cases} \tag{4}$$

Drucker-Prager yield criterion with parabolic yield surface is expressed as

$$\begin{cases} F = \sqrt{l_0^{\,2} + q^2} - p\tan\beta - d' = 0 \\ l_0 = d'|_0 - p_t|_0 \tan\beta \end{cases} \tag{5}$$

$$d' = \begin{cases} \sqrt{l'^2_0 + \sigma_c^2} - \dfrac{\sigma_c}{3}\tan\beta & \text{uniaxial compressive test} \\ \sqrt{l_0^2 + \sigma_t^2} + \dfrac{\sigma_t}{3}\tan\beta & \text{uniaxial tesile test} \\ \sqrt{l'^2_0 + d^2} & \text{direct shear test} \end{cases} \tag{6}$$

Drucker-Prager yield criterion with exponent yield surface is expressed as

$$F = aq^b - p - p_t = 0 \tag{7}$$

Where a and b are model parameters and can be determined from Table 2.

$$p_t = \begin{cases} a\sigma_c^b - \dfrac{\sigma_c}{3} & \text{uniaxial compressive test} \\ a\sigma_t^b + \dfrac{\sigma_t}{3} & \text{uniaxial tensile test} \\ ad^b & \text{direct shear test} \end{cases} \tag{8}$$

Table 2. Relationship between parameters in Mohr-Coulomb yield criterion and parameters in Drucker-Prager yield criterions (Associated flow rule).

Linear Drucker-Prager	$\tan\beta = \dfrac{\sqrt{3}\sin\varphi}{\sqrt{1+\frac{1}{3}\sin^2\varphi}}$	-----	$\dfrac{d}{c} = \dfrac{\sqrt{3}\cos\varphi}{\sqrt{1+\frac{1}{3}\sin^2\varphi}}$
Parabolic Drucker-Prager	$\tan\beta = \left(\dfrac{\sqrt{3}\sin\varphi}{\sqrt{1+\frac{\sin^2\varphi}{3}}}\right)$		$\sigma_c = \dfrac{c\sqrt{3}\cos\varphi}{\left(1-\frac{\sin\varphi}{\sqrt{3+\sin^2\varphi}}\right)\sqrt{1+\frac{1}{3}\sin^2\varphi}}$
Exponent Drucker-Prager	$a = \dfrac{\sqrt{1+\frac{1}{3}\sin^2\varphi}}{\sqrt{3}\sin\varphi}$	$b=1$	$\sigma_c = \dfrac{c\sqrt{3}\cos\varphi}{\left(1-\frac{\sin\varphi}{\sqrt{3+\sin^2\varphi}}\right)\sqrt{1+\frac{1}{3}\sin^2\varphi}}$

When parameters in Mohr-Coulomb yield criterion are known, parameters in Drucker-Prager yield criterions can be determined.

3. Practical application of Drucker-Prager yield criterions to slope stability analysis

Some cross section of Bijia mountain, located in Jinzhou city, Liaoning province, is shown in Figure 1. The slope angles are 50° and 60° in left and right sides, respectively. The height of FEM model is 61.5 m. The widths of FEM model are 20m and 106.8m in upper and bottom. Main mechanical parameters of rock materials are listed in Table 1.

The shear strength reducing method is used to simulate rock slope stability and determine the safety factor of rock slope. Three kinds of constitutive models for Drucker-Prager yield criterion are used to investigate rock slope stability and evaluate the safety factor of rock slope. The relationship between reduced shear strength of rock materials and the safety factor of rock slope is expressed as

$$C_r = \frac{C}{F_S} \tag{9}$$

Where F_s denotes the safety factor of rock slope, C_r is cohesion of rock materials after reducing when considering safety factor F_s.

$$\tan\varphi_r = \frac{\tan\varphi}{F_S} \tag{10}$$

Where φ_r is internal friction angle of rock materials after reducing. Distributions of plastic zone in rock slope in limited state with different Drucker-Prager yield criterions are shown in Figure 2, 3 and 4.

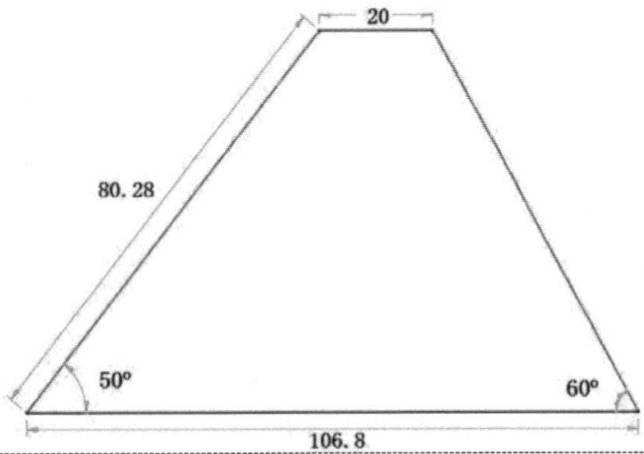

Figure 1. Geometry sizes of a rock slope on Bijia mountain

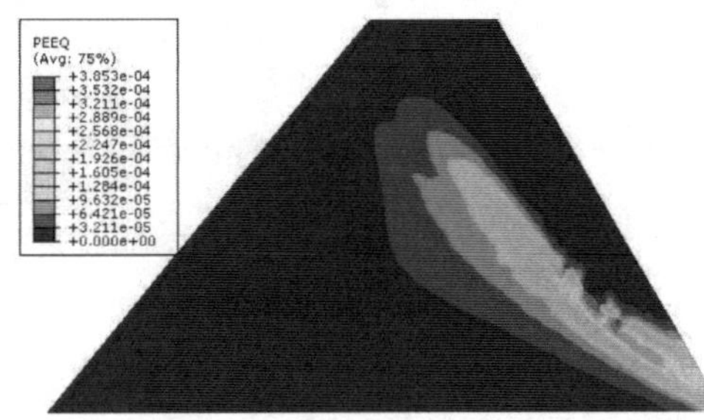

Figure 2. Distribution of plastic zone in rock slope in limited state with Drucker-Prager yield criterion with linear yield surface (Fs=1.412).

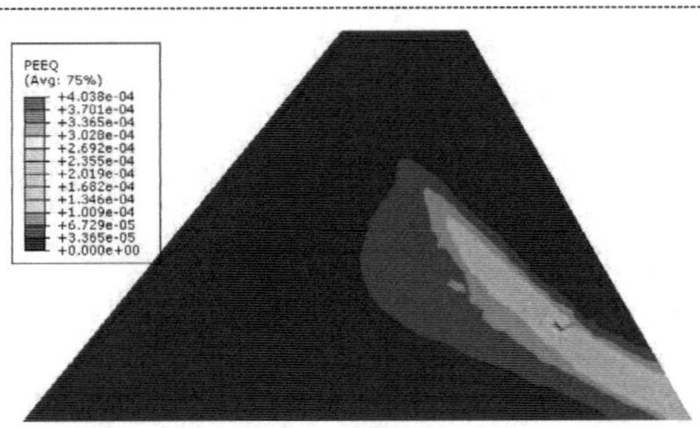

Figure 3. Distribution of plastic zone in rock slope in limited state with Drucker-Prager yield criterion with parabolic yield surface (Fs=1.466).

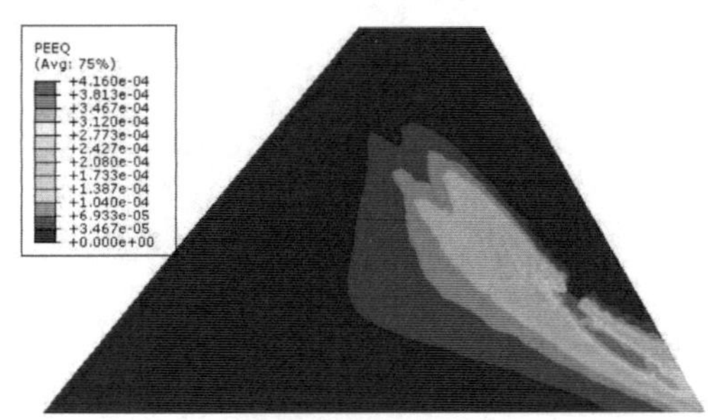

Figure 4. Distribution of plastic zone in rock slope in limited state with Drucker-Prager yield criterion with exponent yield surface (Fs=1.559).

It can be observed from Figure 2, 3 and 4 that the shapes of distribution of plastic zones in rock slope in limited state for three kinds of Drucker-Prager yield criterions are identical, and the safety factors of rock slope with different Drucker-Prager yield criterions are 1.412, 1.466 and 1.559, respectively. The maximum relative error is less 10% by using different Drucker-Prager yield criterions. The safety factors of rock slope with Drucker-Prager yield criterion with linear yield surface is minimum by comparing with parabolic yield surface and exponent yield surface.

4. Conclusions

Finite element method in conjunction with Drucker-Prager yield criterions is validated to be a reliable numerical procedure for evaluating the factor of safety of rock slopes. This procedure can conveniently compute the factor of safety of rock slopes based on the evaluating criterions of slope stability with unconverged FE solution or appearing inflexion for deformation of FEM model by reducing shear strength of rock materials. The investigation shows that the safety factors of rock slope with different Drucker-Prager yield criterions are slightly different and the maximum relative error is less 10% by using different Drucker-Prager yield criterions.

Acknowledgments

The research described in this paper was financially supported by the National Key Basic Research and Development Program of China (Grant No. 2015CB057804).

References

[1] Raghuvanshi T K 2019 *Journal of King Saud University – Science* **31** p 101
[2] Elmo D, Donati D and Stead D 2018 *Engineering Geology* **245** p 81
[3] Schlotfeldt P, Elmo D and Panton B 2018 *Journal of Rock Mechanics and Geotechnical Engineering* **10** p 72
[4] Tang S B, Huang R Q and Tang C A 2017 *Engineering Failure Analysis* **79** p 999
[5] Lv Q C, Liu Y R and Yang Q 2017 *Engineering Geology* **228** p 39
[6] Belghali M, Saada Z and Garnier D 2017 *Journal of Rock Mechanics and Geotechnical Engineering* **9** p 659
[7] Chen T, Deng J and Sitar N 2017 *Engineering Geology* **221** p 70
[8] Stead D, Wolter A 2015 *Journal of Structural Geology* **74** p 1
[9] Wong L, Wu Z 2014 *Engineering Fracture Mechanics* **119** p 1

Journal of Physics: Conference Series **1419** (2019) 012029 doi:10.1088/1742-6596/1419/1/012029

On a new SIMVW epidemic model with vaccination and endemic testing

J. Fernandez-Pereda[1], M. De la Sen[1,2] and S. Alonso-Quesada[1]

[1]Institute of Research and Development of Processes, Po Box-644, Bilbao, Spain

[2]E-mail: manuel.delasen@ehu.eus

Abstract. A new epidemic model referred to as SIMVW is proposed which includes the following subpopulations: susceptible human, infectious human, susceptible vector, infectious vector and vaccinated human. The transmission vector is the anopheles mosquito for the malaria disease. The disease-free equilibrium points concerning the absence of infection in humans and /or mosquitoes are defined and discussed from a stability point of view. A set of simulated examples are performed mainly related to typical endemic situations which are tested for a malaria case study in the absence and in the presence of vaccination efforts. The vaccination efforts delete susceptible numbers which are transferred to the vaccinated subpopulation and include online feedback information on the susceptible subpopulation to adjust the control effort. The vaccination effects are mainly to improve the convergence rates to the equilibrium point, to decrease the numbers of susceptible individuals while increasing those of recovered ones and to reduce the basic reproduction number so that the range of stability of the disease-free equilibrium points is larger compared to the case of absence of vaccination.

1. Introduction

In this work, a study of the stability and equilibrium points of an epidemic model has been carried out by extending a previous one in [1]. The new proposed epidemic model is referred to as an SIMVW epidemic model which includes the subsequent subpopulations: susceptible human, infectious human, susceptible vector, infectious vector and vaccinated human. The transmission vector in our investigation is the anopheles mosquito for the malaria disease. The model is analyzed and discussed using theoretical results of [2], that is, the numerical value that influences the evolution of the populations and determines whether or not the disease will spread. In particular, the values of the parameters used in the epidemic model have been based on known data on the malaria disease so that there are two populations involved: the mosquito vectors and the humans, each of them with the relevant subpopulations integrated within the whole epidemic model [3-6] and controlled via feedback of the susceptible subpopulation as it has been proposed in the literature for previous epidemic models [7-12]. In order to take into account a potential vaccination, the model used in the reference [1] has been modified by implementing in the basic differential equations a vaccination feedback effort and adding a new population to the model, namely, the subpopulation of the vaccinated humans. These facts imply the modification of the equilibrium points of the system related to the vaccination-free case, and, therefore, the need to calculate the new equilibrium points and the evaluation of their associated influence in the model dynamics. The vaccination controls involve the use of proportional feedback information on the susceptible so that the susceptible variation rate is decreased to improve in parallel that of the recovered by the same amount. Also, the basic reproduction number is below unity for larger values of the coefficient transmission rate related to the case of absence of vaccination so that the disease-free equilibrium points have a larger stability domain for higher values of such a coefficient and small numbers at the equilibrium point.

2. The model

The basic model of Kermack and McKendrick is well- known given by:

$$\begin{cases} \frac{dS(t)}{dt} = -\beta S(t)I(t) \\ \frac{dI(t)}{dt} = \beta S(t)I(t) - \frac{1}{\tau}I(t) \\ \frac{dR(t)}{dt} = \frac{1}{\tau}I(t) \end{cases}$$

where $\beta > 0$ is the disease transmission rate and $\frac{1}{\tau} > 0$ is the recovery rate, with all the variables having non-negative values under non-negative initial conditions. The so-called SIMVW epidemic model that we propose is given below while it is an extension of that discussed in [1] including vaccination with feedback information:

$$\begin{cases} S' = \frac{dS(t)}{dt} = B(N)N - \mu S - b\beta VS + \gamma I - u(t), S(0) = S_0 \geq 0 \\ I' = \frac{dI(t)}{dt} = b\beta VS - (\mu + \gamma + d)I, \quad I(0) = I_0 \geq 0 \\ M' = \frac{dM(t)}{dt} = F(T)T - \varepsilon M - b\alpha IM, \quad M(0) = M_0 \geq 0 \\ V' = \frac{dV(t)}{dt} = b\alpha IM - \varepsilon V, \quad V(0) = V_0 \geq 0 \\ W' = \frac{dW(t)}{dt} = u(t) - \mu W, \quad W(0) = W_0 \geq 0 \end{cases} \tag{1}$$

where S, I, M, V y W are respectively, the subpopulations: susceptible human, infectious human, susceptible vector, infectious vector and vaccinated human subpopulations. Those subpopulations define the acronym SIMVW used to name the epidemic model, and the prime prime subscripts of the model denote time-derivatives. N is the total human population ($N = S + I + W$) and T is the total vector population ($T = M + V$). $B(N)$ and $F(T)$ are the non-linear so-called Ricker birth functions of human and vector which together with the remaining parameters ($\mu,\varepsilon,d,\alpha,\beta,\gamma,b,u(t)$) are:

- $B(N)$: nonlinear birth rate of humans
- $F(T)$: nonlinear birth rate of vectors
- μ : natural human mortaslity rate
- ε : natural vector mortality rate
- b : per-capita amount of contacts of vectors to humans
- α : transmission probability from infectious human to non-infectious vector
- β : transmission probability of infectious vector to non-infectious
- d : human disease-due mortality rate
- γ : recovery rate of infectious human
- W : vaccinated human subpopulation
- $u(t) = gS(t)$: vaccination feedback control effort
- g: control gain parameter
- $k = \frac{\mu}{\mu+g}$: auxiliary control gain

Summing-up separately the two subpopulation groups time-derivatives leads to
$N' = B(N)N - \mu N - dI, T' = F(T)T - \varepsilon T$
The nonlinear birth rates satisfy the subsequent conditions:

- $B(N) > 0; F(T) > 0$.
- $B(N),\ F(T)$ are continously- differentiable with $B'(N), F'(T) < 0$.

- $B(0^+) > \mu > B(\infty); F(0^+) > \varepsilon > F(\infty)$.

In particular, we will use the particular functions commonly referred to as Ricker functions:

$$B(N) = c_1 \cdot e^{-a_1 N}\ ;\ F(T) = c_2 \cdot e^{-a_2 T}$$

where c_1, c_2, a_1 y a_2 are real constants subject to $c_1 > \mu$; $c_2 > \varepsilon$; $a_1 > 0$; $a_2 > 0$. The above functions offer a consistent population model since the populations sizes are taken into account while taking care of an uncontrolled growth of them and their possible incoherent negative values. The flow diagram of the model is shown in Figure 1.

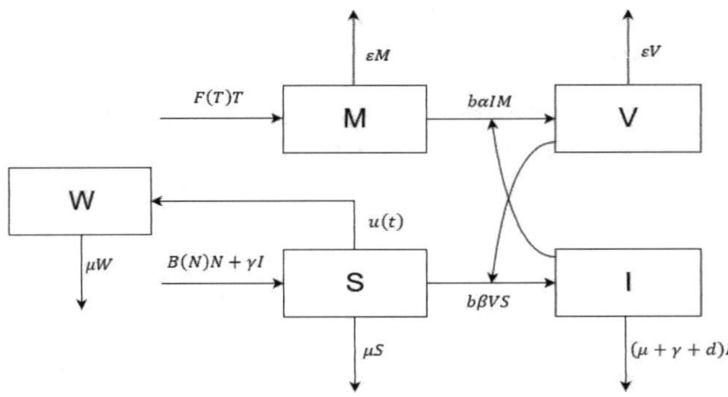

Figure 1. Flow diagram of the epidemic model.

This model is able to describe, the malaria. Now, we proceed to establish the basic reproduction number R_0 which depends on the disease parameters and the control gain g. Consider the following functions for secondary infections (F) and for the transfer rate of individuals ($\mathcal{V}$):

$$\mathcal{F} = \begin{pmatrix} \mathcal{F}_{I'} \\ \mathcal{F}_{V'} \end{pmatrix} = \begin{pmatrix} b\beta VS \\ b\alpha IM \end{pmatrix}; \mathcal{V} = \begin{pmatrix} \mathcal{V}_{I'} \\ \mathcal{V}_{V'} \end{pmatrix} = \begin{pmatrix} (\mu+\gamma+d)I \\ \varepsilon V \end{pmatrix}$$

Since $N = S + I + W = \left(1 + \frac{g}{\mu}\right) S = N^c$, then one gets $S = \frac{\mu N^c}{(\mu+g)}$ and $T = M + V = M = T^c$, where N^c and T^c are, respectively, the human and vector disease-free equilibrium values. After using a linearization around such an equilibrium point, one gets:

$$F = \frac{\partial F_i}{\partial x_j} = \begin{pmatrix} \frac{\partial F_{I'}}{\partial I} & \frac{\partial F_{I'}}{\partial V} \\ \frac{\partial F_{V'}}{\partial I} & \frac{\partial F_{V'}}{\partial V} \end{pmatrix} = \begin{pmatrix} 0 & b\beta \frac{\mu N^c}{(\mu+g)} \\ b\alpha T^c & 0 \end{pmatrix}; V = \frac{\partial \mathcal{V}_i}{\partial x_j} = \begin{pmatrix} \frac{\partial \mathcal{V}_{I'}}{\partial I} & \frac{\partial \mathcal{V}_{I'}}{\partial V} \\ \frac{\partial \mathcal{V}_{V'}}{\partial I} & \frac{\partial \mathcal{V}_{V'}}{\partial V} \end{pmatrix} = \begin{pmatrix} \mu+\gamma+d & 0 \\ 0 & \varepsilon \end{pmatrix}$$

leading to

$$FV^{-1} = \begin{pmatrix} 0 & \frac{b\beta\mu N^c}{\varepsilon(\mu+g)} \\ \frac{b\alpha T^c}{\mu+\gamma+d} & 0 \end{pmatrix}$$

and the basic reproduction number becomes:

$$R_0 = \rho(FV^{-1}) = \sqrt{\frac{b^2 \alpha\beta\mu N^c T^c}{\varepsilon(\mu+\gamma+d)(\mu+g)}}$$

The value of this equation is of vital importance since it indicates the total number of secondary cases that will occur in a population of humans from a single infectious vector. In particular, an infectious individual generates, in average, less than another infectious individual if $R_0 < 1$. However, if $R_0 > 1$ then an infectious individual generates more than new one infectious individual. Note also the relevance of the vaccination control gain g (zero in the vaccination- free case) to the value of the basic reproduction number.

3. Equilibrium points

The equilibrium points of the form $P_i = (S,I,M,V,W)$ are calculated by zeroing the time-derivatives of the various subpopulations of the whole model, that is:

$$\begin{cases} B(N)N - \mu S - b\beta VS + \gamma I - gS = 0 \\ b\beta VS - (\mu + \gamma + d)I = 0 \\ F(T)T - \varepsilon M - b\alpha IM = 0 \\ b\alpha IM - \varepsilon V = 0 \\ gS - \mu W = 0 \end{cases}$$

The simplest equilibrium point is given by $S = I = M = V = W = 0 \rightarrow P_0 = (0,0,0,0,0)$. The disease-free equilibrium points occur when the infectious subpopulations are zeroed, that is,

$$\begin{cases} B(N)N - (\mu + g)S = 0 \\ F(T)T - \varepsilon M = 0 \\ gS - \mu W = 0 \end{cases} \tag{2}$$

By using the Ricker function $B(N)$ into the first above equation and using $N = S + I + W = S + W = \left(1 + \frac{g}{\mu}\right)S = N^c$ where N^c is such that $B(N^c) = \mu$ holds, one gets:

$$c_1 \cdot e^{-a_1 N^c} N^c - \mu N^c = 0 \rightarrow e^{-a_1 N^c} = \frac{\mu}{c_1} \rightarrow N^c = -\frac{1}{a_1}\ln\left(\frac{\mu}{c_1}\right) \geq 0$$

In order to reach non-negative subpopulations, $\ln\left(\frac{\mu}{c_1}\right) < 0$ so that $c_1 > \mu$ and $a_1 > 0$. A similar procedure is performed for the second equation by replacing $F(T)$ by its Ricker-function and noting that $T = M + V = M = T^c$, where T^c satisfies $F(T^c) = \varepsilon$, so that:

$$c_2 \cdot e^{-a_2 T^c} T^c - \varepsilon T^c = 0 \rightarrow e^{-a_2 T^c} = \frac{\varepsilon}{c_2} \rightarrow T^c = -\frac{1}{a_2}\ln\left(\frac{\varepsilon}{c_2}\right) \geq 0$$

Again for non-negativity, $c_2 > \varepsilon$ and $a_2 > 0$. As a result, we get the four subsequent disease- free equilibrium points:

$$P_0 = (0,0,0,0,0),\quad P_1 = \left(\frac{\mu N^c}{(\mu+g)},0,0,0,\frac{gN^c}{(\mu+g)}\right),\ P_2 = (0,0,T^c,0,0),\quad P_3 = \left(\frac{\mu N^c}{(\mu+g)},0,T^c,0,\frac{gN^c}{(\mu+g)}\right)$$

where P_0 describes extinction of both human and vector populations, P_1 has uniquely susceptible and vaccinated human subpopulations, P_2 has only vector susceptible subpopulation while P_3describes the coexistence of human subpopualtions (both susceptible and vaccinated) and a susceptible vector subpopulation. The reach ability and stability of those points is now investigated. Remember that $N = S + I + W = I + (1 + \frac{g}{\mu})S$ and $T = M + V$. After some calculations using the model equations yields:

$$\begin{cases} N' = S' + I' + W' = B(N)N - \mu N - dI \\ I' = b\beta V(N - I)\frac{\mu}{(\mu+g)} - (\mu + \gamma + d)I \\ T' = M' + V' = F(T)T - \varepsilon T \\ V' = b\alpha I(T - V) - \varepsilon V \\ W' = g(N - I) - (\mu + g)W \end{cases} \tag{3}$$

- If $B(0^+) \leq \mu$ y $F(0^+) \leq \varepsilon$ then P_0 is locally asymptoically stable in R_+^5.
- If $B(0^+) \leq \mu$ y $F(0^+) > \varepsilon$ then P_2 is locally asymptotically stable.
- If $B(0^+) > \mu$ and $F(0^+) \leq \varepsilon$ then P_1 is locally asymptotically stable.
- For the case when $B(0^+) > \mu$ and $F(0^+) > \varepsilon$ our system can have any of the equilibria $(P_0, P_1, P_2\ y\ P_3)$.

It is now studied how the disease-free equilibrium point behaved for values of R_0 larger than and less than unity by assuming that $B(0^+) > \mu$ and $F(0^+) > \varepsilon$. To this end, the following set of differential equations, which is equivalent to that of the basic model, is used:

$$\begin{cases} S' = B(N)N - (\mu + g)(N - I - W) - b\beta VS + \gamma I \\ I' = b\beta V(N - I - W) - (\mu + \gamma + d)I \\ M' = F(T)T - b\alpha IM - \varepsilon(T - V) \\ V' = b\alpha IM - \varepsilon V \\ W' = g(N - I) - (\mu + g)W \end{cases} \tag{4}$$

By calculating the Jacobian matrix of (4) with respect to the equilibrium point P_3, we get the Jacobian $J_{P_3}(S,I,M,V,W)$ and its eigenvalues:

$$\begin{pmatrix} B'(N^c)N^c & B'(N^c)N^c + \mu + \gamma + g & 0 & -kb\beta N^c & B'(N^c)N^c + \mu + g \\ 0 & -(\mu + \gamma + d) & 0 & kb\beta N^c & 0 \\ 0 & -b\alpha T^c & F'(T^c)T^c & F'(T^c)T^c + \varepsilon & 0 \\ 0 & b\alpha T^c & 0 & -\varepsilon & 0 \\ 0 & -g & 0 & 0 & -(\mu + g) \end{pmatrix}$$

$\lambda_1 = B'(N^c)N^c < 0;\ \lambda_2 = F'(T^c)T^c < 0, \lambda_3 =- (\mu + g) < 0,$

$$\lambda_{4,5} = \frac{-(\varepsilon+x)\pm\sqrt{(\varepsilon+x)^2-4\varepsilon x(1-(R_0)^2)}}{2}, \lambda_4\lambda_5 = \varepsilon x(1 - (R_0)^2)$$

If $R_0 > 1$ then $\lambda_4\lambda_5 < 0$. Thus, one of the eigenvalues has a positive real part. Therefore, the equilibrium point P_3 is hyperbolically unstable. If $R_0 < 1$ then $\lambda_4 < 0$ y $\lambda_5 < 0$ and P_3 is locally asymptotically stable. On the other hand, by summing-up all the equations (1) $F(T) = \varepsilon$, with $F(0^+) > \varepsilon$, $F(\infty) < \varepsilon$ and $F'(T) < 0$. Then, $T = T^c$ since T has a unique positive root. By noting that $N = S + I + W$ by summing -up the equations referred to the human subpopulations), one gets:

$$S = \frac{(\mu+d-B(N))N}{d}, I = \frac{(B(N)-\mu)N}{d} \tag{5}$$

where $\mu < B(N) < \mu + d$ to ensure that $S,I > 0$. By taking into account the Ricker functions $B(N)$, one gets $z < N < N^c$, where $z = \frac{-1}{a_1}\ln\left(\frac{\mu+d}{c_1}\right)$. After some simple calculations, one gets also for $T = T^c$:

$$M = T^c - V, V = \frac{(\mu + d + \gamma)(B(N) - \mu)}{b\beta(\mu + d - B(N))}$$

$$f(N) = (\mu + d + \gamma)[(B(N) - \mu)Nb\alpha + d\varepsilon] - T^c b^2\alpha\beta(\mu + d - B(N))N = 0$$

$$f'(N) = [(\mu + d + \gamma)b\alpha + T^c b^2\alpha\beta]B'(N)N - \frac{(\mu + d + \gamma)\varepsilon d}{N} < 0$$

since $f'(N) < 0$. To get the solution of (1), one considers the positive range of solutions (z,N^c):

$$f(z) = \begin{cases} (\mu + d + \gamma)d[\varepsilon + b\alpha z] > 0, si\ B(0^+) > (\mu + d) \\ (\mu + d + \gamma)d\varepsilon > 0, si\ B(0^+) \le (\mu + d) \end{cases}, f(N^c) = (\mu + d + \gamma)d\varepsilon(1 - R_0)$$

for $R_0 < 1, f(N^c) > 0$. Therefore, there is no positive root in the interval (z,N^c). That means that the system has no positive equilibrium point of infectious population. If $R_0 > 1$ then $f(N^c) < 0$. This implies that the system has a unique positive root in the interval (z,N^c). Therefore, there is a unique equilibrium point which is given by $P_4 = (S^*,I^*,M^*,V^*,W^*)$, where S^*,I^*,M^*,V^* y W^* are:

$$S^* = \frac{(\mu+d-B(N^*))N^*}{d}, I^* = \frac{(B(N)-\mu)N^*}{d}, W^* = \frac{g(\mu+d-B(N^*))N^*}{\mu d}, M^* = T^c - V^*, V^* = \frac{(\mu+d+\gamma)(B(N^*)-\mu)}{b\beta(\mu+d-B(N^*))}$$

Note that the equilibrium values of the various subpopulations depend on the control parameter g. Therefore, they are dependent on the existence of vaccination effort ($g \neq 0$) or not ($g = 0$).

4. Simulated examples

The subsequent simulated study is based on the endemic malaria disease evolution. The model parameters are given in $days^{-1}$. In particular, $\mu = \frac{1}{83\cdot365} \approx 3.30\cdot10^{-5}$ which corresponds to an average life expectation of 83.09 years (Spain, 2017). There are more than two thousands of mosquito species. In average, they live from one week to a month. We consider here, an average of two weeks of life for a mosquito leading to $\varepsilon = \frac{1}{2\cdot 7} \approx 7.14\cdot 10^{-2}$. One chooses the following transmission probabilities from human infectious to non-infectious vector (α) and from infectious vector to non-infectious human (β): $\alpha = 0.50, \beta = 0.50$. For instance in the dengue case, the typical values would be $\alpha = \beta = 0.75$. The amount of per-capita contacts are in the range $b \in (0.2, 0.8)$. The chosen value is $b = 0.80$. Some simulatons have been performed by varying this value. The infectious-human mortality rate (d) and infectious-human recovery rate (γ), the last one varies in the range $\gamma \in (0.01, 0.05)$. The used values are $d = 0.002$ and $\gamma = 0.05$. The relative values of mosquito to human have been taken from two to one. The initial conditions are $S(0) = 1$, $I(0) = 0$ and $W(0) = 0$. That is, we do not have initially infectious and vaccinated individuals. The initial subpopulations of mosquito are $M(0) = 1.99$ with a single infectious individual, $V(0) = 0.01$. In order that the growing human and mosquito rates $B(N)$ and $F(T)$ agree with values found in the background literature. For mosquito the birth rates are $F(T) \in (0.036, 42.5)$ per day. For humans, the fertility rate in Spain, according to data published by the World Health Organization (WHO), Spanish acronym: OMS, for 2017, was 1.3 per person and year. Then, c_1, c_2, a_1 and a_2 are chosen such that the population model is appropriate so that they approach initially $B(N)$ and $F(T)$ to their respective estimated rates, so that $c_1 = 0.003$; $c_2 = 0.537$; $a_1 = a_2 = 1$. The ranges of values of the control gain from no vaccination to full vaccination effort is $g \in [0,1]$. Table 1 gives the values of parameters in en $days^{-1}$ and the initial conditions:

Table 1. Parameter values and initial conditions.

subpopulation numbers	**Value**
susceptible human S(0)	1
Infectious human I(0)	0
susceptible vector M(0)	1.99
infectious vector V(0)	0.01
vaccinated human W(0)	0
human birth rate B(N)	$1.10\cdot 10^{-3}$
vector birth rate F(T)	$7.27\cdot 10^{-2}$
human mortality rate μ	$3.30\cdot 10^{-5}$
vector mortality rate ε	$7.14\cdot 10^{-2}$
transmission probability: infectious human–healthy vector α	0.50
transmission probability: infectious vector healthy-human β	0.50
human recovery propability γ	0.05
infectious human mortality probability d	0.002
per capita contacts b	0.80
Vaccination gain g	Varying

The values of $B(N)$ and $F(T)$ in Table 1 are the initial ones. Then, they follow their expressions accordingly to the evolution of the subpopulations. The simulations are performed via the Scilab package. Unless some particular specification be given, the parameters and initial values are those of Table 1 and the time range is ($t = 0{:}0.01{:}100$) implying 10^4 iterations. In Figure 2, there is no vaccination, i.e. $g = 0$. We observe that for a single infectious vector, the human/vector infectious grow very fast. This is expected since the basic reproduction number is $R_0 \approx 19.74 > 1$. In order to observe the values of the equilibrium subpopulations, the simulation is repeated with more iterations, (10^6). See Figure 3. It is seen that both infectious subpopulations reach endemic equilibria, the human equilibrium population been small due to the low values of $B(N)$ comapred to the mortality rate d. Next, we apply vaccination effort with $g = 0.05$. See Figure 4. Note that the infectious human subpopulations reduce significantly. By comparing Figures 2 and 3, note that the infectious vector subpopulation also becomes reduced.

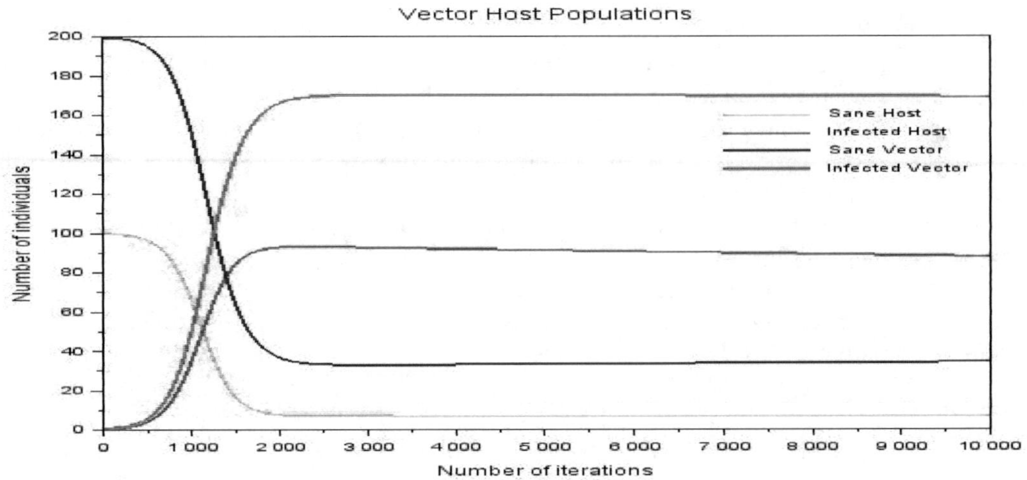

Figure 2. Subpopulations evolution in the vaccination-free case.

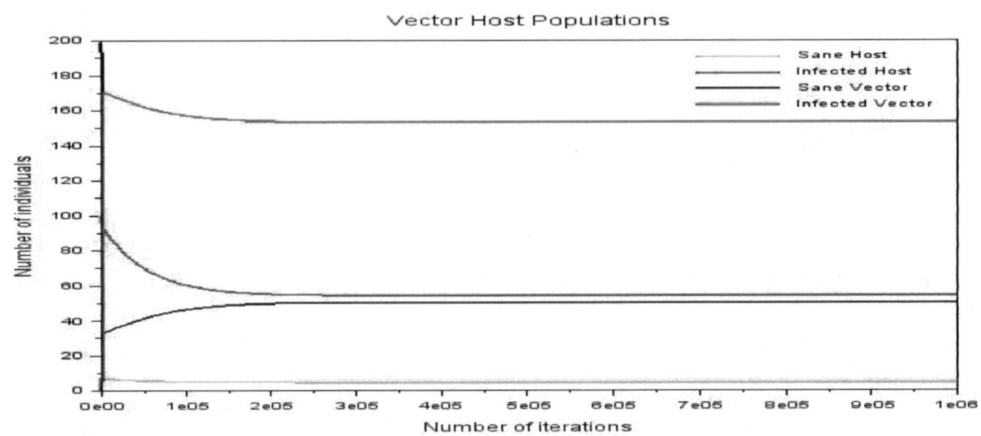

Figure 3. Subpopulations evolution in the vaccination-free case for a larger observation time.

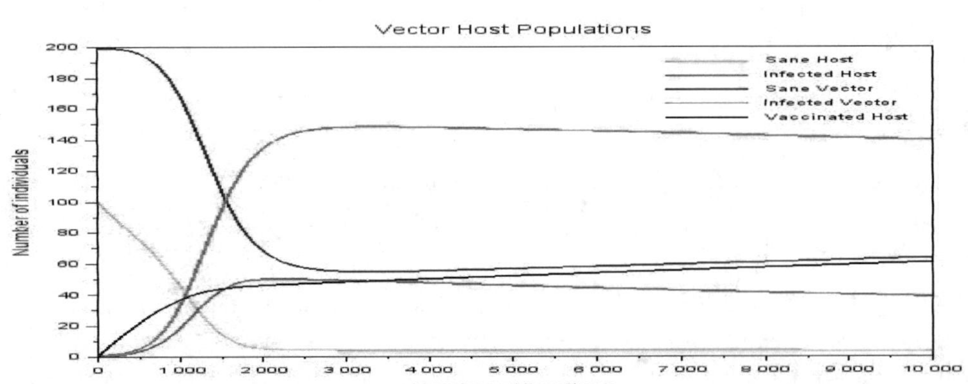

Figure 4. Subpopulations evolution with a vaccination effort of 5%.

Figure 5 is appropriate to compare the subpopulations of vector and host without and with vaccination. R_0 decreases with a vaccination effort of 5% to yield $R_0 \approx 1.60 > 1$. Still the disease becomes endemic in this case but with a lower intensity exhibition. See Figure 5 (continuous portraits).

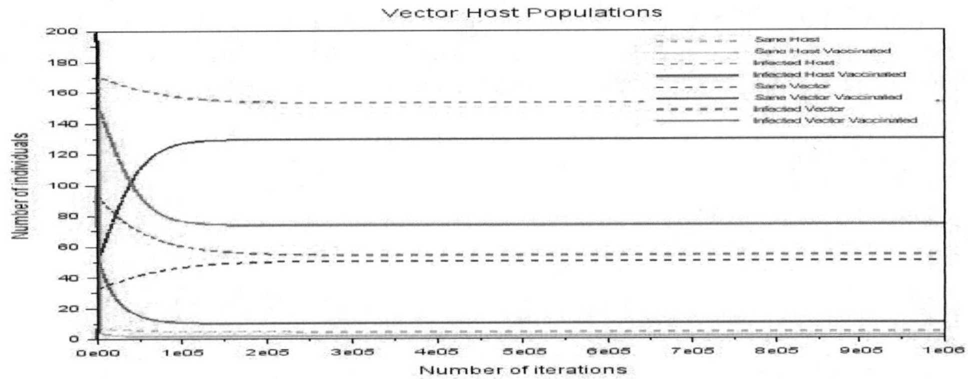

Figure 5. Subpopulations evolution with a vaccination effort of 5% (continuous portrait) and in the vaccination-free case (discontinuous portrait).

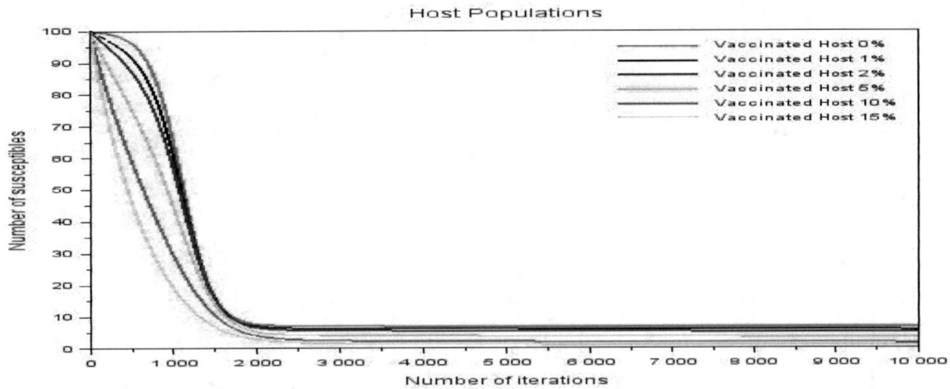

Figure 6. Human susceptible subpopulation versus vaccination effort.

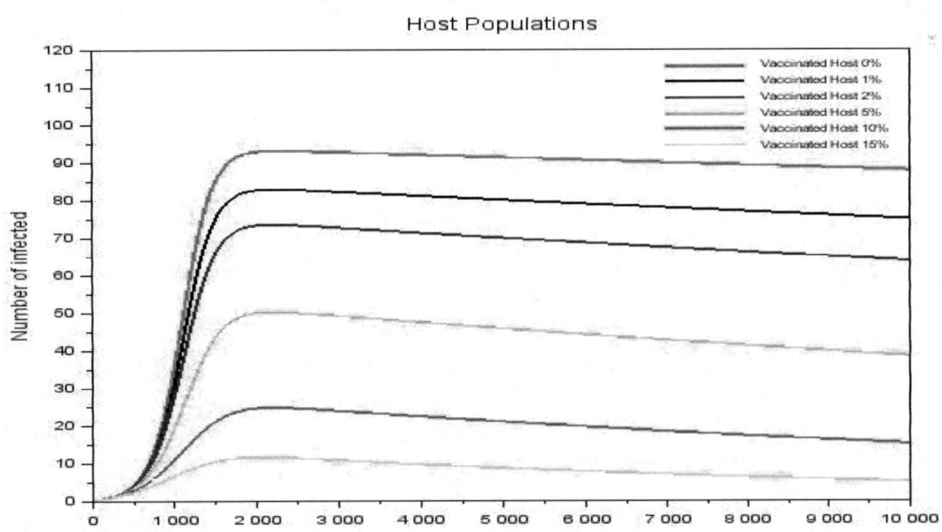

Figure 7. Human infectious subpopulation versus vaccination effort.

In the subsequent Figures 6, 7 and 8 the evolution of the percentages of the subpopulations is represented versus the vaccination effort in the endemic situation.

In Figure 6, it is seen how the susceptible numbers decrease faster as the vaccination effort increases to get smaller equilibrium numbers at the equilibrium steady-state. Figure 7 indicates that larger vaccination efforts give smaller amount of infectious humans as expected. Figures 8 and 9 display, respectively, the growing rates of the human vaccinated populations versus vaccination efforts and some mixed human- vector subpopulations depending on the vaccination efforts.

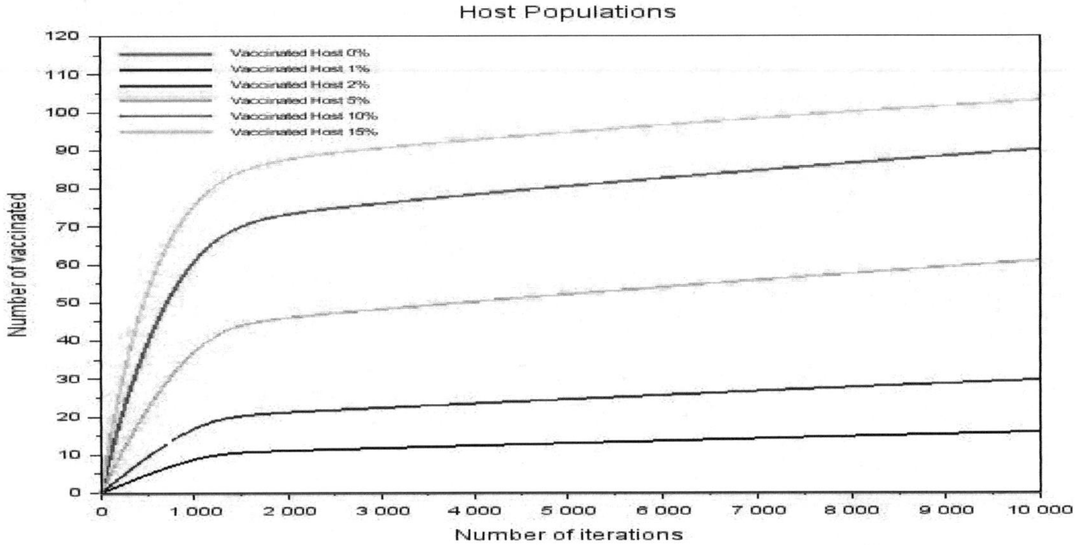

Figure 8. Human vaccinated subpopulation versus vaccination effort.

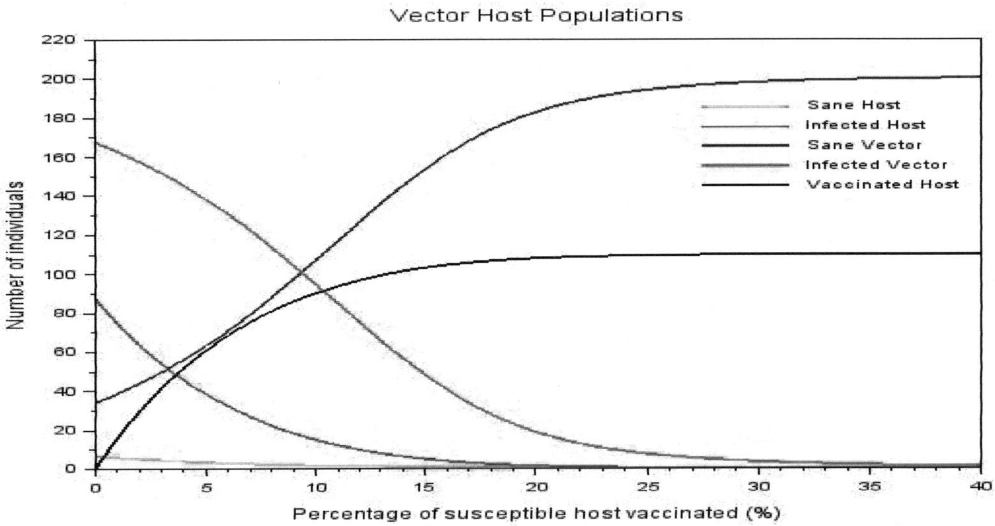

Figure 9. The various subpopulations versus the vaccination effort.

As it has been observed, at higher values of vaccination we would see populations of healthy mosquitoes increasing, since the transmission of infection between vectors will be slower. This occurs since the human population is rapidly vaccinated and a large number of infectious humans are not generated. In Figure 10, the case of convergence to a disease- free steady- state is displayed In this case the reproduction number is less than unity and in particular the coefficient transmission rate takes values under a critical values which corresponds to a basic reproduction number equalizing unity.

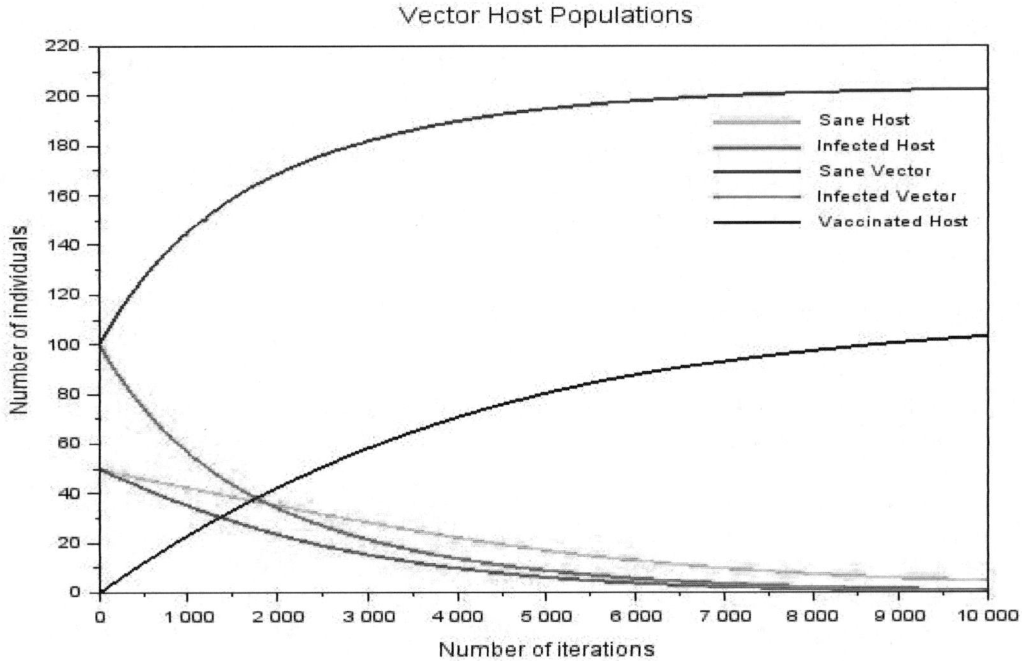

Figure 10. Convergence to the Disease- free equilibrium with =b/2=0.1, g=0.05.

5. Results and Discussion

The basic reproduction number of our model has been calculated while demonstrating that it depends on the vaccination control used, its value being lower the greater the control parameter used. This fact allows the infection to be eliminated asymptotically with higher tolerated virulence rates of the infection if it is vaccinated than if it is not. It has been concluded that, if both vector and human mortality rates are greater than their respective birth rates then a stable equilibrium is reached in which populations disappear. It has also been demonstrated analytically that a disease-free equilibrium will be reached as long as the number of reproduction is less than than one that leading to the disease-free equilibrium point to be asymptotically stable. The latter has also been tested experimentally. For values of this number exceeding the unity value, it has been found that there is a positive balance since a single infectious individual will be replaced by more than one new infectious case. This endemic equilibrium state exists and it is stable, presenting non-zero values of infectious and susceptible populations in equilibrium. For small coefficient transmission rates under a certain critical value which corresponds to the basic reproduction number equalizing unity, the deqauiolibrium state is disease-free and the infectious number reach the zero value.

6. Conclusion

In this paper, we have studied a deterministic epidemic model that allows to representing the transmission dynamics of infectious diseases between two species. In particular, the relationship between human populations and vectors against the disease of malaria has been analyzed mainly in the endemic steady-state. A vaccine-dependent function of the population of susceptible humans has also been incorporated into the model which has a term of proportional feedback information to the susceptible. The proportionality constant is a control parameter which can be used for design so as to improve the equilibrium numbers and the convergence rates. The simulation results have displayed endemic equilibrium situations with nonzero numbers of infectious subpopulations if the basic reproduction number exceeds unity and disease- free equilibrium steady- states if such a number is below unity.

Acknowledgments

This research was supported by Spanish Government and Basque Government through the respective Grants RT12018-094336-B100 (MCIU/AEI/FEDER, UE) and IT1 207/19.

References

[1] Qiu Z 2008 *Computers and Mathematics with Applications* **56** p 3118
[2] Driessche P and Watmough 2002 J *Mathematical Biosciences* **80** p 29.
[3] TumwiineT, Mugisha J Y T and Luboobi L S 2007 *Applied Mathematics and Computation* **189** p 1953
[4] Chitniss N et al. 2010 J.*Trop. Med. Hyg.* **83** p 230
[5] Chitniss N et al. 2008 *J. Biological Dynamics* p 259
[6] Jong M C H, Diekmann O and Heesterbeek H 1994 How does transmission of infection depend on population size? *Research Gate*
[7] Nistal-Riobello R 2015 On some new mathematical models for infective disease: analysis, equilibrium, positivity and vaccination controls *Ph. D. Thesis, University of Basque Country, ZTF-FCT*
[8] De la Sen M, Nistal R, Alonso-Quesada S and Ibeas A 2018 A switched multicontroller for an SEIADR epidemic model with monitored equilibrium points and supervised transients and vaccination costs, *Advances in Difference Equations* Article number 390
[9] De la Sen M and Alonso-Quesada S 2010 On vaccination control tools for a general SEIR epidemic model. *In 18th Mediterranean Conference on Control & Automation*, 23-25th June 2010, Marrakech 1322
[10] De la Sen M, Ibeas A, Alonso-Quesada S and Nistal R 2017 On a new epidemic model with asymptomatic and dead-infective subpopulations with feedback controls useful for Ebola disease *Discrete Dynamics in Nature and Society* Article ID 4232971
[11] Alonso-Quesada S, De la Sen M and Nistal R 2018 On vaccination strategies for a SISV epidemic model guaranteeing the nonexistence of endemic solutions *Discrete Dynamics in Nature and Society* Article ID 9484121
[12] Saralegui U, De la Sen M and Alonso-Quesada S 2016 A feedback vaccination law for an SIR epidemic model: A case study. *In 2016 UKACC 11th International Conference on Control (Control)*, Belfast, UK, 31st August-2nd September

Expert strategic management systems in precision farming

Ilya M. Mikhaylenko[1,3], Valery N. Timoshin[2]

[1]Pulokovskoye chose, house 22, apt., 196158, Sankt-Petersburg, Russia
[2]Krzhizhanovskogo street, house 3, building 5, apt., 21193231, Sankt-Petersburg, Russia

[3]E-mail: ilya.mihailenko@yandex.ru

Abstract. Evaluating the modern market of the Internet of things, one should consider equipment connected into a single network, solutions, applications along the entire chain of product creation, including the end user. In this paper, such a binding is based on cloud information technologies. Through these technologies, the intellectualization of agrotechnology management is implemented through the creation of expert management decision support systems (DSS). The aim of the work is to consider the methodology for constructing DSS of strategic management in precision farming systems, where this type of management has not been implemented to date. To this kind of management, the task of forming strategies for the application of mineral fertilizers and ameliorants of prolonged action for all the years of various types of crop rotations is included. To solve the problem, an algorithm for the formation of optimal strategies for the application of mineral fertilizers and ameliorants, which is implemented in an analytical automated control system of agricultural technologies, through which a knowledge base (KB) is transmitted from the cloud system to local DSS, is substantiated. To select the best option from the knowledge base, the pattern recognition method is used. The technique was tested on arbitrary initial conditions of local systems, including extreme combinations of initial conditions. Based on the analysis of the loss of optimality associated with the discrepancy of the initial conditions on the local DSS and the KB, a method for controlling the formation of the BR, aimed at reducing these losses, is substantiated.

1. Introduction

The concept of the digital economy appeared in the last decade of the last century. Nicholas Negroponte, an informatics specialist, founder of the Media Labs at the Massachusetts Institute of Technology (MIT), formulated the fundamental principles of the digital economy. Back in 1995, he spoke about the shortcomings of classic goods (weight, raw materials, transport) and the advantages of the new economy (lack of weight of goods, virtuality, instant global movement) [13].

Today, agriculture is on the verge of the Second Green Revolution. Experts believe that, thanks to the new information technologies based on the Internet of Things, a surge in yields of such magnitude that humanity did not see even at the time of the appearance of tractors, the invention of herbicides and genetically modified seeds, could follow. Technologies evolved, became cheaper and advanced to such a level that for the first time in the history of the industry it became possible to obtain data on each agricultural object and its environment, mathematically calculate the algorithm of actions and predict the result. Digitalization and automation of the maximum number of agricultural processes is included as a conscious need for a development strategy for this sector of the economy [1].

The task of informatization and digitalization is the maximum automation of all stages of the production cycle to reduce losses, increase business productivity, and optimal resource management.

Further automation represents a higher level of digital integration, which affects the most complex organizational changes in the business. However, their implementation can dramatically affect the profit and competitiveness of products and the company as a whole. Integrating data with various intelligent IT applications that process them in real time provides a revolutionary shift in decision making for the farmer, providing analysis of multiple factors and a rationale for subsequent actions. In this case, the more sensors, sensors and field controllers are connected to a single network and exchange data, the more intelligent the information system becomes and the more useful information for the user it can provide.

The initial base of digitization and intellectualization of agriculture was provided by the scientific and technical progress of the 20th and early 21st centuries, which provided new and unique possibilities for actually reducing the risk of agricultural production. They include new robotic machines, measuring devices, computing equipment and modern mathematical base. The rapid development of information technologies enables researchers and developers to combine measuring instruments of various physical natures with computer equipment and automated agricultural machines into a single, controlled complex. The natural response to such opportunities was the development of a new agrarian-technological direction, known as "precision farming" or "precision agriculture" (PA). At its core (PA), it implies the solution of the problem of increasing the manageability of crop production by solving a set of tasks for managing agrotechnologies [9].

At the same time, it should be borne in mind that a high scientific and technical level of agrotechnology management systems requires the availability of highly qualified personnel for their operation, which is a difficult problem to solve in conditions of agricultural production. This problem can be solved by switching to cloud computing technologies, when computer resources and capacities are provided to the user as an Internet service [7,8,10,14]. Cloud technologies have become possible due to the rapid development of hardware. Processor power is growing day by day, multi-core architecture is developing and hard drive volumes are increasing, and high-bandwidth Internet channels are widely used. The capabilities of cloud technologies allow implementing the full range of procedures necessary for making management decisions.

The situation is greatly simplified in the transition to expert systems in which control decisions are made directly from the input information, bypassing complex multi-step computational procedures. The main information core of expert systems is knowledge bases (KB), which are the source of information for making management decisions. To form a KB, the principle of coding the knowledge of a human expert cannot be used, no matter how highly qualified he may be. An expert system with such a KB is not a modern breakthrough information technology and does not allow optimizing the control decisions made. However, the high efficiency of expert systems can be ensured if, instead of a human expert, the source of information will be software and hardware complexes of analytical systems for managing agrotechnologies. Through such complexes, a large number of optimal solutions are formed for various soil and climatic conditions of growing crops. Each such case is an element of the knowledge base and is the optimal control solution for the given conditions, ensuring the maximization of the result in a controlled system.

The formation of the KB optimal solutions for the given conditions is a serious foundation that provides a high scientific and technical level of the management process. But it remains unclear how the user himself can make management decisions on the KB and the signals of his information-measuring system. One of the most effective means of making such decisions is the pattern recognition method, when the current situation is searched for the best option, taken as a class or image [6]. Considering the fact that there are many such images in the knowledge base, the recognition algorithm can be constructed based on the methods of conditional probabilities of classes or the “nearest neighbor”.

The purpose of this work is to present the methodology for constructing expert systems for strategic management in TK systems. To this level of management, we include the task of optimizing the doses of application of mineral fertilizers of prolonged action in crop rotations of various types. Currently, there is no theory for building such systems, and the results are presented for the first time [3-5].

2. Algorithm Strategic Management

The problem to be solved is formulated as follows: for the sequence of crops in the adopted crop rotation, denoted by the indices j = 1,2,3 ..., N, it is necessary to find a strategy for introducing the main nutrients and ameliorants in a given field, ensuring the achievement of the management goal, namely: minimizing the yield losses of all crop rotation, with minimal consumption of resources and given technological constraints.

To formalize the problem, we introduce into consideration a vector of parameters of the chemical condition of the soil of a given field that is averaged over the area of the field: V = [3x1], with components: $v_1 = pH, v_2 = P, v_3 = K$; *pH*- acidity, P - phosphorus, K - potassium. We will also need a vector of crop cultivation conditions unregulated by this strategy: F = [4x1], with components: f_1 - is the seasonal sum of temperatures; f_2 - is the seasonal total precipitation; f_3 - is the total flux of Phyto active radiation; f_4 - is the annual consumption of available forms of nitrogen.

We accept the assumption that for each crop of the used crop rotation the optimum content of the main nutrients and the optimum value of soil acidity are known. Any deviation from these optimal values will lead to yield losses [5]. Taking into account the fact that all the above indicators of the chemical state of the soil act simultaneously, we will consider the following forms of yield loss patterns for each j-th crop in crop rotation

$$\Delta \mathrm{u}_j(\mathrm{T}) = \mathrm{k}_{1j}^{\mathrm{T}}(\mathrm{V}_j^* - \mathrm{V(T)}) + (\mathrm{V}_j^* - \mathrm{V(T)})^{\mathrm{T}}\,\mathrm{K}_{2j}(\mathrm{V}_j^* - \mathrm{V(T)}) \tag{1}$$

where: V* is the optimal value of the vector of the chemical state of the soil in a given field for the j-th crop of the crop rotation; $\Delta u_j(T)$ - yield losses for the j-th crop rotation due to the deviation of the vector of the chemical state of the soil from the optimal value;

$\mathrm{k}_{1j}^{\mathrm{T}} = \begin{bmatrix} k_1 & k_2 & k_3 \end{bmatrix}_j$ – matrix-row parameters of the linear part of the model,

$\mathrm{K}_{2j} = \begin{bmatrix} k_4 & k_5 & k_6 \\ 0 & k_7 & k_8 \\ 0 & 0 & k_9 \end{bmatrix}_j$ – matrix of parameters of the quadratic part of the model.

At the same time, in order to identify model (1), we need its observed yield - the amount of yield loss due to the deviation of the chemical state parameters from the optimal values - $\Delta u_j(T)$, which is formed by comparing the potential yield for the j-th crop for given cultivation conditions determined by the F vector and the real or projected yield $u_j(T)$ for the same conditions

$$\Delta u_j(T) = \mathrm{B}_{\mathrm{j}}^{\mathrm{T}}\mathrm{F}(T) - u_j(T) \tag{2}$$

$\mathrm{B}_j^{\mathrm{T}} = \begin{bmatrix} b_1 & b_2 & b_3 & b_4 \end{bmatrix}_j$ – vector of parameters of a linear model of potential yield.

Optimization of strategies for the introduction of agrochemicals and ameliorants is possible only if there are predictions of the chemical state of the soil, by which it is possible to estimate the total yield loss in crop rotation. For this purpose, a dynamic model of all components of the chemical state of the soil of the following type is proposed

$$\begin{aligned} \dot{v}_{1j} &= a_{11}v_{1j}(T) + b_1 d_{Ca}(T) + c_1 f_2(T), \\ \dot{v}_{2j} &= a_{22}v_{2j}(T) + b_2 d_P(T) + c_2 f_2(T) + d_2 \hat{u}_j(T), \\ \dot{v}_{3j} &= a_{33}v_{3j}(T) + b_3 d_K(T) + c_3 f_2(T) + d_3 \hat{u}_j(T), \end{aligned} \tag{3}$$

where: $\hat{u}_j(T) = \mathrm{B}_{\mathrm{j}}^{\mathrm{T}}\mathrm{F}(T) - \Delta u_j(T)$ – crop yield taking into account losses; $d_P(T)$, $d_K(T)$, $d_{Ca}(T)$ – doses of nutrients and improver for years of crop rotation (elements of strategy); a_{11}–a_{33}, b_1–b_3, c_1–c_3 – model parameters.

The scalar form of the model (5) is convenient for its line-by-line identification, but to form an optimal strategy for introducing agrochemicals and ameliorants, the canonical vector-matrix form expanded

$$\begin{bmatrix} \dot{v}_1 \\ \dot{v}_2 \\ \dot{v}_3 \end{bmatrix} = \begin{bmatrix} a_{11} & 0 & 0 \\ 0 & a_{22} & 0 \\ 0 & 0 & a_{33} \end{bmatrix} \begin{bmatrix} v_1(T) \\ v_2(T) \\ v_3(T) \end{bmatrix} + \begin{bmatrix} b_1 & 0 & 0 \\ 0 & b_2 & 0 \\ 0 & 0 & b_3 \end{bmatrix} \begin{bmatrix} d_P(T) \\ d_K(T) \\ d_C(T) \end{bmatrix} + \begin{bmatrix} c_1 \\ c_2 \\ c_3 \end{bmatrix} f_2(T) + \begin{bmatrix} 0 \\ d_2 \\ d_3 \end{bmatrix} u(T); \quad (4)$$

and compact character

$$\dot{V} = AV(T) + BD(T) + cf_2(T) + du(t) \quad (5)$$

The introduced designations and models allow us to form an optimal solution criterion for the problem, adequate to the goal

$$I = M\{\sum_{T=1}^{N}[(V^* - V(T))^T G(V^* - V(T)) + c_{Tu}\, \Delta u_T(T) + C_d^T D(T)]\} \quad (6)$$

where: M - the operation of the mathematical expectation of the area of the field; G – matrix of weight coefficients, by which the ratio of the parameters of the chemical state of the soil varies; c_{Tu} - crop unit price in the T-year of crop rotation, C_d^T - the vector of prices for mineral fertilizers for each nutrient.

Criterion (6) has the meaning of the average risk of crop underperformance in crop rotation and over-consumption of fertilizers spent on its production [5,11,12].

Let us consider the Hamiltonian of the system (7), (8) [2]

$$\begin{aligned} H(T) = {} & [(V^* - V(T))^T G(V^* - V(T)) + c_u[K^T(V^* - V(T)) + \\ & +(V^* - V(T))^T H(V^* - V(T))] + \\ & + C_d^T D(T)] + \lambda^T[AV(T) + BD(T) + cw(T) + du(t)], \end{aligned} \quad (7)$$

where λ = is the vector of conjugate variables associated with the vector of the chemical state as follows:

$$\begin{aligned} & \dot{\lambda} = \frac{\partial H(T)}{\partial V} = [G + H](V^* - V(T) + c_u K + \mathbf{A}^T \lambda(T), \\ & T \in (N,0),\ \lambda(N) = 0. \end{aligned} \quad (8)$$

The partial derivative of the Hamiltonian in the vector of doses of agrochemicals

$$g(T) = \frac{\partial H(T)}{\partial D} = C_d + B^T \lambda(T). \quad (9)$$

The optimal strategy for introducing agrochemicals for all years of crop rotation is a sequence of dose vectors D (T) and is found by performing the following computational procedure:

$$D_n^*(T) = D_{n-1}^*(T) - D_n[C_d + B^T I_n(T)],\ D_1 £ D_n^*(T) < D_2 \quad (10)$$

if $D_n^*(T) < D_1$, that $D_n^*(T) = 0$, if $D_n^*(T) > D_2$, that $D_n^*(T) = D_2$.

3. Expert Strategic Management System

Figure 1 shows a block diagram of an automated agrotechnology management system with training from a data center for use in cloud information technologies. Here in the information cloud is the data center of the automated control system, which implements the above-described algorithm of strategic management. Through this algorithm, a knowledge base (KB) is formed, in which for different variants of the initial values of the parameters of the chemical state of the soil and different climatic conditions for each type of crop rotation, optimal strategies for the application of mineral fertilizers and ameliorants are formed.

The approbation of the system consisted in presenting a local expert DPRS of an arbitrary version of the initial conditions for choosing the optimal strategy from the BK. When using the pattern recognition approach, it is advisable to use the "nearest neighbor" method, when class membership is estimated by the minimum distance between the vectors

$$\rho_i(T) = \sqrt{\sum_{j=1}^{n} (z_{ji}(T) - y_j(T))^2}, \; j = \overline{1,J}, \tag{11}$$

where: i=1,2,…I – record numbers in the BK, y_j ,z_j - factors for making decisions on the local DSS and on KB, j=1,2...J – indices and the total number of components of the combined vector of decision-making factors.

The choice of the best variant of the optimal strategy from the BR for a local DSS is taken according to the condition

$$I^* = \arg\min_i \rho_i(T) \tag{12}$$

Were chosen arbitrary initial conditions for the formation of optimal strategies are shown in Table 1.

Table 1. Initial conditions of local DSS.

Initial conditions	Acidity kg·ha^{-1}	Phosphorus kg·ha^{-1}	Potassium kg·ha^{-1}
	4,30	30,40	41,80

To assess the possible loss of optimality for these conditions, an optimal strategy has been formed for ameliorating and mineral fertilizers, shown in Fig. 2. The effectiveness of such a strategy is characterized by the value of the optimality criterion of 534.55 rub.·ha^{-1}.

In accordance with rule (13), the closest variant of the strategy was chosen in the BK, which is characterized by the following initial conditions.

Table 2. The closest initial conditions from the BK.

Initial conditions	Acidity kg·ha^{-1}	Phosphorus kg·ha^{-1}	Potassium kg·ha^{-1}
	5,10	40,20	42,40

This option corresponds to the optimal strategy shown in Fig. 3, the effectiveness of which is estimated at 492.44 rub.·ha^{-1}. Comparison of the results shows that the discrepancy between the initial conditions in the KB and the local DSS leads to a loss of optimality of the strategy by 8.5%.

The low level of loss of optimality in the considered example indicates that arbitrary initial data in the local DSS is chosen from the range of possible values close to optimal. Of interest are cases far from this area. Such a case is presented in table 3.

Table 3. The case of extreme initial conditions.

Initial conditions	Acidity $kg \cdot ha^{-1}$	Phosphorus $kg \cdot ha^{-1}$	Potassium $kg \cdot ha^{-1}$
	4,10	20,00	25,00

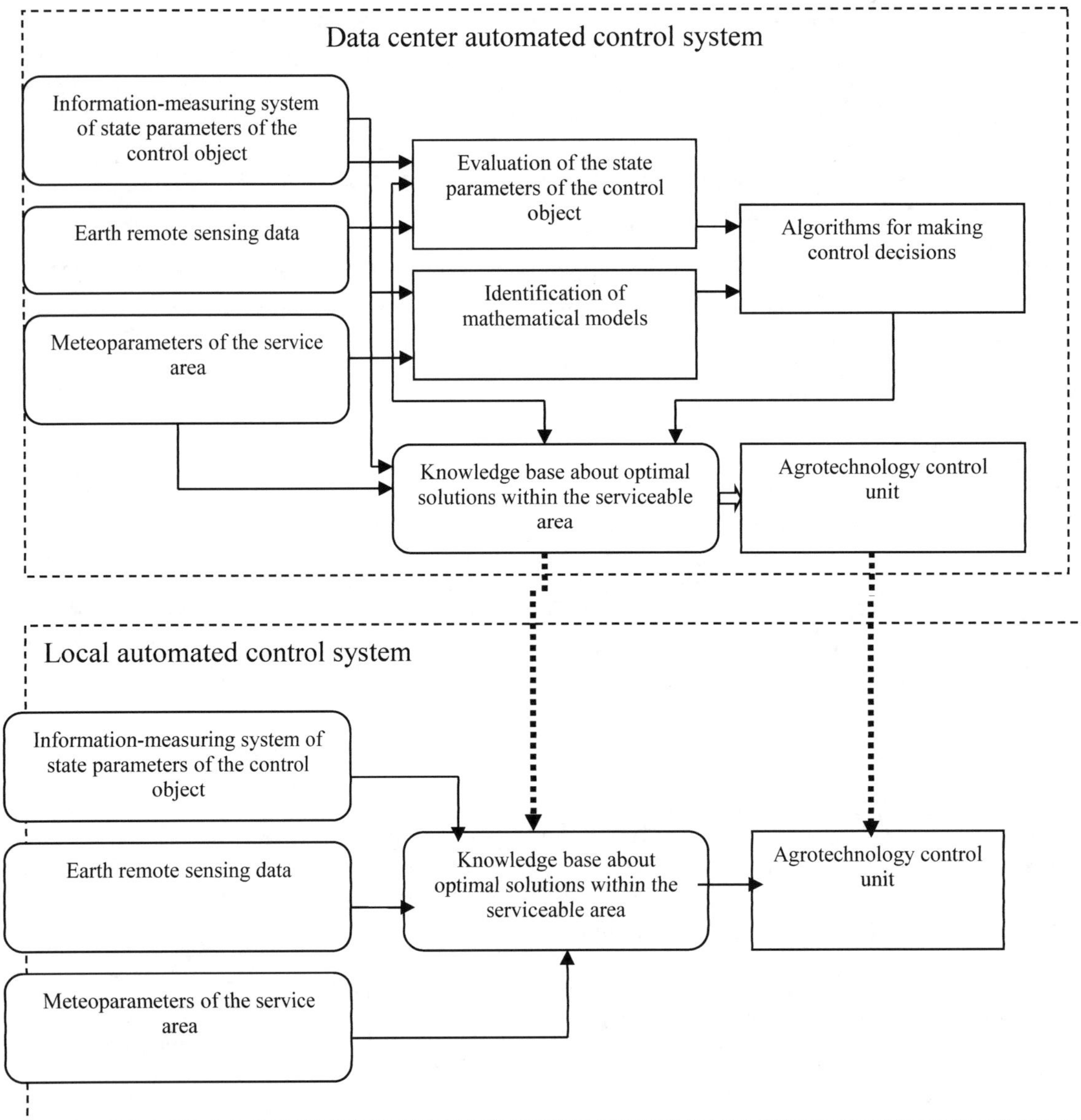

Figure 1. Block diagram of the automated process control system with training from the data center.

Figure 4 presents the optimal strategy for the application of improver and mineral fertilizers for extreme initial conditions. The effectiveness of this variant of the strategy is characterized by the value of the optimality criterion of 1,158.10 rub.$\cdot ha^{-1}$.

In accordance with rule (13), the closest variant of the strategy was chosen in the BR, which is characterized by the following initial conditions:

Table 4. The closest initial conditions from the BR.

Initial conditions	Acidity kg·ha^{-1}	Phosphorus kg·ha^{-1}	Potassium kg·ha^{-1}
	4,40	33,30	29,60

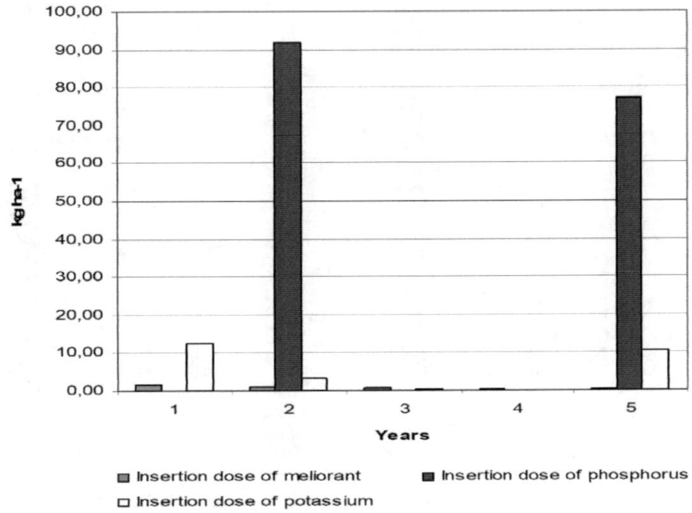

Figure 2. The optimal strategy for the introduction of improver and mineral fertilizers for arbitrary initial conditions of the DSS.

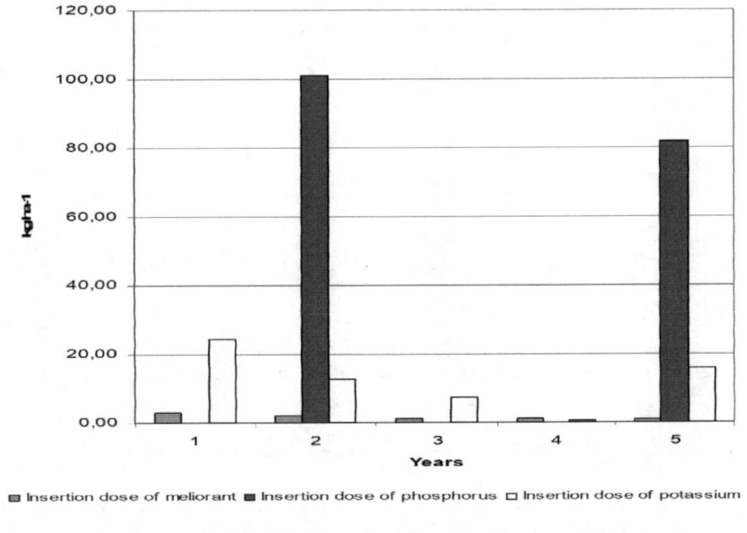

Figure 3. The optimal strategy for the application of improver and mineral fertilizers for the closest case of the BK.

This option corresponds to the strategy presented in Figure 5, the effectiveness of which is estimated by the value of the optimality criterion of 587.81 rub./ha. In this case, the loss of optimality due to the inconsistency of the initial conditions is 99%.

Such a high loss of optimality of the strategies for amelioration and mineral fertilizers makes it necessary to pay serious attention to the formation of the BR, which uniformly covers the entire multidimensional region of possible values of the initial conditions for rotation of crops. To replenish the BR can use two sources. In the first case, new variants of optimal strategies are generated through the above-described strategic management algorithm, and in the second case, real data from local DSS can be used during their centralized service in the cloud information system.

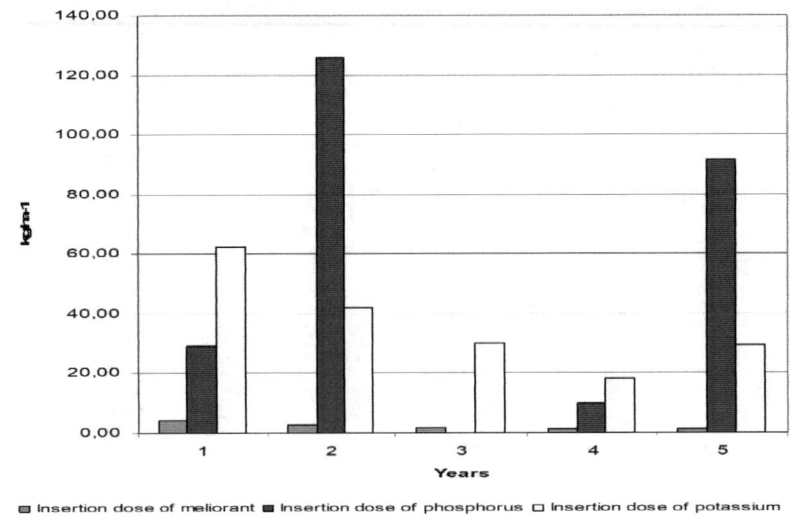

Figure 4. The optimal strategy for applying improver and mineral fertilizers for extreme initial conditions.

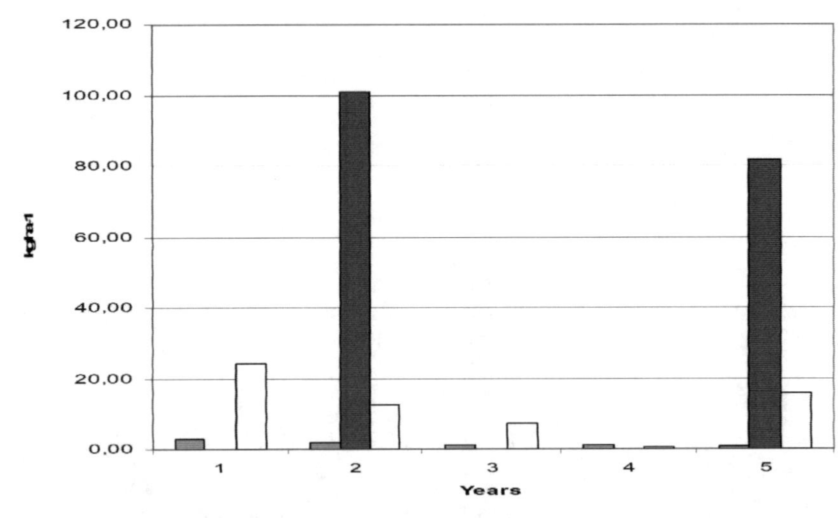

Figure 5. The optimal strategy for the introduction of improver and mineral fertilizers for the closest case of KB and extreme initial conditions.

At the same time, to decide on the inclusion of the next variant of the optimal strategy in the BR, the same criterion of proximity of options (12) can be used, for which its threshold value Δ is established, and the decision rule is applied:

a new option is included in the BK, if $\min\rho \geq \Delta$,

a new option is not included in the BK, if $\min\rho < \Delta$.

A significant method of increasing the reliability of expert DSS with centralized maintenance of local systems is the segmentation of KBs by soil and climatic conditions and crop varieties.

When forming KB segments according to above rule, only differences in the parameters of mathematical models (1) - (7) used in the strategic control algorithm will remain the main source of optimality losses. Losses from such differences do not exceed 20-25%, and their elimination is possible only by adapting models according to the operational monitoring of the state of crops in analytical ASMCs tied to specific control objects.

4. Conclusion

The method of building expert management decision support systems (DSS) in precision agriculture is proposed. The method involves the use of cloud information technologies, in which a knowledge base (KB) is transmitted from the cloud to local DSS, by means of which the closest optimal strategies for applying mineral fertilizers and ameliorants in various types of crop rotations are selected. An algorithm for the formation of optimal strategies has been developed for the formation of the cloud BS, minimizing the risk of yield losses for all crops of crop rotation and over-expenditure of mineral fertilizers and ameliorants. Based on the results of testing the methodology, a methodology for managing the formation of a KB, aimed at reducing the loss of optimality of strategies associated with a mismatch of initial conditions on local DSS and knowledge base of an information cloud, is substantiated.

Acknowledgments

The work was carried out with the partial support of the Russian Foundation for Basic Research (project No. 18-016-00008).

References

[1] Averyanov M A, Evtushenko S N, Kochetkova E Yu 2016 Digital Society: New Challenges *Economic Strategies* **7** (141): 90-91
[2] Kazakov I Ye 1987 *Optimization Methods for Stochastic Systems*. Moscow: Nauka p 354
[3] Mikhaylenko I M, Timoshin V N 2016 Optimization of soil chemical parameters in field crop rotations *Agrochemistry* **3** p 3-10
[4] Mikhaylenko I M, Timoshin V N 2018 Soil chemical state management based on Earth remote sensing data *Tavrichesky herald of agricultural science* 1 (13): 65-76
[5] Mikhaylenko I M 2017 *Theoretical foundations and technical implementation of agrotechnology management*. Ed. St. Petersburg State Technical University p 250
[6] Mikhaylenko I M 2019 Intellectualization of agrotechnology management *Bulletin of Russian agricultural science* **2** p 24-28
[7] Orlov S 2012 Cloud computing *Journal of Network Decisions* No. 01 [Electronic resource] http://www.osp.ru/lan/2012/01/13012475/
[8] Razumnikov S V 2014 Risk assessment modeling using cloud IT services *Basic research* **5** (1): 39-43
[9] Zakharenko A V, Yakushev V P 2009 *Precision Agriculture* ed. D. Shpaar, SPb. VIZR p 397
[10] Shirokova E A 2011 Cloud technologies *Int. Conf. on Modern trends in technical sciences: materials* (Ufa, October 2011) p 30-33
[11] Roudier P, Tisseyre B, Poilve H and Roger J M 2011 A technical opportunity index adapted to zone-specific management *Precision Agriculture* **12** p 130-145
[12] Kim K, Chavas J P 2003 Technological change and risk management: An application to the economics of corn production *Agricultural Economics* **29** p 125-142
[13] Neogronte N 1995 New York: Alfred A. Knopf (Paper edition 1996, Vintage Books)
[14] Gillam. Lee 2010 Cloud Computing: Principles, Systems and Applications *Nick Antonopoulos* (Computer Communications and Networks) p 379

Structural design and development of a water transfer sample

Yongguo Chen[1], Yonghua Chen[2,3] and Qingkui Liu[2]

[1]Guilin University of Technology, Guilin 541004, China
[2]Institute of Oceanology, Chinese Academy of Sciences, Qingdao 266071, China

[3]E-mail: yonghuachen@aliyun.com

Abstract. Understanding rare suspended particulate matter (including planktonic microorganisms) in the deep ocean is important but difficult to sample, for which we deigned a large-sized in-situ water sampler (200-1000L), i.e., a large volume water transfer system sampler. The system comprises of a deep-ocean pump, a filtering device, data acquisition and control devices, a supporting frame, a power supply component, a connecting mechanism, and a flow meter, etc. The filtering device integrates membranes for three-graded filtration at 0.22, 1, and 5μm, and an additional membrane aperture upon the need. The control device adopts two working modes: depth trigger and time trigger. The system can be added with a CTD and a fluorescence meter, etc. In addition, for a stationed site survey, several sets (such as 3 sets) of the system can be fixed at an interval on the suspension cable to work at different depths simultaneously. The system can be applied for high-level multi-layer filtration to obtain suspended particles. The system has been applied in several marine expeditions. The development of this system provides an ideal tool for deep-ocean in-situ sampling technology in China. The applications of the large volume water transfer system sampler for in-situ sampling in deep-ocean have been launched for a few times. Some interesting phenomena have been discovered and the advantages of in situ sampling are proved.

1. Introduction

Accurate measurements of deep-ocean suspended particles concentration (including planktonic microorganisms and suspended sediments) are of great significance to the study of ocean material transport and characteristics of water elements [1]. The traditional measurement method is considered to be the most accurate method. Through on-site water sampling (three or six-point method), water samples are filtered, weighed and suspended particles concentration mass concentration are calculated. However, only a few layers of depth and a large time interval of suspended particles concentration data can be obtained, which is time-consuming and labor-consuming. Modern measurement methods including using optical, acoustic, density, dielectric constant sensors to indirectly observe suspended particles concentration have been developed. Their characteristics are high efficiency, continuous acquisition, high spatial and temporal resolution of suspended particulate matter information can be obtained, but their measurement accuracy is low, and these observation methods need to be calibrated before use, and are limited by the applicable depth [2]. Therefore, how to accurately measure suspended particles concentration in multi-layer water body with time-saving and labor-saving is the key technology to be solved urgently in marine material research.

The research and development of seawater in-situ filtration technology began in the 1980s. The Woods Hole Institute of Oceanography firstly started the research and development of seawater direct current pump and seawater in-situ filtration technology. In the 1990s, Institute for Marine Research at the University of Kiel also developed a suspended particles concentration in-situ filtration device with a working depth of 4000 meters. In addition, the Woods Hole Institute of Oceanography and Mclane Company developed a large volume water transfer system (WTS-LV) together, and formed the product, as shown in the figure below. WTS-LV is a large-capacity in-situ micro-porous filter sampler for continuous extraction of water. It can collect suspended and soluble particulate matter in water through membrane filter paper or adsorption filter cartridge in the filter holder [3]. A series of sampling studies have been carried out using WTS-LV at home and abroad. The main function part of WTS-LV sampler is piston pump. The main body of WTS-LV sampler is made of stainless steel and titanium alloy. It can withstand the maximum water pressure of 5500 meters. It can be used in many water bodies, such as oceans, lakes, rivers and reservoirs, for sampling plankton samples, trace metal samples, sediment particles, etc. In addition, based on the need of research, some researchers have integrated sensors to acquire environmental parameters (CTD, fluorometer, turbidimeter, etc.) in deep-sea microbial sampling [4].

Since 2013, the Institute of Oceanography, Chinese Academy of Sciences, has carried out the deep-ocean in-situ filtration sampling technology, and a sampling system for obtaining suspended particles concentration by staged filtration in deep-ocean environment was successfully developed, which has been applied in many scientific voyages [5]. In the first part of this paper, the structure and design criteria of high flux deep ocean water sampling and grading filtration system are introduced in detail, and the depth meter of this system is compared with high precision sensors widely used in the world in the second part. The deep ocean sampling experiment is discussed in the third part.

2. System Design

As shown in Fig. 1, the main parts are the filter holder (1), the fixing clamp (2), the flow meter (3), the electronics and power pack housing (4), the integrated deep-ocean pump (5) and the mounting frame (6).

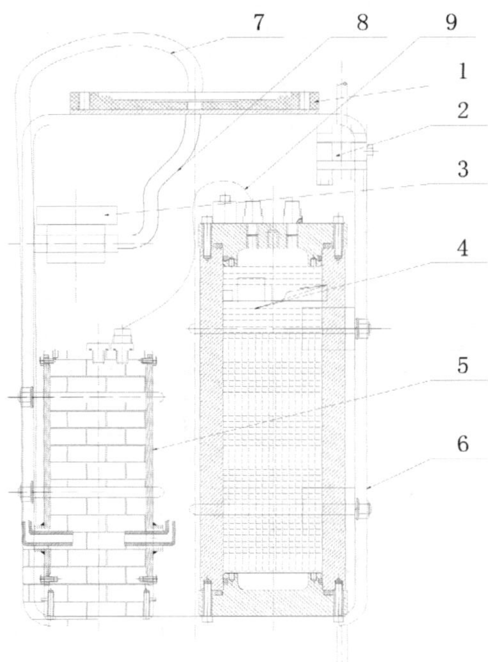

Figure 1. The sketch of the large volume water transfer system sampler.
1 - filter holder, 2 - fixed clamp, 3 – flow rate counter, 4 - control and power pack housing, 5 - integrated deep-ocean pump, 6 - mounting frame, 7 - pressure pipe, 8 - water pipe, 9 - watertight cable.

Technical parameters of the large volume water transfer system sampler are shown in the table below.

Table 1. Technical parameters of the large volume water transfer system sampler.

Parameter	**Value**
Working depth	3000m
Working mode	depth threshold trigger/time trigger
Filter layers	≥3
Filter diameter	200mm
Continuous sampling time	>26H
Pressure measurement accuracy	0.1% (FSO)
Flow rate	~5L/min(first class seawater)
Flux error	<5%
Size	(90×57×30)cm3
Weight	70kg(in air)

Note: FSO: Full Scale Output

The operation of the unit is controlled by software (PC) and all essential data are recorded for later evaluation. The built-in computer controls and records flow rate as well as the times and duration of pumping. The energy is supplied by battery packs with Li-cells storage batteries. Up to six battery packs are optional. The number of battery packs determines how much water can be extracted. Filter clogging may, however, be the limiting factor in turbid waters.

The system has the following details:

As shown in Fig. 2, the filter holder includes the top cover (102), the acquisition distribution layer (103), the fixtures (108), the flow diffusion (109), the bottom support plate (110) and the multi-stage support network, which are located between the top cover and the bottom support plate and are connected to form the closed cavity through the fixtures, and the bottom support plate is fixed on the mounting frame. There are three layers of filters and support nets stacking up and down, which are the class I filter and support net (104), the class II filter and support net (105) and the class III filter and support net (106) respectively. The fixing bolts are passed through the bottom support plate, the support nets, the acquisition distribution layer and the top cover respectively, and are locked and fixed by nuts with sealing achieved by O-rings (107) between them. The inlet (101) is screwed on the outside of the top cover and the outlet (111) is screwed on the outside of the bottom plate.

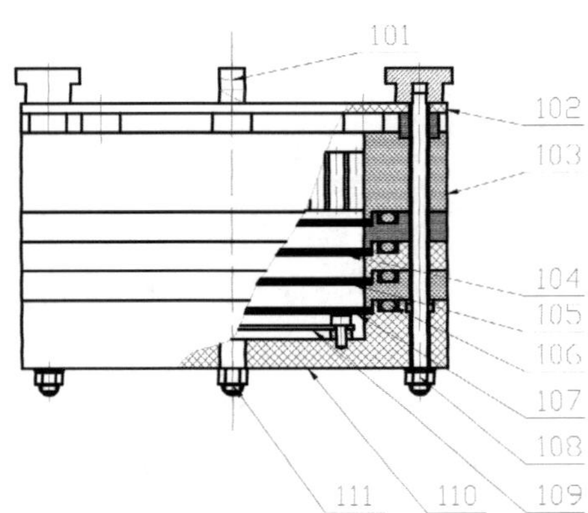

Figure 2. Cross-sectional view of the filter holder.

As shown in Fig. 3, the fixed clamp (2) includes the upper press plate (201), the upper splint (202), the fixed bolt (203), the rotating shaft A (204), the supporting rod (205), the lower splint (206), the lower press plate (207), the rotating shaft B (208) and the fastening screw. The lower press plate is fixed on installation frame. The upper press plate and the lower press plate are connected together by two supporting rods passing through the rotating shaft A . The upper splint is fixed on the upper pressure plate through four fixed bolts, and the lower splint is fixed on the lower press plate through another four fixed bolts. The other end of the press plate is articulated with the fastening screw. The fastening screw includes the fastening bolt (209) and the fastening nut (210). One end of the fastening bolt is articulated with the rotary shaft B, and the other end is connected with the fastening nut. The upper splint and the lower splint are combined with the rotation of the upper and lower pressure plates respectively, and the cable is between them, which makes the whole device be clamped on the cable.

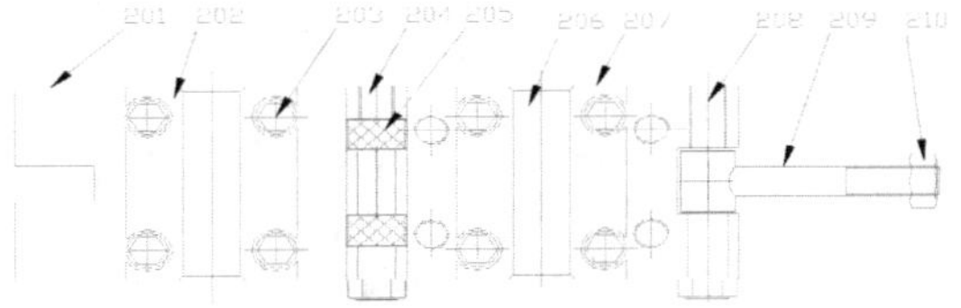

Figure 3. The sketch of the fixing clamp.

As shown in Fig. 4, the electronics and power pack housing is controller (401) and power packs (402) in the sealed cabin. The sealed cabin includes cylinder (408), left cover (405), right cover (409), sacrificial anode (403), Water-tight connector (404), positioning bolt (406) and O-ring (407). The shell of the sealed cabin is composed of cylinder, the left cover and the right cover, and two covers are installed at each end of the cylinder respectively with O-rings to make the left cover and the right cover sealed with the cylinder respectively, which forms a cylindrical sealing cabin. Water-tight connector and sacrificial anode are installed on left cover.

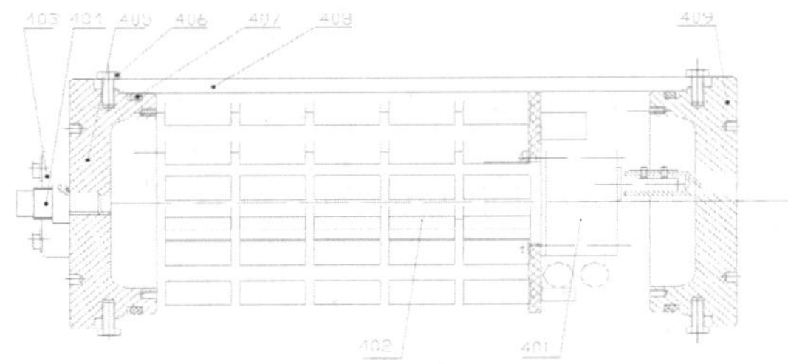

Figure 4. The sketch of the electronics and power pack housing.

As shown in Fig. 5, the integrated deep-ocean pump includes the sealing house, the DC motor (513), the pressure buffer, the connecting shaft (518), the pump head (519) and the deep-ocean water-tight connector (501). The sealing house includes the cylinder (512), the left end cover (504) and the right end cover (516), and the left end cover and the right end cover are fixed at the left and right ends of the cylinder with two O-rings separately. The sealing house is filled with filling oil (514). The DC motor is placed in the sealing house through the rod (511) fixed on the right cover. The pressure buffer comprises two rods (502) with hole, two clamps (507) and a rubber hose (508). The rods are installed on the left end cover filled with the O-ring A (503) and the rubber hose (508) wrapped on them inside the house with two clamps, which make the rods and rubber hose stick together. The inner side of the rubber hose is connected with the outer side of the sealing house through the holes on two rods to maintain the same pressure inside and outside the sealing house. A skeleton oil seal (517) is placed on the right end cover, and the connecting shaft is inserted across the skeleton oil seal. The outer side of the right end cover is provided with a fixed screw hole for fixing the pump head along the axis. Thus the connecting shaft is sealed on the right end cover, which makes the pump head connect with DC motor. The deep-ocean pump water-tight connector is installed on the left end cover and is electrically connected with the DC motor.

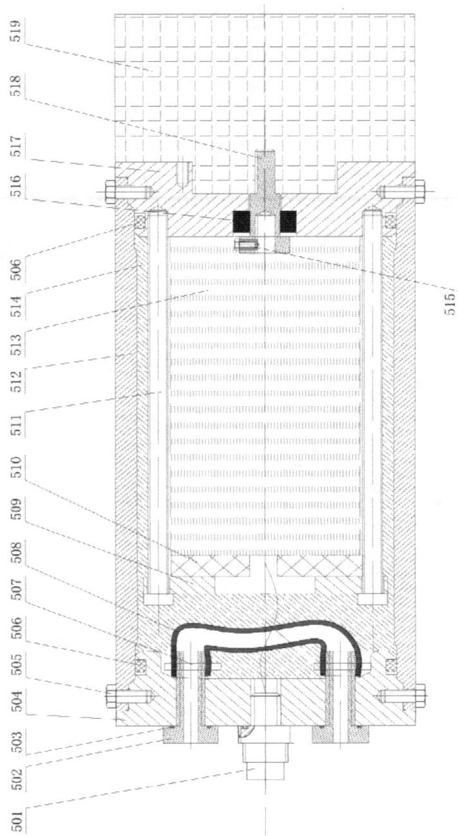

Figure 5. The sketch of the integrated deep-ocean pump.

The mounting frame is composed of the support frame, the fixing plate and some clips. The support frame is the rectangular frame, and the fixing plate is welded to the inner side of the support frame, while the clips are U-shaped. Flow meter, electronics and power pack housing and integrated deep-ocean pump are fixed on mounting frame with the clips respectively. The outlet of integrated deep-ocean pump is connected with the inlet of the filter holder through the pressure water pipe (7), and the outlet of the filter holder is connected with the inlet of the flow meter through the water pipe (8). The filter holder is equipped with multi-layer filter membranes to achieve graded filtration; and the electronics and power pack housing are integrated in the cylindrical sealed cabin and connected with the integrated deep-ocean pump through the watertight cable (9).

The electronic control has an external clock module to effectively control the working time, and the multi-channel analog to digital converter (ADC) sampling module is used to realize the measurement of various marine environmental parameters and the attitude of the filtering system, so as to accurately restore the changes of the marine environment in the sampling process. The electronic control system adopts two working modes: depth trigger and time trigger. The workflow of time-triggered and depth-triggered is shown in Fig. 6. The software of electronic control calibrates and controls the pumping rate, controls the times of start and end of pumping, records the volume sampled, records flow in user-selected time intervals and documents the battery condition at all times.

The Software is written in Turbo C for Windows systems. Pumping is stopped at a battery power below a preset value (watchdog/voltage control). This allows the preservation of all data in the case of power failure. Data exchange takes place through the RS-232C interface. Software control is menu driven, data are stored on SD card, data can be converted into any desired format. The programme allows setting of real time on the PC, the calibration factor of the rate counter, the times of start and end of pumping, the pumping rate, and the voltage below which pumping must be stopped. The selected configuration can be checked prior to deployment. After recovery of the unit, the measured data are transferred to a personal computer.

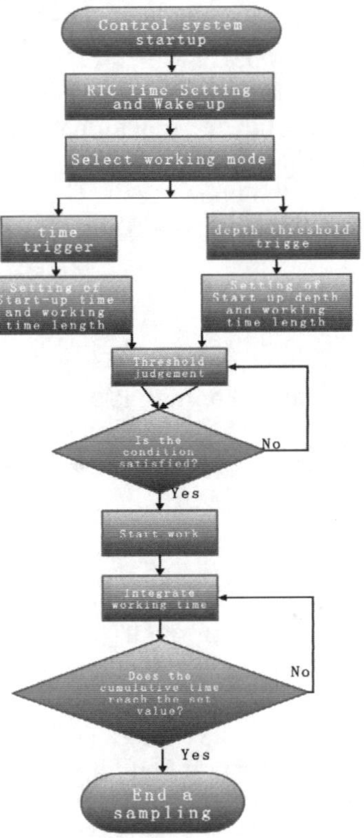

Figure 6. Flow chart of the sampling.

3. System Test

The design and the construction of prototypes were achieved at the Institute of Oceanology, Chinese Academy of Sciences. The system has been tested and applied successfully in subsurface and deeper waters (down to 2000 m) of the Western Pacific Ocean (As shown in Tab.2), including the extraction of more than 170 dm3 water within half an hour.

Table 2. Test Position of the System.

Test station	Position
TW2-2	121.85E, 22.88N
TW2-3	122.10E, 22.76N

Prior to deployment of the unit, preparations are made on deck. The flow controller is calibrated. The filter is inserted and the sampling tube is brought into position. The required time of pumping, the pumping rate and the safety voltage are entered. Each unit is attached to the hydrographic wire by means of clamps, the tubes are opened and the inlet is positioned about 1 m from the unit by means of a rod, extending into a direction away from the hydrographic wire. Thus, resampling of water exiting from the filtered water is avoided. The unit is then lowered to the desired depths. After retrieval of the unit on deck, the filters are removed and a PC is used to read the data. The unit is ready for use again after the battery packs have been recharged. The quality of the batteries is checked by software during and after the charging. The filters are kept in aluminium foil at -20°C until analysis in the home laboratory.

4. Summary

The equipment is flexible because it can be programmed: the user is assisted by a menu, allowing any change in the program to be realised in a very short time, say 1 minute. It is mechanically stable, even during launching and recovery operations at rough sea conditions. The systems have been tested down to 3000 m, the various stainless steel housings have been designed to resist pressures down to 6000 m. The challenge in their application lies in the fact that all essential operational steps are carried out under total exclusion from the ship's atmosphere: contamination can be kept at an extremely low level.

The units can also be applied in shallow waters. The limiting factor may be the much larger suspended matter concentrations than in the open ocean. The filters may get clogged before sufficient dissolved material has been collected. We are presently working on a version that should reduce the extent of this problem.

Acknowledgments

We wish to acknowledge assistance from senior engineering Li Siren and financial support from Key research and development of science and technology Program (Grant No. 2016YFC1402602 and No. 2017YFC1403406), the major scientific research equipment R & D projects of Chinese Academy of Sciences (Grant No. YZ201522 and No. YZ201625) and the Science Fund of the National Natural Science Foundation of China (Grants No. 41776109).

References

[1] Johnson R L, Pankow J F, Cherry J A 1987 Design of a ground-water sampler for collecting volatile organics and dissolved gases in small-diameter wells *Groundwater* **25** (4): 448-454

[2] Rai A K, Kumar A 2015 Continuous measurement of suspended sediment concentration: Technological advancement and future outlook *Measurement* **76** p 209-227

[3] Morrison A T, Billings J D, Doherty K W 2000 The McLane WTS-LV: a large volume, high accuracy, oceanographic sampling pump. In: *Conf. Of OCEANS 2000 MTS/IEEE* Providence, RI, USA, USA: IEEE **2** p 847-852

[4] Yu Yali, Wang Yi, Zou Tiancheng, et al. 2013 Design of New Portable Sampler for Water Quality at Multiple Depths *Northwest Hydropower* (2): 8-11

[5] Chen Yonghua, Yu Fei, Li Xiao-long, et al. 2017 The Large Volume Water Transfer and Graded Filtration System for in-Situ Sampling in Deep-sea *Oceanologia et Limnologia Sinica* **48** (6):1465-1470

A quantitative analysis pattern for regional risk conduction

Jie Su[1], Xinyue Zhang[1], Ling Zhou[2,4] and Yue Yin[3]

[1]School of Management Science and Engineering, Central University of Finance and Economics, Beijing, China
[2]School of Social Development and Public Policy, Beijing Normal University Beijing, Beijing, China
[3]China Minsheng Bank Corp., Ltd., Beijing, China

[4]E-mail: www.yueeaya@163.com

Abstract. Risk conduction refers to a certain risk can lead to the spread of risk accidents, causing other risks through many ways. Conductivity is a basic characteristic of risk, especially when the research object is regional risk. Aiming at the conductive characteristics of regional risk in scenic spots, based on Bayesian rule and total probability formula, this paper explored a regional risk conduction pattern. Firstly, risk management system consists of risk carrier, risk incentive and risk performance. The consequences of one risk may also be the risk environment of another risk, so the correlation between risks can be established. Secondly, construct regional risk transmission network with three layers of variables. Starting with the specific risks, analyzing the possible risk consequences, judging whether these risk consequences will lead to secondary risks and calculating the conditional probability of secondary risks based on conditional probability and full probability formula. Last, according to the conduction probability between risks, the comprehensive severity of the consequences is calculated after considering risk transmission. The conduction probability between risks and seriousness of risk after considering conduction can be obtained through this model. It provides a quantitative analysis tool for regional risk management and is helpful in two aspects: providing monitoring focus before risk occurs, and cutting off conduction path to prevent further risk expansion when risk occurs.

1. Introduction

With the continuous growth of China's economic development level and the improvement of people's quality of life, tourism has become the first vacation choice for thousands of families. However, in recent years, the frequency and scale of security incidents in scenic spots have been expanding. For risk accidents in scenic area, it is very likely that multiple safety incidents will occur in a short time. For example, in the summer of 2013, due to the high temperature in the past few days, the Xingshan rafting scenic spot in Hubei Province was overcrowded. Due to the chaotic work order of the management commanders, tourists rushed to log in to the kayak, causing vicious incidents such as fighting conflicts and eventually leading to stampedes. Such risk-conducting accidents are common in the safety management of scenic spots. When certain risks occur, the various regional routes lead to the spread of risk accidents in the scenic area, resulting in more serious consequences.

At present, a lot of scholars' research on regional risk of scenic spots has been relatively mature and have some safety management systems of scenic spots. Arbel's [1] scenic risk management article published in 1980 first studied the scenic area risk management and proposed a planning model for tourism scenic area risk management.

Since then, some scholars have studied the regional risk of scenic spots from tourism and crisis management, disaster planning [2-4], tourism risk management [5,6], tourism risk assessment [7].

In China, domestic research on regional regional risks is still in its infancy. Although there is a small amount of quantitative analysis, most of them are macro-analysis of regional risk of scenic spots from a qualitative point of view, using AHP, expert scoring, GIS technology and other methods. These studies studied the regional risk of scenic spots from the perspectives of the management of tourism safety and mechanism [8,9], safety risk index system [10-12], and scenic spot risk evaluation [13-16].

Research on risk transmission is currently concentrated in natural disaster systems. Shi Peijun (2002) defined the disaster chain as a series of disasters caused by a disaster, and proposed the common typhoon-storm chain, cold wave disaster chain, drought disaster chain and earthquake disaster chain in China [17]. Zhang Weixing (2013) established a risk assessment conceptual model of natural disaster chain based on the risk measurement method of supply chain system [18]. In addition to the natural disaster system, there are also risk transmission studies on software systems [19] and supply chain systems [20].

It can be seen that the current research on the regional risk of the scenic spot pays more attention to the establishment of the scenic area risk management system, risk index system, and risk assessment system, but less attention is paid to the risk transmission in the scenic area. For articles about risk transmission, the focus is on natural disaster transmission, work chain risk transmission, etc., and lack of research on risk transmission in a certain region.

In order to make up for the problem that most researches of scenic area risk management are qualitative analysis, this paper systematically quantifies the regional risk relationship and deeply studies the transmission relationship of regional risk. Combined with the regional security risk transmission process and management characteristics, we are going to establish a regional risk conduction quantitative pattern.

2. Introduction of mathematical methods and formula derivation

The mathematical theory used in this paper includes Bayesian rule and total probability formula. The Bayesian rule is used to describe the relationship between two conditional probabilities. The mathematical expression is in equation (1):

$$P(B_i|A) = \frac{P(B_i)P(A|B_i)}{\sum_{i=1}^{n} P(B_j)P(A|B_j)} \quad (1)$$

The total probability formula can transform the probabilistic solution of complex event A into the summation of the probabilities of simple events in different situations. If events B_1, B_2, B_3... B_n constitutes a complete event group, that is, they are incompatible with each other, and their sum is a complete set, so formula (2) holds for any event.

$$\mathrm{P(A)} = \sum_{i=1}^{n} P(A, B_i) \quad (2)$$

According to above two formulas, we can deduce the conditional probability between events A and B.

$$\mathrm{P(A|B)} = \frac{P(A,B)}{P(B)} = \frac{\sum_{C_i} P(A,B,C_i)}{P(B)} = \frac{\sum_{C_i} P(A,B|C_i) * P(C_i)}{P(B)}$$

Suppose that A and B are independent under the condition of C, the above formula can be further simplified.

$$\mathrm{P(A|B)} = \frac{\sum_{C_i} P(A,B|C_i) * P(C_i)}{P(B)} = \frac{\sum_{C_i} P(A|C_i)P(B|C_i) * P(C_i)}{P(B)}$$

$$= \sum_{C_k} \frac{P(A|C_i) * P(C_i)}{P(B)} * \frac{P(C_i|B) * P(B)}{P(C_i)}$$

$$= \sum_{C_k} P(A|\, C_i) * P(C_i|B) \quad (3)$$

According to formula (3), the conditional probability between two events can be written as the product of two conditional probabilities.

3. Quantitative analysis pattern of regional risk conduction

This chapter will specifically introduce the quantitative analysis method of regional risk transmission, which is divided into three parts: establishment of risk management system, establishment of transmission network and calculation of comprehensive risk severity.

3.1. Regional risk management system

When a risk occurs, the consequences of the risk, together with other factors in the environment, lead to other risks in the same region. This is the basic process of risk transmission. In order to study this process, we must clearly understand important factors in the process and construct regional risk management system. This paper constructs the regional risk management system from three perspectives: risk carrier, risk environment and risk performance (Figure 1).

Risk carriers, represented by symbol B, refer to disaster-bearing items in risk areas. They can be roughly divided into natural environment carriers, regional facilities carriers, item carriers, disaster prevention system and personnel carriers.

Risk environment, also known as risk incentives, is represented by symbol C. It refers to incentives that may lead to risks. It can be divided into four aspects: the property of items, natural factors, man-made environment and safety management.

Risk performance, expressed by symbol R, refers to the specific manifestation of risk, which can be divided into four categories: natural disaster risk, accident risk, public hygiene event and personnel risk.

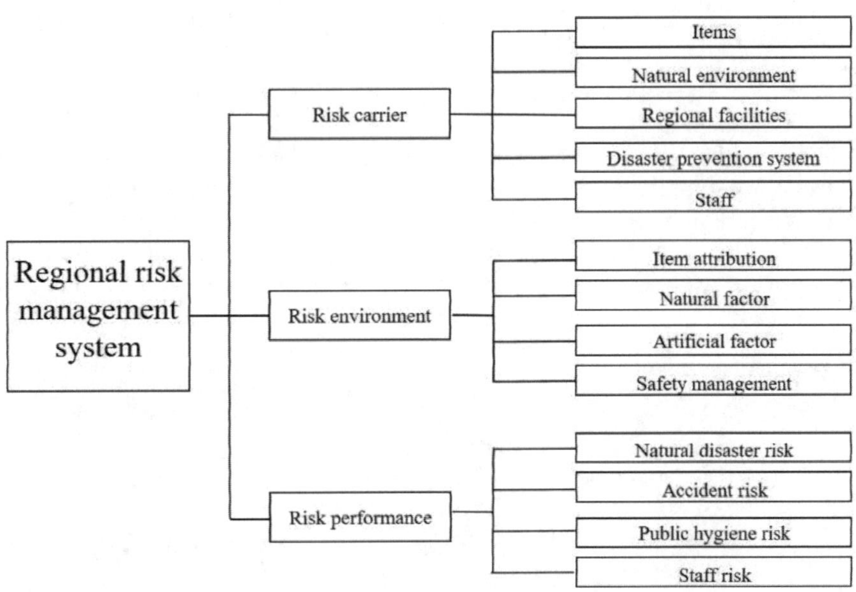

Figure 1. Regional risk management system.

3.2. Regional risk conduction network

Generally speaking, regional risk transmission can be summarized as a network model with input layer, intermediate layer and output layer. Taking the observed risk performance as the input layer of conduction pattern, the risk consequence which is also the risk incentive for other risk as the intermediate layer, and the secondary risk that first occurs as the output layer, the regional risk conduction network model is established. Figure 2 shows the results of regional risk conduction relationship.

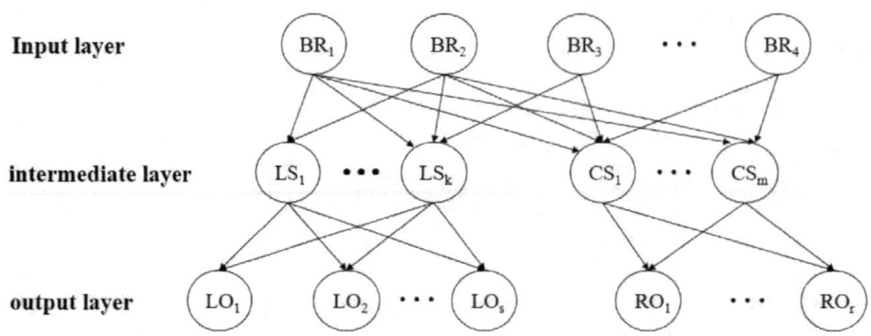

Figure 2. Regional risk conduction network.

The meaning of the variables in the figure is as follows:

Input variable BR: the starting point of regional risk conduction relationship, it's the observed risk performance.

State variable S: represents the risk consequences of the first risk, including loss state LS, which is directly expressed as loss, and incentive state RS, which forms the risk environment of other risks.

Output variable O: includes loss LO based on loss state and secondary risk performance RO based on incentive state.

After establishing the above regional risk conduction network, it is necessary to define two types of conditional probabilities for quantitative analysis: $P(C_i|BR_j)$ and $P(BR_i|C_j)$. $P(C_i|BR_j)$ indicates the probability of environmental factor C_i based on the risk BR_j, and $P(BR_i|C_j)$ indicates the probability of the risk BR_i based on environmental factor C_j.

Then, according to equation (3), the probability of risk BR_i based on risk BR_j can be calculated by formula (4).

$$P(BR_i|BR_j) = \sum_{C_k} P(BR_i|\, C_k) * P(C_k|BR_j) \tag{4}$$

Because the focus of this paper is the risk transmission relationship within a certain region, the geographical scope is limited, and the number of risk carriers in the region is limited, too, we can exhaust all risk in this region. Through risk questionnaires, on-the-spot investigation, expert opinions and other methods, all risk carriers, risk incentives, and risk performances in the region can be enumerated and graded, so the conditional probabilities mentioned above can be calculated.

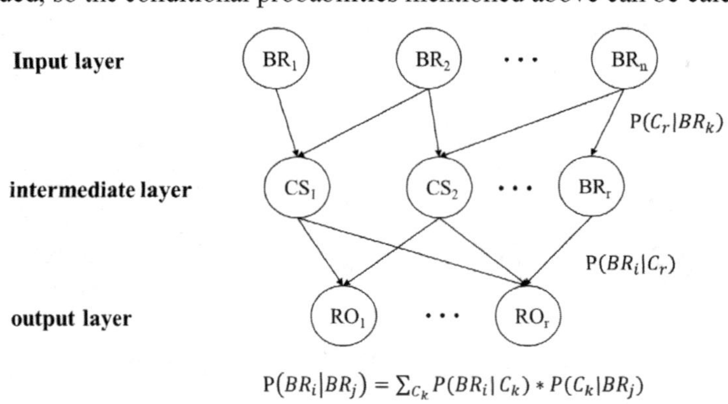

Figure 3. Regional risk conduction network with probability.

3.3. Comprehensive seriousness of risk taking risk transmission into account

According to the risk transmission network in last section, we can know when a risk occurs, which risk is more likely to be triggered, and we can accurately calculate the probability of being triggered. Knowing this probability, the severity of the secondary risk can be taken into account when assessing the severity of the risk.

Represent the original consequence severity of risk BR_i with the symbol Φ_i, the severity of risk BR_i after considering risk conduction is γ_i, which can be calculated by formula (5).

$$\Upsilon_i = \sum_k \phi_k * P(BR_k|BR_i) + \phi_i \tag{5}$$

4. Example: risk conduction relation of cultural relics in exhibition hall based on XX scenic spot

This chapter applies the regional risk conduction analysis method proposed in the previous chapter to the risk management of XX scenic spot, and illustrates the application of the above theory in the actual management.

4.1. Data preparation

The data of this part is from the questionnaire of the XX scenic spot management personnel and second key interview.

The respondents were the management staff of 13 administrative departments in scenic spots. The questionnaire contains three types of questions, namely, whether the risk environment of scenic spots plays a role in risk performance, the possibility of risk occurrence, and the severity of risk consequences. 58 valid questionnaires were collected. Then the results of the questionnaire survey were sorted out, and the experts of the relevant conclusions were interviewed to collect opinions. Finally, the data of XX scenic area risk management system is obtained. According to the questionnaire and interview, XX scenic area risk management system contains 16 types of risk carriers, 42 types of possible risk performance, and 69 kinds of risk incentives. As mentioned before, B_i and C_i respectively represent risk carriers, risk performance, and risk incentives. BR_i is used to indicate a specific risk on a carrier.

The number of valid questionnaire is 58, so number of sample is N = 58. Use variable θ_i to indicate the seriousness of risk BR_i, θ_i = 0,1,2,3,4,5, 0 means no impact, 1 means the impact is very small, 2 means risk BR_i may cause some impact but not serious, 3 means the impact is a little serious, 4 means serious impact and 5 means the impact is very serious. Use binary variable η_{ij} to indicate whether the incentive C_j has an effect on BR_i. Define the conditional probability $P(C_i|BR_j)$ and the consequence severity of risk BR_i.

$$\phi_i = \frac{\sum \theta_i}{N},\ P(C_i|BR_j) = \frac{\eta_{ij}}{N} \tag{6}$$

The other kind of conditional probability $P(BR_i|C_j)$ is given by risk experts in combination with specific environment and risk manifestations.

4.2. Data analysis

Taking the fire of cultural relics in the exhibition hall of XX scenic spot as an example, the above-mentioned regional risk conduction management model is used for risk conduction analysis.

According to the expert assessment, when the exhibition hall's cultural relics get on fire, it may cause several risk incentives, including damage of safety monitoring facilities, open fire, crowd crowding and inadequate managing workers, the probability of damage to safety monitoring facilities is 12.8%, the probability of open fire is 91.5%, the probability of crowd crowding is 77.1%, and the probability of insufficient managing workers is 19.6%.

According to the results of the questionnaire survey, when the security monitoring facilities are damaged, it may lead to the loss of exhibition cultural relics, and the probability of this risk is 32.7%. The probability of building fire caused by open fire is 1.7%. When crowds are crowded, cultural relics may be lost and stampede may occur, the probability is 5.1% and 20.6% respectively. Insufficient management may lead to a variety of risks, the probability of building on fire is 1.7%, the probability of cultural relics loss is 13.6%, and the probability of stampede is 6.8%.

Substituting possible risk environment, secondary risk and probability mentioned above into the regional risk conduction network. Result is showed in figure 4.

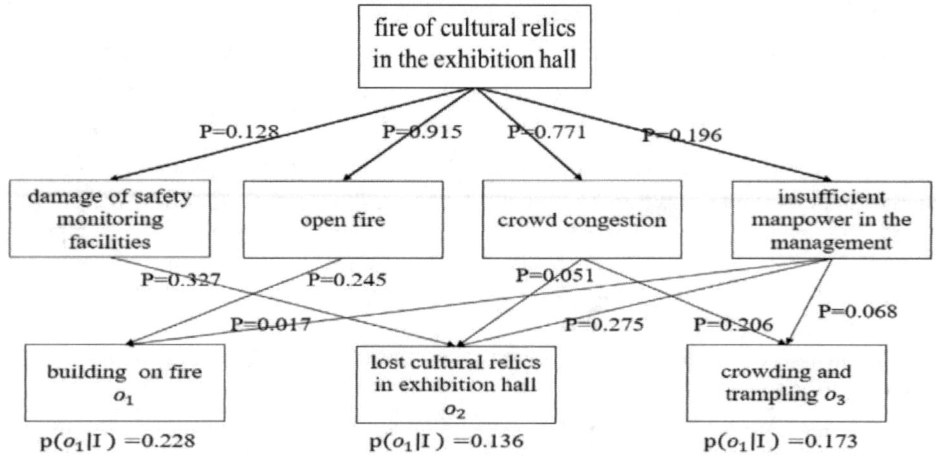

Figure 4. Regional risk conduction relationship based on fire of cultural relics in the exhibition hall.

According figure 4, we can see that when the fire happens in the exhibition hall of scenic spots, the risk consequences include the damage of safety monitoring facilities, open fire, crowd congestion and insufficient manpower in the management. These consequences result in the risk environment of the building on fire, the loss of cultural relics in exhibition hall and the occurrence of stampede event. Through the calculation, the conditional probability of building on fire is 0.228, the conditional probability of the loss of cultural relics in exhibition hall is 0.136, and the conditional probability of crowding and trampling risk is 0.173.

Next, according to the conditional probability of three kinds of risk and equation (5), the comprehensive seriousness of cultural relics fire in exhibition hall can be calculated. The severity of fire is showed in figure 5.

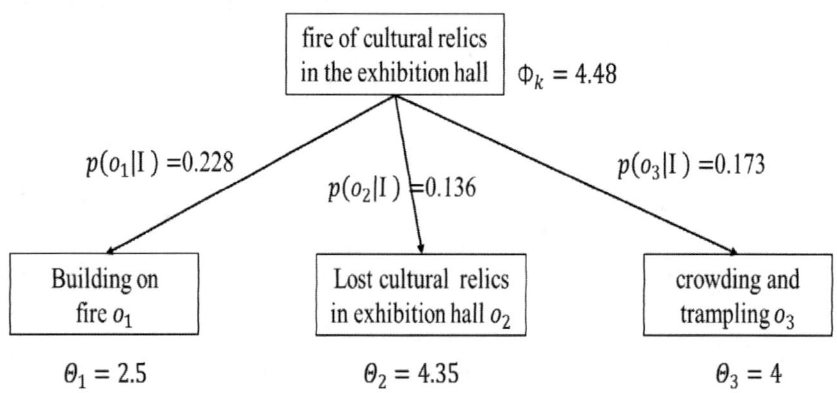

Figure 5. Severity of risks.

After calculation, the comprehensive risk consequence seriousness of the fire risk of cultural relics in exhibition hall is 6.33.

$$\gamma_k = 4.48 + 2.5 * 0.228 + 4.35 * 0.136 + 4 * 0.173 = 6.33$$

So, in the actual work of scenic area risk management, once the fire risk of cultural relics in the exhibition hall of scenic area occurs, it may lead to many other risks, among them, the most likely secondary risk is stampede. In addition, after knowing the conditional probability of risk transmission, risk conduction factors should be taken into account when evaluating the seriousness of risk. Considering the secondary risk, the severity of cultural relics fire in exhibition hall increased to 6.31.

5. Conclusion and contribution

In a limited area, the occurrence of risk often has the nature of conduction, and one risk may lead to other risks. Based on this, this paper explored a regional risk conduction pattern. The regional risk conduction pattern consists of three parts.

First, establish regional risk management system, have an intimate knowledge of specific environment from the risk carrier, risk incentives and risk performance. Second, establish regional risk conduction network, calculate conditional probability of secondary triggered by first risk. Based on this, we can know accurately which risk is more likely to be triggered with specific probability. Last, according to conduction probability between risks, calculating comprehensive severity of risk taking conduction factor, which can reflect possible losses more thoroughly.

This model can help regional risk management from two aspects.

First, providing monitoring focus before risk occurs. When assessing the consequence of a risk, we should consider not only the seriousness of the impact caused by the risk itself, but also its possible secondary risk consequences. Update risk consequence severity assessment according to conditional probability of secondary risk and consequence severity of secondary risk. The improved risk consequence assessment can reflect the seriousness of risk consequence more comprehensively, help to distinguish the focus of pre-management.

Second, cutting off conduction path to prevent further risk expansion when risk occurs. By calculating the conduction probability, managers can have a better control of risk conduction chain. The staff should deal with the direct loss, and at the same time, effectively deal with the transmission chain of secondary risk which may be caused by the fire, to avoid risk spreading. Knowing which risk is more likely to triggered by risk consequence help managers to know what risk and risk incentive need to be check preferentially and make risk management work more orderly when risk occurs.

Acknowledgments

This work was supported by National Key R&D Program of China (Grant No. 2018YFC0809900) and The National Social Science Fund of China (16ZDA054).

References

[1] Arbel A, Bargur J 1980 A planning model for crisis management in the tourism industry *Eurorean Journal of Operarional Research* **5** p 77-85
[2] Bayley M R 1989 Tourism and disaster planning *Geographical Review* **79** p 36-46
[3] Lehrman C K 1986 When Fact and Fantasy Collide: Crisis Management in the Travel Industry. *Public Relations Journal* **42** p 25-8
[4] Scott R 1988 Managing crisis in tourism: a case study of Fiji *Travel and Tourism Analyst*
[5] Faulkner B 2001 Towards a framework for tourism disaster management *Tourism Management* **22** p 135-47
[6] Russell R, Faulkner B 1999 Movers and shakers: chaos makers in tourism development *Tourism Management* **20** p 411-23
[7] Sheng-Hshiung T, Gwo-Hshiung T and Wang G Q 1997 Evaluating tourist risks from fuzzy perspectives *Annals of Tourism Research* **24** p 796-812
[8] Xiao Y B 2010 Thoughts on Tourism Safety in Scenic Spots *The Merchandise and Quality* p 13-23
[9] Xie Z W 2010 Study on the construction of tourism security pre-warning system in China *China Safety Science Journal (CSSJ)* **20** p 171-6
[10] He G P 2014 Studying of Guarding System of Safety Risk in Scenery Area Tourism *Journal of Lanzhou University of Arts and Science(Social Science Edition)* **30** p 19-23
[11] Wen J 2005 Study on the Risks and Countermeasures of China's Tourism in the Globalization *Problems of Forestry Economics* **25** p 334-8
[12] Zhu X, Wang J and Wu G Q 2010 Research on the Construction and Application of Indexes for Early Warning of Tourism Security *Technoeconomics and Management Research* p 133-136
[13] Cai R, Liu F G, Huang Y M, et al. 2014 The natural environment influence on tourism safety in Qinghai lake basin *Journal of Arid Land Resources and Environment* p 5-35

[14] Yang S S, Zhu R N 2013 The risk evaluation model of Sun Island scenic area in Harbin *Science-Technology and Management* **15** p 115-9
[15] Zhou L J 2012 *Study on Risk Assessment and Management of Tourism Safety in Mountain Scenic Areas* (Northeast Normal University)
[16] Zou Y G, Zheng X M 2011 Safety Evaluation Model and Empirical Study for Tourism Scenic Spots *China Safety Science Journal* **21** p 156-162
[17] Shi P J 2014 Complexity of Disaster System and Comprehensive Disaster Prevention and Reduction *Disaster Reduction in China* p 20-21
[18] Zhang W X, Zhou H J 2013 Conceptual Model for Risk Assessment of Disaster Chain: A Case Study of Wenchuan 5.12 Earthquake *Progress in Geography* **32** p 130-8
[19] Meng J, Jiang L M, Li Q M and Zhang H 2011 Research on Risk Correlation Based on the Module Coupling *Logistics Technology* **32** p 217-24
[20] Geng X F, Liu K and Wang D Z 2007 Fuzzy Compr ehensive Evaluation of SC Risk *Logistics Technology* **26** p 164-167

Journal of Physics: Conference Series **1419** (2019) 012033 doi:10.1088/1742-6596/1419/1/012033

Prevention and control on urban NIMBY caused by nuclear facilities in China

Jiu Liu

Nantong Street No.145, Nangang District, City of Harbin, Heilongjiang Province, 150001, China

E-mail: nine699@sina.com

Abstract. In order to improve the energy structure and the air quality, China has made every effort to develop civil nuclear industry and establish some nuclear facilities. Due to the potential radioactivity of nuclear facilities, the location and construction of the facilities will cause dissatisfaction and panic of the surrounding residents, resulting in "Not in My Backyard (NIMBY)" incidents. This paper will analyze the reasons for the NIMBY incidents, which are inadequate and not timely disclosure of information for residents, the lack of relevant education and compensation mechanism with legal analytical methods. Therefore, the prevention and control methods of such NIMBY incidents should be put forward: Governmental departments and operators of nuclear facilities should adhere to scientific and democratic decision of location selection and construction, and establish proper compensation mechanism for the residents nearby.

1. Introduction

The civil nuclear industry in China started in the early 1980s. So far, China has been one of the few countries with a complete system of civil nuclear industry and nuclear industry in China has been ranked the fourth in the world. By 2020, the scale of the civil nuclear industry of China will have been ranked second in the world [1]. Nuclear industry has played an important role in economic and social development in China and still keeps a good record of nuclear safety and security. However, it cannot be neglected that nuclear facilities and materials have potential radiological hazards. The nuclear accidents happened in Three Mile Island of the United States, Chernobyl of the Soviet Union and Fukushima of Japan made it quite difficult to decrease people's anxiety and increase people's acceptance for nuclear facilities. Therefore, the residents living near the nuclear facilities always worry about the influences on their health, environmental conditions and property values. So, when the nuclear facility may be constructed nearby, the residents will carry on highly emotional boycotts and even fight against the construction decisions, which always cause the NIMBY incidents.

The NIMBY (Not in My Backyard) movement or incident was initially raised by O'Hare, an American scholar, who defined "NIMBY" that "the establishment and operation of some facilities would benefit the citizens across wide areas but the citizens near the facilities will suffer from relevant negative influences [2]. Therefore, they will be sick of the facilities. When they change their resenting attitude into boycotts, it will be a collective conflict and lead to NIMBY." In the simple words, NIMBY means the citizens near the facilities insist that "we need this facility, but not in my backyard". The occurrence of concept was relevant to the incident that the residents nearby boycotted the decision of the setting of facilities like garbage furnace and nuclear facilities of the developed countries in the 1970s. The event that citizens of Lianyungang boycotted the construction of spent fuel reprocessing factory in 2016 is typical NIMBY.

In China, it is common for the government to make a compromise and abandon the candidate location of NIMBY nuclear facilities to temporarily solve the confliction. Actually, this method will waste much social and economic cost, still fail to reduce people's fear of nuclear, and quite bad to improve people's acceptance of nuclear facilities.

This paper proposes the following research questions: Why do the NIMBY incidents caused by nuclear facilities happen? And how can the government prevent and control such NIMBY incidents by nuclear facilities? In order to solve the NIMBY problems in China, this paper adopts qualitative and case methodologies to summarize and analyse the current legislation and cases pertaining to nuclear facilities and environmental pollution in China and put forward the reasons of such incidents and solutions to prevent the incidents and increase people's phycological acceptance of nuclear facilities.

2. The reasons of NIMBY incidents by nuclear facilities

In the profound investigation on reasons for Nuclear-fear NIMBY Events with the incident of Lianyungang citizens boycotted the irradiated fuel reprocessing factory, it was learned that the citizens nearby insisted that the use of facility would lead to potential radiological hazards and threat or harm their own interests. From the legal perspective, the insufficient publicity of setting decision made by the operator of nuclear facilities and local governments, to the greatest extent, neglected people's rights of knowing the details. In addition, they failed to conduct the popularization of science for local citizens and haven't collected the suggestions and opinions from local people. Moreover, they haven't promised to provide relevant payment for local people. All the elements led to the occurrence and conflicts in such events.

Table 1. The reasons of NIMBY incidents by nuclear facilities.

·NIMBY citizens are lack of rights of disclosure and knowing details
·NIMBY citizens do not have sufficient relevant promotional education
·Government is unable to develop relevant compensation mechanism for NIMBY citizens

2.1. Unable to provide NIMBY citizens rights of knowing details

The construction of nuclear facilities leads to great benefits but they also bring out regional hazards. So, in some sense, the cost will be only paid by citizens near the plants and not shared by the beneficiaries, which is unfair for the citizens nearby. Therefore, it will cause complains and boycotts from citizens who take the risks. Besides, the imbalanced information and failure of delivering opinions will aggravate the existing confliction. In selection and construction of the nuclear facilities, the intensification of public learning the details and participation may increase the cost and influence the efficiency. However, fear is usually from something unknown and understanding is from general publicity. In many similar events, as the relevant information is not revealed from mainstream channels, there is no chance for people to deliver their own opinions and safeguard their own benefits, which are the main reasons for citizens to boycott the setting of NIMBY facilities.

In China, there are legal requirements to provide people sufficient rights to be informed and rights of discourse in legislation and regulation. According to Article 53 and 56 in The Environmental Protection Law as well as Article 21 and 23 in Evaluation Law of Environmental Influence, firstly, the operators involved in special projects like nuclear facilities shall submit the documents relevant to environmental influences of construction projects. Before approving the documents, it is necessary to give the explanation for citizens who may be influenced and collect the views from relevant operators, experts and citizens by holding public hearings or other ways of announcement. Secondly, chief environmental protection departments of governments at all levels shall release environmental information to the public according to the law and establish sound mechanism of people's participation. Thirdly, the citizens, according to the law, have the rights to obtain environmental information and participate in and oversee environmental protection. Meanwhile, according to Article 11 and 15 in Regulations on the Administration of Construction Project Environmental Protection issued by State Council, construction operators of nuclear facilities should compile the environmental impact reports and demand the examination and approval from the competent department of environmental protection administration under the State Council, in addition, the report compiling should solicit the views of the operators and residents concerned of the locality. As for the laws and regulations relevant to nuclear facilities and radiation, according to The Prevention and Control of Radioactive Pollution Law, Chinese safety supervision of nuclear facilities and radioactivity, prevention and control of radioactive pollution should be mainly administrated and supervised by governments and relevant departments, in the meantime, relevant operators and individuals have rights of reporting and accusing the conducts of radioactive pollution. Therefore, according to the laws and regulations mentioned above, governmental institutions like the operators of nuclear projects and environmental protection departments have obligation of issuing relevant information and solicit the opinions from the public who also have the rights of gaining information and put forward their views relevant to their own direct interests and participate in reporting, supervising and accusing the conducts of radioactive pollution. Only when the residents concerned of the locality gain sufficient information may they escape from exaggeration and panic of the hazards and reduce the impulse of opposing the NIMBY facilities.

However, in NIMBY Events like the incident of Lianyungang citizens boycotted the irradiated fuel reprocessing factory, as the final locality hasn't been selected, relevant operators neglected the aforementioned laws and regulations and simply revealed in some enterprise news with a few words to explain the reason for selecting Lianyungang as the locality of the plant. In addition, they never let local residents know the details of the news, which restricted the local residents from knowing the details and participation. However, according to Environmental Protection Law, local environmental protection departments, for all the time, have the obligation of "disclosing environmental information with the accordance of law and completing the procedure of people's participation." As it was in selecting the locality of spent fuel reprocessing factory, the relevant selections haven't made real impact on the local residents and then the government neglected local residents' rights to know and disclose. Therefore, without gaining real and correct information and candidate locality, local residents were misunderstood by We-media and then became panic, which led to NIMBY phenomenon.

2.2. Lack of relevant promotional education

In dealing with NIMBY, besides making public a variety of information, American government also organized and trained professional staffs to help local residents learn professional standards and data relevant to NIMBY facilities in order to improve people's reasonable judgement and eliminate panic emotions. In China, The Prevention and Control of Radioactive Pollution Law also regulates that "the people's government at the county level or above shall arrange for and carry out the pertinent publicity and education on the prevention and control of radioactive pollution, and make the public know about the relevant information and scientific knowledge on the prevention and control of radioactive pollution." However, in Lianyungang NIMBY Incident, as it was in the period of site selection, the governments and enterprises neglected sufficient promotion and education and never revealed relevant information timely making the local residents lack relevant knowledge and information. In addition to the characteristics of Network Era, some fake information spread through various We-media rapidly, which made the people misunderstood and become panic.

As for the construction of a spent fuel processing plant, from scientific and legal perspective, it is suitable to implement popularization of science on the knowledge of "spent fuel" and "spent fuel processing". For example, in science, "spent fuel" refers to the used nuclear fuel generally produced in the reactors of nuclear plants. In nuclear reactors, nuclear fuel has nuclear reactions with neutron bombardment. After a period of time, it is removed from the reactors. Meanwhile, the fuel contains many unused proliferating materials including Uranium 238 or Thorium 232. As the Uranium contents of fuel are reduced, it cannot continue with nuclear reactivity. Therefore, the fuel is called the spent fuel. From the perspectives of laws and international conventions, according to Joint Convention on the Safety of Spent Fuel Management and on the Safety of Radiation Waste Management, "spent fuel" means nuclear fuel that has been irradiated in and permanently removed from a reactor core. Therefore, according to Article 2 in Regulations for Radioactive Waste Management issued by State Council, there is the definition of radioactive waste that the spent fuel is not nuclear waste. And, it can deal with the resource limits of nuclear fuel and reduce nuclear waste and optimize the management of waste after processing of the spent fuel. In addition, if such conducts are in line with Article 40 in Environmental Protection Law, "The state shall promote cleaner production and resource recycling" and relevant regulations in Promotional Law of Recycling Economy, the residents won't believe the fake information and consider it as processing of nuclear rubbishes. They just think it is harmful and dangerous and refuse to agree on the construction of establishing the local spent fuel plant for fear of hurting their own health and economic interests. However, in this incident, nuclear facility operator CNFR (CNNC Nuclear Fuel Reprocessing Co., Ltd) hasn't carried out the aforementioned science popularization for the citizens of Lianyungang, who lives in the possible site of the spent fuel plant. When responding the relevant issues, they just focus on the importance of project instead of answering the people's puzzles directly and reply people's needs, which lead to more misunderstandings and resentful emotions.

2.3. Unable to develop relevant compensation mechanism for NIMBY citizens

According to the research of O'Hare Michael, an important failing of current practice in sitting locally noxious facilities is the strategic problem which results from failure to pay compensation to neighbors who suffer costs (loss on property values or less measurable amenity costs) not found to be a taking under law. If the residents in nuclear facility areas can be provided with relevant financial compensation for their loss from potential risks and great reduction in possible added value of permanent assets like houses, it will bring out relevant economic interests and improve the enthusiasm of construction for the residents in nuclear facility areas. In addition, it is also good for reducing people's hostility and improving their recognition and acceptance of nuclear risks. However, there are no relevant systems in the fields of nuclear law in China. Take the relevant spent fuel processing system as an example, there are regulations in Temporary Measures for the Use and Management of the Spent Fuel Treatment and Disposal Fund for Nuclear Power Plants Issued by MOF (the Ministry of Finance), NDRC and MIIT (Ministry of Industry and Information Technology) and Management Measures for the Spent Fuel Treatment and Disposal Fund for Nuclear Power Plants issued by SASTIND that the spent fuel processing fund can be used for of spent fuel's transportation, storage, later processing and the processing of highly radiation waste from later processing of spent fuel, spent fuel processing plant's construction, operation, reconstruction and retirement as well as other outcomes of spent fuel processing. Meanwhile, they haven't regulated clearly that local residents can be provided with financial compensation. Local governments usually provide relevant land compensation for the residents of NIMBY areas rather than the residents whose lands haven't been requisitioned. For example, in construction of high-speed railroad, people with relevant interests in relocation can be provided with compensation but other people in the project are the exception of the compensation. Therefore, it has been one of the reasons for NIMBY phenomenon in site selection and construction of nuclear facilities.

3. Prevention and control of NIMBY caused by nuclear facilities

Law is the most efficient solution in dealing with NIMBY incidents caused by nuclear facilities. As the stable and predictable rules, law can make all sides with relevant interests participate in construction and policy-making equally and fairly. And, it will make the people with loss of personal interests gain the compensation so as to prevent the frequent happening of NIMBY events and solve such events properly. Then, it may reduce the economic loss from the events and safeguard social harmony and stability and win the people's confidence. As a result, our relevant governmental departments and enterprises should make good use of law to prevent and solve the NIMBY phenomenon in the frame of laws and regulations.

For maintaining social stability and making Chinese nuclear industry with smooth and sustainable development, the government's security departments, and environmental protection department, local governments of candidate areas, relevant enterprises and institutions must make site selection of nuclear facilities and construction policy-making scientific and democratic. Science refers to the evaluation of various safety issues related to the selection of such great nuclear facilities. In addition, it is also needed to implement sustainable supervision and examination of the construction and the later processing to prevent from the risks. As for being democratic, it means to establish public participation mechanism and conduct profound communication with the residents and provide popularization for residents and collect the public opinions and focus on public rights to know and supervise.

Table 2. Methods of prevention and control of NIMBY caused by nuclear facilities.

·Government and operators of nuclear facilities should adhere to scientific location selection and construction
·Government and operators of nuclear facilities should adhere to democratic site selection and policy-making of construction
·Legislature should establish financial compensation mechanism for residents nearby the nuclear facilities

3.1. Adhere to scientific nuclear facilities' decision of location selection and construction

In order to make scientific location selection of nuclear projects， Prevention and Control of Radioactive Pollution Law in China provides that "The competent department for nuclear facilities under the State Council, together with the competent environmental protection administrative department under the State Council, shall draw up, based on the environmental impact assessment, a solid radioactive waste disposal siting plan on the basis of geological factors and solid radioactive waste disposal needs. The plan shall be implemented after being reported to and approved by the State Council." [3] Joint Convention on the Safety of Spent Fuel Management and on the Safety of Radioactive Waste Management issued by IAEA also stipulate that Each Contracting Party shall take the appropriate steps to ensure that procedures are established and implemented for a proposed spent fuel management facility: to evaluate all relevant site-related factors likely to affect the safety of such a facility during its operating lifetime; to evaluate the likely safety impact of such a facility on individuals, society and the environment; to make information on the safety of such a facility available to members of the public; to consult Contracting Parties in the vicinity of such a facility, insofar as they are likely to be affected by that facility, and provide them, upon their request, with general data relating to the facility to enable them to evaluate the likely safety impact of the facility upon their territory, meanwhile Each Contracting Party shall take the appropriate steps to ensure that: the design and construction of a spent fuel management facility provide for suitable measures to limit possible radiological impacts on individuals, society and the environment, including those from discharges or uncontrolled releases; at the design stage, conceptual plans and, as necessary, technical provisions for the decommissioning of a spent fuel management facility are taken into account; the technologies incorporated in the design and construction of a spent fuel management facility are supported by experience, testing or analysis [4].

For making site selection more scientific, procedural and systematic, according to Nuclear Safety Law, the top legislative law in nuclear field in China, which officially implemented at the beginning of 2018, it regulates: "The state shall develop a nuclear facility safety licensing system. A nuclear facility operating entity that conducts such activities as siting, construction, operation and decommissioning of a nuclear facility shall apply for licensing to the nuclear safety supervision and administration department of the State Council... A nuclear facility operating entity shall make scientific assessment of geology, earthquake, weather, hydrology, environment, population distribution and other factors, submit a nuclear facility site safety analysis report to the nuclear safety supervision and administration department of the State Council on the premise of conforming to the requirements of nuclear safety technical appraisal, and obtain the nuclear facility site selection review opinions after the review finds it conform to the requirements for nuclear safety." [5] It is believed that with the aforementioned scientific site selection, the operating party of nuclear facilities and relevant departments of local governments will do well in prevention of risks and accidents to safeguard the safety and interests of local residents.

3.2. Adhere to democratic nuclear facilities' site selection and policy-making of construction

As for democratic nuclear facilities' site selection and policy-making of construction, International Atomic Energy Agency (IAEA) suggests, with the development of the peaceful uses of nuclear energy, public understanding of and confidence in the technology have required that the public, the media, legislatures and other interested bodies be provided with the fullest possible information concerning the risks and benefits of using various nuclear related techniques for economic and social development. Therefore, the transparency principle requires that bodies involved in the development, use and regulation of nuclear energy make available all relevant information concerning how nuclear energy is being used, particularly concerning incidents and abnormal occurrences that could have an impact on public health, safety and the environment [6].

Relevant French legislation and regulations also require nuclear operators to popularize the public with knowledge of nuclear power and open to the public with nuclear facilities on construction or operation in order to strengthen people's safety confidence of nuclear projects. In administrative permission of project approval and site selection, Japan has taken consideration of national safety, nuclear development plan and regional economy. In addition, it has also considered the rights of public expression in system and procedure. The operator should try to make the information public and ensure transparency of business activities and establish the platform to strengthen the communication with regional society. For example, at Kashiwazaki and East Sea Village, together with local residents, Tokyo Electric Power established Transparency Organization of Kashiwazaki Kariwa Nuclear Power Plant and Proposal Organization of East Sea Village's Environment and Nuclear Energy Safety [7].

In order to improve the democracy of decision-making and prevent the NIMBY Events, China has always focused on the cultivation of public rights of knowing the details and awareness of participation. Since 1996, State Council has issued Decision on Some Issue of Environmental Protection and initially put forward public participation mechanism. In 2004, Administrative Permission Law stipulated the public participation's procedures, methods and conditions. As for nuclear projects and safety, Statement of Nuclear Safety Culture Policy issued by National Nuclear Safety Administration, National Energy Administration put forward to establish harmonious public relations. "Through ways of public communication like information opening, public participation, science popularization, it is needed to ensure the public rights to know, participate and supervise. The departments of decision and administration should learn different opinions in different ways and with open mind. In addition, they should try to deal properly with various demands of relevant people." [8] According to 12th Five-year Plan and 2020 Long-rang Objective of Nuclear Safety and Protection of Nuclear Radioactive Pollution, China will "Establish an open and transparent platform of communication to improve transparency of industry; establish the system to open information of nuclear facilities and make clear governmental departments and operational sections' scope, duty and procedure in releasing the information; improve the public participation in site selection, construction, operation and retirement. In basic education, it is needed to add the popularization knowledge of nuclear and radioactive security. In addition, the effective educational and promotional system of nuclear security must be established to meet the public demands of relevant nuclear security information so as to improve people's confidence and understanding in the safe application of nuclear and technology [9].

The Chapter Five of Nuclear Safety Law has also provided "Information Disclosure and Public Participate" for nuclear facilities. Firstly, the relevant departments of the State Council and the departments designated by the people's governments of provinces, autonomous regions and municipalities directly under the Central Government where nuclear facilities are located shall publicly disclose the information relating to nuclear safety within their respective functions and responsibilities in accordance with the law. The nuclear safety supervision and administration department of the State Council shall publicly disclose under the law administrative licensing relating to nuclear safety and such information as reports on safety supervision and inspection of activities regarding nuclear safety, overall safety situation, radiation environment quality, and nuclear accidents. The State Council shall regularly report nuclear safety situation to the Standing Committee of the

National People's Congress. Secondly, the nuclear facility operating entities shall publicly disclose such information as its nuclear safety management system and relevant documents, safety situations of the nuclear facility, radiation monitoring data of effluents and surrounding environment, and annual nuclear safety reports. The nuclear safety supervision and administration department of the State Council shall develop specific measures. And the nuclear facility operating entities shall solicit the opinions of interested parties on major nuclear safety matters involving public interests through questionnaire survey, hearing, discussion meeting, or symposium or by any other means and provide feedback in an appropriate form. The people's government of the province, autonomous region or municipality directly under the Central Government where the nuclear facility is located shall solicit the opinions of interested parties on major nuclear safety matters impacting public interests by holding hearing, discussion meeting, symposium or any other means and provide feedback in an appropriate form. The entities shall take measures to conduct nuclear safety publicity campaigns, such as opening the nuclear facility to the general public in an orderly manner on the premise of ensuring the safety of the nuclear facility; cooperating with schools to conduct nuclear safety knowledge education campaigns for students; establishing nuclear safety publicity places and printing and issuing nuclear safety publicity materials and etc. Moreover, nuclear safety information required to be publicly disclosed by the law shall be so disclosed in a timely manner by government announcements, websites or any other means facilitating public knowledge. Citizens, legal persons or any other organizations may, in accordance with the law, apply for access to the information relating to nuclear safety to the nuclear safety supervision and administration department of the State Council and the department designated by the people's government of the province, autonomous region or municipality directly under the Central Government where the nuclear facility is located and they have the right to report any act with hidden nuclear safety risks or in violation of nuclear safety laws and administrative regulations to the nuclear safety supervision and administration department or any other relevant department of the State Council [10].

Besides, the upcoming Nuclear Power Management Regulations (draft for approval) drafted by NDRC (National Development and Reform Commission) and National Energy Bureau regulates in general principles that in Article 8, “the state's relevant departments and units shall strengthen nuclear power and its safety, nuclear radioactive protection and popularization and education of nuclear emergency knowledge and gradually make them into national compulsory educational system. Establish the sound system of nuclear power information disclosure and public participation. According to the law, it is needed to disclose relevant information of nuclear plant construction, operation and emergency management of nuclear accidents. In addition, it also leaves a special chapter (Chapter 10) to regulate information disclosure and public participation.

Through issuing and implementing the aforementioned laws, it is believed that China’s nuclear information should be released more openly, apparently and instantly, which will reduce the people’s doubt and fear. In addition, it will prevent and control the happening of NIMBY incidents.

3.3. Establish financial compensation mechanism for residents nearby the nuclear facilities

Some financial compensation provided for local residents can avoid the happening of NIMBY events. For example, Regulations on Site Setting of Final Disposal Facilities for Low Radioactive Waste issued in Taiwan regulates that the rewarding system for local residents to gain feedback fund directly in order to improve the residents’ acceptance of final settlement plant. Mainland China’s law has similar regulations. On the basis of such concept, Environmental Protection Law regulates that “the state shall establish and improve an ecological compensation mechanism.” [11] The reason why financial compensation is provided for local residents is because NIMBY facilities will do harm to the health of relevant residents. In addition, as for the cost and benefits, it will devalue the building properties surrounding the facilities and reduce the house occupancy as well as the rent. As a result, financial interests of the residents surrounding NIMBY facilities will be damaged.

Therefore, some relevant departments in China should develop some compensation mechanism and make up such procedures, methods and standards to provide the financial compensation for the residents around nuclear facilities. Therefore, it will reduce their rejections for fear of nuclear facilities and reduction of housing prices and foreign income. Then, the residents’ recognition and acceptance of nuclear facilities will be improved to reduce the happening of NIMBY incidents. In the ways of compensation, it's suggested that the government should try many ways, comprehensive benefit compensation modes. Apart from money, they can provide to present the health insurance, the reduction of taxes of small business, the exemption from the expenditure of relevant residents' water, electricity and gas. Therefore, it will relieve NIMBY emotions of residents around the facilities.

4. Conclusion

Due to insufficient and delayed disclosure of information, lack of promotion of knowledge on nuclear facilities safety and failure of establishing relevant compensation mechanism for NIMBY residents, NIMBY incidents happened some time. In order to protect and control the happening and spreading of such NIMBY public events, this thesis has made an analysis based on legislation and regulations in China. It's suggested that China's government departments and operators involved in nuclear facilities should adhere to scientific and democratic decision of location selection and construction and establish proper compensation mechanism for the residents nearby to reduce the possibility of radioactive damage and relieve anxiety and opposition of residents so as to reduce the possible loss.

Acknowledgments

This research was funded by National Social Science Fund of China, project number is 18CFX049.

References

[1] http://www.iaea.org/PRIS/World Statistics/operational reactors by country.aspx
[2] O'Hare Michael 1977 Not on My Block You Don't: Facility Sitting and the Strategic Importance of Compensation *Public Policy* **25** (4): 407-458
[3] Prevention and Control of Radioactive Pollution Law of the People's Republic of China: Article 44
[4] IAEA: Joint Convention on the Safety of Spent Fuel Management and on the Safety of Radioactive Waste Management: Article 6 and Article 7.
[5] Nuclear Safety Law of the People's Republic of China: Article 22 and Article 23
[6] International Atomic Energy Agency (IAEA) "Handbook on Nuclear Law", IAEA, Vienna 2003 p 12
[7] ZHAO, Wei 2017 *The Atomic Energy Law Symposium* CUPL Press September 2017 p 363
[8] http://www.cnpe.cc/tabid/852/InfoID/88376/frtid/870/Default.aspx
[9] http://news.bjx.com.cn/html/20121016/394681.shtml
[10] Nuclear Safety Law of the People's Republic of China: Chapter 5.
[11] Environmental Protection Law of the People's Republic of China: Article 31.

Financial deepening, inclusive finance and economic growth: analysis based on endogenous growth theory

Jun He, Yue Zhang*, Datang Zheng and Hongyan Wan

School of Management, University of Science and Technology of China, Hefei, Anhui, China

E-mail: yuezha@mail.ustc.edu.cn

Abstract. This article studies the relationship between financial development and economic growth from the perspectives of financial deepening and inclusive finance. The theoretical research expands the financial endogenous economic growth model, constructs the Hamiltonian function, using the first-order optimization conditions to obtain explicit relationship between economic growth and financial development under balanced path. Theoretical research shows that financial deepening and inclusive finance are significant and stable in promoting economic growth.

1. Introduction and literature review

The relationship between financial development and economic growth has always been an issue of concern and discussion by economists, the long-term research on the level of financial development is mainly based on the perspective of financial deepening, due to the difference between theoretical models and statistical methods, there are also differences in the conclusions of research on the effects of financial development to economic growth and its conduction paths. Piano (1993) earlier proposed that financial development promoted economic growth by increasing the marginal productivity of social capital. Levine (2005) proposed five possible channels that financial development influenced economic growth: (I) Providing sound investment information and enhancing capital distribution efficiency. (II) Supervising enterprises to improve management level. (III) Reducing innovation risks and enhancing enterprises' enthusiasm for technological innovation. (IV) Improving savings-investment conversion efficiency. (V) Promoting trade in goods and services. Han Tingchun (2003) mainly studied the role of financial deepening in promoting economic growth from the perspective of savings-investment conversion efficiency. James (2011) embedded the level of financial development into the R&D model and concluded that efficient financial service level promoted the emergence of new technologies. In the related literature, as a unique case, China has caused various countries' scholars to explore the relationship between financial development and economic growth. Hasan, etc. (2006) used China's inter-provincial macro data to discover that financial development had a negative impact on local economic growth, and he attributed the negative impact to the huge amount of banking credit used to support state-owned loss-making enterprises. Shao Yihang, etc. (2015) used the systematic GMM regression method to analyze the impact of financial development on economic growth in eastern and central China, the empirical research found that financial development had different effects on economic growth due to regional innovation differences.

In 2005, the United Nations officially proposed the concept of inclusive finance, which means providing effective and sustainable financial services to all society sectors with financial services needs, and reducing the financial services gap between regions and sectors. Focusing on the purpose of inclusive finance, economists try to scientifically calculate the inclusive financial index, Beck, etc. (2007) proposed an inclusive financial index from the availability, usable and actual use status of financial services. According to the United Nations Development Programme Human Development Index (HDI), Sarma (2008) adopted the coefficient of variation method by selecting a number of financial service indicators to measure the comprehensive development index of inclusive finance in the given region. Domestic scholars Wang Jing and Hu Guohui (2013), Chen Sanmao and Qian Xiaoping (2014) used the calculation method of foreign scholars to calculate the inclusive financial index of China and various provinces and regions.

Traditional finance has gathered in economically developed regions, leading to the widespread financial exclusion for vulnerable groups. Inclusive finance emphasizes the coverage and sustainability of financial services, helps the bottom of the society to gain the ability to put into production, and offers financial services to weak enterprises which have the ability to technological innovation but also have financing constraints. At present, the research literature on the relationship between inclusive finance and economic growth is still relatively inadequate. Tita and Aziakpono (2017) found that inclusive finance in Africa played an active role in promoting poverty reduction and reducing income gap. Fan and Zhang(2017) found that inclusive finance reduced entrepreneurial financing constraints, information asymmetry and transaction costs between commercial partners, and its development would promote the formation of entrepreneurs. However, domestic scholars Li Tao, etc. (2016) used the inclusive financial panel data from 2005 of more than 180 economies in the world, the regression analysis showed that part of the inclusive financial indicators had a significant negative impact on regional economic growth, the other indicators selected had no significant impact on economic growth.

Based on the above literature, we find that domestic research on the impact of inclusive finance on economic growth is still relatively inadequate, financial structure changes will have an impact on economic growth, existing researches lack a comparison between financial deepening and inclusive finance on economic growth. The systematic endogenous growth model of financial development is not perfect enough, and the model fails to explain the inherent logic of financial development affecting economic growth. Therefore, this paper constructs a complete endogenous growth model, and uses dynamic optimization conditions to solve the relationship between economic growth rate and financial development level.

2. Theoretical model

Financial market plays an important role in the modern economy and is closely related to economic growth. The endogenous growth theory is a reliable theoretical method for studying financial development and sustainable economic growth. The model extends the endogenous growth models of James (2011) and Chou (2007), the main innovation of the model lies in the views of scholars such as Luintel and Khan (1999), Sarma and Jesim (2011), which proposed that socio-economic level promotion could stimulate financial development. Thus the model considers endogenous factors of financial development, holds the opinion that economic level and material capital accumulation have an impact on financial development. In addition, the model includes the human capital development department and uses dynamic optimization tools to solve the explicit relationship between variables. Considering a four-sector closed economy consisting of a production department, a financial department, a technology innovation department and a human resources department, in which human capital H is allocated to each department according to a certain proportion.

2.1. Final product department

The production function follows the Cobb-Douglas production function form, it could be expressed as:

$$Y=F_Y K^{\alpha}(Au_Y H)^{1-\alpha} \tag{1}$$

Where Y is the output level, F_Y is a fixed coefficient, K is the physical capital, A is the technical level, and the human capital invested in the production sector is $u_Y H$.

According to the theoretical models of Han Tingchun (2003) and Chou (2007), the improvement of financial development level would increase savings-investment conversion efficiency. Therefore, the accumulation equation of physical capital is:

$$\dot{K} = \xi S - \delta K, \quad \xi = G\tau \tag{2}$$

ξ represents savings-investment conversion efficiency, which is positively related to the financial development level τ , S represents savings, its relationship with output and consumption is $S = Y - C$, and G is a fixed coefficient, δ indicates depreciation.

2.2. Financial department

The Financial Deepening Index (DFI) and the Inclusive Financial Index (IFI) represent the depth and breadth of financial development, respectively, $\tau = \tau(DFI, IFI)$, and $\partial\tau/\partial DFI > 0$, $\partial\tau/\partial IFI > 0$. Some literature models directly set the financial development level τ to be exogenous, however, as mentioned above, financial development level is closely related to the socio-economic level and financial innovation technology. Therefore, this paper proposes that financial development, technical level A and material capital accumulation K are related.

$$\dot{\tau} = F_\tau (u_\tau H)^\lambda \tau^\varphi A^\varepsilon K^\upsilon \tag{3}$$

F_τ is a fixed coefficient, the human capital occupied by financial innovation development is $u_\tau H$, each coefficient satisfies $0 < \lambda, \varphi, \varepsilon, \sigma < 1$.

2.3. Technology innovation department

Financial development, especially inclusive finance, could improve the efficiency of enterprises' technological innovation, on the other hand, reduce the capital pressure and uncertainty of application of new technologies, and accelerate the generation of new technologies. Embedding financial development into the technological innovation model, F_A is a fixed coefficient, $0 < \eta, \beta, \theta < 1$, the human capital invested in the R&D innovation sector is $u_A H$.

$$\dot{A} = F_A (u_A H)^\eta \tau^\beta A^\theta \tag{4}$$

2.4. Human resources department

The human capital allocated to the human resources department is $u_H H$, where $u_Y + u_\tau + u_A + u_H = 1$, F_H is a given exogenous coefficient.

$$\dot{H} = F_H (u_H H) \tag{5}$$

2.5. Consumer budget constraint

The consumer chooses the current consumption C_t for maximizing the utility in a lifetime, $U(C)$ is the instantaneous utility function, ρ indicates the time preference, and σ is the relative risk aversion coefficient, the maximize utility function is:

$$\Omega = \int_0^\infty U(C) e^{-\rho t} dt, \quad U(C) = \frac{C^{1-\sigma} - 1}{1-\sigma} \tag{6}$$

2.6. Competitive equilibrium solution

According to the above constraints, the dynamic optimization solution system of the economy at consumption level C is:

$$\begin{cases} \Omega = \int_0^\infty U(C)e^{-\rho t}\,dt \\ \dot{K} = \xi S - \delta K \\ \dot{\tau} = F_\tau(u_\tau H)^\lambda \tau^\varphi A^\varepsilon K^\upsilon \\ \dot{A} = F_A(u_A H)^\eta \tau^\beta A^\theta \\ \dot{H} = F_H(u_H H) \end{cases} \tag{7}$$

Constructing the Hamiltonian functional with equation (7), where λ_1、λ_2、λ_3、λ_4 are Hamilton multipliers:

$$\Psi = U(C) + \lambda_1\dot{K} + \lambda_2\dot{\tau} + \lambda_3\dot{A} + \lambda_4\dot{H} \tag{8}$$

Optimizing first-order conditions:

$$\frac{\partial\Psi}{\partial C}=0,\quad \frac{\partial\Psi}{\partial u_Y}=0,\quad \frac{\partial\Psi}{\partial u_\tau}=0,\quad \frac{\partial\Psi}{\partial u_A}=0 \tag{9-12}$$

Euler equation:

$$\frac{\partial\Psi}{\partial K}=\lambda_1\rho-\dot{\lambda}_1,\quad \frac{\partial\Psi}{\partial \tau}=\lambda_2\rho-\dot{\lambda}_2,\quad \frac{\partial\Psi}{\partial A}=\lambda_3\rho-\dot{\lambda}_3,\quad \frac{\partial\Psi}{\partial H}=\lambda_4\rho-\dot{\lambda}_4 \tag{13-16}$$

Based on the equation (9) — (16), the economic growth rate under the balanced growth path is:

$$g_Y = g_C = \frac{1}{\delta}[G\tau(Au_YH)^{1-\alpha}\alpha F_Y K^{\alpha-1} + G\tau(Au_YH)^{1-\alpha} F_Y \frac{(1-\alpha)\upsilon u_\tau}{K\lambda u_Y} - \frac{F_\tau(u_\tau H)^\lambda A^\varepsilon K^\upsilon}{\tau^{1-\varphi}} - \delta - \rho] \tag{17}$$

The relationship between economic growth rate, financial deepening and inclusive finance is:

$$\frac{\partial g_Y}{\partial\tau} = \frac{1}{\delta}[F_Y G(Au_YH)^{1-\alpha}(\alpha K^{\alpha-1} + \frac{(1-\alpha)\upsilon u_\tau}{K\lambda u_Y}) + (1-\varphi)\frac{F_\tau(u_\tau H)^\lambda A^\varepsilon K^\upsilon}{\tau^{2-\varphi}}] > 0$$
$$\frac{\partial g_Y}{\partial DFI} = \frac{\partial g_Y}{\partial\tau}\cdot\frac{\partial\tau}{\partial DFI} > 0,\qquad \frac{\partial g_Y}{\partial IFI} = \frac{\partial g_Y}{\partial\tau}\cdot\frac{\partial\tau}{\partial IFI} > 0 \tag{18}$$

So far, we have solved the theoretical model and found that the two dimensions of financial development namely financial deepening and inclusive finance promote sustainable economic growth. There are two channels for financial development to influence economic growth in the model: one is to stimulate savings-investment conversion efficiency, and the other is to promote technical research and development efficiency.

3. Conclusion and suggestion

The level of financial development includes two dimensions: depth and breadth. This paper selects financial deepening and inclusive finance to study the driving effect of financial development on economic growth. By constructing the endogenous growth model of four sectors and solving it by Hamilton function, obtains the relationship between economic growth rate and financial development level. It shows that financial deepening and inclusive finance both have significant positive effects on economic growth.

According to the conclusions of theoretical model, we propose the following suggestions for the future financial development direction and policy formulation: First, continue to promote financial deepening, for mature financial market is an important component to ensure steady economic growth. Actively carry out financial innovation, and excavate the role of financial deepening in efficiently adjusting resource allocation, improving savings-investment conversion efficiency. Second, financial deepening and inclusive finance have their own advantages and emphasis, should have coordinate sustainable development. Third, the concept of inclusive finance has not yet been perfectly constructed, but the basic idea is to provide effective and sustainable financial services for all sectors of society, fully tap the superiority of inclusive finance in solving the financing constraints of SMEs and promoting technological innovation, and provide a sound financing environment for innovation and entrepreneurship. Fourth, the development of internet technology is on the rise, internet finance and big data science have begun to play a unique role in financial innovation and inclusive finance. Technology helps reduce information asymmetry and the cost of financial services. We should make full use of internet technology to improve financial services and financial inclusion.

References

[1] Piano M. Financial markets and growth: An overview[J]. European Economic Review. 1993, 37(2-3): 613 -622.

[2] Levine. Financial and growth: Theory and evidence[J]. Handbook of Economic Growth. 2005(1): 865-934.

[3] Han Tingchun. The endogenous mechanism of financial development and economic growth[J]. Journal of Tsinghua University(Philosophy and Social Sciences) . 2003(18): 80-85.

[4] James B A. Financial development, liberalization and technological deepening[J]. European Economic Review. 2011, 55(5): 688-701.

[5] Hasan I. Wachtel P. and Zhou M. Institutional development, financial deepening and economic growth: Evidence from China[J]. Journal of Banking & Finance. 2006, 33(1): 157-170.

[6] Shao Yihang, Liu Shibao, Zhang Chaoyang. Financial development model and economic growth under innovation difference: Theory and evidence[J]. Management World. 2015(11): 29-39.

[7] Beck T. Demirguc-Kunt A and Peria MSM. Reaching out: Access to and use banking services across countries[J]. Journal of Financial Economics. 2007(85): 234-266.

[8] Sarma M. Index of financial inclusion[R]. ICRIER Working Paper. 2008, No. 215.

[9] Wang Jing, Hu Guohui. Analysis on the development evaluation and influential factors of China's inclusive finance[J]. Finance Forum. 2013(6): 31-36.

[10] Chen Sanmao, Qian Xiaoping. Financial inclusion index and its measurement of China's provinces[J]. Finance Forum. 2014(9): 3-8.

[11] Tita A F, Aziakpono M J. The effect of financial inclusion on welfare in sub-Saharan Africa: Evidence from disaggregated data[R]. ERSA Working Paper 679, May 2017.

[12] Fan Z, Zhang R. Financial inclusion, entry barriers, and entrepreneurship: Evidence from China [J]. Sustainability. 2017, 9(2): 203.

[13] Li Tao, Xu Xiang, Sun Shuo. Inclusive finance and economic growth[J]. Journal of Financial Research. 2016(4): 1-16.

[14] Yuan K Chou. Modeling financial innovation and economic growth: Why the financial sector matters to the real economy [J]. Journal of Economic Education. 2007, 38(1): 78-91.

[15] Luintel K B, Khan M. A quantitative reassessment of the finance–growth nexus: Evidence from a multivariate VAR[J]. Journal of Development Economics. 1999, 60(2): 381– 405.

[16] Sarma M, Jesim P. Financial inclusion and development[J]. Journal of International Development. 2011(23): 613-628.

Acknowledgments
Thanks to the National Natural Science Foundation of China (71573240) for its strong support for this

Journal of Physics: Conference Series **1419** (2019) 012034 doi:10.1088/1742-6596/1419/1/012034

article.

Robust beamforming optimization for magnetic resonance coupling wireless power transfer systems

Wuxing Lin, Zhimeng Xu and Jinsong Hu

College of Physics & Information Engineering, Fuzhou University, 350108, Fuzhou, China

E-mail: zhmxu@fzu.edu.cn

Abstract. Magnetic Resonance Coupling (MRC) wireless power transfer (WPT) is a promising technique for portable devices and Internet of Things applications, which can transfer power to energy-constrained terminals in mid-range with relatively high efficiency. To enhance the transfer performance, by applying more coils in transceivers, multiple-input single-output (MISO) MRC-WPT system has attracted much attention in recent years. In this paper, a MISO MRC-WPT system with multiple transmitters (TXs) and a single receiver (RX) is studied and its robust magnetic beamforming problem is investigated. The optimal robust beamforming design of the currents of TXs is investigated to maximize the power received by RX with the mutual inductance error between TXs and RX, whereas the total power required is less than a certain value and each TXs has a certain limited transmitting power. By using certain transformation techniques, the optimization problem is converted into a relaxed semidefinite programming (SDP) problem which can be solved efficiently. We further prove that the solution of the relaxed SDP problem is always rank-one, which indicates that the relaxation is tight, and the optimal solution for the problem was obtained. Numerical results demonstrate that the proposed beamforming design can achieve up to 97% transfer efficiency by reducing the measurement errors.

1. Introduction

Recently, Wireless power transfer (WPT) [1-2] has attracted widely attention because it make possible a reliability, convenience and safety power supply method for the portable devices and Internet of Things applications. For the technology of wireless power transfer, magnetic resonance coupling (MRC) has been considered as a promising technique since it can provide longer transmission distance, higher transmission efficiency and wide working frequency band [3-6].

In 2014, J. Jadidian and D. Katabi of the Massachusetts Institute of Technology proposed the Magnetic multiple-input multiple-output (MIMO) theory [7], applying multiple power transmitters (TXs) and one power receiver (RX) to improve the charging efficiency and charging flexibility. Base on the MIMO structure, magnetic beamforming is anticipated to enhance the transfer performance of WPT system. In [8], magnetic beamforming for a MISO-WPT system with multiple TXs and a single RX is investigated, and the receiving efficiency is significantly improved by optimizing the TXs currents. However, the beamforming design in [8] based on the perfect knowledge of the mutual inductance between TXs and RX. In practical scenarios, perfect knowledge of the mutual information between TXs and RX may not be available due to the influence of environment and positional

Journal of Physics: Conference Series **1419** (2019) 012035 doi:10.1088/1742-6596/1419/1/012035

deviation. Therefore, the robust beamforming design over the mutual inductance error for MRC-WPT system needs to be investigated.

In this paper, we assume that the measured mutual inductance has a norm bounded error, then consider an MRC-WPT system with one RX and multiple TXs whose source current (or corresponding source voltage) can be adjusted so that the induced magnetic field can be optimally superimposed at RX to achieve maximum power transfer. The robust beamforming is designed by jointly optimizing the currents of all TXs to maximize the received power while under the limit of total transmit power constraints and the power constraints of each transmitter circuit. According to convex optimization theory, the proposed optimization problem is a non-convex quadratically constrained quadratic programming (QCQP) [9]. By means of inequality method and S-procedure method, the original optimization problem is transformed into a semi-positive definite programming problem with rank 1. Then, by applying semidefinite relaxation (SDR), we show that the optimal solution of the semidefinite programming (SDP) problem is always rank-one, which means that the relaxation is tight, and the optimal solution of the original problem can be obtain.

2. MISO-WPT system model

The system model discussed in this paper is shown in Figure 1, A MISO MRC-WPT system with multiple power transmitters (TXs) and one power receiver (RX) is considered. For each TXs and RX, an air-core induction coils is applied. Specially, the transceiver circuit comprise a series RLC resonant circuits and the coil is considered as the inductor. $L_n > 0$, $C_n > 0$, $n = 1,\dots,N$ and $L_r > 0$, $C_r > 0$ denote the self-inductance and the capacitance in the resonant circuit of the *n*-th TX and the RX respectively. The capacitances in each resonant circuit are set to make the all circuit resonant at the resonance angular frequency $\omega = \omega_0 = 1/\sqrt{LC}$. Moreover, the *n*-th TX is connected to a stable power source supplying sinusoidal voltage over time given by $\tilde{v}_{tx,n}(t) = \mathrm{Re}\{v_{tx,n}e^{j\omega t}\}$, where $v_{tx,n}$ denoting the complex voltage and $\omega > 0$ denoting the operating angular frequency. Thus, the steady-state current flowing through the *n*-th TX is defined as $\tilde{i}_{tx,n}(t) = \mathrm{Re}\{i_{tx,n}e^{j\omega t}\}$, with the complex current $i_{tx,n}$. This current produces a time-varying magnetic flux in the transmitter's electromagnetic coil, which passes through the RX coil and induces time-varying currents in it. We denote $\tilde{i}_{rx}(t) = \mathrm{Re}\{i_{rx}e^{j\omega t}\}$, with complex current i_{rx}, as the steady state current at receiver.

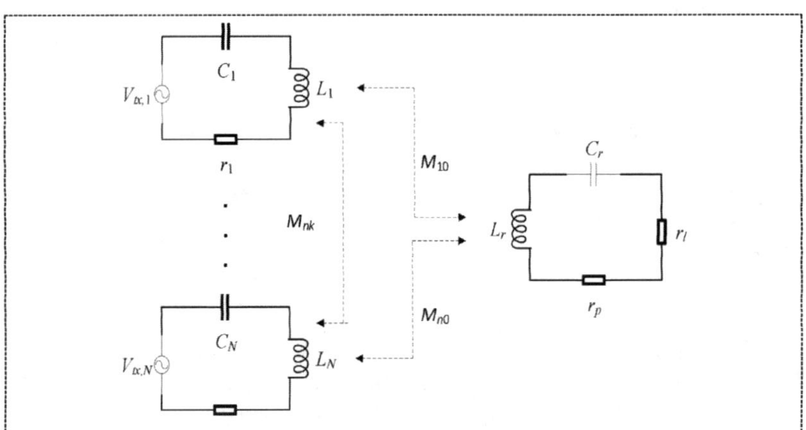

Figure 1. MISO-WPT system model.

Define M_{n0} as the mutual inductance between the *n*-th TX coil and the RX coil, and M_{nk} ($n \neq k$) as the mutual inductance between the *n*-th TX coil and the *k*-th TX coil respectively. The

mutual inductances is a positive or negative real number, which depend on the physical characteristics of each pair of TX and RX coils such as their relative distance, orientations, etc.
The source resistance of the n-th TX is denoted by $r_{tx,n} > 0$. Then the diagonal resistance matrix is defined as $\mathbf{R} \quad \mathrm{diag}\{r_{tx,1}, r_{tx,1}, \cdots, r_{tx,N}\}$. Similarly, the total resistance of RX is denoted by r_{rx} , which is given by $r_{rx} = r_{rx,p} + r_{rx,l}$, where $r_{rx,p} > 0$ is the parasitic resistance and $r_{rx,l} > 0$ is the load resistance. It should be noted that $r_{rx,l}$ is a real value resistor and $r_{rx,l} \quad r_{rx,p}$.

Assume that there is a controller applied in the MRC-WPT system, thus the information of system pre-measured parameters can be obtained. Define the complex TX current $i_{tx,n}$ as the design variable, by optimization, the controller can realize magnetic beamforming.

By applying Kirchhoff's circuit law, the coupled current i_{rx} of to the RX is given by

$$i_{rx} = \frac{j\omega}{r_{rx}} \sum_{n=1}^{N} M_{n0} i_{tx,n} . \tag{1}$$

Define the vector of all TXs currents as $\mathbf{i} = [i_{tx,1}, \cdots, i_{tx,N}]^T$, and the vector of mutual inductances between the RX coil and TXs coils as $\mathbf{m} = [M_{10}, \cdots, M_{N0}]^T$. From (1), the receiving power of RX can be given by

$$p_r = \frac{1}{2}\left|i_{rx}\right|^2 r_{rx,l} = \frac{\omega^2 r_{rx,l}}{2 r_{rx}^{\ 2}} \mathbf{i}^H \mathbf{m}\mathbf{m}^T \mathbf{i} . \tag{2}$$

Similarly, by applying Kirchhoff's circuit law to each TX, the source voltage of the n-th TX can be expressed as

$$v_{tx,n} = \left(r_{tx,n} + \frac{M_{n0}^{\ 2}\omega^2}{r_{rx}} \right) i_{tx,n} + \sum_{k \neq n} \left(j\omega M_{nk} + \frac{M_{k0} M_{n0} \omega^2}{r_{rx}} \right) i_{tx,k} . \tag{3}$$

Based on (3), the total transmitted power of TXs can be expressed as

$$p_t = \frac{1}{2} \mathrm{Re}\left\{ \sum_{k=1}^{N} v_k \, i_k^{*} \right\} = \frac{1}{2}\left(\mathbf{i}^H \mathbf{R} \mathbf{i} + \frac{\omega^2}{r_{rx}^{\ 2}} \mathbf{i}^H \mathbf{m}\mathbf{m}^T \mathbf{i} \right) . \tag{4}$$

3. Problem formulation

In this section, we design the MRC beamforming to maximize the power delivered to the RX load over transmit currents vector $\mathbf{i}$, subject to the following constraints: the total source power should meet a maximum level $\beta > 0$,i.e., $p_t \le \beta$ and each TX n should meet the largest sustainable power value p_n . This is to ensure the normal operation of the transmitter and avoid damage due to excessive power. In practical application scenarios, the receiver load is expected to achieve the maximum energy receiving efficiency under the limited power of the transmitter.

3.1. Norm-bounded model

Define $\mathbf{Q}_n(i,j)$ as $\mathbf{Q}_n(i,j) = \begin{cases} r_{tx,n}, & i = j = n \\ 0, & \text{otherwise} \end{cases}$, based on the discussion above, the problem P1 is formulated as follows

$$\max_{\mathbf{i}} \quad \frac{\omega^2 r_{rx,l}}{2 r_{rx}^{\ 2}} \mathbf{i}^H \mathbf{m}\mathbf{m}^T \mathbf{i} , \tag{5}$$

$$s.t. \quad \frac{1}{2}\left(\mathbf{i}^H \mathbf{R} \mathbf{i} + \frac{\omega^2}{r_{rx}} \mathbf{i}^H \mathbf{m}\mathbf{m}^T \mathbf{i} \right) \le \beta , \tag{6}$$

$$\frac{1}{2} \mathbf{i}^H \mathbf{Q}_n \mathbf{i} \le p_n , n = 1,2,...N . \tag{7}$$

For the optimization problem P1, perfect channel information is assumed. However, in the practical scenario, the mutual inductances cannot be measured precisely, thus the measurement errors must be considered in the magnetic beamforming design. Generally, there are two error models for mutual inductance measurement, namely a norm boundary model and a probability model. The norm-bounded error model generates a worst case of robust design, which can provide a performance boundary of the system. In this paper, the actual mutual inductance between the TXs and RX coils are assumed to be two parts: pre-measured mutual inductance and errors. By applying norm-bounded model to mutual inductance measurement, the actual mutual inductance vector m can be expressed as

$$\mathbf{m} = \mathbf{m}_q + \Delta\mathbf{m}, \tag{8}$$

$$\|\Delta\mathbf{m}\| \le \varepsilon, \tag{9}$$

where $\mathbf{m}_q$ is the pre-measured mutual inductances vector between TXs and RX, $\Delta\mathbf{m}$ is the error vector of the mutual inductances, and ε is the radius of the uncertainty region.

To take the error vector of mutual inductance measurement into account, the optimization problem P1 based on worst-case criterion can be formulated as problem P2, which is given by

$$\text{P2:} \quad \max_{\mathbf{i}} \quad \frac{\omega^2 r_{rx,l}}{2r_{rx}^{\ 2}}\mathbf{i}^H\left(\mathbf{m}_q+\Delta\mathbf{m}\right)\left(\mathbf{m}_q+\Delta\mathbf{m}\right)^T\mathbf{i}, \tag{10}$$

$$\text{s.t.} \quad \frac{1}{2}\left(\mathbf{i}^H\mathbf{R}\mathbf{i}+\frac{\omega^2}{r_{rx}}\mathbf{i}^H\left(\mathbf{m}_q+\Delta\mathbf{m}\right)\left(\mathbf{m}_q+\Delta\mathbf{m}\right)^T\mathbf{i}\right)\le\beta, \tag{11}$$

$$\frac{1}{2}\mathbf{i}^H Q_n\mathbf{i}\le p_n, \quad \text{n=1,2,...,N}, \tag{12}$$

$$\|\Delta\mathbf{m}\|\le\varepsilon. \tag{13}$$

According to the convex optimization theory, (P2) is in general a non-convex QCQP problem [9] due to the objective function in (10). It is well known that the general nonconvex QCQP problem is NP-hard and thus, intractable. However, as we will show in the following, due to the special structure of the objective function and the constraints, problem P2 can be reformulated as a convex SDP problem and solved optimally.

3.2. *Reformulation and optimization*

Before optimizing, the problem P2 is transformed into a more tractable form. For the objective function of P2 in (10), we obtain

$$\left\|\left(\mathbf{m}_q+\Delta\mathbf{m}\right)^T\mathbf{i}\right\| = \left\|\mathbf{m}_q^{\ T}\mathbf{i}+\Delta\mathbf{m}^T\mathbf{i}\right\| \ge \left\|\mathbf{m}_q^{\ T}\mathbf{i}\right\| - \left\|\Delta\mathbf{m}^T\mathbf{i}\right\|. \tag{14}$$

Then applying the Cauchy-Schwarz inequality to the second term in the right-hand-side of (14), we have

$$\left\|\Delta\mathbf{m}^T\mathbf{i}\right\| \le \left\|\Delta\mathbf{m}^T\right\|\left\|\mathbf{i}\right\| \le \varepsilon\left\|\mathbf{i}\right\|. \tag{15}$$

Applying (15) to (14) and defining $\mathbf{I}$ as the identity matrix, we then have that

$$\begin{aligned}\left\|\left(\mathbf{m}_q+\Delta\mathbf{m}\right)^T\mathbf{i}\right\|^2 &\ge \left(\left\|\mathbf{m}_q^{\ T}\mathbf{i}\right\|-\varepsilon\left\|\mathbf{i}\right\|\right)^2 = \left\|\mathbf{m}_q^{\ T}\mathbf{i}\right\|^2 - 2\varepsilon\left\|\mathbf{m}_q^{\ T}\mathbf{i}\right\|\left\|\mathbf{i}\right\| + \varepsilon^2\left\|\mathbf{i}\right\|^2 \\ &\ge \left\|\mathbf{m}_q^{\ T}\mathbf{i}\right\|^2 - 2\varepsilon\left\|\mathbf{m}_q^{\ T}\mathbf{i}\right\|\left\|\mathbf{i}\right\|^2 + \varepsilon^2\left\|\mathbf{i}\right\|^2 = \mathbf{i}^H\left(\mathbf{m}_q\mathbf{m}_q^{\ T}+\varepsilon\left(\varepsilon-2\left\|\mathbf{m}_q^{\ T}\mathbf{i}\right\|\right)\mathbf{I}\right)\mathbf{i}.\end{aligned} \tag{16}$$

For the constraint function of (11), it can be solved with the help of the S-procedure [9]. The S-procedure can be presented as the following lemma.

Lemma1: Let $\mathbf{A}_1$, $\mathbf{A}_1 \in \mathrm{h}\ \mathrm{H}^n$, $\mathbf{b}_1$, $\mathbf{b}_2 \in \mathrm{C}^{n\times 1}$, c_1, $\mathrm{c}_2 \in \mathrm{C}$, and suppose there exists an $\mathbf{x}$ with $\mathbf{x}^H\mathbf{A}_1\mathbf{x}+2\mathrm{Re}\{\mathbf{b}_1^{\ T}\mathbf{x}\}+\mathrm{c}_1<0$. Then, the following relationship:

$$\mathbf{x}^H\mathbf{A}_1\mathbf{x}+2\mathrm{Re}\{\mathbf{b}_1^{\ T}\mathbf{x}\}+\mathrm{c}_1\ge 0, \tag{17}$$

$$\mathbf{x}^H\mathbf{A}_2\mathbf{x}+2\mathrm{Re}\left\{\mathbf{b}_2^T\mathbf{x}\right\}+\mathrm{c}_2\geq 0, \tag{18}$$

holds if and only if there exists a $\lambda>0$ such that

$$\begin{bmatrix}\mathbf{A}_1 & \mathbf{b}_1\\ \mathbf{b}_1^T & \mathrm{c}_1\end{bmatrix}-\begin{bmatrix}\mathbf{A}_2 & \mathbf{b}_2\\ \mathbf{b}_2^T & \mathrm{c}_2\end{bmatrix}\pm 0 \tag{19}$$

Define Hermitian matrix $\mathbf{X}=\mathbf{i}\mathbf{i}^H$. By expanding equation (11) and equation (13), we can obtain that

$$\begin{cases}\frac{r_{rx}}{\omega^2}Tr(\mathbf{RX})+\mathbf{m}_q^T\mathbf{X}\mathbf{m}_q+2\mathrm{Re}\left\{\mathbf{m}_q^T\mathbf{X}\Delta\mathbf{m}\right\}+\Delta\mathbf{m}^T\mathbf{X}\Delta\mathbf{m}-\frac{2\beta r_{rx}}{\omega^2}\leq 0\\ \|\Delta\mathbf{m}\|\leq\varepsilon\end{cases}. \tag{20}$$

Applying Lemma 1, we then equivalently reformulate

$$\Phi(\mathbf{X},\lambda,\varepsilon)=\begin{bmatrix}\lambda\mathbf{I}-\mathbf{X}-\mathbf{X}\mathbf{m}_q\\ -\mathbf{m}_q^T\mathbf{X}-\mathbf{m}_q^T\mathbf{X}\mathbf{m}_q-\frac{r_{rx}}{\omega^2}Tr(\mathbf{RX})-\lambda\varepsilon^2+\frac{2\beta r_{rx}}{\omega^2}\end{bmatrix}\pm 0\ . \tag{21}$$

In summary, problem P2 can be equivalently reformulated to problem P3-SDP as follow

$$\max_{\mathbf{X}}\quad \frac{\omega^2 r_{rx,l}}{2r_{rx}^2}\mathrm{Tr}(\mathbf{MX}), \tag{22}$$

$$\text{s.t.}\quad \Phi(\mathbf{X},\lambda,\varepsilon)\pm 0, \tag{23}$$

$$\mathrm{Tr}(\mathbf{Q}_n\mathbf{X})\leq 2p_n,\quad n=1,2,...N, \tag{24}$$

$$\mathrm{Rank}(\mathbf{X})=1,\mathbf{X}\pm 0, \tag{25}$$

where $\mathbf{M}=\mathbf{m}_q\mathbf{m}_q^T+\varepsilon\left(\varepsilon-2\left\|\mathbf{m}_q^T\mathbf{i}\right\|\right)\mathbf{I}$.

By ignoring the rank-one constraint in (25), we obtain the SDR of problem P3−SDP, denoted by (P3−SDR), which becomes a linear optimization problem that can be solved by existing convex optimization tools. At this point, an important question is that whether the optimal solution of (P3−SDR), denoted as $\mathbf{X}^*$, is rank-one. If the solution $\mathbf{X}^*$ to (P3−SDR) satisfies $\mathrm{Rank}(\mathbf{X})=1$, $\mathbf{X}^*$ is also the optimal solution to (P3−SDP). In the following part, we will prove that the rank of the solution of (P3−SDR) is exactly 1.

Theorem 1: The optimal solution $\mathbf{X}^*$ for problem (P3−SDR) is rank-one.

Proof: Let $\mathbf{A}_q\pm 0,\boldsymbol{\rho}=[\rho_1,...,\rho_N]^T\geq 0$ and $\mathbf{S}\pm 0$ be the dual variables corresponding to the constraints given in (23), (24), and (25), respectively. Thus, the Lagrangian of (P3−SDR) is written as

$$L(\mathbf{X},\mathbf{A}_q,\boldsymbol{\rho},\mathbf{S})=-\frac{\omega^2 r_{rx,l}}{2r_{rx}^2}\mathrm{Tr}(\mathbf{MX})+\mathrm{Tr}(\mathbf{A}_q\Phi(\mathbf{X},\lambda,\varepsilon))+\sum_{n=1}^N\rho_n(\mathrm{Tr}(\mathbf{Q}_n\mathbf{X})-2p_n)-\mathrm{Tr}(\mathbf{SX}). \tag{26}$$

Moreover, let $\boldsymbol{\Omega}_q=[\mathbf{I}\mathbf{m}_q]$, the Karush−Kuhn−Tucker (KKT) conditions [9] of (P3−SDR) are given by

$$\nabla_{\mathbf{X}}L(\mathbf{X}^*,\mathbf{A}_q^*,\boldsymbol{\rho}^*,\mathbf{S}^*)=-\frac{\omega^2 r_{rx,l}}{2r_{rx}^2}\mathbf{M}+\boldsymbol{\Omega}_q\mathbf{A}_q\boldsymbol{\Omega}_q^H+\sum_{n=1}^N\rho_n\mathbf{Q}_n-\mathbf{S}=0\ , \tag{27}$$

$$\mathbf{S}^*\mathbf{X}^*=0. \tag{28}$$

Next, by multiplying (27) by X* on both sides and substituting (28) into the obtained equation, we have

$$-\frac{\omega^2 r_{rx,l}}{2r_{rx}^2}\mathbf{MX}^*+\left(\boldsymbol{\Omega}_q\mathbf{A}_q\boldsymbol{\Omega}_q^H+\sum_{n=1}^N\rho_n\mathbf{Q}_n\right)\mathbf{X}^*=0. \tag{29}$$

Then, we obtain

$$\mathrm{rank}\left(\left(\boldsymbol{\Omega}_q\mathbf{A}_q\boldsymbol{\Omega}_q^H+\sum_{n=1}^N\rho_n\mathbf{Q}_n\right)\mathbf{X}^*\right)\leq\mathrm{rank}(\mathbf{M})=1 \tag{30}$$

Since the matrix $\mathbf{\Omega}_q\mathbf{A}_q\mathbf{\Omega}_q^H \pm 0$ and $\mathbf{Q}_n$ must have full rank. Hence, $\left(\mathbf{\Omega}_q\mathbf{A}_q\mathbf{\Omega}_q^H + \sum_{n=1}^{N}\rho_n\mathbf{Q}_n\right)$ has full rank. Together with (30), $\mathrm{Rank}(\mathbf{X}^*) = 1$ can be proved.

4. Numerical results

In this section, we consider the MRC-WPT system with N=5 TX coils and one RX coil, each of which has 100 turns and a radius of 0.1 meter and cooper wire with radius of 0.1 millimeter was used. The resistance of all TXs is set identically as 0.336Ω. For the RX, its parasitic resistance and load resistance are $r_{rx,p} = 0.336\ \Omega$ and $r_{rx,l} = 50\ \Omega$, respectively. The self and mutual inductances are given in Table 1.

Table 1. Mutual/Self inductances (μH).

	TX1	**TX2**	**TX3**	**TX4**	**TX5**
TX1	47700	2.2970	0.8074	2.2970	6.5741
TX2	2.2970	47700	2.2970	0.8074	6.5741
TX3	0.8074	2.2970	47700	2.2970	6.5741
TX4	2.2970	0.8074	2.2970	47700	6.5741
TX5	6.5741	6.5741	6.5741	6.5741	47700

The load power of RX versus different levels of uncertainty bounds is given in Figure 2. It can be seen from the figure that the receiving power decreases gradually as the uncertainty bounds increases, therefore, the uncertain boundary should not be too large to guarantee the load power.

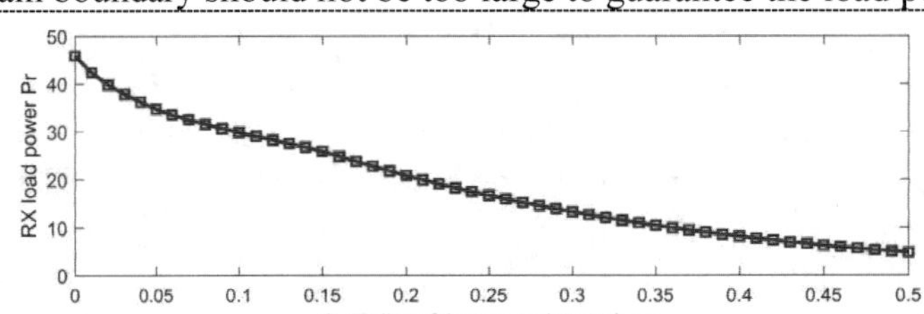

Figure 2. RX load power versus error region.

Figure 3 and Figure 4 show the total source power efficiency and load power of RX under the uncertainty bounds. It can be seen from Figure 3 that, the beamforming design can achieve 97% efficiency when uncertainty region ε=0 , however, the efficiency decreases significantly with the increase of ε . The results show that the beamforming design can work well only when the measurement error is small. Furthermore, as shown in Figure 3, when the total transmitted power increase to a certain extent, the efficiency will decrease. The reason of this phenomenon is that when the when the total transmits power increases, the maximum limit of each transmitter power becomes the main constrains of the beamforming design. As shown in Figure 4, when the total TXs power is small, the received power in RX increase with the increase of the total TXs power, this mean that the maximum total TXs power is the main constrain of the beamforming design under this condition. Like the results in Figure 3, when the total transmits power increases to a certain extent, the received power tends to be flat, which means the maximum limit of each transmitter power becomes the main constrains.

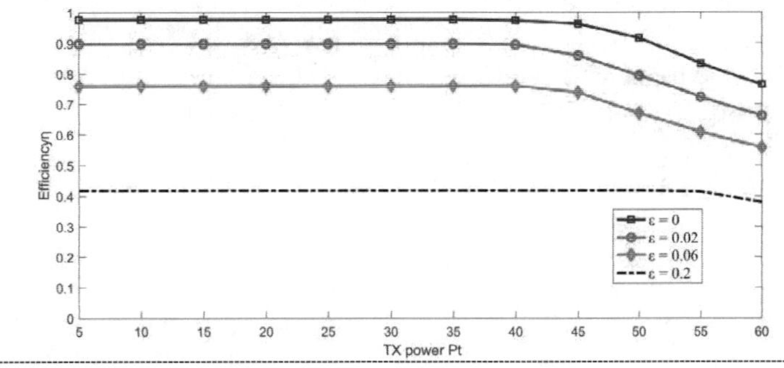

Figure 3. Efficiency versus TX power.

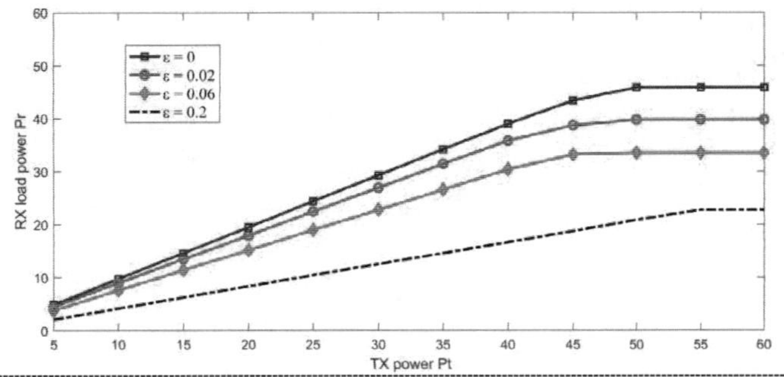

Figure 4. RX load power versus TX power.

5. Conclusion

In this paper, we investigated the robust beamforming design for the MISO MCR-WPT system with the mutual inductance error between TXs and RX. The problem of maximize the power received by RX while limit the total power required and the transmitting power of each TXs is solved by means of semidefinite relaxation and thus the optimal solution of the problem is obtained. Simulation results show that the beamforming design can achieve up to 97% transfer efficiency and the measurement error mutual inductances significantly reduce the beamforming performance. Therefore, in order to guarantee the beamforming performance, mutual inductances estimation errors should be minimized as possible.

References

[1] C. Kim, D. Seo, J. You, J. Park, and B. H. Cho, "Design of a contactless battery charger for cellular phone," IEEE Trans. Ind. Electron., vol. 48, no. 6, pp. 1238–1247, Dec. 2001.

[2] J. Garnica, R. A. Chinga, and J. Lin, "Wireless power transmission: From far field to near field," Proc. IEEE, vol. 101, no. 6, pp. 1321–1331, Jun. 2013.

[3] A. Kurs, A. Karalis, R. Moffatt, J. D. Joannopoulos, P. Fisher, and M. Soljacic, "Wireless power transfer via strongly coupled magnetic resonances," Science, vol. 317, no. 83, pp. 83–86, Jul. 2007.

[4] Z. Fei, S. A. Hackworth, F. Weinong, L. Chengliu, M. Zhihong, and S. Mingui, "Relay effect of wireless power transfer using strongly coupled magnetic resonances," IEEE Trans. Magn., vol. 47, no. 5, pp. 1478–1481, May 2011.

[5] J. Shin et al., "Design and implementation of shaped magnetic-resonancebased wireless power transfer system for roadway-powered moving electric vehicles," IEEE Trans. Ind. Electron., vol. 61, no.3, pp. 1179–1192, Mar. 2014.

[6] L. Chen, Y. C. Zhou, and T. J. Cui, "An optimizable circuit structure for high-efficiency wireless power transfer," IEEE Trans. Ind. Electron., vol. 60, no. 1, pp. 339–349, Jan. 2013.

Journal of Physics: Conference Series **1419** (2019) 012035 doi:10.1088/1742-6596/1419/1/012035

[7] J. Jadidian and D. Katabi. “Magnetic MIMO: how to charge your phone in your pocket,” in Proc. 20th Annu. Int. Conf. Mobile Comput. Netw. (ACM), Sep. 2014, pp. 495–506.
[8] G. Yang, M. R. V. Moghadam, and R. Zhang, “Magnetic beamforming for wireless power transfer,” in Proc. IEEE Int. Conf. Acoust., Speech Signal Process., Mar. 2016, pp. 3936–3940.
[9] S. Boyd and L. Vandenberghe, Convex Optimization. Cambridge University Press, 2004.

Acknowledgments
This work was supported in part by the National Natural Science Foundation of China under Grant 61401100, the Natural Science Foundation of Fujian Province under Grant 2018J01805, and the Scientific Research Foundation of Fuzhou University under Grant GXRC-18083.

Optimization of helicopter blade section based on multi-objective optimization algorithm

Shuai Wang and Guohua Wang

Department of Instrumentation and Optoelectronic Engineering, Beijing University of Aeronautics and Astronautics, Beijing, China

E-mail: 851423771@qq.com

Abstract. This paper aims to find a better performance multi-objective optimization algorithm to optimize the helicopter blade section structure, so that the optimized blade section characteristics are better. Optimization objectives include the section quality linear density of the blade and the section stiffness characteristics of the blade. In this paper, three multi-objective optimization algorithms including non-dominated sorting genetic algorithm II (NSGA-II), non-dominated sorting genetic algorithm III (NSGA-III) and multi-objective evolutionary algorithm based on decomposition (MOEA/D) algorithm are used to optimize the blade section. The performance evaluation results of the three algorithms are evaluated by inverted generation distance(IGD). The calculation results show that the MOEA/D algorithm has better convergence, uniformity and extensiveness, and the comprehensive performance is better than other algorithms in high-dimensional multi-objective optimization. After optimization by the algorithm, the quality linear density of the section is reduced by 1.8%, the torsional stiffness of section is increased by 2.1%., the shimming stiffness of section is decreased by 1.9%, the waving stiffness of section is decreased by1.7%, and the tensile stiffness is increased by 0.9 %.

1. Introduction

Among the various components of the helicopter, the rotor system is the key moving part of the helicopter [1]. It consists of blades and hubs. The rotation of the blades around the center of the hub produces the helicopter's lift, forward force and maneuvering force, thus achieving helicopter flight. The structure of the blade directly affects the flight quality, reliability, safety and other performance of the helicopter. The performance of blade’s design has a very important impact on the performance of the helicopter. Compared with other structures, the blade structure dynamics problem (the problem of blade structure characteristics) is particularly prominent. In addition, because it works under continuous elastic vibration, it causes rotor fatigue problems and helicopter vibration problems that are

Journal of Physics: Conference Series **1419** (2019) 012036 doi:10.1088/1742-6596/1419/1/012036

highly valued in helicopter development. Therefore, the optimization design of the blade structure is very important in the development of the helicopter. The optimization of the section characteristics of the blade is an important optimization goal. The section characteristics of the blade mainly include the quality linear density of the section, the torsional stiffness of the section, the tensile stiffness of the section, the waving stiffness of section, and shimming stiffness of section. There are mainly direct analytical methods and finite element analysis methods for the analysis of section characteristics of blade. This paper uses direct analytical method [2] to construct the section analysis model of blade. In the optimization, the section characteristics model is taken as the optimization target and the geometric parameters of the components in section of blade are used as design variables to optimize the section of the blade.

For blade structure optimization, the commonly used optimization methods can be divided into the following three types [3]: numerical optimization algorithm, direct search method and global exploration method. The numerical optimization algorithm includes MMFD [4] (Modified Method of Feasible Directions) algorithm and NLPQL [5] (Sequence Quadratic Programming) algorithm, which are mainly used to solve problems with nonlinear and continuous characteristics. The advantage of this method is that it can effectively explore the local area around the initial design point. The disadvantage is that it has a strong dependence on the initial design point and is easy to fall into the local optimal solution. The direct search method includes the Hooke-Jeeves algorithm [6], etc., which has the advantage of being able to efficiently search for the area around the initial point, and can search for a broader design space than the numerical optimization algorithm. The disadvantage is that the dependence on the initial design point is relatively strong. And it is possible to fall into the local optimal solution. The global exploration method mainly includes multi-objective optimization algorithms such as genetic/evolutionary algorithms [7,8] and simulated annealing algorithms [7]. The advantage is that there is a large probability to find the global optimal solution. The disadvantage is that the computational complexity is relatively large. In the optimization design, it is necessary to fully consider the characteristics of the actual optimization problem and select the appropriate optimization algorithm.

In engineering practice like the blade optimization, the optimization problems encountered are mostly multi-objective optimization problems—allowing multiple sub-goals to achieve optimal problems at the same time. In most cases, these sub-goals are often mutually exclusive, and improvements in one sub-goal may lead to deterioration of other sub-goals. In response to the above problems, designers generally compromise the goals and try to optimize each sub-goal. In recent years, researchers have proposed many excellent multi-objective optimization algorithms, including NSGA-II (non-dominated sorting genetic algorithm II), NSGA-III (non-dominated sorting genetic algorithm III), MOEA/D (multi-objective evolutionary algorithm based on decomposition) and other algorithms to solve the multi-objective optimization problem. NSGA-II [9] was proposed by Kalyanmoy Deb et al. in 2002. It is an improvement of the NSGA. Compared with NSGA, the improvement of NSGA-II has two main points: (1) Propose a fast non-dominated sorting, making Pareto The time complexity of dominating sorting is reduced. (2) A crowded distance is proposed to measure the distribution of solutions, and based on this selection of suitable individuals in the population. At present, NSGA-II is a mainstream multi-objective optimization algorithm. In particular, its framework based on Pareto dominance relationship has been adopted by many algorithms, such as NSGA-III, etc. At the same time, NSGA-II is good for low-dimensional multi-objective optimization problems, but for high-dimensional multi-objective optimization problems, the first problem is that the selection pressure is too small due to its Pareto dominance relationship, and secondly That is, the crowded distance is not applicable in high-dimensional space, and the computational complexity is also relatively high. NSGAIII [10] was proposed by Kalyanmoy Deb in 2014. The algorithm adapted the order of crowding to maintain population diversity by introducing widely distributed reference points. MOEA/D [11] was proposed by Mr. Zhang Qingfu and others in 2008. MOEA/D converts a multi-objective optimization problem into multiple scalar quantum problems, and each sub-problem consists of a uniformly distributed weight vector. For each new solution generated, the solution near the subproblem is replaced based on the aggregate function.

In the optimization of the blade section in this paper, there are five optimization objectives, which belong to the high-dimensional multi-objective optimization problem. In this paper, We will use inverse generation distance [12] (IGD) to compare the effects of NSGA-II, NSGA-III and MOEA/D multi-objective optimization algorithms on blade section optimization to find a more suitable solution to the problem proposed in the paper. The optimized blade section structure is obtained by the better algorithm, and the improvement of the section characteristics of the optimized blade is calculated.

2. Modelling of blade section characteristics

2.1. Parametric description of blade section

This paper mainly studies the section structure of C-spar blade. Fig.1 is the schematic diagram of the C-spar blade section. It consists of four parts: skin, spar, beam and trailing edge.

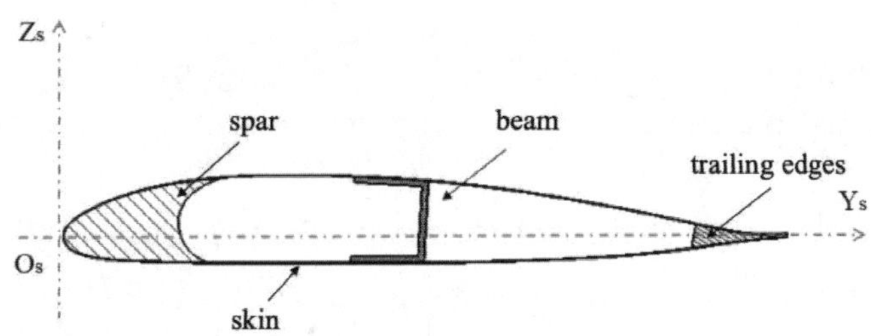

Figure 1. Schematic diagram of the C-spar blade section.

For parametric description of blade section structure, we need to define the type of airfoil selected and the chord length of the airfoil (C). On this basis, four components are parameterized. For the skin, this paper uses the form of full skin layer, so it only needs to define the thickness (h_i, i=1,...n) and quantity of the layer (n). For C-spar, trailing edges and beam, their parametric descriptions are shown in Fig. 2, Fig. 3, Fig. 4.

As shown in fig.2. The section of C-spar consists of inner and outer contour lines. The outer contour line can be obtained by cutting the inner airfoil contour of the skin layer by the points P_u and P_d. P_u is the point on the outer contour of the spar, and its chordwise distance to the leading edge point is Y_u. P_d is the lower end point of the outer contour of the spar, and its chordwise distance to the leading edge point is Y_d. The inner contour line can be determined by the coordinates of the Pu point, the Pd point, and the intermediate point $\{P_1, P_2, ..., P_i, ..., P_m\}$. Therefore, the section parameters of the spar can be described by Y_u, Y_d, Y_i^p and Z_i^p (i = 1, 2, ..., m).

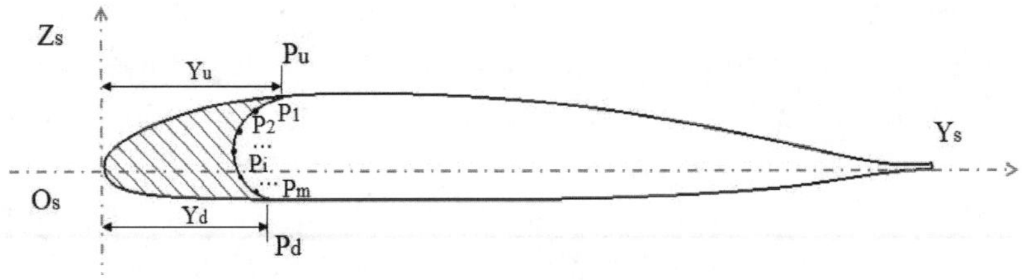

Figure 2. Parametric description of C-spar.

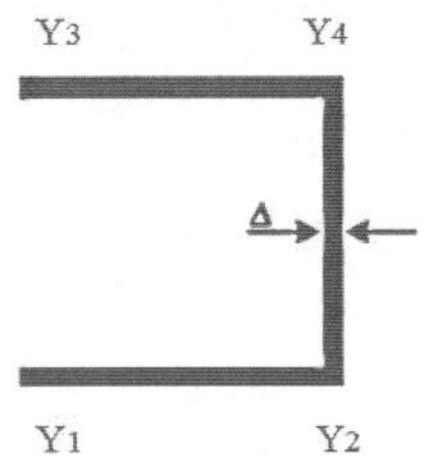

Figure 3. Parametric description of beam.

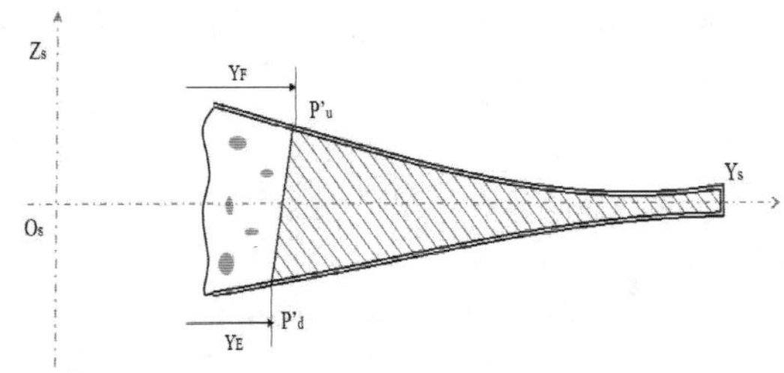

Figure 4. Parametric description of trailing edge.

As shown in fig.3. The profile of the beam can be represented by the chordwise distances of the four endpoints to the leading edge point $\{Y_1, Y_2, Y_3, Y_4, Y_5\}$ and the thickness Δ of the beam. As shown in fig.4. The section of the trailing edge is described by the two parameters of the lower edge endpoint P_u' to the leading edge point chordwise distance Y_F and the upper edge endpoint P_d' to the leading edge point chordwise distance Y_E.

2.2 Modelling of blade section characteristics

The blade section characteristics are determined by the airfoil of the blade section, the distribution of the components, and the material properties of the component. It mainly includes section quality linear density M_p , tensile stiffness E_S, waving stiffness E_{IB}, pendulum stiffness E_{IT} and torsional stiffness Γ. This paper uses direct analytical method[2] which proposed by Yang jianling to calculate the section characteristics of the C-spar blade.

Torsional stiffness of the Blade section Γ_{TOT} is mainly provided by torsion boxes consisting of skins (or consisting of skins and beam) and spar. In engineering applications, the torsional stiffness of the blade section is approximately equal to the sum of the torsion stiffness of the torsion box Γ and the torsional stiffness of the beam section Γ_{Spar}.

$$\Gamma_{TOT}=\Gamma+\Gamma_{Spar} \tag{1}$$

Calculation formula for section torsional stiffness of single torsion box:

$$\Gamma = \frac{4\Omega_1^2}{\alpha_1} \tag{2}$$

Calculation formula for section torsional stiffness of double torsion box:

$$\Gamma = \frac{4(\Omega_2^2\alpha_1 + \Omega_1^2\alpha_2 + (\Omega_1 + \Omega_2)^2\alpha_1^R)}{\alpha_1\alpha_2 + \alpha_1\alpha_1^R + \alpha_2\alpha_1^R} \tag{3}$$

In Eq.(2) and Eq.(3), Ω_i is the area enclosed by the torsion box. α_i is the ratio of the perimeter and thickness of the skin layer constituting the torsion box i (i = 1, 2). α_i^R is the ratio of the perimeter and the thickness of the beam region in the torsion box.

The blade C-spar can usually be regarded as solid beams, so the torsional stiffness calculation formula[] of the solid beam can be applied to obtain the torsional stiffness of the spar section:

$$\Gamma_{Spar} = \frac{G \cdot S^4}{4\pi^2 I} \tag{4}$$

In Eq.(4). S is the cross-sectional area of the spar, and I is the moment of inertia of the spar. They are all obtained by the blade geometry parameter extraction algorithm mentioned later; and G is the shear modulus of material.

Calculation formula of waving stiffness E_B and shimmy stiffness E_T in a Cartesian coordinate system :

$$E_B = \sum_{e=1}^{g} E_e \cdot Z_N^2 \cdot S_e \tag{5}$$

$$E_T = \sum_{e=1}^{g} E_e \cdot Y_N^2 \cdot S_e \tag{6}$$

Tensile stiffness Es is an important parameter for strength checking. Its physical meaning refers to the ability of an object to resist tensile deformation. The calculation of the tensile stiffness of the blade section in the project is also calculated by accumulating the tensile stiffness of the components of the blade section:

$$E_S = \sum_{e=1}^{g} E_e \cdot S_e \tag{7}$$

The quality linear density Mp of the blade section is a parameter that can describe the mass distribution of blade and calculate the blade weight. The blade is a combination of components which made up of a variety of different materials, so the quality linear density of the blade section is obtained by algebraic superposition of the mass linear density of the components:

$$M_P = \sum_{e=1}^{g} \rho_e \cdot S_e \tag{8}$$

In Eq.(5-8), Y_N and Z_N represent the coordinates of the center of gravity of the components in the Cartesian coordinate system. E_e is the Young's modulus of the e-th component material. P_e is the density of the e-th component. S_e is the area of the e-th component material

3. Optimization of blade section characteristics

3.1. Optimization model

In this paper, there are five main objective of section optimization in this article:

- Minimize $M_p(X)$—finding the minimum of mass linear density of section
- Maxmize $\Gamma_{TOT}(X)$—finding the maximum of torsional stiffness of section
- Maxmize $E_S(X)$—finding the maximum of tensile stiffness of section
- Minimize $E_T(X)$—finding the minimum of shimmy stiffness of section
- Minmize $E_B(X)$—finding the minimum of waving stiffness of section

In this paper, based on the previous parametric description of the blade profile, the design variables are selected from the parameters defined. The results are as follows:

Design variable: $X=[C\ h_i\ Y_u\ Y_d\ Y_i^p\ Z_i^p\ Y_u'\ Y_d'\ Y_1\ Y_2\ Y_3\ Y_4\ \Delta]$

Blade section optimization is generally performed given the specific parameters required. However, it is difficult to ensure that the blade profile characteristics are equal to the given value during design. Therefore, according to the design experience (the tensile stiffness and torsional stiffness are as large as possible. mass linear density, the shimmy stiffness and the waving stiffness are as small as possible) and the given values, the following inequalities are obtained. The combination of constraints:

$$\begin{cases} G_1(X) = M_p \geq \overline{M}_p \\ G_2(X) = \Gamma_{TOT} \geq \overline{\Gamma}_{TOT} \\ G_3(X) = E_B \leq \overline{E}_\mathrm{B} \\ G_4(X) = E_T \leq \overline{E}_T \\ G_5(X) = E_s \geq \overline{E}_s \end{cases} \tag{9}$$

3.2. Optimization method

According to the characteristics of blade optimization, it is very important to select the appropriate solution method for the optimization of the blade section structure. In the optimization of the blade section, there are many design variables, the numerical changes are discrete, and the constraints have contradictory characteristics. For example, an increase in torsional stiffness causes an increase in the waving stiffness, while in the blade design it is desirable to have a greater torsional stiffness and a lower waving stiffness. In addition, there is a mutual constraint relationship between the blade optimization targets, which belongs to the multi-objective optimization problem. Therefore, it is necessary to choose a multi-objective optimization algorithm that can adapt to discrete design variables and constraint associations. At present, more applications in blade optimization are NSGA-II algorithms. This algorithm is good for low-dimensional multi-objective optimization problems, but for high-dimensional multi-objective optimization problems, it first faces the problem of too small selection pressure due to its Pareto dominance relationship. Secondly, the crowded distance is not applicable in high-dimensional space, and the computational complexity is also relatively high. Because there are five optimization goals in this paper, it belongs to the high-dimensional multi-objective optimization problem. It has the following three questions:

- As the number of optimization targets increases, the proportion of non-dominated solutions in the population also increases, which leads to a slow search process;
- For high-dimensional target space, the computational complexity of indicators for maintaining diversity is too high, and the neighboring elements of the solution are difficult to find;
- For high-dimensional target spaces, the search ability of the recombination operator is too inefficient.

Therefore, in order to solve the above problem, a predefined target search can be used, which can be divided into two types: the first one is a predefined set of search directions spanning the entire pareto front, and the representative of this method is MOEA/D; The other is to predefine multiple reference points, such as the algorithm NSGAIII.

In this paper, we use NSGA-II, NSGA-III, and MOEA/D algorithms to optimize the profile. We evaluate the optimization effects of the three algorithms through the evaluation indicators of the multi-objective optimization algorithm to find the optimal optimization algorithm. The basic method of blade profile optimization is shown in the fig.5.

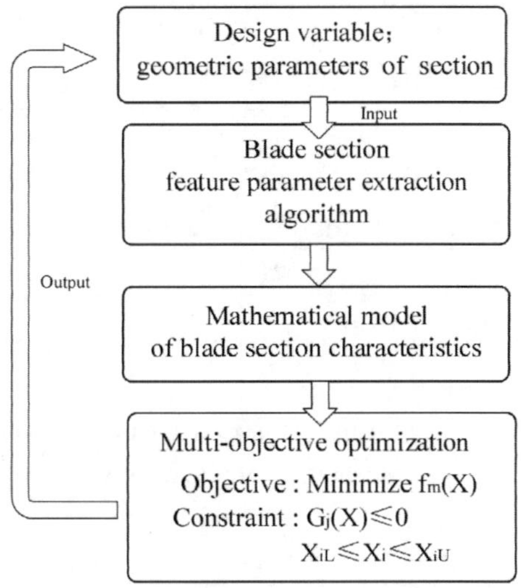

Figure 5. Blade section optimization method.

First you need to enter design variables and divide the population according to the design variables. The blade geometry feature extraction algorithm is used to calculate the geometric feature parameters of the component, such as area S, perimeter L, and moment of inertia I. This step translates the input design variable parameters into parameters that can be used to calculate the profile characteristics of the blade. The mathematical model of the blade profile is then used as the optimization target and constraints, and optimized by the multi-objective optimization algorithm. Finally, the optimized results are output.

The section feature extraction algorithm mainly considers the section of each component of the blade as a complex polygon, and converts the component area, arc length, moment of inertia, and extraction of the center of gravity into polygon area, arc length, moment of inertia and center of gravity. Each side of the polygon is determined by the set of imported airfoil coordinate points and the component geometry.

4. Result and discussion

This paper selects the OA212 as airfoil for optimization, the chord length is 350mm, and the airfoil contains four components: skin, C-spar, beam and trailing edge. In order to reduce the amount of calculation, the number of skin layers is n=1, and the number of intermediate points in the inner contour of the beam is m=1. In order to compare the optimized effects, the profile geometry parameter values must be set before optimization, as shown in table.1.

Table 1. Pre-optimization parameter list.

Component name	Component parameter				
skin	material	ply number	h/mm	ply type	
	C3228	1	1	full coverage	
spar	material	Y_u/mm	Y_d/mm	Y_1^P/mm	Z_1^P/mm	
	V3233	40	30	36	0	
beam	material	Y_1/mm	Y_2/mm	Y_3/mm	Y_4/mm	Δ/mm
	V3233	150	180	150	180	1
trailing edge	material	Y_u' /mm	Y_d' /mm			
	V3233	280	280			

The optimization goal of section optimization is to minimize the mass line density, maximize torsional stiffness, maximize tensile stiffness, minimize shimmy stiffness, and minimize waving stiffness. The constraint conditions for section optimization are shown in Table 2.

Table 2. Blade section optimization constraints.

Constraint	E_S (N)	E_B (N·m²)	E_T (N·m²)	Γ_{TOT} (N·m²)
Range	$\geq 8\times10^7$	$(1.5\times10^4, 2\times10^4)$	$(5\times10^5, 6\times10^5)$	$\geq 2\times10^4$

The value range of the design variables is shown in Table 3.

Table 3. Range of design variables.

variable	h (mm)	Y_u, Y_d (mm)	Y_1^P (mm)	Z_1^P (mm)	Y_1 ,Y_2 ,Y_3 ,Y_4 (mm)	Δ (mm)	Y_u', Y_d' (mm)
Range	(0, 1)	(10, 70)	(10 ,70)	(-10,10)	(100,250)	(0.5,1)	(280,320)

NSGA1-II, NSGA-III, and MOEA/D were used as optimization algorithms for optimization. For high-dimensional multi-objective optimization, it is difficult to show its pareto frontier, so we evaluate the optimization effect by the convergence performance and the distribution performance of the solution set.

The Inverted Generational Distance (IGD) is a comprehensive performance evaluation indicator. It evaluates the convergence performance and distribution performance of the algorithm by calculating the minimum distance between each point on the real Pareto frontier surface and the individual set obtained by the algorithm. The smaller the value, the better the comprehensive performance (convergence performance and distribution performance) of the algorithm.

In this paper, we set the population size to 400 and the evolutionary algebra to 2000, and use three algorithms to optimize the blade profile. Finally, we get the evolution algebra-IGD value relationship shown in Fig.6-8.

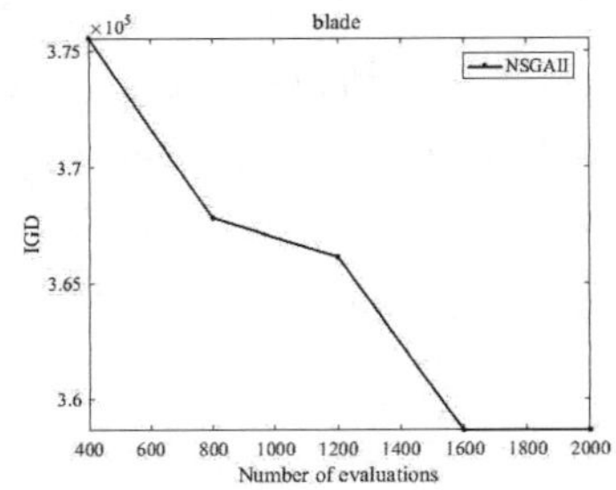

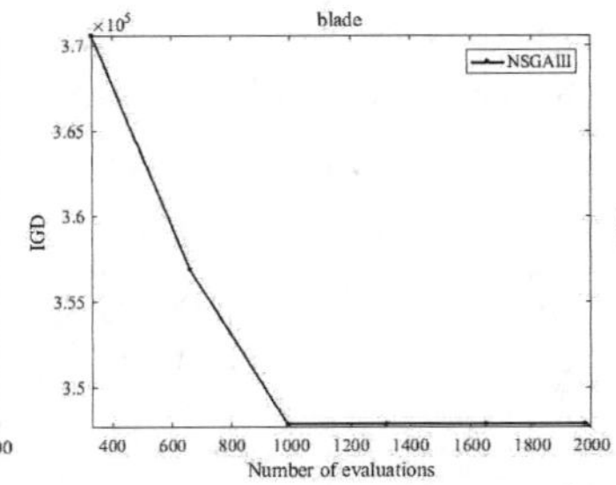

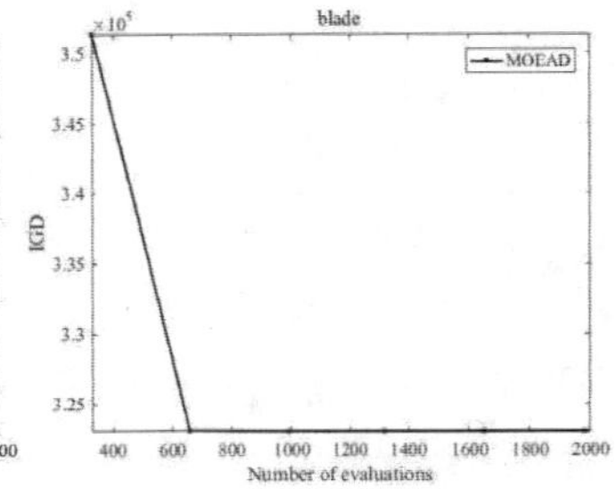

Figure 6. EA-IGD (NSGA-II). **Figure 7.** EA-IGD (NSGA-III). **Figure 8.** EA-IGD (MOEA/D).

As shown in Fig.6-8, we can see that with the increase of evolutionary algebra, the value of IGD is decreasing, and the value of MOEA/D algorithm is decreasing faster, followed by NSGA-III algorithm, and last one is NSGA-II algorithm. It indicate MOEA /D algorithm can shorten the distance between the resulting solution and the real solution faster. In addition, it can be seen from the IGD values of 2000th generation that the IGD values (NSGA-II, NSGA-III, and MOEA/D) are 3.53×10^5, 3.43×10^5 and 3.21×10^5 respectively. The IGD value of MOEA/D is smaller, indicating that the obtained solution set is closer to the real Pareto frontier, the distribution of the solution set is more uniform, and the number of solution sets is also more abundant.

Since the multi-objective evolution/genetic algorithm has different solution sets for each calculation, in order to test the above conclusions, we repeat the experiment 30 times and calculate the mean and standard deviation of the IGD values obtained by each algorithm. We get the table as shown in table.4

Table 4. IGD mean and standard deviation of NSGA-II, NSGA-III, MOEA/D.

	Population size	evolutionary algebra	Run times	Average value of IGD	Standard deviation of IGD
NSGA-II	400	2000	30	3.51×10^5	8.16×10^3
NSGA-III	400	2000	30	3.42×10^5	7.01×10^3
MOEA/D	400	2000	30	3.27×10^5	7.51×10^3

The grayed-out part of the table 4 is the average and standard deviation of the IGD obtained after optimizing the section 30 times using the MOEA/D algorithm. The average IGD of MOEA/D is the smallest among the three algorithms. This proves the previous conclusion——MOEA/D can get better optimization results.

The optimized design variables obtained by the MOEA/D algorithm are shown in Table 5.

Table 5. Pre-optimized and optimized design variables.

/mm	h	Y_u	Y_d	Y_1^p	Z_1^p	Y_1	Y_2	Y_3	Y_4	Δ	Y_u'	Y_d'
before	1	40	30	36	0	150	180	150	180	1	280	280
after	0.62	45.9	45.2	30.2	0	191	221	191	221	0.7	288.3	287.7

According to the model of blade section characteristic, the parameters of the blade profile characteristics before and after optimization of the blade are calculated. The change situation of optimized section characteristics are obtained, as shown in Table 6.

Table 6. rate of change of blade section characteristic.

section characteristic	Mp (kg • m)	Es (N)	E_B (N • m^2)	E_T (N • m^2)	Γ_{TOT} (N • m^2)
	-1.8%	1.3%	-1.7%	-1.9%	2.1%

5. Conclution

Through the IGD evaluation index, the optimization effects of NSGA-II, NSGA-III and MOEA/D algorithms in the optimization of blade profile were compared. The results show that the NSGA-II algorithm performs poorly in the high-dimensional multi-objective optimization problem, while the NSGA-III and MOEA/D algorithms show better comprehensive performance (convergence performance, diversity performance). Among them, MOEA/D works best.

After using the optimized design variables to calculate the blade structure characteristics, the results show that the section quality linear density is reduced by 1.8%, the section torsional stiffness is increased by 2.1%, the tensile stiffness is increased by 1.3%, and the swing stiffness and the shimmy stiffness are respectively reduced. 1.9% and 1.7%.

References

[1] Chen Ming, Xu Guanfeng, Zhang Lei. Key Technologies of Transmission System and Rotor System of helicopter[J]. Aviation Manufacturing Technology, 2010(16): 32-37.

[2] Yang Jianling. Design and optimization of helicopter blade structure based on integral parameter definition[D]. Nanjing University of Aeronautics and Astronautics, 2013.

[3] Lai Yuyang, Jiang Xin. Isight parameter optimization theory and examples[M]. Beijing University of Aeronautics and Astronautics Press, 2012.

[4] Bazaraa M S, Sherali H D, Shetty C M. Methods of feasible directions:[M]. Elsevier Pub. Co. 1960.

[5] Schittkowski K NLPQLP: A New Fortran Implementation of a Sequential Quadratic Programming Algorithm for Parallel Computing[J]. 2002.

[6] R Hooke, T A Jeeves, "Direct search" solution of numerical and statistical problems. Journal of the Association for Computing Machinery (ACM), 8 (2), 212–229, 1961.

[7] Davis L. Genetic algorithms and simulated annealing[M]. Morgan Kaufmann Publishers, 1987.

[8] Fogel D B. Evolutionary algorithms in theory and practice.[J]. Complexity, 1997, 2(4):26–27.

[9] Deb K, Pratap A, Agarwal S, et al. A fast and elitist multiobjective genetic algorithm: NSGA-II[J]. IEEE Transactions on Evolutionary Computation, 2002, 6(2):182-197.

[10] Deb K, Jain H. An Evolutionary Many-Objective Optimization Algorithm Using Reference Point-Based Nondominated Sorting Approach, Part I: Solving Problems With Box Constraints[J]. IEEE Transactions on Evolutionary Computation, 2014, 18(4):577-601.

[11] Zhang Q, Li H MOEA/D: A Multiobjective Evolutionary Algorithm Based on Decomposition[J]. IEEE Transactions on Evolutionary Computation, 2008, 11(6):712-731.

[12] A Zhou, Y Jin, Q Zhang, B Sendhoff, and E Tsang, "Combining model-based and genetics-based offspring generation for multi- objective optimization using a convergence criterion," in Proceedings of the 2006 IEEE Congress on Evolutionary Computation, 2006, pp. 892–899.

Evolutionary dynamics of Chinese tourism market extremely low-priced strategies based on wright-fisher processes

Wanqing Lv, Yijie Wang and Xiangli Ni

School of Business and Tourism Management, Yunnan University, Kunming, China

E-mail: kamiolily@icloud.com

Abstract. Extremely low-priced of package tours is an important cause of Chinese tourism market price disorder. Regulating the price of package tours in Chinese tourism market, can solve the problem of Chinese tourism market price disorder. In order to research the reason of extremely low package tours price, consider the stochastic dynamics of N homogeneous mixed travel agency pricing strategies, and each travel agency could choose Extremely low-pricing or regular pricing. Based on Wright-Fisher processes, we obtain the dominant strategy of Extremely low-pricing and regular pricing under the weak selectivity in the package tours price setting, when the travel agency groups are finite. And then further verify these conditions by numerical simulation.The simulation shows that the Price elasticity of tourism demand, the transparency of the tourism market and the degree of government supervision strength have important effects on the stochastic evolution of package tours price setting. Reducing the demand elasticity of tourism market, increasing the transparency of the tourism market and the degree of government supervision strength can make travel agency choose regular pricing and solve chaotic of Chinese tourism market further. Therefore, this paper puts the suggestion of "two increases and one decrease" for the development of Chinese tourism market.

1. Introduction

Extremely low-priced of package tours is a chronic disease of Chinese tourism industry, which first appeared in the 1990s. In recently, with the occurrence of "compulsive shopping" and "abusive tourist" and other tourism complaints due to extremely low-priced of tours market [1]. The public is starting to concerned about the tourism market price disorder caused by extremely low-priced of package tours. For a time, extremely low-priced of package tours were pushed to the forefront of public opinion and led to declining the credibility of China's tourism market. Therefore, extremely low-priced of package tours has seriously affected the image of Chinese tourism market and inhibited the development of Chinese tourism industry.

Extremely low-priced of package tours is a tourism market price disorder, which violated the objective economic law. Pricing strategies of tourist product providers is an important factor that caused tourism market price disorder. In the mass tourism market, travel agencies usually are the main suppliers of the package tours products. The pricing strategy of travel agencies has a direct impact on the price of products in the tourism market. Thus, the travel agency is the liability subject of Chinese tourism market price disorder, and the low-priced strategy of travel agency is the important factor which led to Chinese tourism market price disorder. So, researching the pricing strategies of travel agencies is significant to solve tourism market price disorder and keep the Chinese tourism industrial consistent developing.

2. Literature review

Nowadays, most of the research on travel agency pricing strategy game in academia is based on the infinite overall and completely rational object model. Long Yong and others studied the pricing strategies of travel agencies when they cooperated with online tourism platforms based on Stackelberg game. Long Yong believed that the pricing of tourism products by travel agencies was mainly affected by the Commission threshold and marginal service cost. In the cooperation between them, online tourism platforms always wanted to keep the prices of tourism products low. Therefore, Online Tourism Platform will be more inclined to choose travel agency cooperation with lower service cost. According to Picazo P [2] and Estrella Díaz [3], the pricing strategy of travel agency is mainly influenced by the marketing strategy of travel agency. The brand image and marketing content of travel agency will affect the final pricing decision of travel agency. [3,4] Furthermore, Joaqun Alegre found that the tourism market environment also has a certain impact on the pricing strategy of travel agencies through comparative study of the tourism market prices in Germany and Britain. When the tourism market is oligopoly, travel agencies tend to implement the low-price strategy because of fierce intra-industry competition.[5] The result is similar, Guanhua studied the pricing strategy of Chinese travel agencies based on Bertrand model, and found that the travel agency industry in China is in an oligopoly competition state.[6] Because of the high homogeneity of Chinese tourism commodities, there exists intense price competition among travel agencies. It is based on the principle that market price equals marginal cost to price tourism commodities. Meanwhile, based on fair preference Lin Qiang analyzed the pricing strategy of travel agency in game with other players in tourism commodity supply chain. Lin believed that the pricing strategy of travel agency in game with other players in tourism commodity supply chain was related to whether the information of game system was symmetrical or not. When the information of game system was asymmetrical, the game system will be a zero-sum game, the travel society adopts a low-price strategy to compete with other main bodies in the tourism commodity supply chain because of its complete rationality.[7] For further explore price competition among the Chinese travel agencies, NingzeQun analyzed the development trend of Chinese travel agency industry from 1997 to 2000. Though the research on the Profit and Loss Statements of Chinese travel agency, Ning found a paradox of Chinese tourism industrial that the number of Chinese travel agencies' entry and their profitability show a reverse trend. For this paradox, Ning put forward a theory of "tourism black box pricing". It is meaning in Chinese tourism market, the process of tourism consumption is a black box, so that travel agencies can make up for their losses due to intense price competition by forcing them to make extra consumption in the process of providing tourism services [8].

Through the analysis of the above literature, we can see that the current studies on the pricing strategy of travel agencies is usually carried out under the scenario of competition and cooperation between travel agencies and other main bodies in the tourism commodity supply chain, and the theoretical model is mainly based on the object model of infinite totality and complete rationality. However, due to the existence of external factors such as national policy, corporate responsibility and market threshold, the number of travel agencies is always limited. Therefore, the basic hypothesis of infinite totality is a common defect in the study of travel agency pricing strategy in the current academic circles. And meanwhile, as the literature shows, the intense price competition of Chinese travel agencies is the main reason of extremely low-priced tourism products. According to Bertrand's price competition model, the excessive competition within the industry has a more significant impact on the emergence of very low prices than excessive competition within other stakeholder in the industrial chains.[9] Compared with other tourism-related acts, competition and game among different travel agents will have more significant impact on the pricing strategy of travel agencies. However, reviewing the literature, we can see that there are many studies on pricing competition between travel agencies and other tourism-related subjects in the academic circles, but there are few studies on pricing competition between different travel agencies.

Under the actual situation, however, the number of travel agencies is limited, and pricing decisions are limited and rational. The final pricing strategy of travel agencies always needs several dynamic games to converge to a stable state. Therefore, the evolutionary game theory has better applicability in the analysis of travel agency pricing decision-making process. So, in this paper, we based on the assumption of limited rationality and limited aggregate, studies the price competition and pricing game among different travel agencies under the scenario of intra-species competition and cooperation, analyzing and describing the evolution process of pricing strategy of travel agencies.

The evolutionary game with finite population has been discussed extensively in the field of biology and behavioral economics [10]. Schaffer analyzed the economic meaning of evolutionary learning from the viewpoint of evolution [11], and extended Smith's concept of infinite population evolutionary stability to finite population [12]. Based on the research of Schaffer, Tanaka and Ania further discussed the evolutionary game characteristics of finite population ,and believed that the evolutionary game dynamics of finite population is a stochastic process [13,14]. Nowadays, The evolutionary game dynamics of a finite population is mainly studied by means of the frequency-dependent Moran and Wright-Fisher processes [15-21]. The former is an asynchronous update process, in which, at each time step, an individual in a population is selected to produce a descendant proportional to its fitness and its offspring replaces a randomly selected individual [22]. The Wright-Fisher evolutionary process is also a stochastic process, but unlike the Moran process, it is synchronous. In each generation, individuals produce offspring proportional to their payoff. The next generation is sampled randomly from this pool of offspring. The total population size is constant. Therefore, every time step in the Wright-Fisher process corresponds to an entire generation of N time steps in the Moran process[23]. In 2006, for the first time, Imhof and Nowak pro-posed that the dynamic evolution of a finite population can be studied by a frequency-dependent Wright-Fisher process and studied the fixation probability in a 2×2 game model [24]. Traulsen pointed out that, in an infinite population, strict Nash equilibrium evolution is always stable, but in a limited population, random effects can cause the system to deviate from strict Nash equilibrium. For the reason [25], Traulsen produced a new evolutionary stability concept and proved that the two equilibrium concepts can be gained from both the frequency-dependent Moran and Wright-Fisher processes.

Analogical ecology shows that price competition among different travel agencies is essentially an intraspecific competition and intraspecific game phenomenon. In this phenomenon, the learning and replacement process of travel agency's population strategy will be displayed, so the Moran process and Wright-Fisher process have better applicability to the research of the evolution of pricing strategy in travel agency's population. In fact, the evolution of strategies among different travel agencies is usually simultaneous and synchronous. Therefore, based on the actual situation, this paper uses frequency-dependent Wright-Fisher process to study the evolution of intraspecies pricing strategy of Chinese travel agencies, in order to explore the evolution mechanism of extremely low-priced of package tours in Chinese tourism market, and solving the problem of tourism market price disorder.

3. Model

3.1. Evolutionary Game Theory

Ever since Smith, Price and Taylor put forward the basic concepts of evolutionary stability strategy and replicating dynamic equation[26], evolutionary game theory has been widely used in various fields. Evolutionary game theory, as the main theory to analyze the evolution process of group strategy, has good applicability to the study of decision evolution in group game. In recent years, evolutionary game theory has been used in the study of ecotourism development and tourism conflict, which involves the interests conflict of multi-party tourism subjects.

In evolutionary game theory, the stable convergence evolution of game process is mainly described by replication dynamic equation. The replication dynamic equation is:

$$\frac{dx_i(t)}{dt}=[u(e^i,x)-\sum_{i=1}^{k}x_i u(e^i,x)]x_i$$

In this equation, each individual can choice strategies i or other strategies, the probability of strategies i is x_i, while the $u(e^i,x)$ refers to the expected payoff of strategies i in the game.

From the replication dynamic equation of evolutionary game, can be found the game population in evolutionary game analysis needs to be assumed as an infinite population, so that the replication dynamic equation can be expressed in differential form. For this reason, when study evolutionary game of finite population, the stochastic process should be introduced to.

3.2. Wright-fisher process with different selection intensities

Studied the evolutionary game dynamic in a mixed homogeneous finite population. Suppose a population that consists of a fixed number of N individuals, each individual can be of one of two types, A or B, where A -type individuals refer to those who adopt strategy A in the game, do B type individuals, and every individual can interact with any other. The state of the population is fully characterized by the numbers i of individuals of using strategy A and $N-i$ of individuals using strategy B. The payoffmatrix for the game is:

$$\begin{array}{cc} & \begin{array}{cc} A & B \end{array} \\ \begin{array}{c} A \\ B \end{array} & \begin{bmatrix} a & b \\ c & d \end{bmatrix} \end{array}$$

In this game, the average payoffs of individuals using strategies A and B are given by:

$$F = \frac{a(i-1) + b(N-i)}{N-1}, G = \frac{ci + d(N-i-1)}{N-1}$$

Under a classical evolutionary game dynamic framework, the expected payoff represents the fitness. However, from the perspective of heredity transmission, the individual reproduction rate is proportional to the corresponding game revenue. Therefore we should consider the difference in selection intensities ω of the two types of individuals are introduced.[27] The fitness of an A and B individual is given by:

$$f = 1 - w + wF, g = 1 - w + wG$$

Suppose that in the current generation i individuals use strategy A. Then, the composition of the next generation is determined through N independent binomial trials. So, the Wright-Fisher evolutionary process is a discrete-time Markov chain with the state space $\{0,1\cdots\cdots N\}$ and transition probabilities

$$P_{ij} = \binom{N}{j}\left(\frac{if_i}{if_i + (N-i)g_i}\right)^j \left(1 - \frac{if_i}{if_i + (N-i)g_i}\right)^{N-j}$$

Where $\binom{N}{j} = C_N^j$.

3.3. Parametric Hypothesis

In actual market, different travel agencies have many strategic choices in pricing game. In order to highlight the evolution of extremely low-priced and simplify the analysis process, this paper assumes that the travel agency population is a limited and homogeneous single population mixed with multiple travel agencies, and that there are only two kinds of pricing strategies within the population, namely " extremely low-priced strategy" and "normal pricing strategy". The following are specific parameter assumptions.

Hypothesis 1. In a travel agency market (M), there are N travel agencies that provide the same goods. These travel agencies have the same market position and are all bounded rational individuals. In this market, the expected payoff of travel agencies just effected by their strategy, instead of their market position. Therefore, the pricing strategy game between each travel agency is symmetrical game.

Hypothesis 2. In M market, the pricing strategy of each travel agency can be of one of two types, T_y $(y = 1,2)$. In here, T_1 strategy denotes the set of extremely low-price pricing strategies and T_2 strategy denotes the set of normal pricing strategies. Among them, we assume the number of travel agencies who select strategy T_1 is i, and the number of travel agencies who select strategy T_2 is j ,and $i + j = N$.

Hypothesis 3. In M market, the total tourism demand is L, the demand elasticity is θ. In T_1 strategy, Travel agencies will set the price of tourism products is P_1.While in strategy T_2, Travel agencies will set the price of tourism products is P_2.Where $P_2 \gg P_1$.

Hypothesis 4. Based on the research of Andrea[28] ,this paper supposes that when the pricing strategy in M market is the same, the tourism demand will be digested by all travel agencies on average. When there are different pricing strategies in M market, part of the tourism demand of the higher-priced side will flow to the lower-priced side. Based on Alfred Marshall's demand elasticity theory, this paper assumes that the change of tourism demand flow is ΔQ. Where ΔQ is a function of θ, $\Delta Q = (1 - \frac{P_1}{P_2})\frac{L\theta}{2}$, and $(0 \le \Delta Q \le \frac{L}{2})$.

Hypothesis 5. Review the research of Chinese tourism Market, when the travel agencies choose extremely low-price pricing strategies, with information dominance, they usually compensate for the loss of extremely low-price pricing strategies by forcing tourists to shop and increasing additional tourism consumption in the process of providing tourism services[29]. For this reason, it is assumed that the travel agencies with strategy T_1 have additional Grey Income (G), and that the additional Grey Income (G) is a function of the pricing difference , $G = k\ (P_2 - P_1)$, where k as a compensation coefficient, which is negatively correlated with the transparency of tourism market.

Hypothesis 6. As well known, the Chinese government's behavior has a great Impact on Chinese tourism market. In order to analyze the influence of government behavior on travel agency pricing game, this paper introduces the parameter of government punishment behavior(F).It is assumed that, in order to ensure market price order, when there are travel agencies in M market that are far below the average market price, they will be punished economically by government(F).

Based on the above hypothesis, the symmetric game payment matrix is obtained as shown in Table 1. Because of the symmetry, this paper only gives the payment matrix of a single entity.

Table 1. Symmetric game payment matrix.

	T_1	T_2
T_1	$\frac{L}{2}P_1 + G$	$\left(\frac{L}{2} + \Delta Q\right)P_1 + G - F$
T_2	$\left(\frac{L}{2} - \Delta Q\right)P_2$	$\frac{L}{2}P_2$

3.4. Stochastic game Model

3.4.1. Fitness calculation

According to the parameter hypothesis and the payment matrix, the expected payoff of travel agencies chooses T_1 and T_2 strategies without considering the individual's own game are

$$F_i = \frac{(\frac{L}{2}P_1 + G)(i-1) + [(\frac{L}{2} + \Delta Q)P_1 + G - F](N-i)}{N-1} \quad (1)$$

$$G_i = \frac{P_2(\frac{L}{2} - \Delta Q)i + \frac{L}{2}P_2(N-i-1)}{N-1}$$

consider the difference in selection intensities ω .Then the fitness of T_1 and T_2 strategies are

$$f_i = 1\text{-}\omega + \omega F_i = 1\text{-}\omega + \omega \frac{(\frac{L}{2}P_1 + G)(i-1) + [(\frac{L}{2} + \Delta Q)P_1 + G - F](N-i)}{N-1}$$
$$g_i = 1\text{-}\omega + \omega G_i = 1\text{-}\omega + \omega \frac{P_2(\frac{L}{2} - \Delta Q)\, i + \frac{L}{2}P_2(N-i-1)}{N-1} \tag{2}$$

Selective intensity ω is a description of the contribution of expected payoff to actual fitness. For a finite population model, however, the parameters ω have an impact on the long-term behavior. If $\omega = 0$, then $f_i = g_i = 1$,for all states i and covers the case of neutral selection. If $\omega = 1$, then $f_i = F_i, g_i = G_i$,the game will be in a state of complete rationality. This case is the so-called strong selection. If $\omega \in \{\omega | 0 < \omega < 1\}$ then the payoffs from the game have only a marginal influence on fitness f_i and g_i, the game will be in a state of bounded rationality. This case is the so-called weak selection. [30] Realistic, the pricing strategies of travel agencies are bounded rationality. Thus, this paper assumes $\omega \in \{\omega | 0 < \omega < 1\}$.

Due to the requirement of non-negative fitness in genetic algorithm, based on the formula (2), the fitness function is modified by adding ∝ correction coefficient. The corrected fitness is

$$f_i = 1\text{-}\omega + \omega F_i = 1\text{-}\omega + \omega \frac{(\alpha + \frac{L}{2}P_1 + G)(i-1) + [\alpha + (\frac{L}{2} + \Delta Q)P_1 + G - F](N-i)}{N-1}$$
$$g_i = 1\text{-}\omega + \omega G_i = 1\text{-}\omega + \omega \frac{[\alpha + P_2(\frac{L}{2} - \Delta Q)]\, i + (\alpha + \frac{L}{2}P_2)(N-i-1)}{N-1} \tag{3}$$

$$\left(\omega \in (0,1);\ f_i > 0;\ g_i > 0\right)$$

3.4.2. Dynamic Equation

Based on the independent binomial distribution characteristics of policy update in Wright-Fisher process.[31] Formula (3) shows that the probability of generating T_1 strategy in each experiment is as follows:

$$P_i = \frac{if_i}{if_i + (N-i)g_i},\quad i \in (0, N) \tag{4}$$

Therefore, the process of generation of T_1 strategy individuals is a N-weight Bernoulli experiment with probability of $P_i = \frac{if_i}{if_i + (N-i)g_i}$. So, the Proportion evolution of T_1 strategy from I to Δi is a discrete-time Markov chain with the state space $\{0,1\cdots\cdots N\}$ and transition probabilities:

$$P_{i\Delta i} = \binom{N}{\Delta i}\left(\frac{if_i}{if_i + (N-i)g_i}\right)^{\Delta i}\left(1 - \frac{if_i}{if_i + (N-i)g_i}\right)^{N-\Delta i} \tag{5}$$

Since the individual strategy updating is synchronized in Wright-Fisher process, it is assumed that the individual strategy updating step is $\Delta t(\Delta t > 0)$. The frequency of T_1 strategy at t-Time is $x(x =$

$\frac{i}{N})$.Thus, When the total number of N is large and finite, the evolution relationship between frequency X and time t of T1 strategy in Wright-Fisher process can be described by Langevin equation[32].

$$\frac{dx}{dt}=m(x)+\sqrt{v(x)}\xi \tag{6}$$

Where, $m(x)=\frac{E(\Delta x)}{\Delta t}$ is drift term, $v(x)=\frac{E[(\Delta x)^2]}{\Delta t}$ is diffusion term, ξ is gauss noise and Δx is the unit step of T_1's increment. Substitute $x=\frac{i}{N}$ into $m(x)$ and $v(x)$,we can obtain the result:

$$\begin{aligned} m(x)&=m(\frac{i}{N})=\frac{E(\Delta(\frac{i}{N}))}{\Delta t}=\frac{1}{N}\frac{1}{\Delta t}\sum_{\Delta i=0}^{N}(\Delta i-i)\binom{N}{\Delta i}(P_i)^{\Delta i}(1-P_i)^{N-\Delta i} \\ &=\frac{1}{\Delta t}[(\frac{if_i}{if_i+(N-i)g_i})-\frac{i}{N}] \\ v(x)&=v(\frac{i}{N})=\frac{E(\Delta(\frac{i}{N})^2)}{\Delta t}=\frac{1}{N}\frac{1}{\Delta t}\frac{if_i(N-i)g_i}{[if_i+(N-i)g_i]^2} \end{aligned} \tag{7}$$

When the total number of N is large and finite, namely $N\to\infty$,

$$\begin{aligned} \lim_{N\to\infty} m(x)&=\lim_{N\to\infty} m(\frac{i}{N})=\frac{1}{\Delta t}[(\frac{\frac{i}{N}f}{\frac{i}{N}f+(1-\frac{i}{N})g})-\frac{i}{N}]=\frac{1}{\Delta t}[(\frac{xf}{xf+(1-x)g})-x] \\ \lim_{N\to\infty} v(x)&=\lim_{N\to\infty} v(\frac{i}{N})=0 \end{aligned} \tag{8}$$

Where $f=1-\omega+\omega\left(x\left(\alpha+\frac{L}{2}P_1+G\right)+(1-x)\left(\alpha+\left(\frac{L}{2}+\Delta Q\right)P_1+G-F\right)\right)$, $g=1-\omega+\omega\left(x\left(\alpha+P_2\left(\frac{L}{2}-\Delta Q\right)\right)+(1-x)\left(\alpha+\frac{L}{2}P_2\right)\right)$.

So, form the dynamic equation of frequency-dependent Wright-Fisher process based on equation (8) is obtained as follows:

$$\begin{aligned} F(x)&=\frac{dx}{dt}=m(x)+\sqrt{v(x)}\xi=\frac{1}{\Delta t}[(\frac{xf}{xf+(1-x)g})-x] \\ &=\frac{1}{\Delta t}\frac{x(1-x)(f-g)}{x(f-g)+g} \end{aligned} \tag{9}$$

4. Strategic stability analysis

4.1. Stability Point discussion

If there is a stable strategy in Wright-Fisher process, the dynamic equation needs to be satisfied $F(x)=0$,Substitute $f = 1-\omega+\omega\left(x\left(\alpha+\frac{L}{2}P_1+G\right)+(1-x)\left(\alpha+\left(\frac{L}{2}+\Delta Q\right)P_1+G-F\right)\right)$, $g = 1-\omega+\omega\left(x\left(\alpha+P_2\left(\frac{L}{2}-\Delta Q\right)\right)+(1-x)\left(\alpha+\frac{L}{2}P_2\right)\right)$ into equation(8):

Thus,

$$F(x)=\frac{1}{\Delta t}\frac{x(1-x)\omega\left(x(F+P_2\Delta Q-P_1\Delta Q)+\frac{L}{2}P_1-\frac{L}{2}P_2+P_1\Delta Q+G-F\right)}{x\omega\left(x(F+P_2\Delta Q-P_1\Delta Q)+\frac{L}{2}P_1-\frac{L}{2}P_2+P_1\Delta Q+G-F\right)+1-\omega+\omega\left(x\left(\alpha+P_2(\frac{L}{2}-\Delta Q)\right)+(1-x)\left(\alpha+\frac{L}{2}P_2\right)\right)} \tag{10}$$

Let $\sigma = \omega\left(x(F+P_2\Delta Q-P_1\Delta Q)+\frac{L}{2}P_1-\frac{L}{2}P_2+P_1\Delta Q+G-F\right)$, $\eta = x\omega(x(F+P_2\Delta Q-P_1\Delta Q)+\frac{L}{2}P_1-\frac{L}{2}P_2+P_1\Delta Q+G-F\Big)+1-\omega+\omega(x(\alpha+P_2\left(\frac{L}{2}-\Delta Q\right)+(1-x)(\alpha+\frac{L}{2}P_2))$.

Thus, Formula (10) can be reduced to

$$F(x)=\frac{1}{\Delta t}\frac{x(1-x)\delta}{\eta} \tag{11}$$

Let $F(x)=\frac{dx}{dt}=0$, there are two solutions can be obtained, $x_1 = 0$, $x_2 = 1$.

If the evolution stable state $x_i(i = 1,2)$ is stable, then x_i is robust to small fluctuation disturbance. Therefore, based on the stability theorem of differential equation, the stability point $x_i(i = 1,2)$ needs to satisfy the conditions $F(x)'=0$ before it can evolve to a stable state, which was introduced by Richard Bellman (1980).Thus, this paper gets definitions as follows.

Definition 1. When $x_1 = 0$ is a stable point of intra-species pricing game for travel agencies, the parameters need to satisfy condition $S_1 = \left\{G+P_2\Delta Q<\frac{L}{2}(P_2-P_1)\right\}$.

Definition 2. When $x_2 = 1$ is a stable point of intra-species pricing game for travel agencies, the parameters need to satisfy condition $S_2 = \left\{P_1\Delta Q+G-F>\frac{L}{2}(P_2-P_1)\right\}$.

Proof.: Derivation of formula (11), thus

$$F(x)'=\frac{1}{\Delta t}\frac{\left((1-x)\delta-x\delta+x(1-x)\delta'\right)\eta-x(1-x)\delta\eta'}{\eta^2} \tag{12}$$

Substitute $x_1 = 0$ and $x_2 = 1$ into equation (12), we obtain

$$F(0)'=\frac{\delta(0)}{\eta(0)}=\frac{1}{\Delta t}\frac{\omega\left(\frac{L}{2}P_1-\frac{L}{2}P_2+P_1\Delta Q+\text{G-F}\right)}{1-\omega+\omega\left(\alpha+\frac{L}{2}P_2\right)}$$

$$F(1)'=-\frac{\delta(1)}{\eta(1)}=\frac{1}{\Delta t}\frac{-\omega\left(\frac{L}{2}P_1-\frac{L}{2}P_2+G+P_2\Delta Q\right)}{1-\omega+\omega\left(\alpha+\frac{L}{2}P_1+G\right)} \quad (13)$$

If $F(0)'<0$,then $\frac{1}{\Delta t}\frac{\omega\left(\frac{L}{2}P_1-\frac{L}{2}P_2+P_1\Delta Q+\text{G-F}\right)}{1-\omega+\omega\left(\alpha+\frac{L}{2}P_2\right)}<0$.From equation（3）, we can see $0<\omega<1$ and $1-\omega+\omega\left(\alpha+\frac{L}{2}P_2\right)>0$.Beacuse of $\Delta t>0$,to make $\frac{1}{\Delta t}\frac{\omega\left(\frac{L}{2}P_1-\frac{L}{2}P_2+P_1\Delta Q+\text{G-F}\right)}{1-\omega+\omega\left(\alpha+\frac{L}{2}P_2\right)}<0$, just let $\frac{L}{2}(P_1-P_2)+P_1\Delta Q+G-F<0$,namely $P_1\Delta Q+G-F<\frac{L}{2}(P_2-P_1)$.

If $F(1)'<0$ then $\frac{1}{\Delta t}\frac{-\omega\left(\frac{L}{2}P_1-\frac{L}{2}P_2+G+P_2\Delta Q\right)}{1-\omega+\omega\left(\alpha+\frac{L}{2}P_1+G\right)}<0$.From equation (3) , we can see $0<\omega<1$ and$1-\omega+\omega\left(\alpha+\frac{L}{2}P_1+G\right)>0$.Beacuse of $\Delta t>0$,to make $\frac{1}{\Delta t}\frac{-\omega\left(\frac{L}{2}P_1-\frac{L}{2}P_2+G+P_2\Delta Q\right)}{1-\omega+\omega\left(\alpha+\frac{L}{2}P_1+G\right)}<0$, just let $-\left(\frac{L}{2}(P_1-P_2)+P_2\Delta Q+G\right)<0$,namely $G+P_2\Delta Q>\frac{L}{2}(P_2-P_1)$.

Assume $S_\epsilon(\epsilon=1,2)$ is the set of conditions, when $x_1=0$ or $x_2=1$ is the stable point. And $U_\gamma(\gamma=1,2)$ is the set of values, when $x_1=0$ or $x_2=1$ is the stable point. Then $\overline{U_\gamma}(\gamma=1,2)$ is the set of values, when $x_1=0$ or $x_2=1$ is not the stable point. Where $U_1=\{P_1\Delta Q+G-F<\frac{L}{2}(P_2-P_1)\}$;$U_2=\left\{G+P_2\Delta Q>\frac{L}{2}(P_2-P_1)\right\}$;$\overline{U_1}=\left\{P_1\Delta Q+G-F>\frac{L}{2}(P_2-P_1)\right\}$;$\overline{U_2}=\{G+P_2\Delta Q>\frac{L}{2}(P_2-P_1)\}$.

Thus, when the function satisfies the condition that $S_1=U_1\cap\overline{U_2}=\left\{P_1\Delta Q+G-F<\frac{L}{2}(P_2-P_1)\right\}\cap\left\{G+P_2\Delta Q<\frac{L}{2}(P_2-P_1)\right\}$, $x_1=0$ is the stable point. While, when the function satisfies the condition that $S_2=U_2\cap\overline{U_1}=\left\{G+P_2\Delta Q>\frac{L}{2}(P_2-P_1)\right\}\cap\left\{P_1\Delta Q+G-F>\frac{L}{2}(P_2-P_1)\right\}$, $x_2=1$ is the

stable point.
From the Hypothesis 3., we can see $P_1 \ll P_2$. And from Hypothesis 6., we can see $F > 0$.Thus, there is $G + P_1\Delta Q - F < G + P_2\Delta Q$.Further, we can get $S_1 = U_1 \cap \overline{U_2} = \left\{P_1\Delta Q + G - F < \frac{L}{2}(P_2 - P_1)\right\} \cap \left\{G + P_2\Delta Q < \frac{L}{2}(P_2 - P_1)\right\} = \overline{U_2}$, and then, Definition 1. has been proofed.

Similarly, we can get $S_2 = U_2 \cap \overline{U_1} = \left\{G + P_2\Delta Q > \frac{L}{2}(P_2 - P_1)\right\} \cap \left\{P_1\Delta Q + G - F > \frac{L}{2}(P_2 - P_1)\right\} = \overline{U_1}$, and then , Definition 2. has been proofed.

4.2 Analysis of Stability Conditions

When the Definition 1. is true, that is the expected payoff function satisfies $S_1 = \{G + P_2\Delta Q < \frac{L}{2}(P_2 - P_1)\}$, then $x_1 = 0$ will be The Stability Strategy of Intra-species Pricing Game for Travel Agencies. In this case, the selection ratio $(\frac{i}{N})$ of T_1 strategy in the evolution game system will be 0. From the Hypothesis 2, where $i + j = N$. So in this case, the selection ratio $(\frac{j}{N})$ of T_2 strategy in the evolution game system will be 1. Thus, When the Definition 1. is true, normal pricing strategies will be the Stability Strategy of Travel Agency Industry in M Market.

From the Hypothesis 3, compared with adopting extremely low-price pricing strategy, when travel agencies adopt normal pricing strategies, the price of package tours will be higher, namely $P_2 \gg P_1$. Thus, when P_1 is small enough to be $P_1 \to 0$, there is $\lim_{P_1 \to 0} S_1 = \left\{\frac{G}{P_2} + \Delta Q < \frac{L}{2}\right\}$. And from the Hypothesis 4 and Hypothesis 5, there are $\lim_{P_1 \to 0} \Delta Q = \frac{L\theta}{2}$, $\lim_{P_1 \to 0} G = k$.Thus, when the Definition 1 is true, under the Hypothesis 3, The evolution and stability of normal pricing strategies of travel agencies are mainly affected by demand elasticity θ and compensation coefficient k.

When the demand elasticity θ is small, tourists are not sensitive to the price of tourism commodities. In the case, Higher pricing strategies of tourism commodities will help travel agencies to obtain more revenue. When the compensation coefficient (k) is low, the transparency of the tourism market is high, tourism market information is symmetrical. Thus, in this case, travel agencies will find it difficult to cross-subsidize extremely low prices by forcing tourists to shop and increasing additional tourism consumption. Therefore, When the Definition 1 is true, the normal pricing strategy is strictly dominant and stable in evolution.

When the Definition 2 is true, that is the expected payoff function satisfies $S_2 = \{P_1\Delta Q + G - F > \frac{L}{2}(P_2 - P_1)\}$. then $x_2 = 1$ will be The Stability Strategy of Intra-species Pricing Game for Travel Agencies. In this case, the selection ratio $(\frac{i}{N})$ of T_1 strategy in the evolution game system will be 1. From the Hypothesis 2., where $i + j = N$. So in this case, the selection ratio $(\frac{j}{N})$ of T_2 strategy in the evolution game system will be 0. Thus, When the Definition 2 is true, extremely low-price pricing strategies will be the stability strategy of travel agency industry in M market.

Similarly, from the Hypothesis 3, when P_1 is small enough to be $P_1 \to 0$, there is $\lim_{P_1 \to 0} S_2 = \left\{G - F > \frac{L}{2} P_2\right\}$. Thus, when the Definition 2 is true, under the Hypothesis 3, The evolution and stability of extremely low-price pricing strategies of travel agencies are mainly affected by government penalties F and additional grey income G. When grey tourism income (G) is higher and government penalty (F) is lower, travel agencies can not only seize the market through extreme low prices, but also cross-subsidize extreme low prices through grey tourism income (G). Meanwhile, because of the low punishment (F) of the government, the illegal cost of tourism enterprises is low, extremely low-price pricing strategy can bring more benefits to travel agencies. Thus, when the Definition 2 is true, the extremely low-price pricing strategy is strictly dominant and stable in evolution.

5. Simulation analysis

To further verify stable conditions, in this section, we will analyze the influence of main parameters on the evolution of travel agency pricing strategy.

5.1. Simulation of Definition 1

For verifying Definition 1, with the condition of $S_1 = \left\{G + P_2 \Delta Q < \frac{L}{2}(P_2 - P_1)\right\}$, this paper sets the parameters as follow: $L = 100, \Delta t = 10$, $\omega = 0.5$, $P_2 = 30$, $P_1 = 1$, $F = 60$, $\theta = 0.5$, $k = 5, \alpha = 60$. And then, Simulate the evolutionary game with initial strategy T_1 ratio is $x\left(\frac{i}{N}\right)$, where $x\left(\frac{i}{N}\right) \in [0.01, 0.99]$. Fig. 1 shows the result.

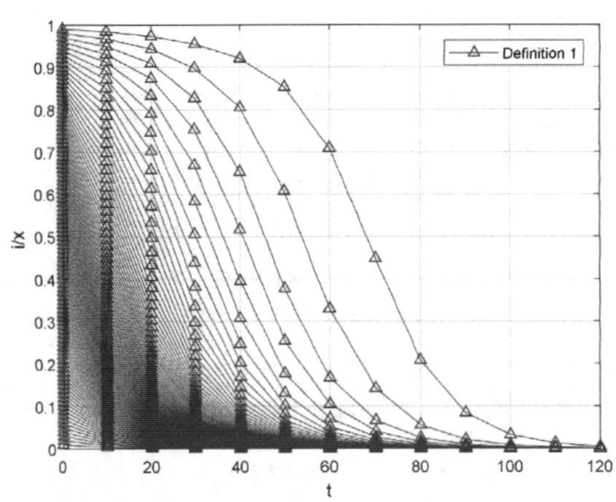

(a)

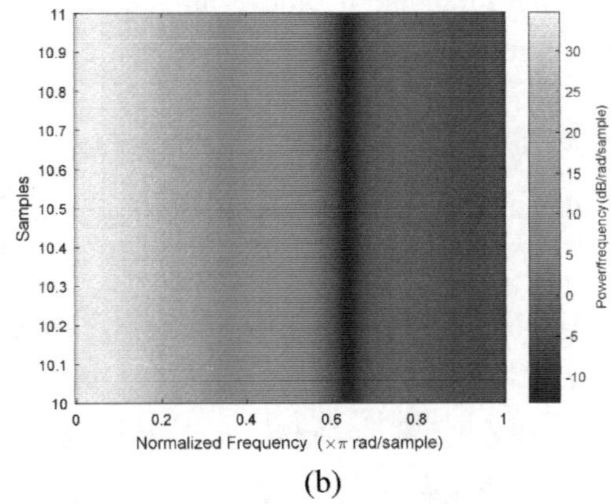

(b)

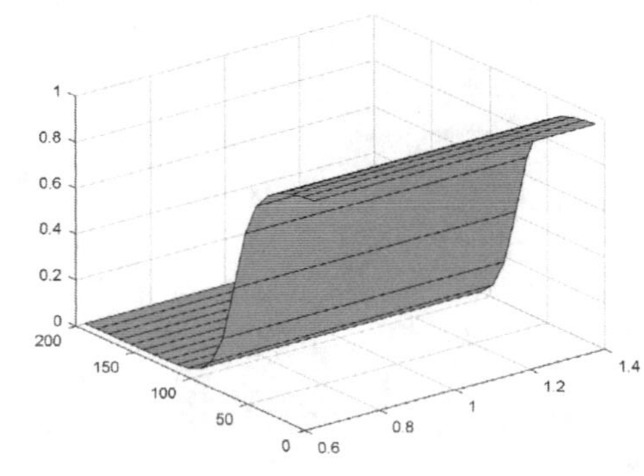

(c)

Figure 1. Simulation of Definition 1.

Fig. 1 (a)–(c) shows that, when Definition 1 is true, the demand elasticity (θ) of tourism market and the compensation coefficient (k) of travel agencies will be lower. The market with low demand elasticity is the market with low demand elasticity, under the circumstances, the rate of price change will be greater than the rate of demand change, so it will be difficult for travel agencies to gain more benefits by low-price competition. Because the compensation coefficient (k) is inversely proportional to the transparency of tourism market, when the compensation coefficient (k) is low, the transparency of travel agency market will be higher, tourists have easier access to tourism market and tourism commodity information. Thus, it is difficulty for travel agencies obtain additional grey tourism income through compulsory shopping and additional tourism consumption.

Therefore, when Definition 1 is true, the fitness of the extremely low-price pricing strategy of travel agencies will be lower than that of the normal pricing strategy, and the normal pricing strategy will strictly dominate in the game. Whatever the initial proportion of T_1 strategy is set, the final selection of the system will evolve to a stable state of $x_1 = 0$. At this time, the selection proportion of T_1 strategy will be 0, and that of T_2 strategy will be 1. Therefore, normal pricing strategy will become a stable pricing strategy for travel agencies in M market.

Although the initial proportion setting has no effect on the result of strategy evolution, it can be seen from Figure 1 that the initial proportion setting has a great influence on the speed of evolution stability. When Definition 1 is true, the rate of stability of M market strategy evolution is inversely proportional to the initial proportion setting, and the longer it takes for M market strategy evolution to be stable when the initial proportion setting is higher.

5.2. Simulation of Definition 2

For verifying Definition 2, with the condition of $S_2 = \left\{P_1 \Delta Q + G - F > \frac{L}{2}(P_2 - P_1)\right\}$, this paper sets the parameters. To ensure the comparability of experiments, comparing with the simulation of Definition 1, only the parameters F, θ and K are changed in the simulation of Definition 2. The parameters of Definition 2 are set as follows: $L = 100, \Delta t = 10, \omega = 0.5, P_2 = 30, P_1 = 1, F = 30, \theta = 1.5, k = 60, \alpha = 60$.Similarly, we Simulate the evolutionary game with initial strategy T_1 ratio is $x\left(\frac{i}{N}\right)$, where $x\left(\frac{i}{N}\right) \in [0.01,0.99]$. And Fig. 2 shows the result.

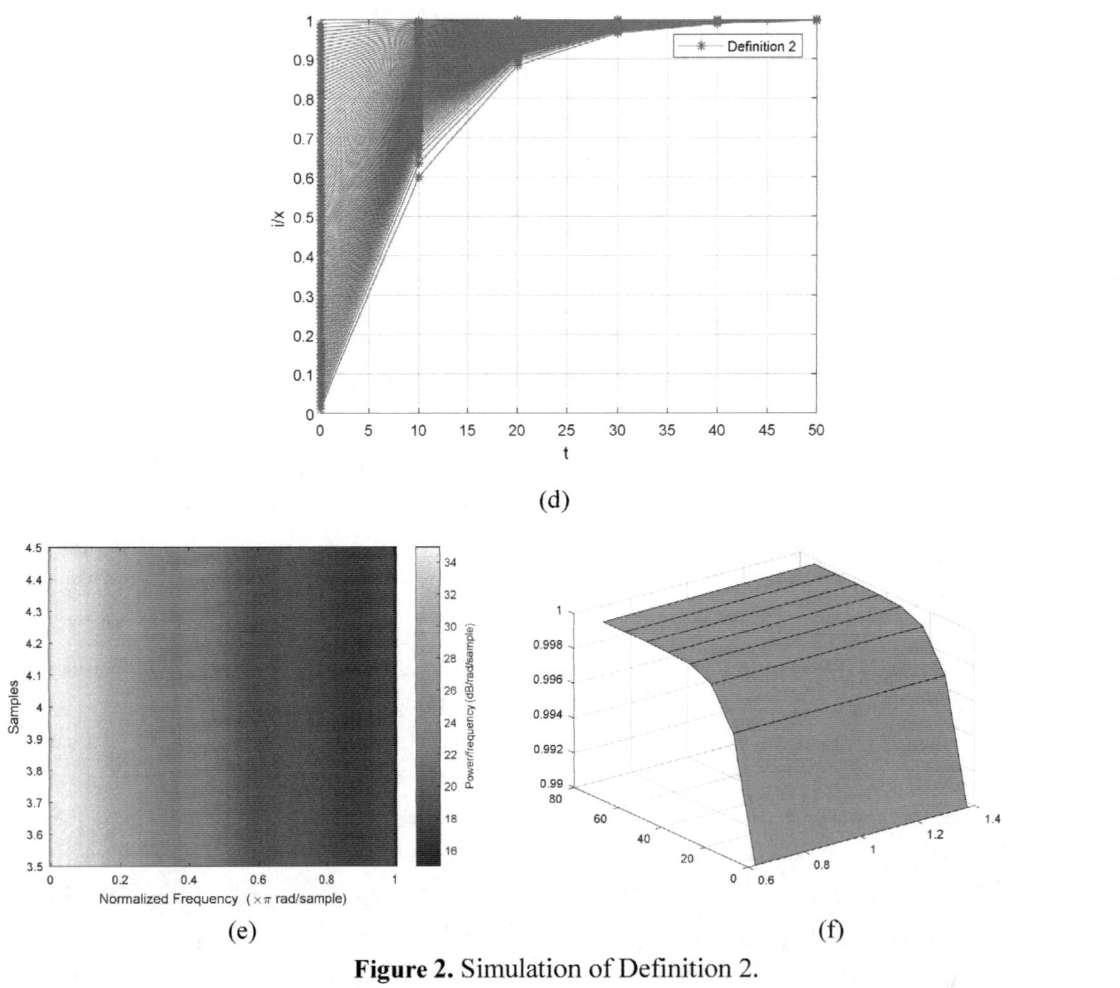

Figure 2. Simulation of Definition 2.

Fig. 2 (d)–(f) shows that, when Definition 2 is true, the demand elasticity (θ) of tourism market and the compensation coefficient (k) of travel agencies will be higher and the government penalties F is lower. The market with high demand elasticity is the market with high demand elasticity,

Journal of Physics: Conference Series **1419** (2019) 012037 doi:10.1088/1742-6596/1419/1/012037

under the circumstances, the rate of demand change will be greater than the rate of price change. From the Hypothesis 4, $0 \le \Delta Q \le \frac{L}{2}$. Thus, when the demand elasticity (θ) is 1.5, there is $\Delta Q = (1-\frac{P_1}{P_2})\frac{L\theta}{2} \approx 72.5 > \frac{L}{2}$. Therefore, under the Definition 2 condition, the change of tourism demand flow ΔQ is 50, namely the demand of travel agencies that make normal pricing strategies will all flow to the travel agencies with extremely low-price strategies. At this time, travel agencies can seize the whole market by making extremely low-price pricing. Meanwhile, due to the compensation coefficient (k) is inversely proportional to the transparency of tourism market, when the compensation coefficient (k) is high, the transparency of travel agency market will be lower. In this case, the tourism market is a black box market, it is easy for travel agencies to obtain grey tourism income by compulsory shopping and increasing additional tourism consumption.

Therefore, when Definition 2 is true, extremely low-price pricing strategy will bring more benefit to travel agency, and because the lower government punishment (F), the illegal cost of travel agencies is low. So, the fitness of the extremely low-price pricing strategy of travel agencies will be higher than that of the normal pricing strategy, and the extremely low-price pricing strategy will strictly dominate in the game. Whatever the initial proportion of T_1 strategy is set, the final selection of the system will evolve to a stable state of $x_1 = 1$. At this time, the selection proportion of T_2 strategy will be 0, and that of T_1 strategy will be 1. Therefore, extremely low-price pricing strategy will become a stable pricing strategy for travel agencies in M market.

Same as the simulation of Definition 1, although the initial proportion setting has no effect on the result of strategy evolution, while the initial proportion setting has a great influence on the speed of evolution stability. When Definition 2 is true, the rate of stability of M market strategy evolution is directly proportional to the initial proportion setting, and the shorter it takes for M market strategy evolution to be stable when the initial proportion setting is higher.

5.3. Simulation of different selective intensity

For further analyze the influence of selectivity intensity ω on the evolution of travel agency pricing strategy under the circumstance of limited rationality. Based on the parameters setting of Definition 2, this paper analyses the evolution of grouped 1 and 2 with different selectivity intensity ω. Where, the parameters setting of grouped 1 are : $L = 100, \Delta t = 10$, $\omega_1 = 0.01$, $P_2 = 30$, $P_1 = 1, F = 30$, $\theta = 1.5$, $k = 60, \alpha = 60$; the parameters setting of grouped 2 are : $L = 100, \Delta t = 10$, $\omega_2 = 0.99$, $P_2 = 30$, $P_1 = 1, F = 30$, $\theta = 1.5$, $k = 60, \alpha = 60$ 。 Simulate the evolutionary game with initial strategy T_1 ratio is 0.1, 0.5 and 0.9. The Fig. 3 shows the result.

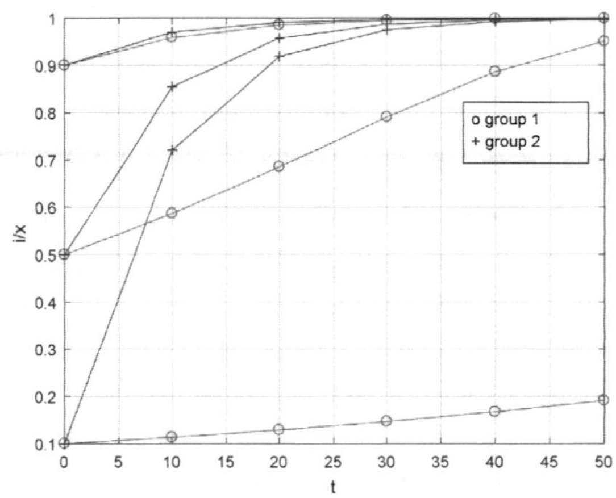

Figure 3. Simulation of different selective intensity.

From Figure 3, the selectivity intensity ω has no effect on the ultimate outcome of evolution game when the game is a bounded rational game, but has a significant effect on the speed of evolution game. The selectivity intensity ω of group 1 is 0.01 and the selectivity intensity ω of group 2 is 0.99. While in Figure 3, the speed of evolution stability in group 2 is always faster than that in group 1, Therefore, it can be concluded that the selectivity intensity ω has a significant influence on the evolution speed of the game in the limited rational game, and the selectivity intensity ω has a positive correlation with the evolution speed of the game.

6. Conclusion

Extremely low-priced of package tours is an important cause of Chinese tourism market price disorder. In this paper, we considered the Chinese tourism market extremely low-priced strategies evolutionary dynamic process of Wright-Fisher in a finite population under weak selection. And found that the change of tourism demand flow (ΔQ), grey additional tourism income (G) and government penalties (F) have direct effects on the evolution of travel agency pricing strategy. The demand elasticity (θ), compensation coefficient (k) and government penalties (F) are the fundamental reasons that affect the choice of travel agency pricing strategy. Therefore, based on the conclusion of the study, this paper proposes a solution to the price disorder in Chinese tourism market as follow.

I. Strengthen the Government's Penalty and Increase the Low Competition Cost of Travel Agencies

In the intraspecific pricing game of travel agencies, the government plays the role of supervisor. Chinese government punishment (F) can effectively increase the cost of vicious price competition and enhance the potential benefits of normal competition. It can be seen from the establishment conditions of Definition 2 that, once the punishment intensity of the government is $F < P_1\Delta Q + G + \frac{L}{2}(P_1 - P_2)$, the extremely low-price pricing strategy of will become the stable pricing strategy of travel agencies. Therefore, strengthening the Chinese government's punishment to maintain the implementation cost of the extremely low-price strategy at a high level will play a positive role in preventing and improving Chinese tourism market price disorder.

However, it should be noted that the supervision and incentive effect of government punishment (F) has a significant effect only when the proportion of travel agencies choose extremely low-price strategy is small. From the proof of Definition 1, the stability of normal pricing strategy of travel agencies is only affected by these parameters G, P_2 and ΔQ ,because there is always $P_1\Delta Q + G - F < G + P_2\Delta Q$. Only when $G + P_2\Delta Q < \frac{L}{2}(P_2 - P_1)$, the pricing strategy of travel agency will be stability at normal pricing. Therefore, once the extremely low-price pricing strategy has been the stable pricing strategy in the tourism market, it will be difficult for the government to make normal pricing become a stable pricing strategy in the tourism market again by means of government penalty (F).

II. Strengthen market transparency and reduce the difficulty of obtaining tourism market information

The grey additional tourism income (G) is an important profitable means, when travel agencies to implement extremely low-price pricing strategy. It is also one of the main case for tourism complaints and chaos in the Chinese tourism market. Reducing the grey tourism additional income (G) of travel agencies can effectively control extremely low-cost travel disorder and reduce the occurrence of complaints and chaos in Chinese market. From the actual situation and related literature, the dominant information of travel agencies is the fundamental reason why travel agencies can obtain tourism grey income (G).Thus, the Chinese government and relevant agencies can further enhance the transparency of the tourism market, make the information between tourists and travel agencies symmetrical, eliminating the driving force of the grey tourism additional income (G) of travel agencies, further reduce the occurrence of travel agencies choosing extremely low-price strategy.

III. Promoting the resident's Tourism consumption concept and Reducing the Elasticity of Chinese tourism Market Demand

From the fitness of T_1 and T_2 strategies, we can find that the fitness of T_1 strategies f_i is positively correlated with tourism demand elasticity θ, while the fitness of T_2 strategies g_i is negatively correlated with tourism demand elasticity θ. Therefore, the higher the demand elasticity θ, the more likely the pricing strategy of travel agencies will be stable in the extremely low-price pricing strategy. Thus, reducing the elasticity of tourism demand θ is of great significance for solving the problem of unreasonable low-price pricing of travel agencies.

From the theory of elasticity, demand elasticity is related to consumption concept. While, in Chinese tourists' consumption concept, tourism product are often non-necessities or even luxury goods, and meanwhile most of Chinese tourists' consumption behavior is uncertain consumption, lack of long-term planning of tourism consumption, short time to adjust tourism demand, tourists are very sensitive to the price of tourism commodities. In addition, because China's tourism market is a typical off-peak season market and the demand for tourism is concentrated, the elasticity of tourism demand in China's tourism market is high. Thus, extremely low-price pricing strategy has a high fitness in China's tourism market.

Therefore, through the promotion of tourists' consumption concept, guiding tourists' consumption habits, making tourism commodities a necessity, making the residents form the habit of long-term planning consumption and off-season travel, can effectively reduce the elasticity of Chinese tourism market demand θ.Further alleviating and solving the price disorder in Chinese tourism market.

7. Discussion

This paper discusses the stability conditions of normal pricing strategy and extremely low-price pricing strategy of Chinese travel agencies, and based on game Analysis and simulation experiments, discusses the main factors affecting the pricing of Chinese travel agencies, and puts forward some suggestions to alleviate the price disorder in Chinese tourism market. The research has certain practical significance and practical value. However, in order to highlight the evolution process of extremely low-price strategy, this paper does not fully reflect the evolution of all pricing strategies of travel agencies. It only focuses on the evolution of normal pricing and extremely low-price pricing of travel agencies, which has some limitations. Therefore, based on the Wright-Fisher process to explore the evolution of intra-species pricing strategy in multi-strategy scenarios is the next research focus of this paper.

References

[1] Hu Fu Sheng. The Research on the Cause and Governance Policies of Unreasonable Low-Price- Tourism[J]. Price Theory and Practice,2017(02):32-34.

[2] Long Yong, Shi Ping ping. Pricing strategies of tour operator and online travel agency based on cooperation to achieve O2O model[J]. Tourism Management,2017,62(05):302-311.

[3] Picazo P, Moreno-Gil, S. Tour operators' marketing strategies and their impact on prices of sun and beach package holidays[J]. JOURNAL OF HOSPITALITY AND TOURISM MANAGEMENT.2018,35(02):17-28.

[4] Estrella Díaz, David Martín Consuegra, Águeda Esteban. Perceptions of service cannibalization : The moderating effect of the type of travel agency[J].Tourism Management, 2015,48(12):329-342.

[5] Joaquín Alegre, Maria Sard. Price strategies by German and British tour operators in Mallorca[J].Journal of Hospitality and Tourism Management,2017,33(09):93-102.

[6] Guan Hua, Zhao Li Ming. Price Competition and Product Differentiation Strategy of Bertrand Travel Agency[J]. Journal of East China Normal University, 2010,42(04):108-111.

[7] LIN Qiang, WEI Guangxing. Pricing-decisions and Contract Coordination of Tourism Service Supply Chain under Fairness Preference[J]. Tourism Tribune, 2018,33(04):59-69.

[8] NING Ze Qun. On the Paradox of the Market Access into China' s Travel Service Market and the Black-Box Pricing Pattern of Travel Service Enterprises Tourism Tribune, 2006(05):80-85.

[9] FANG Zhong-xiu, DONG Liang. Research on Questionable Patent, Firm Behavior and Patent System—Analysis Based on the Bertrand Competition[J]. Soft Science, 2018,32(06):15-20.

[10] Traulsen A, Hauert C. Stochastic evolutionary game dynamics[J].Reviews of nonlinear dynamics and complexity,2009,2: 25-61.

[11] M E, Schaffer. Evolutionarily stable strategies for a finite population and a variable contest size[J]. Journal of Theoretical Biology.1988,12:29-45.

[12] Smith J M, Price G R. The Logic of Animal Conflict[J]. Nature. 1973,246:15-18.

[13] Yasuhito Tanaka. Stochastically stable states in an oligopoly with differentiated goods: equivalence of price and quantity strategies[J].Journal of Mathematical Economics.2000,34:235-253.

[14] Ania A B. Evolutionary stability and Nash equilibrium in finite populations, with an application to price competition[J]. Journal of Economic Behavior and Organization. 2006 (3),65:472-488.

[15] Nowak, M.A., Sigmund, K, 2004. Evolutionary dynamics of biological games. Science 303 (5659), 793–799.

[16] Traulsen A, Pacheco, J.M, Nowak, M.A, 2007. Pairwise comparison and selection temperature in evolutionary game dynamics. J. Theor, Biol. 246 (3), 522–529.

[17] Wu, B, Altrock P M, Wang, L, et al, 2010. Universality of weak selection. Phys. Rev. E 82 (2), 46106.

[18] Ewens, W.J, 2004. Mathematical population genetics: theoretical introduction. Springer-Verlag.

[19] Nowak, M.A., 2006. Evolutionary Dynamics: Exploring the Equations of Life. Harvard University Press.

[20] Chen, J.H, Liu, X.N, 2011. Fixation probabilities in evolutionary dynamics with a Wright-Fisher process in finite diploid populations. J. Southwest Univ. 33 (9), 40–49.

[21] Traulsen A, Pacheco, J.M, Nowak, M.A, 2007. Pairwise comparison and selection temperature in evolutionary game dynamics. J. Theor, Biol. 246 (3), 522–529.

[22] Traulsen A, Hauert C. Stochastic evolutionary game dynamics[J]. Reviews of nonlinear dynamics and complexity,2009,2: 25-61.

[23] Corradi V. Continuous approximations of stochastic evolutionary game dynamics[J]. Journal of Economic Theory,2000,94:163-191.

[24] Nowak, M.A, 2006. Evolutionary Dynamics: Exploring the Equations of Life. Harvard University Press.

[25] Traulsen, A, Pacheco, J.M, Imhof, L.A, 2006a. Stochasticity and evolutionary stabil- ity. Phys. Rev. E 74 (1), 21905.

[26] Taylor C, Fudenberg D, Sasaki A, et al. Evolutionary game dynamics in finite populations[J]. Bulletin of Mathematical Biology,2004,66: 1621-1644.

[27] Nowak, M.A, 2006. Evolutionary Dynamics: Exploring the Equations of Life. Harvard University Press.

[28] Andrea Guizzardia, Flavio Maria Emanuele Ponsa, Ercolino Ranieri. Competition patterns, spatial and advance booking effects in the accommodation market online[J]. Tourism Management, 2019,71: 476-489.

[29] Lin Long Fei, Hu Kun. Research on Gray Income of Tour Guide—Take shopping discounts as an example[J]. Journal of Hunan Institute of Engineering, 2018,28(02):21-25.

[30] Imhof L A and Nowak M A. Evolutionary Game Dynamics in a Wright-Fisher Process[J]. Journal of Mathematical Biology,2006,52:667-681.

[31] WangXian-jia, GuCui-ling, QuanJi. Evolutionary game dynamics of the Wright-Fisher process with different selection intensities[J].Journal of Theoretical Biology. 2019,465:17-26.

[32] Ohtsuki H, Nowak M A, Pacheco J M, 2007. Evolutionary graph theory: Breaking the symmetry between interaction and replacement. J. Theor Biol. 246 (4), 681–694.

Grid-search wolf pack optimization algorithm for two-dimensional OTSU image segmentation

Ziyi Lv[1], Kang Liu[1,2], Dongxing Wang[1,3] and Xu Qian[1]

[1]China University of Mining & Technology, Beijing, China

[2]E-mail: kangliu@cumtb.edu.cn
[3]E-mail: wangdongxing85@163.com

Abstract. Compared with One-Dimensional OTSU, Two-Dimensional OTSU includes the relationship about the pixels of the image and improves the anti-noise ability of the algorithm. However, as a result of too much computation caused by considering the relationship about pixels of the image to be processed, Two-Dimensional OTSU always takes much time. To keep the good performance of original Two-Dimensional OTSU as well as reduce the time spent; this paper proposes Grid-Search Wolf Pack Optimization Algorithm (GSWA) for Two-Dimensional OTSU Image Segmentation. Firstly, it is based on the thought of adaptive shrinking grid search chaos wolf optimization algorithm with adaptive standard-deviation updating amount (ASGS-CWOA) that traditional wolf pack optimization algorithm was improved to enhance its performance, which includes Grid Search and Opposite-Middle Raid(OMR). Grid Search can enhance the wolf pack's summon and siege capability. Moreover, discrete step-size was adopted to adapt to discrete solution space of Image-Segmentation. Experimental results indicate that GSWA-OTSU not only reduces the segmentation time, but also takes the segmentation quality into account.

1. Introduction

Image segmentation is based on some specific similarity criteria, which divides the image into a number of non-overlapping regions with special properties. In the same region, it satisfies certain similarity criteria of texture, grayscale, colour, etc. In different regions, the image characteristics are significantly different. The common methods used in image segmentation are Threshold method [1], watershed [2], multi-scale morphological method [3], fuzzy clustering [4], deep learning-based image segmentation [5] and Bayesian [6,7].

In the paper [9], the authors proposed a new wolf swarm algorithm by Yang and others in 2007.The wolf swarm algorithm has been developed and improved continually, and is also used in many fields, such as: in the paper [10], the authors proposed a method for the chaotic systems. In the paper [11], the authors proposed a method for path planning. In order to solve a problem in mathematical field, the authors of paper [12] proposed a method. In the paper [13], the authors proposed a new method that adds a radius and a parameter to the old method. In order to increase the ability of hunting, the authors of paper [14] proposed a method that the action of the wolves is obeying the command of the leader wolf. In order to solve the electrical problem, Gupta and Saurabh [15] proposed a method. For the purpose of solving the global optimization problems, Jitkongchuen et al. [16] proposed a method. A new wolf pack algorithm to solve problems was proposed in paper [17] by Saurabh and Gupta. Yao et al. [18] proposed a new wolf pack algorithm which is applied to solve the uncertain bi-level knapsack problem.

Summarizing the research results of predecessors, especially the paper [19], this paper proposes a grid search wolf optimization algorithm. This paper uses this improved wolf group algorithm combined with two-dimensional OTSU algorithm. This algorithm not only can ensure the quality of image segmentation, but also can make the running time shorter.

2. Related works

2.1. OTSU

OTSU method is an important threshold method [8] which proposed by Scholar OTSU in 1979. OTSU threshold method is an automatic, nonparametric, unsupervised threshold selection method, which is based on discriminant measure criterion. When the discriminant measure criterion gets maximum, we get the best threshold. OTSU is global threshold selection method which widely used due to its high efficiency and simplicity.

One-dimensional OTSU is a simple method, which requires a grey level histogram. However, the histogram can't show obvious troughs and peaks in practical applications because of the noises and other factors. So One-Dimensional OTSU doesn't achieve good segmentation effect.

2.2. Two-dimensional OTSU

Since one-dimensional OTSU did not achieve perfect segmentation, two-dimensional OTSU method was proposed. The main method of the two-dimensional OTSU algorithm is to apply the pixel gray value and the average gray value of the pixel neighbourhood. The two-dimensional histogram containing the gray level of the pixel and its neighbourhood mean is designed, and the optimal threshold is obtained by the two-dimensional histogram to realize the automatic threshold selection.

2.3. Wolf Pack algorithm

2.3.1. Original wolf pack algorithm. Inspired by the group intelligence algorithm, the wolf group algorithm appeared. The specific process of original wolf group algorithm is: predatory wolf send several wolves to search for prey, wolves continue to search for prey by smell; then, when the prey is discovered, the wolf closes to the prey as soon as possible and summons companion through howling to lurk the prey in all directions; next, when the time is proper, the wolves catch the prey; in addition, the prey is distributed to the wolves according to their contribution, and the weak wolf will be eliminated.

2.3.2. ASGS-CWOA. In the paper [14], the authors proposed a new wolf pack algorithm which named Leader wolf algorithm. When the wolf pack prepares to get food, the wolves which have the better ability of siege will compete to be the leader wolf. The wolves of the pack except the leader wolf will obey the rules which made by the leader wolf. This system is good to the survival of the wolf pack in the harsh environment.

The COWA is proposed which originated from Leader wolf algorithm. The improvements of the COWA are as follows:

First, the chaos optimization is introduced to the algorithm. The equation of logistic map is following, which is from [20].

$$Chaos_{k+1} = \mu * Chaos_k * \left(1 - Chaos_k\right) \qquad Chaos_k \in [0,1] \tag{1}$$

Second, a new search which includes migration step size and siege step size is introduced to the algorithm. The new search step size is adaptive variable. For the step size of siege, CWOA adopts an adaptive adjustment factor α into the improved equation following.

$$step\_a_{new} = \alpha * step\_a_0 \ , \alpha = 1 - \left(\frac{t-1}{T}\right)^2 \tag{2}$$

And the latter is about the step size of siege. The equation is following.

$$step\_c_{new} = step\_c_0 * rand(0,1) * [1 - \left(\frac{t-1}{t}\right)^2] \tag{3}$$

In the paper [19], the ASGS-COWA is proposed which originated from COWA.

The new ideas of the algorithm are as follows:

The first of the innovations is ASGS. In the migration of the wolf pack, the wolf can choose any directions for its step. In the LWPS and CWOA, the wolf only chooses a direction in the H dimensions, and a point is generated. If the wolf chooses a better direction, the algorithm convergence speed will be increased. As shown in figure 1 and figure 2, there are (2*k+1) points generated in H dimensions, which reflect the neighbourhood range of the current position. The each point of the grid is got by the following equation:

$$[k, x_{ih-new}] = x_{ih} + step\_a_{new} * k(k = -K, -K+1, ..., 0, ..., K-1, K; i = 1, 2, ..., Q; h = 1, 2, ..., H) \quad (4)$$

So a point from the grid can be got by the following equation:

$$x_{i-new} = \{[k, x_{i1-new}], [k, x_{i2-new}], ..., [k, x_{iH-new}]\}(i = 1, 2, ..., Q; k = -K, -K+1, ..., 0, ..., K-1, K) \quad (5)$$

There are $(2*K+1)^H$ points produced in the grid during any migration which will be new position of the searching wolf, and the position with best fitness will be chosen as the leader wolf's position. As the step_anew become smaller, the searching grid becomes smaller and smaller. Grid search which has $(2*K+1)^H$ points is better than an isolated point in each iteration. This means is good to find the optimal threshold in the local neighbourhood which includes $(2*K+1)^H$ points.

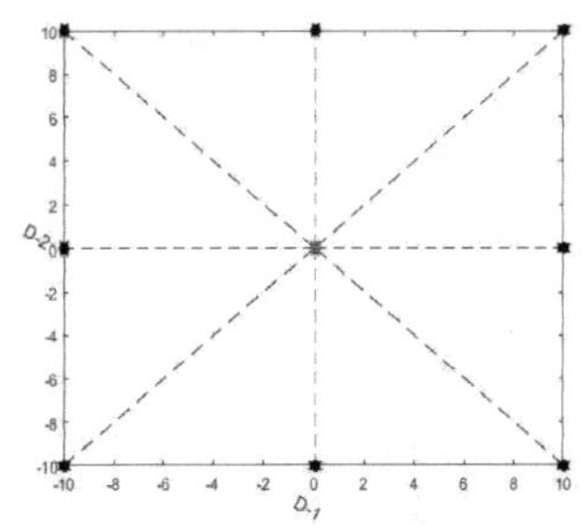

Figure 1. Locations of searching wolves (dimension=2)

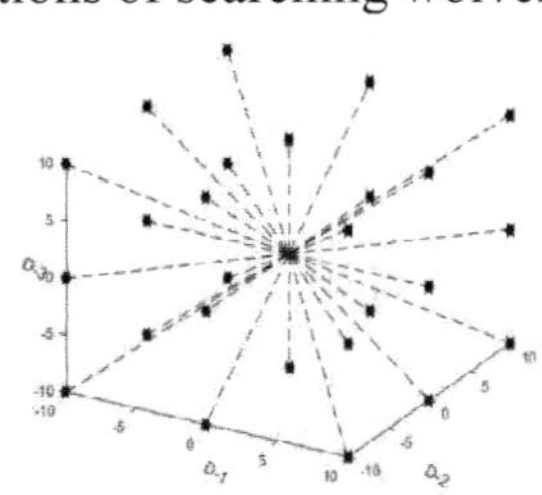

Figure 2. Locations of searching wolves (dimension=3)

Figure 1 shows the range of locations of searching wolves in ASGS-CWOA when dimension is 2 [19]; Figure 2 shows the range of locations of searching or competing wolves in ASGS-CWOA when dimension is 3, where step_a = 10 or step_c = 10 and K = 1 [19].

This strategy is also introduced into the process of siege. But the different is that the siege step size is shorter than the migration step size. In the end, the wolf with the best fitness will be chose of the neighbourhood range points.

The second of the innovations is OMR. When the leader wolf send the signal of summon, the other wolves try to close to the leader wolf as soon as possible and find the preys on the way. The OMR is about the opposite position of the current position relative to the leader wolf. Its position is calculated by the following equation:

$$X_{id-opposition} = 2 * X_{ld} - X_{id} \quad (i = 1, 2, ..., Q; d = 1, 2, ..., D) \quad (6)$$

To get the better position, we can change the current position to the opposite position based upon the fitness of opposite position is better than the fitness of the current position. If the wolves don't get the better position, the wolves except the leader wolf get the new position by the following equation.

$$X_{i-m\_d}=(X_{i\_d}+X_{1\_d})/2,$$
$$X_i=Bestfitness([X_{i\_1},X_{i\_2},...X_{i\_(D-1)},X_{i\_D}],[X_{i\_1},X_{i\_2},...X_{i\_(D-1)},X_{i-m\_D}],...[X_{i-m\_1},X_{i-m\_2},...X_{i-m\_(D-1)},X_{i-m\_D}]) \quad (7)$$
$$(i=1,2,...Q;d=1,2,...D)$$

3. OTSU based on grid search wolf pack algorithm

3.1. Improvement of traditional wolf pack optimization algorithm

In the algorithm ASGS-COWA, the Adaptive Shrinking Grid Search (ASGS) is proposed. In the process of migration and siege, the step size is getting smaller and smaller, and the step size may become decimal. This method is suitable for solving complicate problems. When we solve the problem of image segmentation, the gray level is from1 to 256, and there is no decimal case, so we don't need to use the adaptive variable step size strategy to generate the gird.

Based on the aforementioned thoughts of grid search, the GSWA is proposed. In the migration of the wolf pack, the wolf can choose any directions for its step. As shown in figure 1 and figure 2,there are (2*k+1) points generated in H dimensions, which reflect the neighbourhood range of the current position, where K represents the number of points taken in the same direction of each dimension. The each point of the grid is got by the following equation:

$$[k,x_{ih-new}]=x_{ih}+step\_a*k(k=-K,-K+1,...,0,...,K-1,K;i=1,2,...,Q;h=1,2,...,H) \quad (8)$$

So a point from the grid can be got by Eq.(5).

During the process of any migration, the position with best fitness from the grid including $(2*K+1)^H$ points will be selected as new position of the searching wolf, and at the end of the migration, the wolf with best fitness will be successful to compete to be the leader wolf of the wolf pack.

There are $(2*K+1)^H$ points generated in the grid during any migration which will be new position of the searching wolf, and the position with best fitness will be selected as the leader wolf's position. The searching grid will not be smaller because the step_a is the same. The process of migration is also used the gird search strategy.

3.2. Steps of GSWA-OTSU

The steps of the GSWA algorithm are as follows:

A. Initialization of population

Set parameters: N, D, range_min, range_max, T, step_a, step_b, step_c.

N is the total number of wolf pack, D is the dimension of solution space, [range_min ,range_max] is the ranges of solution space , T is the maximum number of iterations, step_a is the initial step size for migration, step_b is the initial step size for summons, step_c is the initial step size for siege. And the N wolves are generated by Eq.(9).

$$x_{id}=range\_\min+rand(0,1)*(range\_\max-range\_\min) \ (i=1,2,...,N;d=1,2,...,D) \quad (9)$$

Where, rand (0, 1) is a random number distributed uniformly in the interval [0, 1]; x_{max} and x_{min} are the upper and lower bounds of the search space respectively.

B. Migration

In order to show the neighbourhood space of the current position, a grid with H dimensions which has $(2*K+1)^H$ points gets by Eq.(5).In the end, the position with best fitness will be selected as the leader wolf's position.

C. Summon

When the leader wolf send the signal of summon, the other wolves try to close to the leader wolf as soon as possible and find the preys on the way. In this process, we can use the innovation OMR of the

ASGS-COWA.

D. Siege

The wolves except the leader wolf wait for opportunities to move where the prey can't see. Select the middle point of the current wolf position and the best wolf position is got by the following equation:

$$x_{i-m\_d} = (x_{i\_d} + x_{1\_d}) / 2 \tag{10}$$

Based on the middle point, the GSWA was used to generate a grid with H dimensions by Eq. (5).

Comparing the fitness of the opposite position and the fitness of the current position, we can get the better position of the current position in the grid. Comparing the fitness of the best position and the fitness of the current position, we can get the better position of the current position in the grid.

The wolf with best fitness will be chosen as the leader wolf at the end of each siege.

E. Regeneration

Following the wolf's survival rules, the food will be distributed according to the wolf's hunting ability, and the wolves with weak hunting ability will be eliminated. The same amount of wolves should be generated along with the weaker wolves has been removed.

F. Loop

The iteration won't stop until the exit condition is ok. Go to B if the exit condition is not ok. Once the loop stops, the global optimal value is found through the best fitness.

3.3. Steps of GSWA-OTSU

Convert an image to M*N points. The gray level is L. The gray level of the neighbourhood smooth image G(X,Y) is also L, and the value of each pixel i in the image and the neighbourhood average gray value j are calculated to form a binary group (i, j).

The steps of the GSWA-OTSU algorithm are described below.

A. Initialization of population

Set parameters: N, D=2, range_min=1, range_max=256, T, step_a, step_b, step_c.

N is the total number of wolf pack, D is the dimension of solution space, [range_min ,range_max] is the ranges of solution space , T is the maximum number of iterations, step_a is the initial step size for migration, step_b is the initial step size for summons, step_c is the initial step size for siege. And the N wolves are produced by Eq.(9).

B. Migration

In order to show the neighbourhood space of the current position, a grid with H dimensions which has $(2*K+1)^H$ points gets by Eq.(5). In the end, the position with best fitness will be selected as the leader wolf's position.

C. Summon

When the leader wolf send the signal of summon, the other wolves try to close to the leader wolf as soon as possible and find the preys on the way. In this process, we can use the innovation OMR of the ASGS-COWA.

D. Siege

The wolves except the leader wolf wait for opportunities to move where the prey can't see. Choose the middle point of the current wolf position and the best wolf position is got by Eq.(10).Based on the middle point, the GSWA was used to generate a grid with H dimensions by Eq.(5).

If the fitness of the best position which in the grid is better than the best wolf position, change the best wolf position to the best position in the grid. If the fitness of the best position which in the grid is better than the current position, change the current position to the best position in the grid.

The wolf with best fitness will be chosen as the leader wolf at the end of each siege.

E. Regeneration

Following the wolf's survival rules, the food will be distributed according to the wolf's hunting ability, and the wolves with weak hunting ability will be eliminated. The same amount of wolves should be generated along with the weaker wolves has been removed.

F. Loop

The iteration won't stop until the exit condition is ok. Go to B if the exit condition is not ok. Once the loop stops, the global optimal value is found through the best fitness.

G. Calculate the segmentation threshold

According to the position of the best wolf, the best segmentation threshold is got by equation:

$$Threshold = (bestwolf(i) + bestwolf(j))/2 \quad (i=1, j=2) \tag{11}$$

4. Experimental and experimental analysis

As shown below, the two pictures are the experimental objects.

Figure 3. Female picture.

Figure 4. Elephant picture.

In order to make the results of the GSWA-OTSU algorithm more convincing, one-dimensional OTSU and two-dimensional OTSU image segmentation algorithm are selected for comparison experiments under the same conditions.

(a) (b) (c) (d)

Figure 5. (a) shows the original grey female picture; (b) shows the picture which processed by one-dimensional OTSU; (c) shows the picture which processed by two-dimensional OTSU; (d) shows the picture which processed by GSWA- OTSU.

It can be seen from the comparison of various algorithms in the above figures that the two-dimensional OTSU algorithm is clearer than the one-dimensional OTSU algorithm because of the addition of the neighbourhood average grey value. The upper left and upper right corners of the figure 5(c) are more clearly contoured, and the female's hair is handled more clearly. It can be seen from the figure6(c) that the outlines of swamps and elephants are handled more clearly.

(a) (b) (c) (d)

Figure 6. (a) shows the original grey elephant picture; (b) shows the picture which processed by one-dimensional OTSU;(c) shows the picture which processed by two-dimensional OTSU; (d) shows the picture which processed by GSWA- OTSU.

Comparing the figure 5(c) and figure 5(d), figure 6(c) and figure 6(d), we can see that the GSWA-OTSU algorithm has no significant difference in image segmentation quality from the traditional two-dimensional OTSU method.

5. Results and Discussion

From table 1, we can see the running time of the one-dimensional OTSU, two-dimensional OTSU and GSWA-OTSU. The unit is seconds. Comparing the running efficiency of the algorithm, the average time of the one-dimensional OTSU segmentation algorithm is about 0.0002-0.0009 seconds, the average running time of the two-dimensional OTSU algorithm is about 43.1-44.1 seconds, and the algorithm of this paper is about 13.8-18.7 seconds. The GSWA-OTSU algorithm takes about 32% of the time of the traditional two-dimensional OTSU method, and the running time is greatly reduced. The image segmentation quality of the two two-dimensional OTSU algorithms has no obvious difference. When the female image and the elephant image are segmented, the outlines of the female and the elephant are better extracted. Although the one-dimensional OTSU algorithm takes less time, its image segmentation quality is poor. Under the premise of efficiency and quality, the GSWA-OTSU algorithm proposed in this paper is an ideal segmentation method.

Table 1. Running time of the three methods

method / time	One-dimensional OTSU	Two-dimensional OTSU	GSWA-OTSU
Female image	0.00092	44.1363	13.791395
Elephant image	0.00021	43.0967	18.73625

Although GSWA-OTSU algorithm performs well in terms of time, this algorithm is not good enough in some aspects. For example， the segmentation quality still needs to be improved. If the segmentation edge is clearer, it is good for the follow-up work.

6. Conclusion

This paper proposes an improved two-dimensional OTSU segmentation method, which greatly improves the segmentation speed compared to the traditional two-dimensional OTSU segmentation method. It's core procedure are initialization the wolf group in the image pixel area, migration, summon, capture the prey, hand out the food and update the wolf pack. The optimal two-dimensional segmentation threshold of the grey image is finally got by the fitness function.

Experiments show that compared with the one-dimensional OTSU algorithm and the traditional two-dimensional OTSU algorithm, the GSWA-OTSU algorithm takes the image segmentation quality and segmentation speed into account. The outlines of the segmented images are more clearly. This segmentation method lays a good foundation for the follow-up work. And, the proposed GSWA-OTSU segmentation method is an effective image segmentation method.

Acknowledgments
This study is supported by the National Key Research and Development Program of China (Grant No. 2016YFB0700502, 2016YFB1001404).Thanks to Professor QIAN for his careful guidance and kind care. Thanks to the classmates for their encouragement and support.

References

[1] Dewan M A A, Ahmad M O and Swamy M N S 2011 Tracking biological cells in time-lapse microscopy: An adaptive technique combining motion and topological features *IEEE Trans. Biomed* **58** p 1637-1647
[2] Li Q, Ni X and Liu G 2007 Ceramic image processing using the second curvelet transform and watershed algorithm *IEEE International Conf. on Robotics and Biomimetics* p 2037-2042
[3] Ullah A, Liu G, Luan J, Li W, Rahman M U and Ali M 2014 Three-dimensional visualization and quantitative characterization of grains in polycrystalline iron *Mater. Charact* **91** p 65-75
[4] Ali M, Le H S, Khan M and Tung N T 2017 Segmentation of dental X-ray images in medical imaging using neutrosophic orthogonal matrices *Expert Syst. Appl.*
[5] Boyuan M, Xiaojuan B, Huang H, Chen Y, Liu W and Zhi Y 2018. Deep learning-based image segmentation for al-la alloy microscopic images *Symmetry* **10** p 107
[6] Comer M, Bouman C A, Graef M D and Simmons J P 2011 Bayesian methods for image segmentation *JOM* **63** p 55-57
[7] Simmons J P, Bartha B, Graef M D and Comer M 2008 Development of bayesian segmentation techniques for automated segmentation of titanium alloy images *Microsc. Microanal.* **14** p 602-603
[8] Vala M H J, Baxi A 2013 A review on Otsu image segmentation algorithm *Int. J. Adv. Res. Comput. Eng. Technol.* **2** p 387-389
[9] Yang C, Tu X & Chen J 2007 Algorithm of Marriage in Honey Bees Optimization Based on the Wolf Pack Search *International Conference on Intelligent Pervasive Computing* vol. 871 p 462-467
[10] Li H, Wu H 2016 An oppositional wolf pack algorithm for parameter identification of the chaotic systems: Optik *International Journal for Light and Electron Optics* **127** (20): 9853-9864
[11] Chen Y B, Mei Y S, Yu J Q, Su X L and Xu N 2017 Three-dimensional unmanned aerial vehicle path planning using modified wolf pack search algorithm *Neurocomputing*
[12] Yang N, Guo D L 2016 Solving polynomial equation roots based on wolves algorithm *Science & Technology Vision*
[13] Hui X B, Guo Q, Ping-Ping W U and Zhao Y 2017 An improved wolf pack algorithm *Control & Decision*
[14] Qiang Z, Zhou Y Q 2013 Wolf colony search algorithm based on leader strategy *Application Research of Computers* **30** (9): 2629-2632
[15] Gupta S, Saurabh K 2017 Modified artificial wolf pack method for maximum power point tracking under partial shading condition *Proc. Int. Conf. Power Embedded Drive Control* p 60-65
[16] Jitkongchuen D, Sukpongthai W and Thammano A 2017 Weighted distance grey wolf optimization with immigration operation for global optimization problems *Proc. IEEE/ACIS Int. Conf. Softw. Eng., Artif. Intell., Netw. Parallel/Distrib. Comput.* p 5-9
[17] Saurabh K, Gupta S 2017 Modified artificial wolf pack optimization for optimal power flow, in *Proc. Int. Conf. Circuit, Power Comput. Technol.* p 1-6
[18] Yao Q, Junjie X, Ying W, Xiangfei M and Maolong L 2017 Binary smart wolf pack algorithm for uncertain bilevel knapsack problem *Proc. 2nd IEEE Int. Conf. Comput. Intell. Appl. (ICCIA)*, Beijing, China p 133-140
[19] Wang D, Qian X, Liu K and Ban X 2019 An adaptive shrinking grid search chaos wolf optimization algorithm with adaptive standard-deviation updating amount (to be published) doi: 10.13140/ RG.2.2.30190.05440
[20] Qinlong G U, Yao M and Zhang R 2003 Design of pid controller based on mutative scale chaos optimization algorithm *Basic Automation*

Journal of Physics: Conference Series **1419** (2019) 012039 doi:10.1088/1742-6596/1419/1/012039

Low-carbon optimization of spatial pattern in Shenfu New District based on genetic algorithm

Peiwen Jia[1], Shilei Fu[1,2], Zhenxing Li[1] and Huan He[1]

[1]College of Architecture and Planning, Shenyang Architecture University, Shenyang, Liaoning Province, China

[2]E-mail: fsl116@foxmail.com

Abstract. The realization of low-carbon development is the key to solving the global climate change problem. China's total carbon emissions share in the world is increasing year by year. Taking the road of low-carbon city development is the inevitable trend of China's city development in the future. Increasing the concentration of urban land space can improve energy utilization and achieve the optimal goal of energy conservation and emission reduction. Genetic algorithm is a method to search the optimal solution by simulating the natural evolution process, providing a scientific solution for low-carbon city planning. This paper starts with the spatial arrangement of land use, minimizes people's spatial travel distance through the global optimization ability of genetic algorithm. It summarizes the low-carbon spatial organization characteristics of Shenyang-Fushun New City, explores the analysis method of urban spatial low-carbon pattern, puts forward the optimization strategies of urban spatial structure, urban transportation network, green space system and low-carbon industrial layout, adjusts the development trend of urban areas, promotes low-carbon living and production behaviours, and provides theoretical basis and practical experience for the planning of low-carbon cities.

1. Introduction

1.1. Research background

China's development of low-carbon eco-city is an inevitable choice to promote new urbanization. The Party's "19th Congress" proposed that China should promote green development, establish and perfect a green and low-carbon circular economic system, and advocate a simple and moderate, green and low-carbon lifestyle [1]. Traditional planning methods mostly use qualitative methods to analyze and plan or the models built are mostly single-level optimization problems, which make it difficult to find the most ideal low-carbon solution. Genetic algorithm is a method to search for the optimal solution by simulating the natural evolution process. It, with better global optimization ability, can automatically obtain and guide the optimized search space, adaptively adjust the search direction, and can provide scientific solutions for low-carbon city planning [2].

The optimization of urban low-carbon spatial pattern includes both quantity distribution and spatial distribution. In the pattern optimization, the key to realize a low-carbon city is to study the effect of the spatial pattern of land use on reducing urban carbon emissions and environmental pressure, and how to combine with economic benefits and other factors to jointly establish a multi-objective optimal allocation model of land use under certain constraints .Under the guidance of low carbon as the goal, this paper introduces genetic optimization algorithm to coordinate multiple goals between urban land use form and low carbon orientation, in order to form an ideal low carbon urban space form.

1.2. Overview of research area

Shenyang-Fushun connection zone is located in the eastern part of Shenyang city and the western part of Fushun city in Liao Ning province, with an area of 605.34 square kilometers. Shenyang-Fushun New District is located in the middle of Shenyang-Fushun connection zone, with obvious advantages in location conditions (Fig. 1). Shenfu New District is adjacent to Shenyang's ecological conservation zone in the west and Fushun's key development zone in the east. Fushun City has built a large number of industrial parks in the new town. There are industries with high pollution and energy consumption, which conflict with Shenyang's development orientation. Ecological protection and economic development are the difficult problems facing the new district. It is urgent to promote the green and low-carbon development of Shenfu New Town and optimize the spatial structure of the new town in an all-round way.

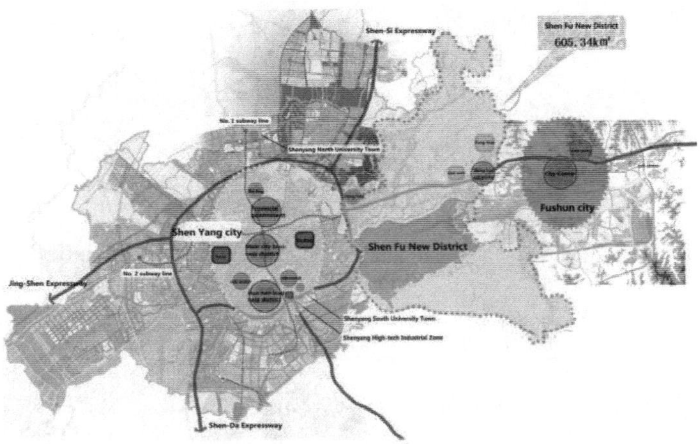

Figure 1. Map of Shenfu New District.

2. Application of Genetic Algorithm in Spatial Optimization of Shenfu New District

2.1. Model Optimization Ideas and Data Acquisition

2.1.1. Optimization Ideas. According to the low-carbon planning objectives, land use status and socio-economic basis of Shenfu New District, a low-carbon spatial optimization model is constructed, the constraints in the model are defined, and decision variables are selected according to the objective function to understand the phenotype of each individual and the scope of solution. According to the low-carbon space optimization model, the target system and the constraint system are determined to determine the fitness value, and the decision-making genetic algorithm is used to optimize the allocation operation to form the low-carbon space optimization model of Shenfu New District [3] (Fig. 2).

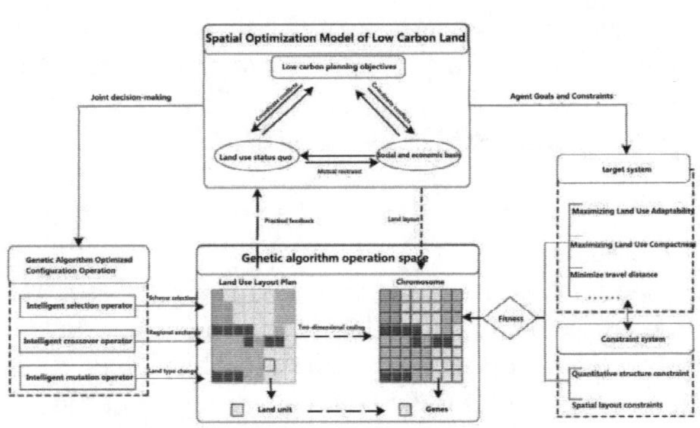

Figure 2. Genetic Algorithm Model Frame Diagram.

2.1.2. Data Acquisition. Based on the remote sensing images of Shenfu New District, the geographic spatial information units in the remote sensing images are identified and rasterized. Each grid is taken as the basic unit in the study, and the study area is divided into 5177 grids with the same specifications by using a grid of 200m×200m, and this is taken as the basic data for empirical analysis. In the optimization process, these small squares are used as optimization units to establish layers and various land numbers. According to the output of current land types, the research obtains the database of current unit land use (Fig. 3) [4].

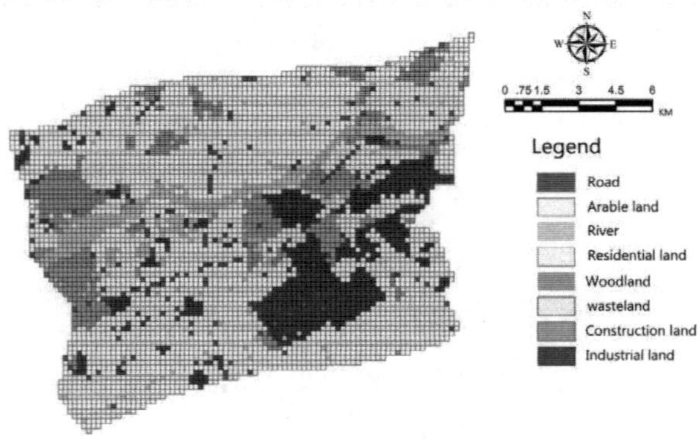

Figure 3. The current land grid map of the study area.

2.2. Construction of Low Carbon Space Optimization Model

2.2.1. Model selection. In this paper, the multi-objective optimization model is selected. The development of many factors in land use is not only restricted by the subjective factors of decision makers, but also influenced by various objective factors, which makes the land use structure more scientific and reasonable. All variables should be comprehensively analyzed, and the subjective and objective factors that may affect the overall objective should be analyzed one by one, so as to improve the overall optimization. The optimization mathematical model can be expressed by formula (1) [5]:

$$\begin{cases} opt\ Z = \alpha f(x_1) * \beta f(x_2) * \gamma f(x_3) * \delta f(x_4) \\ s.t\ \ g(x) \geq B \\ x \geq 0 \end{cases} \tag{1}$$

In the formula: Opt-Fitness z value; f(x) -objective function; β -Weight of objective function 2; γ-Weight of objective function 3; g(x)-Constraint function; x-decision variable.

2.2.2. Construction of Objective Function. The objective of this paper is to construct the objective function of low-carbon land use structure optimization in Shenfu New District by integrating low carbon, ecology and current situation. For the sake of simplification, we actually set 5 types of land in the model, of which road land is a linear element and cannot be reflected in grid blocks, so it is divided equally into other land units. Leisure land includes green land and square land and land for public management and public service facilities. The working land includes three types of land: industrial land, storage and logistics land and public facilities land. The corresponding relations between codes and land types are: 1 - residential land, 2 - commercial land, 3 - working land, 4 - leisure land, 5 - unused land, and 0-non-calculation object.

In this paper, the targets shown in equations (2) to (5) are selected as optimization targets.

The objective of maximizing land use suitability is to make the spatial arrangement of land more suitable for the current geological, topographic, hydrological and other natural conditions through the extraction of land use suitability, to find out the suitable arrangement areas for each type of land, and to standardize the rationality of optimal arrangement.

$$maxZ = \sum f_{(x1)} + \sum f_{(x2)} + \sum f_{(x3)} + \sum f_{(x4)} \quad (2)$$

In the formula: maxZ is the land use suitability value; $f_{(x)}$ is the land suitability value corresponding to the decision variables.

Maximizing land use compactness target is to increase the spatial concentration of industries, realize labor market sharing, improve energy utilization rate, and achieve the optimal goal of energy conservation and emission reduction. This paper sets the goal of maximum land compactness to reflect the utility of land spatial agglomeration in low-carbon cities.

$$\max \sum_{i=1}^{N} \sum_{j=1}^{M} \sum_{k=1}^{K} \frac{\sum_{i',j' \in Nei} X_{i'j'k} \exp\left(-{dist_{ij,i'j'}}/{\lambda}\right)}{\sum_{dist_{ij,i'j'}} \exp\left(-{dist_{ij,i'j'}}/{\lambda}\right)} X_{ijk} \quad (3)$$

Nei represents the neighborhood of the cell (i, j); Dist i j, i'j' represents Manhattan distance of two units (i, j) and (i', j') of land use type k; λ is the adjustment coefficient.

Maximizing the land mix target is that the land mix can be reflected by travel distance, which reflects the low-carbon degree of the city, as well as by the distance between housing and the nearest work place, shopping center and leisure center. Exponential distance attenuation function is adopted in the model to express its spatial attraction to position.

$$E_{distance} = c_1 A_1 e^{-B_1 D_{work}} + c_2 A_2 e^{-B_2 D_{shop}} + c_3 A_3 e^{-B_3 D_{leisure}} \quad (4)$$

$E_{distance}$ is low-carbon trip distance evaluation. c1, c2 and c3 are the weight coefficients of each distance influence factor, c1+c2+c3 = 1, C1 > C2 > C3. A1, A2 and A3 are the intensity coefficients of spatial influence. B1, B2 and B3 are the attenuation coefficients of spatial influence, and Dwork is the distance to the work center. Dshop is the distance to the shopping center. Middleware is the distance to the Dleisure center.

Minimize the goal of land conversion cost: spatial optimization of urban land will result in a change in the current use of some units of urban land. The goal of establishing the cost function of urban land type conversion is to minimize the conversion cost as much as possible.

$$\min \sum_{i=1}^{N} \sum_{j=1}^{M} \sum_{k=1}^{K} C_{ijk} X_{ijk} \quad (5)$$

C_{ijk} indicates the cost of arranging K types of land use at the locations of units (i,j); X_{ijk} is a variable. If the land use type of units (i,j), (i',j') is K, its value is 1, otherwise it is 0; K represents the land use type of unit (i,j).

2.2.3. Construction of Constraints. In this paper, economic benefits and ecological benefits are placed in the system of constructing objective function, and low-carbon benefits are taken as the main constraint conditions. Respectively from the land quantity, the population quantity, the present situation condition and unifies the Shenfu New District land plan and the actual situation and so on 11 aspects carries on the restraint to the Shenfu New District land.

The number of each type of land use must meet a predefined structural ratio.

$$\sum_{i=1}^{N} \sum_{j=1}^{M} X_{ijk} = X_k, \forall k \quad (6)$$

Every space unit must have a land use.

$$\sum_{k=1}^{K} X_{ijk} = 1, \forall ij \quad (7)$$

One land unit is at least adjacent to another.

$$S_{L_{ij,i'j'}} \geq 1, \forall (ij, i'j') \quad (8)$$

Spatial adjacency constraints of the same type of land use units. It is better to form a complete plot of land for units with the same land use. This is a planning need, so that each type of land will not be fragmented in spatial distribution and will not cut off the close interrelation within a certain type of land. If the distance of 2 units (i1j1) and (i2j2) of land use type K is expressed.

$$m_{i_1j_1k,\, i_2j_2k} = 1 \quad (9)$$

Distance constraint. It is necessary to keep a certain distance between certain land use types. If the Euclidean distance between the unit (i_1 j_1) with land use type k1 and the unit (I_2J_2) with land use type k2 is required to be greater than D, the distance constraint can be expressed as follows:

$$d_{i_1j_1k,\, i_2j_2k} > D \quad (10)$$

Direction constraint. For example, factories with serious air pollution should choose the downwind direction in densely populated cities. If B_{ij} represents a set of units in a plot unit (ij) forbidden area:

$$i \notin B_{ij} \quad \text{and} \quad j \notin B_{ij} \quad (11)$$

The total area of the land is about 10,000 square meters. TL is the total land area of 42.55 (hm^2).

$$\sum_{i=1}^{8} x_i = TL \quad (12)$$

Minimum area constraint of residential land: $1 \geq TBF$, where TBF is the land area 14 (hm^2) calculated according to the per capita residential area.

Minimum area constraint of management service facility land: $2 \geq TLt$, where TLt is 2.75 (hm^2) calculated according to the area of service facility per capita.

Constraint on green square area: $4 \leq UJL$, where UJL is the land area of 5(hm^2) calculated according to the per capita green square area.

Land conversion constraints: set constraints in basic farmland protection areas, rivers and lakes ecological protection areas and other areas to prohibit inappropriate land use conversion. Ω indicates a space control area; P(Xij) represents the conversion probability of plots (i, j).

$$P(X_{ij}) = 0,\ \text{if}(i, j\epsilon\Omega) \quad (13)$$

2.3. Operation Flow of Genetic Algorithm

In the case of selecting the initial population, the target solution is gradually obtained by continuously selecting, crossing and mutating individuals in the population on the premise of a certain optimization target [6] (Fig. 4).

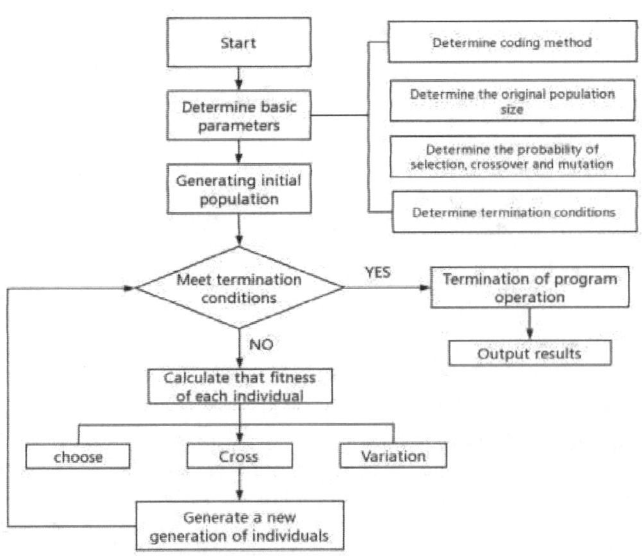

Figure 4. Genetic algorithm flow chart.

2.3.1. Initial Population Setting. The optimal layout of land space is mostly based on the current utilization, and there are many hard restrictions. For example, the current land use for water areas and water conservancy facilities will not be easily changed in general. Therefore, this paper must satisfy certain prior knowledge when generating the initial group. It is used in the optimization configuration, and it will not be considered in the unit optimization process where the purpose is clear and changes are not allowed, so they will not be coded, thus reducing the scale of the problem and improving the efficiency of the algorithm.

2.3.2. Determination of model fitness function. When constructing the fitness function, a dynamic penalty function method is introduced (i.e. the penalty factor changes continuously in the process of convergence). Its essence is:

For equality constraint cτ (x) = 0, constructor

$$\varphi(c_\tau(x)) = \begin{cases} 0,, \quad \text{if } -\varepsilon \le c_\tau(x) \le \varepsilon \\ \tau = 1,, 2,, \ldots,, s \\ |c_\tau(x)|,, \quad \text{else} \end{cases} \tag{14}$$

For inequality constraint hω(x) ≤ 0, construct function

$$\Psi(h_\omega(x)) = \begin{cases} 0,, \quad \text{if } h_\omega(x) \le -\varepsilon \\ \omega = s+1,, s+2,, \ldots,, M \\ |h_\omega(x)|,, \quad \text{else} \end{cases} \tag{15}$$

Where: S is the number of equations in the constraint condition; M is the number of constraints; ε is Archimedes infinitesimal, that is, a number less than any positive number and greater than zero. Then the fitness function of genetic algorithm for minimizing constrained optimization problem is

$$\max f(x, \sigma_\theta) = -f_0(x) - \sigma_\theta\left[\sum_{\tau=1}^{s} \varphi^2(c_\tau(x)) + \sum_{\omega=s+1}^{M} \Psi^2(h_\omega(x))\right] \tag{16}$$

Where: f0(x) is the original objective function; for the penalty factor, =1/ = C, c ∈ (0,1), as evolution progresses, the penalty factor gradually increases, making the solution group more feasible.

2.3.3. Output of Optimization Results. The GA optimization module has an initial population size of 100, an optimized individual of 10, a population size of 600,000, a crossover probability of 0.75, a mutation probability of 0.1, and an evolution algebra of 1,000 generations. In this paper, Visual Studio software is selected as the platform to write the spatial configuration module of genetic algorithm. During the process of program operation, the evolution of the program gradually slows down and tends to be stable as time progresses. In the optimization module, the data obtained from the final results of optimization are re-imported into Arc Map for spatial display.

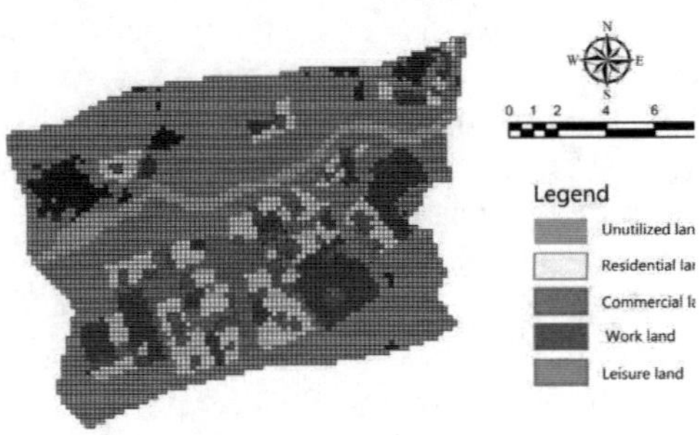

Figure 5. Simulation optimization scheme.

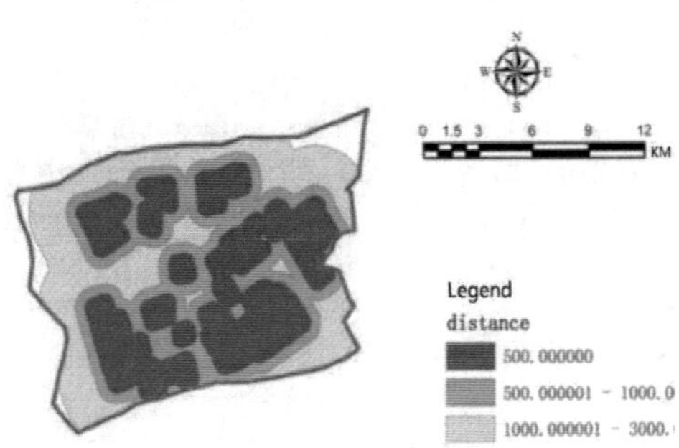

Figure 6. Scope of work site services.

The simulation optimization in this paper has output a number of results, and the most representative scheme among them has been selected for in-depth study (Fig. 5). The optimization results show that within the working travel radius of 500 meters, the construction land basically covers most areas, the 1000-meter range circle covers all areas except some leisure land, and the 3000-meter working travel range circle covers all planned land as shown in Fig. 6. The 500-meter travel radius of commercial land covers most of the built-up area, and the 1,000-meter range circle realizes the full coverage of residential land, for example, the 3,000-meter range circle realizes the full coverage of construction land, for example (Fig. 7); the 500-meter travel radius of leisure land covers most of the built-up area, and the 1,000-meter range covers all the construction land as shown in (Fig. 8). Therefore, the optimization results of the plan are in line with the low-carbon plan results expected in the previous period.

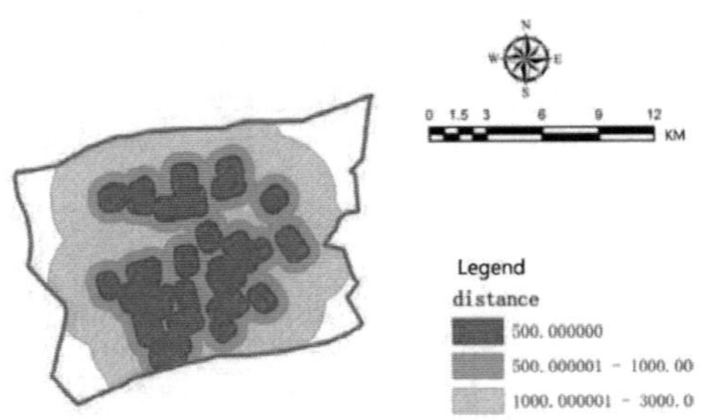

Figure 7. Scope of commercial land services.

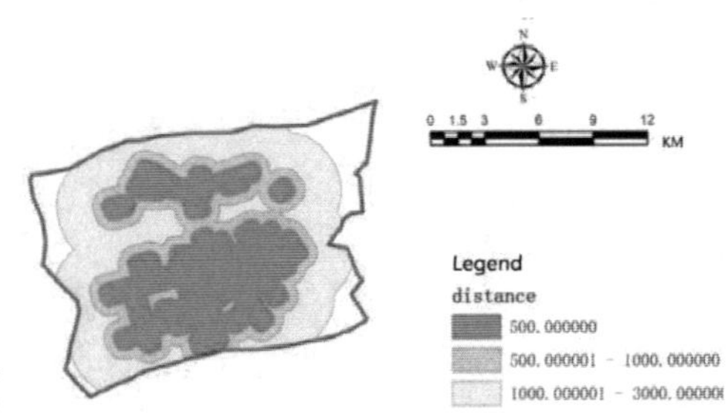

Figure 8. Scope of leisure land services.

3. Low carbon spatial pattern optimization strategy

Based on the low-carbon land use zoning method, the optimized low-carbon land use structure in Shenfu New District is spatially distributed to achieve the goal of rational allocation of land resources and full realization of low-carbon utilization. The spatial layout is optimized and analyzed from four aspects: low-carbon land use spatial structure, low-carbon transportation system, low-carbon green space system and low-carbon industrial system [7].

3.1. Low-carbon Optimization of Urban Land Use Structure

In the planning scheme, centralized industrial layout is adopted, and the original industrial construction land is fully reserved in the optimization, thus reducing the planned land area of the urban built-up area and adjusting the planned population capacity (Fig. 9).In this plan, some residential land and commercial land in the north of Hunhe River will be adjusted to forest land or other land. There is a large area of wetland landscape in the south of Hunhe River, which has been preserved and renovated in the planning, forming a large area of ecological wetland. The industrial land of the whole land has retained the centralized layout mode, but the industries of different natures have been distributed to balance the travel distance of residents.The commercial area has been distributed in a scattered way [8], and the green space has been arranged to form a network layout mode. Broaden the ecological corridors on both sides of Hunhe River, together with the two ecological corridors in the north-south direction, thus strengthening the links between the north and south sides of Hunhe River.

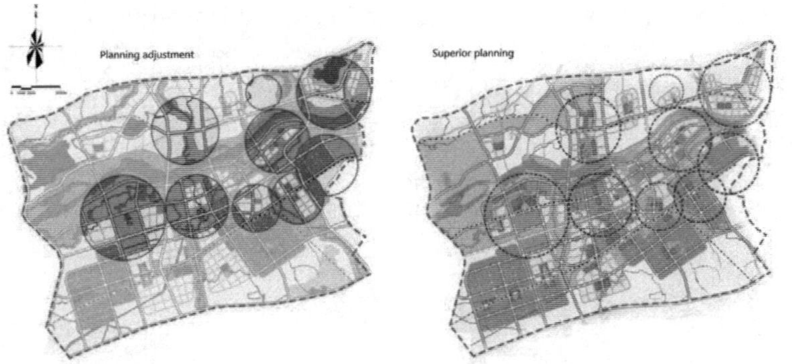

Figure 9. Adjustment of Low Carbon Planning Plan.

3.2. Low-carbon Optimization of Urban Transportation System

Based on people's need for green travel, the plan is guided by the "B+R" system. By arranging bicycle parking facilities at bus stops and setting up a comfortable and convenient greenway network, people are encouraged to connect buses by riding or walking to reach their final destination and improve the coverage of the bus network [9]. In this paper, aiming at the green commuting preference and exercise travel needs of specific citizens, slow-moving networks such as riverside greenways, urban greenways and country greenways are planned and set up to meet the personalized travel needs of citizens(Fig. 10).

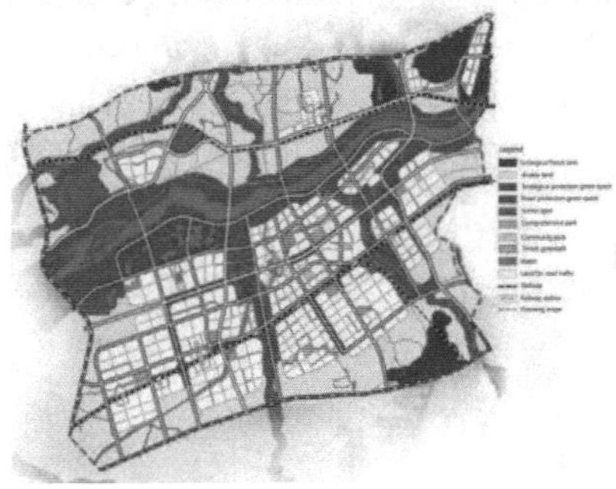

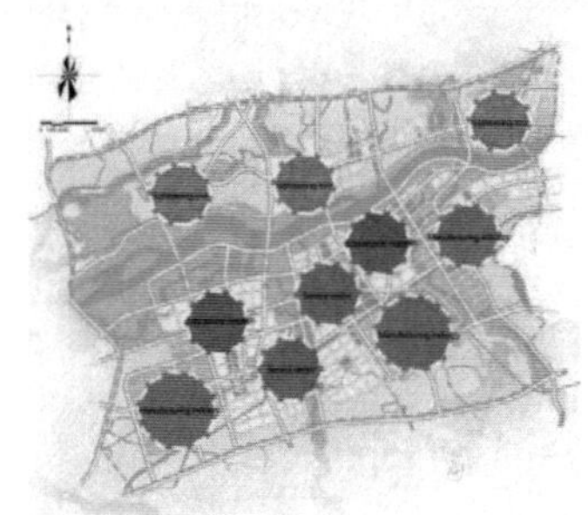

Figure 10. Low carbon greenway planning.

Figure 11. Green space system plan.

Figure 12. Industrial zoning layout.

3.3. Low-carbon Optimization of Urban Green Space System

This paper adopts a spatial layout mode composed of three low-carbon layout modes of "oxygen source green space", "near source green space" and "carbon source green space" to form a low-carbon urban layout. The spatial layout mode of "large dispersion and small aggregation" will produce a spatial dispersion arrangement of carbon sources with large carbon emission, with a large area of oxygen source green space distributed on the periphery, forming a large-scale natural carbon sink absorption barrier around the city to prevent carbon dioxide from continuing to diffuse outward.

However, the local space adopts the layout mode of combining three levels of carbon source space, with near-source carbon sink greenbelts arranged between carbon sources to absorb carbon emissions from the local space nearby, which is also in line with the development direction of local "compact" cities [10] (Fig. 11).

3.4. Low-carbon Optimization of Urban Industrial System

Based on the industrial development structure of the connecting belt in Shenfu New District, this paper designs the spatial layout of the low-carbon industry in the plan. The industries in the plan are centrally arranged in the south of Hunhe River and distributed in the north of Hunhe River. The north area is mainly divided into three regions, and the development of eco-agriculture, forestry and tourism experience industry is considered; the area south of Hunhe River is mainly divided into seven areas.

In order to reduce the impact on living, low-carbon scientific research industry and low-carbon service industry are arranged close to the urban area. The eastern area is updated and transformed into a high-tech industrial park. The southern area is transformed from the original status quo into a low-carbon equipment processing and manufacturing industrial park. A low-carbon automobile logistics industrial park is added to the western area.The adjustment of the whole industry has basically satisfied the residents' travel proximity and reached the low-carbon arrangement to guide residents' travel [11] (Fig. 12).

4. Conclusion

Low-carbon city is the inevitable choice to promote new urbanization. The spatial structure of land use and the rational layout of space are the foundation to realize the construction of low-carbon city in Shenfu New District. In this paper, genetic algorithm combined with spatial information technology is introduced to build a spatial layout model of urban land based on spatial travel distance, which guides urban residents to choose low-carbon travel modes spontaneously, and conducts low-carbon target-oriented simulation optimization on land use in Shenfu New District. In this paper, a multi-objective optimization model based on genetic algorithm is established to optimize and analyze the spatial layout of land use in Shenfu New District. The low-carbon spatial layout results obtained are used to guide the adjustment of the overall planning of Shenfu New Distric.

Acknowledgments

First of all, I would like to thank my graduate advisor Professor Shilei Fu. In the process of writing my thesis, Mr. Fu promptly gave careful guidance to the difficulties and doubts I encountered and put forward many useful suggestions for improvement.

This paper is based on the National Natural Science Foundation presided by Professor Tiemao Shi (51878418) and thanks to the data support provided by Professor Tiemao Shi in the process of writing my paper. At the same time, I would also like to thank senior Zhenxing Li and Huan He who provided a lot of help in the process of writing this dissertation. In addition, I would also like to thank the authors in the references for their research articles, which have enabled me to have a clear idea of the research topic.

References

[1] Meirong S U 2012 Hot thoughts on China's low-carbon cities: current situation, problems and trends *China population resources and environment* **3** p 48-56

[2] Zhuo Yu and Zhihua Wu 2008 Research on urban spatial growth model based on Genetic *City planning* **5** p 83-88

[3] Fei Chen, Dajian Zhu 2009 Connotation, Model and Target Strategy Determination of Low-carbon City Research *Journal of Urban Planning* **4** p 7-13

[4] LIAO G C 2012 Intergrated isolation niche immune genetic algorithm for solving bid-based dynamic economic dispatch *International Journal of Electrical Power & Engery System* **42** p 264-275

[5] Lili Zhang, Feng Zhang 2009 Complex constraint multi-objective optimization genetic algorithm based on population classification *Software guide* **12** p 38-42

[6] Yang Zhou,Chang Gong and Pingping Xu 2017 Cloud computing workflow scheduling method based on dynamic target Genetic Algorithm *Natural science journal of Xiangtan University* **39** p 123-6

[7] Haixiao Pan, Yi Tang and Jingyu Wu 2008 China's "Low Carbon City" Spatial Planning Strategy *Journal of Urban Planning* **6** p 57-64

[8] Jiansheng Wu, Na Xu and Xiwen Zhang 2016 Evaluation and spatial pattern analysis of low-carbon cities in China *Progress in Geographical Science* **35** p 204-13

[9] Jianqiang Hu, Jiaxiong Xu, Xianglin Jiang and Yaoting Zhu 2017 Research on Evaluation of Green Transportation Development Based on PSR Model *Traffic Energy Saving and Environmental Protection* **13** p 1-5
[10] Fangzheng Li, Xiong Li 2014 Research on Development Strategy of Low-carbon City Based on Greenway Network Construction *Journal of Theoretical Studies* **9** p 108-11
[11] Jian Li, Zhenzhen Hao 2014 Research on Low-carbon City Construction Based on Industrial Impact and Spatial Reconstruction *China population resources and environment* **24** p 65-72

Journal of Physics: Conference Series **1419** (2019) 012040 doi:10.1088/1742-6596/1419/1/012040

The coordinated development between the internal audit guidance by the government audit departments and internal audit units: applying the principal component analysis and Monte-Carlo simulation

Hong Yu Guo[1, 4], Yi Cen Ding[2], Qian Chang[3], Chang Ping Zhu[4], Ying Zhao[5], Qing Zhang[6]

[1]Tianjin University of Finance and Economics, 300222 Tianjin, China
[2]Audit Bureau of Tianjin Hexi District, 300060 Tianjin, China
[3]Tianjin Municipal Water Affairs Bureau, Tianjin,China
[4]Tianjin Audit Bureau, 300200 Tianjin, China
[5]Tianjin Foreign Studies University, 300204 Tianjin, China
[6]Tianjin University of Science and Technology, 300222 Tianjin, China

E-mail: guohongyu@foxmail.com, yicending@163.com

Abstract. In the new era, reform of the audit management of Official Audit Organs need systems thinking. This paper focus on the guidance and supervision of internal audit， which is the important content of the audit management system. Through the analysis of the present situation from various perspectives, aiming at the outstanding problems that still need to be paid attention to under the new situation. It explores a feasible way to coordinate the development of internal audit guidance and internal audit institutions, and tries to exemplify the application section on the basis of empirical analysis. To quantify the relevance between internal audit guidance and internal audit institutions through scientific method, introducing Principal Component Analysis and Monte Carlo Method, also provides some reference ideas for practical application.

1. Introduction

The 19th National Congress of Communist Party of China has proposed the deployment of reforming the audit management system. Guiding and supervising the internal audit unit according to the law is the important responsibilities of our government audit departments and paying high attention to the guidance on the internal audit is the practical measures to implement the spirit of the 19th National Congress of Communist Party of China. After "The Regulation on the Internal Audit Practice Issued by the National Audit Office of the People's Republic of China" in 2018, the National Audit Office has released "the Opinions On Strengthening the Professional Guidance and Supervision on the Internal Auditing Work" recently, which highlights the importance of the guidance on the internal audit and indicates at the same time that there are still some underlying problems between the internal audit guidance and internal audit unit.

2. The coordination and linkage between the government audit and internal audit

The coordination and linkage between the government audit and internal audit are the basis to achieve a full coverage of audit supervision. The Third Session of the 19th CPC Central Committee in 2018 made comprehensive deployment of deepening the reform in party and government institutions. Seizing the opportunity of deepening the reform and making good use of bonus from reforms to find a solution for the coordinated development of internal audit unit and internal audit guidance on it; preventing the systematic risks with the help of internal audit guidance and supervision and levering the transformation of the internal audit in to the endogenous power to safeguard the high-quality development are the innovative development directions in the new ear and under the situation.

3. The concept and method are to be updated

It's worth noting that some internal audit units in the internal audit guidance range of government audit departments such as grassroots administrative organs and public institutions still follow the tradition to focus on the internally auditing the compliance of financial affairs because of their own special attributes like small scale and single layer. Their business is single and comparatively limited comparing with the rapidly developing market economy and they are lack of communication channels to the new methods and motivation to face the new situation. All the above mentioned are becoming gradually the short slabs of the innovative development of the internal audit in our country.

3.1. Law and regulations construction

Legislative guarantee is the priority among priorities, which means to formulate the legal and regulatory framework with complete levels and in accordance with the Socialism with Chinese Characteristics for the internal audit guidance and to build a legal environment to promote the development of internal audit. The matters which need adjusting through legal means include: organizational structure, rights and duties of internal audit guidance, application of the internal audit achievements, the use of social auditing efficacy, purchase of social auditing service to implement the internal audit and so on.

3.2. The establishment of a dynamic data base system

Setting up the dedicated data base for internal audit and paying high attention to the completeness and integrity of basic data could provide data support for the application of internal audit achievements and macro policy making. The dynamic data base platform could be used to promote the internal audit guidance and the auditing method update of the internal audit unit through the adaptation to the development tendency of big data and cloud technology. Additionally, the disclosure measures of the dynamic information about the internal audit need to be formulated to improve the information transparency, eliminate in steps the restriction on the internal audit caused by the asymmetric information and promote the result and efficacy of internal audit.

3.3. The construction of a relevance measure mechanism

The problems about relevance have not yet gained enough attention. However, the quantization of relevance is an avoidable step not only for the internal audit guidance but also for the scientific transformation of public governance. The relevance and interaction between the audit measuring departments' guidance and internal audit unit's work need to be clarified. The quantized indexes could be used to define clearly the fields that the internal audit of the auditing bodies should regulate and could regulate. The guiding policy should be formulated prudently. The influenced range should be predicted in advance and the policy should be adjusted according to the feedback information. What's more, the process of measuring the factors consisting of the relevance and discriminating and selection results of influencing factors could be further used in the construction of assessment system for the quality of internal audit units' work, which will form the assessment index spontaneously.

4. Explore the systematic development planning under the scientific methodology

Taking the scientific method as the guidance and strengthening the supporting system to ensure the sustainable and long-term coordinated development of internal audit guidance and internal audit unit.

Journal of Physics: Conference Series **1419** (2019) 012040 doi:10.1088/1742-6596/1419/1/012040

To expound the path for a scientific and coordinated development of the relevance between internal audit guidance and internal audit unit, the following ideas are proposed based on the living examples:

- Clarify the internal audit guidance boundary of the government audit departments and influence factors of internal audit unit with the application of the statistical and experimental simulation methods.
- Along with advancement of the internal audit guidance in the new mode and in the new era, more and more data will be added, so we must be able to use the factor analysis to discriminate the influence factors and explore further the internal relationships and structure of each part.
- The interactional model of the internal audit guidance could be set up through regression analysis and it could be verified and modified via real-time data. Finally it could be used for prediction.
- Once the model is matured, analytic hierarchy process could be used to generate an index and weight system which could be used for assessing the internal audit units. In addition, from the perspective of state level, the data will be more, so the data mining technology could be considered to generate more detailed indexes and reference information for making macro policy.

5. The empirical exploration on the relevance between internal audit guidance and internal audit unit

At present, there are in fact lots of matured statistical approaches to scientize the internal audit guiding methods of government audit departments. What we need to prepare is the construction of the scientific public governance idea and the collection of the essential data.

5.1. Using the Principal Component Analysis (PCA)

Principal component analysis is used because it seeks to suitable rotation strategy. Data were analysed with the aid of the SPSS version 12 statistical software application.

Table 1. Total Variance Explained.

Component	Initial Eigenvalues			Extraction Sums of Squared Loadings		
	Total	% of Variance	Cumulative %	Total	% of Variance	Cumulative %
1	2.39	34.136	34.136	2.39	34.136	34.136
2	1.98	28.289	62.425	1.98	28.289	62.425
3	1.033	14.762	77.186	1.033	14.762	77.186
4	0.897	12.813	89.999	0.897	12.813	89.999
5	0.37	5.281	95.281			
6	0.193	2.752	98.033			
7	0.138	1.967	100			

Extraction Method:Principal Component Analysis

Table 2. Component Matrix.

	Component			
	1	2	3	4
Type of internal audit unit	.297	.872	.222	.152
Nature of internal audit unit	-.154	-.394	.230	.875
Full-time internal audit staff	.868	-.286	.131	-.068
Part-time internal audit staff	.073	.916	.271	.116
Independent accounting subordinate units	.876	-.232	.223	-.057
Non-Independent accounting subordinate units	.224	.302	-.867	.246
Internal audit system	.838	-.003	-.197	.161

Extraction Method:Principal Component Analysis.

Table 1&2 provides a summary of the correlations between the component of internal audit unit questionnaire topics which serve as the basis of the factor analysis using principal component procedure. Principal component analyses suggest four factor solutions where the factor represents independent character of internal audit unit, subordinate character of internal audit unit, character of internal audit guidance and scale of the internal audit unit.

5.2. Introducing the Monte-Carlo simulation

The above is only an example of the guidance of internal audit and the research methods of internal audit institutions under the scientific concept. In reality, if historical data is scarce, mature experimental techniques such as simulation can also be selected.

Linking fiscal budget revenue and expenditure, using Monte Carlo simulation method, combining with macro-control data, measuring the expected Value at Risk (VAR) scientifically, focusing on key funds and units, comprehensively determined the national audit annual plan and the internal audit guidance or supervision plan, forming a dynamic closed-loop of full audit coverage under the macro-prudential thinking.

Monte Carlo method is selected to avoid the lack of empirical data and the lack of simulation data in scenario analysis. Due to the research object, internal audit risk and risk loss occurrence events are typical skew thick-tailed problems, non-normal problems. Monte Carlo method can deal with such problems well. A few domestic researchers have applied Monte Carlo method in the field of approximation in practice (Dongxia Liu, 2016).

The main steps of applying Monte Carlo method can be summarized as follows:

- determining variables
- determining the probability distribution of variables
- randomization and bootstrap
- conducting simulation
- finding approximate solutions and making corrections

6. Conclusion

The coordinated development and mutual promotion between the internal audit guidance by the government audit departments and internal audit units are good for the realization of the simultaneous improvement in public governance of national government audit departments, government sectors and state-owned enterprises and institutions and also beneficial to the public sectors' and its officials' abilities in financial and audit to satisfy the social development needs and play the leading roles in the field at home and abroad. A clear target, the selection of proper mode and a complete supporting system which could form collaboratively a systematic and scientific organic and a closed loop, should be able to serve as reference path for the coordinated development of internal audit guidance by the government audit departments and internal audit units.

References

[1] Badara M S, Saidin S Z. Internal audit effectiveness: Data screening and preliminary analysis[J]. Asian Social Science, 2014, 10(10): 76-85.

[2] Sovinová H, Csémy L. The Czech audit: internal consistency, latent structure and identification of risky alcohol consumption[J]. Central European journal of public health, 2010, 18(3).

[3] Thomas Schillemans,Mark van Twist,Martijn van der Steen,Ilsa de Jong. New development: Breaking out or hanging on? Internal audit in government[J]. Public Money & Management,2018,38(7).

[4] Laurence Ferry,Zamzulaila Zakaria,Zarina Zakaria,Richard Slack. Watchdogs, helpers or protectors? – Internal auditing in Malaysian Local Government[J]. Accounting Forum,2017,41(4).

[5] Robyn Pilcher. Role of Internal Audit in Australian Local Government Governance: A Step in the Right Direction[J]. Financial Accountability & Management,2014,30(2).

[6] Dean Neu,Jeff Everett,Abu Shiraz Rahaman. Internal Auditing and Corruption within Government: The Case of the Canadian Sponsorship Program[J]. Contemporary Accounting Research,2013,30(3).

[7] Jeremy Raemaekers. Corporate environmental management in local government: A review of action programmes, internal audits, and state of the environment reports[J]. Planning Practice and Research,1993,8(3).

[8] Zvi Haimon. Internal auditing in local government in Israel: status vs effectiveness?[J]. Israel Affairs,2003,9(4).

[9] Stephen Kwamena Aikins. Government Internal Audits: The Determinants of Quality Supervisory Review of Audit Documentation[J]. International Journal of Public Administration,2013,36(10).

[10] Mary Bowerman,Shirley Hawksworth. Local government internal auditors' perceptions of the Audit Commission[J]. Managerial Auditing Journal,1999,14(8).

[11] Pulina T e tiana V.. The Analysis of Internal Audit Methods to Assess the System of Quality Management in the Government Authorities[J]. Bìznes Inform,2017,8(475).

[12] Novita Puspasari,Meutia Karunia Dewi. The effect of government internal auditors moral reasoning and situational pressure on the tendency to commit fraud when auditing: An experimental study[J]. Journal of Economics, Business & Accountancy,2015,18(3).

[13] Greg Jones,Claire Beattie. Local Government Internal Audit Compliance[J]. Australasian Accounting, Business and Finance Journal,2015,9(3).

[14] Erny Prasetyaningsih,Gufratul Yuhalifiyah,Heri Susanto. Internal Control System Weakness and Non-Compliance to the Provision of Legislation in Practicing Audit of Local Government in Indonesia[J]. Open Journal of Political Science,2014,04(04).

[15] Soratana K, Marriott J. Increasing innovation in home energy efficiency: Monte Carlo simulation of potential improvements[J]. Energy and Buildings, 2010, 42(6): 828-833.

Acknowledgments

This project is supported by the Tianjin Finance Bureau & Accounting Society of Tianjin, Key Accounting Scientific Research Project of 2019-2020 (Y190202).

Compared Analysis of Calculation and Simulation on Heat Transfer Characteristics in PCM Heat Exchange Unit, Part I: Thermal Storage Process

Xiao Hu[1], Pengxiang Song[1], Zhanfeng Deng[1], Guizhi Xu[1], Qingqi Zhao[2], Meng Tang[2] and Changnian Chen[3,*]

[1]State Key Laboratory of Advanced Transmission Technology (Global Energy Interconnection Research Institute Co.,Ltd), Changping District,Beijing 102209, China
[2]State Grid Liaoning Electric Power Co., Ltd., Shenyang 110006, China
[3]School of Energy and Power Engineering, Shandong University, Jinan 250061, China

Email: chen.cn@sdu.edu.cn

Abstract. Based on the temperature change and thermal storage processes of PCM phase change energy storage unit, the PCM channel number of which is 39, the plate spacing is 0.01 m, the air channel is 40 with its plate spacing of 0.003 m, and the unit length is total of 1.62 m. MATLAB solution about the above PCM energy storage unit was performed according to the model and physical property data, and the numerical simulation was carried out by using Fluent software. The theoretical calculation and numerical simulation were compared and analyzed. It is found that only in the case of smaller or larger flow rate working conditions, the solution results of MATLAB for the increase rate of the average temperature of the phase change material inside the system is quite different from the FLUENT simulation result; the comparison of the phase change process degree of advancement is found that at a higher flow rate, the difference between the two results is small, but the difference is larger when the flow rate is smaller; the comparison result of the PCM liquid rate is different under the small flow rate condition. However, the comparative analysis of heat storage shows that the maximum error does not exceed 10%. It can be seen that the flow rate conditions in the model and working conditions are the obvious factors affecting the two results, but the calculation and simulation results can more accurately describe the heat storage characteristics of the PCM phase change energy storage unit, which was calculated in nearly 90% ranges of working conditions, which can prove that the results of MATLAB and FLUENT are consistent.

Journal of Physics: Conference Series **1419** (2019) 012041 doi:10.1088/1742-6596/1419/1/012041

Nomenclature

C1	Solid specific heat, J/(kg K)
C2	Liquid specific heat, J/(kg K)
h	Enthalpy, J/kg
Δh	Enthalpy increase, J/kg
K1	Solid conductivity, W/(m K)
K2	Liquid conductivity, W/(m K)
q	External heat flux, W/m^2
T1	Solid temperature, K
T2	Liquid temperature, K
T_f	Fluid working temperature, K
V	Velocity vector, m/s
ρ	Density, kg/m^3
Subscripts	
M	Material
a	Air
0	Inlet state

1. Introduction

With the rapid development of today's world economy, energy and environment have become two major topics attracting worldwide attention. Especially, energy storage technology has been widely concerned because of its obvious advantages and functions [1]. Energy storage technology and its application have become one of the key areas to make full use of energy and effectively utilize energy. Among them, the application of phase change energy storage materials in energy storage has gradually attracted the attention of scholars all over the world. It has been widely used in water heaters, air conditioning, medical engineering, even military, Aerospace Engineering and other fields [2-5]. Using PCMs, thermal energy can be stored in the material in the form of latent heat by phase change, and then converted into other forms of energy when used. The output temperature and energy are stable and the storage density is high. The storage heat per unit volume is 5-14 times that of sensible heat storage. The advantage is obvious [6-7]. But at the same time, the phase transition process is more complex and difficult to analyze [8-12].

In this paper, the enthalpy method is used to calculate and analyze the performance of PCM phase change energy storage unit by MATLAB, and the numerical simulation is carried out by Fluent 12.0 software. The results of the two methods are compared, and the characteristics of the heat storage of PCM phase change energy storage unit are mainly analyzed, as well as the rationality and validity of the two methods for phase change process analysis.

2. Physical and mathematical model analysis

2.1. Physical model of PCM heat exchange unit

The layout and physical model of the phase change energy storage device used is shown in Figure 1. The structure of the energy storage device is a kind of plate heat exchanger, in which the composite phase change material is encapsulated between staggered parallel plates. The heat transfer fluid (air) flows through the outer parallel channel and exchanges heat with the internal phase change material to achieve the purpose of thermal energy storage and release. In the heat storage process, the heat transfer fluid enters from the inlet above the device and flows out through the lower outlet; in the exothermic process, the flow direction of the heat transfer fluid is opposite to that of the heat storage process, and enters from the lower part and flows out from the upper part. The structure dimension of energy storage device is as follows: veneer area is 0.81 m^2, plate size is 1.6 m *0.505 m, steel plate is as thin as possible, i.e. about 1.0 mm, plate spacing is asymmetric channel, PCM channel is 0.010 mm, air channel is 0.003 mm, corresponding single channel cross-sectional area PCM channel is 0.00505 m, air channel is 0.001515 m^2. A summary of the unit structure parameters and material physical properties has been listed in Table 1 and Table 2.

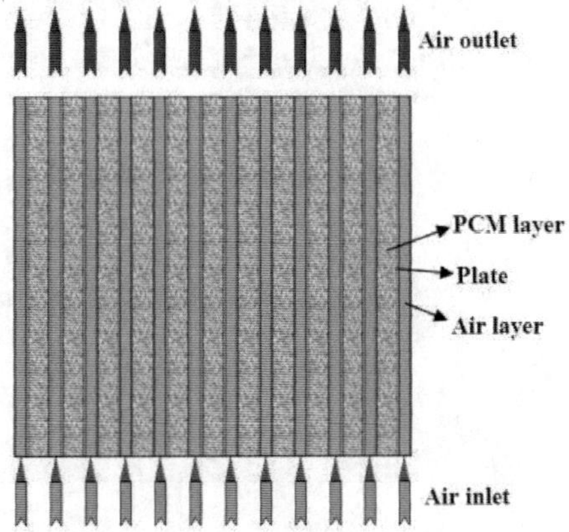

Figure 1. Structure of the PCM heat exchange unit.

In order to obtain more accurate simulation result, the following assumptions are made: 1) the heat conductivity and density of phase change media are constant and do not change with temperature; 2) the temperature of phase change is constant; 3) the heat transfer in phase change media is dominated by heat conduction, ignoring the effect of natural convection; 4) the initial temperature of phase change materials is close to the melting temperature [13-15].

Table 1. The PCM unit structure parameters.

No.	Structure parameters	Values
1	Air flow plate spacing/ m	0.003
2	PCM spacing/ m	0.01
3	Heat transfer area of veneer/ m^2	0.81
4	Cross section area of single channel/ m^2	0.001515
5	Section area of PCM in single channel/ m^2	0.00505
6	PCM channels number	39
7	Total heat transfer area/ m^2	63

Table 2. The PCM physical properties.

No.	Performance parameters	Values
1	Phase change temperature/ ℃	138
2	Energy storage density of materials/ kJ/kg	300
3	Density / kg/m^3	~1000
4	Thermal conductivity/ W/(mK)	~1
5	Specific heat / J/(kgK)	1.9
6	PCM quantity/ kg	200
7	Heat storage capacity /kWh	16.7
8	PCM volume / m^3	0.200
9	Thermal resistance / mK/W	0.005

2.2. Mathematical model of PCM heat exchange unit

For the plate energy storage unit shown in the figure above, because the heat transfer along its thickness direction is very small compared with the flow direction, its internal heat storage and discharge area can be simplified to a two-dimensional structure for simulation calculation [16-18]. In the process of simulation calculation, the average temperature of the PCM area and the outlet temperature of the heat transfer fluid are taken as the criteria, that is, when the average temperature of the PCM area is consistent with the temperature of the heat transfer fluid, or when the outlet temperature of the heat transfer fluid is consistent with the inlet temperature, the PCM inside the device is considered to be completely melted or solidified, other words, the heat energy storage or release are complete.

According to the air exothermic process, a coordinate system with the air inlet position as the origin and the x-axis direction parallel to the air flow direction has been established, as shown in Figure 2.

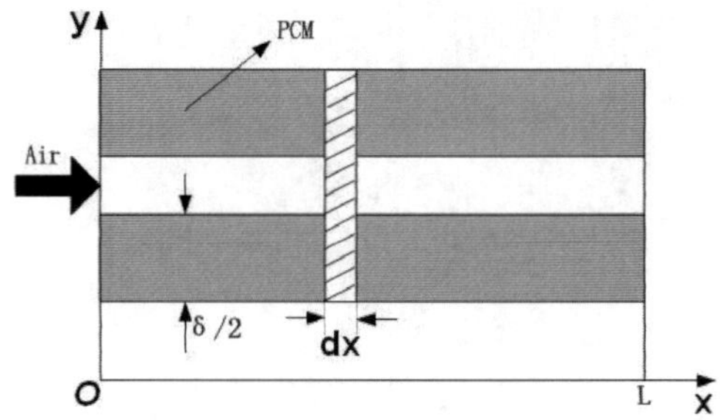

Figure 2. Coordinate system and infinitesimal analysis for the PCM unit.

Therefore, the following heat conduction equation can be established by analyzing the energy balance of dx volume from the origin of x.

$$S\rho_s c_{p,s}\partial T_s / \partial \tau = S\lambda_s \partial^2 T_s / \partial x^2 + KU(T_a - T_s) \tag{1}$$

The ideal gas equation of state is described as follows:

$$p = \rho_a R_M T_a \tag{2}$$

And the mass conservation equation is:

$$\rho_a u_a = \rho_{a,0} u_{a,0} \tag{3}$$

Energy equation:

$$\frac{\partial(\rho h)}{\partial t} + \mathrm{div}(\rho u h) = \mathrm{div}(\alpha\,\mathrm{grad} h) + S_h \tag{4}$$

$$\frac{\partial(\rho h)}{\partial t} = \lambda \nabla^2 T = \frac{\partial}{\partial t}\left(\lambda \frac{\partial T}{\partial x}\right) + \frac{\partial}{\partial t}\left(\lambda \frac{\partial T}{\partial y}\right) \tag{5}$$

The boundary and initial conditions are listed as bellows:

For boundary conditions,

$$T_a(0,\tau) = T_{a,0} \tag{6}$$

$$\partial T_s(0,\tau)/\partial x = \partial T_s(L,\tau)/\partial x = 0 \tag{7}$$

For initial conditions,

$$T_s(x,0) = f(x) \tag{8}$$

Considering the non-steady phase change process for heat storage and discharge in device, it is difficult to obtain the analytical solution under this problem [19-20]. Therefore, numerical method is used to simulate the heat storage and discharge process of heat storage device, time and space are discretized in the solution area, and partial differential mathematical model equation is approximated by forward difference.

3. Results and discuss

3.1. Analysis of air temperature change rate

Taking the inlet velocity range of heat transfer fluid from 30% to 300% of the design value (3.14 m/s), i.e. 1-9 m/s, the MATLAB discrete solution results are compared with the simulation results of FLUENT under the same conditions. The air temperature rise curve is compared with the time change in Fig. 3. It can be seen that within the calculation range of heat transfer fluid velocity, the average temperature increase rate of phase change material side at the initial stage of overall heat storage process is greater than that at the later stage of heat storage. This is consistent with the previous analysis, because of the large temperature difference in the initial stage of heat storage, the corresponding heat transfer rate is larger; in the later stage of heat storage, with the increase of the temperature of phase change material, the temperature difference between the phase change material and the heat transfer fluid gradually reduces, and then the heat transfer rate in the system gradually decreases, while the average temperature of phase change material increases at the initial stage of heat storage.

Two points with different slope (heating rate) on the air temperature distribution curve of Fig.3 are analyzed. At a certain time during the heat transfer process, two points with different positions in the air passage are analyzed. The slope of these two points, i.e. the heating rate, is obtained by the simulation results at that time. Then using the discrete mathematical analytic method of equation and using MATLAB to program and solve, the heating rate of these two points at the corresponding time is obtained, and then the two points are compared and analyzed.

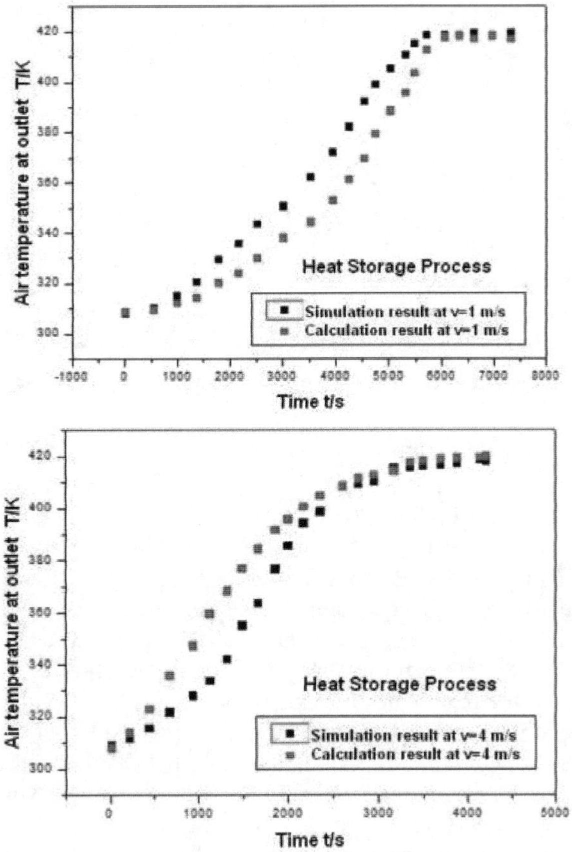

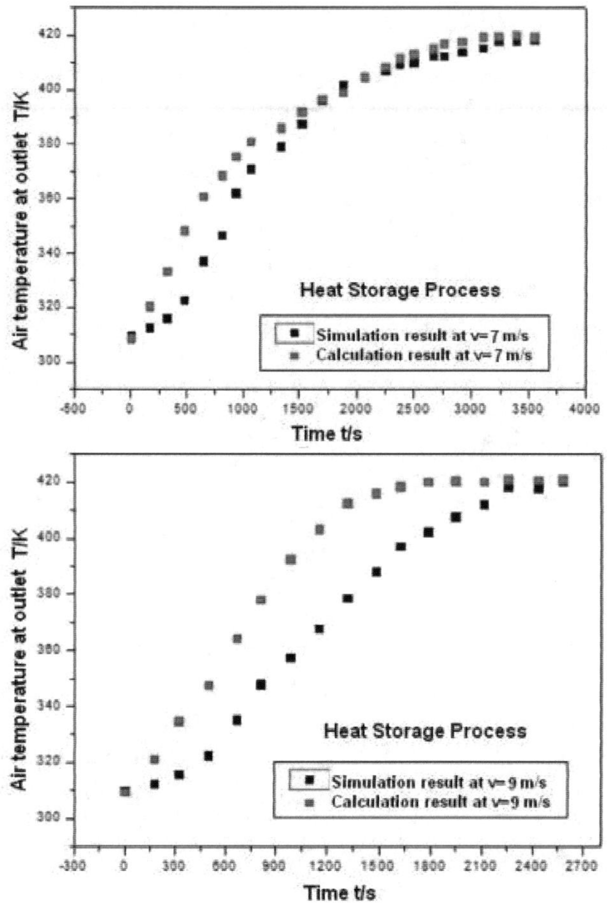

Figure 3. Air temperature change rate at outlet.

It can be seen from Fig.3 that the average temperature increase rate of phase change material in the system increases with the increase of the inlet velocity of heat transfer fluid. The results of MATLAB calculation show that the heat storage time is shorter than that of FLUENT simulation, especially when the flow rate is smaller and larger [21]. The difference of heat storage time between 1 m/s and 9 m/s is more than 200 s and 500 s, respectively, while at medium flow rate. The difference is small. Under the condition of 7 m/s, the difference of heat storage time is less than 100 s. See Fig. 3. Overall, the results show that the thermal storage time of FLUENT simulation results is longer when the flow rate is small, such as phase change thermal storage time more than 7500s under 1m/s condition.

When the velocity of flow is 1 m/s, 4 m/s and 9 m/s, the results of MATLAB and FLUENT simulation show great difference. The average errors of calculation are 16.2%, 13.9% and 17.0% respectively. When the velocity of flow is 7 m/s, the calculated results are in good agreement with the simulation results, the curve coincidence is high, and the average error is reduced to less than 10%. At the same time, it can be seen that the storage time and the temperature rise range of the two algorithms are both reasonable, and the reference velocity can be chosen.

3.2. Analysis of liquid phase ratio of PCM material

Since PCM material exists in two-phase state (liquid phase and solid phase) during the process, the ratio of the mass of PCM material in liquid phase state to the mass of PCM material in general is defined as the liquid phase ratio of PCM material. With the continuous heat transfer process, the liquid phase ratio of PCM material is also changing continuously, that is, from 0% to 100%. By calculating the liquid phase ratio of PCM material at a certain time, the simulation results and design results can be compared and analyzed.

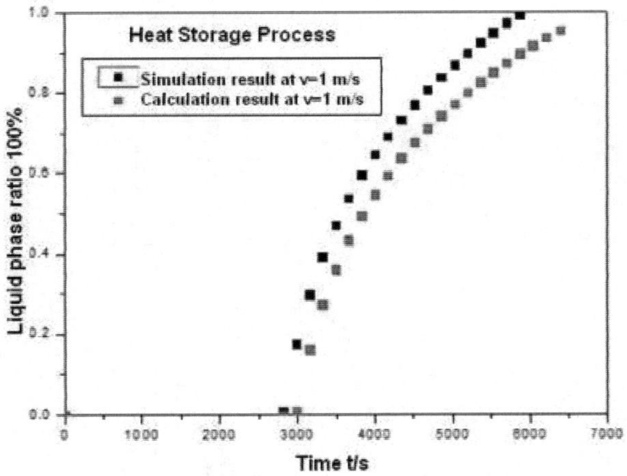

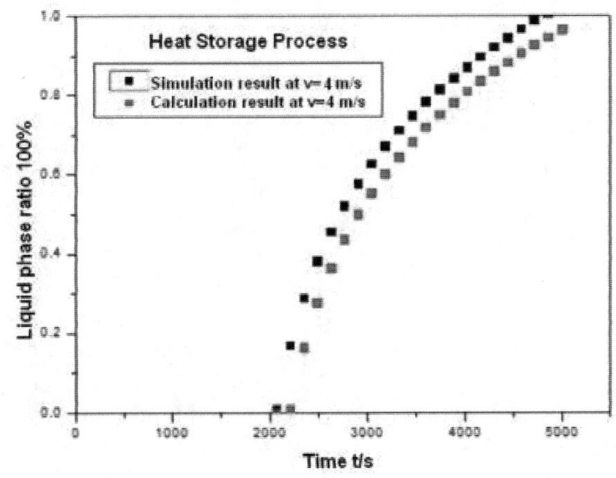

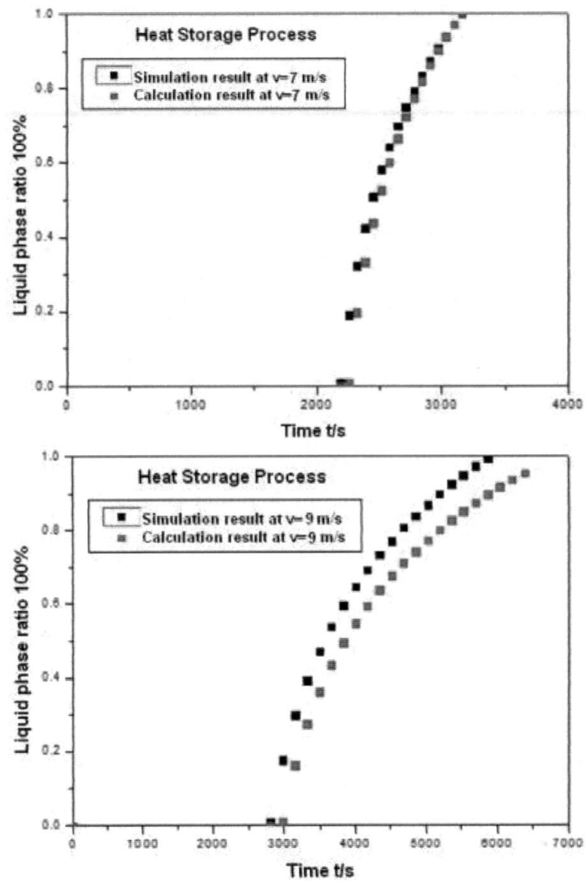

Figure 4. Liquid phase ratio of PCM material.

From Fig.4, it can be seen that the overall liquid phase change rate trend of the simulation results and the design results is the same, but comparing the four different flow rate conditions, it is found that the difference between the two results is smaller at higher flow rate, but larger at smaller flow rate.

3.3. Heat storage amount of PCM materials

With the constant total heat storage amount of PCM, the residual heat storage of PCM materials continue to change with the continuous heat transfer process. The numerical value can be calculated at a certain time in the heat transfer process, and then compared with the simulation value.

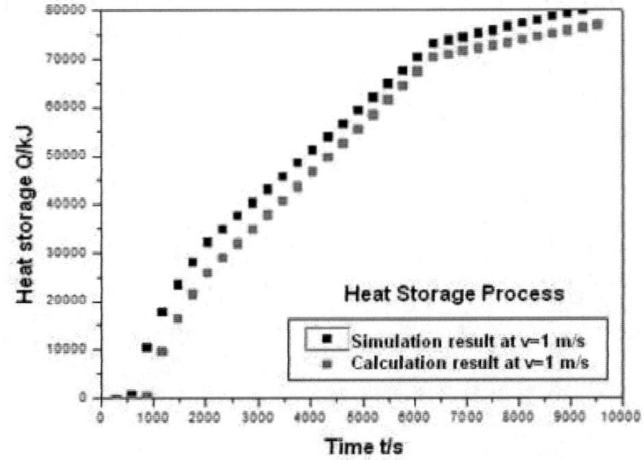
Heat Storage Process
Simulation result at v=1 m/s
Calculation result at v=1 m/s
Heat storage Q/kJ
Time t/s

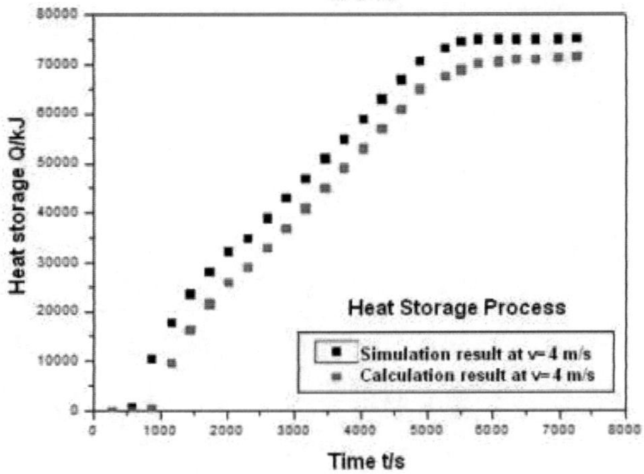
Heat Storage Process
Simulation result at v=4 m/s
Calculation result at v=4 m/s
Heat storage Q/kJ
Time t/s

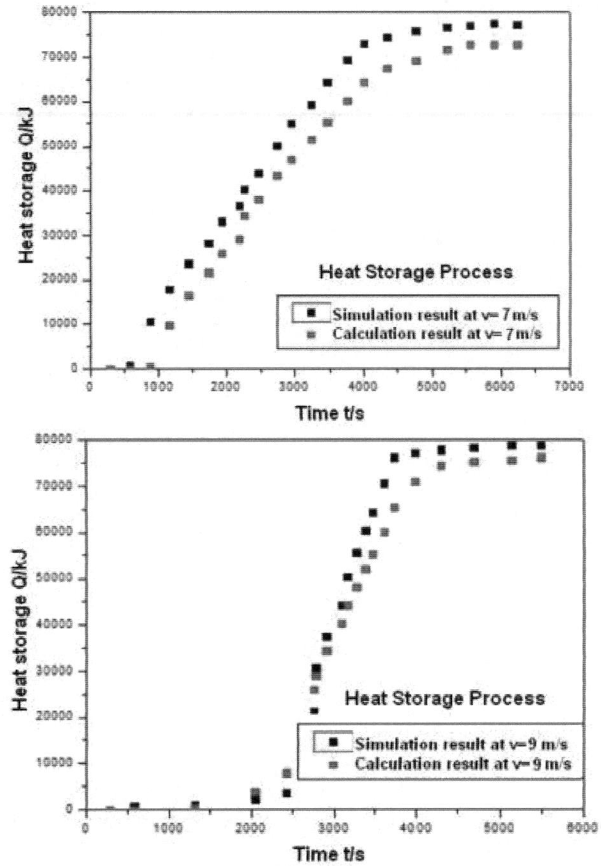

Figure 5. Heat storage amount during the phase change process.

From Fig. 5, it can be seen that the overall heat storage change trend of simulation results and design results is consistent. Four comparisons show that the growth rate of heat storage is relatively small at the beginning stage of heat storage. When phase change occurs, it increases approximately linearly, and the growth rate is relatively large. Until the phase change heat storage stage ends, the change trend of heat storage decreases; 1 m/s, 4 m/s, 7 m/s and 9 m/s flow rate. Under the condition, the time period for the increase of storage heat is gradually shortened, which is about 3000s, 2600s, 1700s and 900s, respectively. The comparison results prove the correctness of the design results, and the maximum check error is not more than 10%, which shows the accuracy of the design results.

4. Conclusions

(1) By comparing and analyzing the change rate of air temperature, it is found that in the calculation range of heat transfer fluid flow rate, the increase rate of the average temperature of phase change material side in the initial stage of overall heat storage process is greater than that in the later stage of heat storage. The increase rate of the average temperature of phase change material inside the system increases with the increase of the inlet velocity of heat transfer fluid. The results show that the errors of air temperature rise rate under different flow velocities are less than 10%.

(2) With the continuous heat transfer process, the liquid phase ratio of PCM material also changes continuously. The trend of the overall liquid phase change rate of the simulation results and the design results is the same, but comparing four different flow velocity conditions, it is found that the difference between the two results is smaller at higher flow velocity.

(3) With the constant total storage heat of PCM and the continuous heat transfer process, the residual storage heat amount of PCM materials are also changing continuously. The analysis and comparison show that the overall storage heat trend of simulation results and design results is the same. The growth rate of storage heat is relatively small at the beginning of thermal storage, and increases linearly when phase change occurs, and the growth rate is similar. It is large until the phase change heat storage stage is over.

References

[1] Y. L. Li, Y. Jin, Y. Huang, etc., Fundamentals of Heat Storage Technology (I) - Basic Principles of Heat Storage and New Trends of Research, Energy Storage Science and Technology, 2013 (01): 69-72.
[2] Z. M. Jiang. Reflections on Energy Issues in China, Journal of Shanghai Jiaotong University, 2008, 42(3): 345-359.
[3] Feldman D, Shapiro M M, Banu D. Organic Phase Change Materials for Thermal Energy Storage [J]. Sol. Energy Mater, 1986, 13(1): 1−10.
[4] Syed M T, Kumar S, Moallem I M K, et al. Thermal Storage Using Form Stable Phase Change Materials [J]. ASHRAE J., 1997, 39(5): 45−50.
[5] Akiyama T, Yagi J. Encapsulation of Phase Change Materials for Storage of High Temperature Waste Heat [J]. High Temp. Mater. P., 2000, 19(4): 219−222.
[6] Yavari F , Fard H R , Pashayi K , et al. Enhanced Thermal Conductivity in a Nanostructured Phase Change Composite due to Low Concentration Graphene Additives[J]. The Journal of Physical Chemistry C, 2011, 115(17):8753-8758.
[7] Vigo T L, Frost C M. Temperature Adaptable Fibers [J]. Text. Res. J., 1986, 56(12): 737−740.
[8] Fan L W , Zhu Z Q , Zeng Y , et al. Transient performance of a PCM-based heat sink with high aspect-ratio carbon nanofillers [J]. Applied Thermal Engineering, 2015, 75:532-540.
[9] Zhang Renyuan. Phase change material and phase change energy storage technology [M]. Beijing: Science Press, 2009: 130-181(in Chinese).
[10] Ge Z , Li Y , Li D , et al. Thermal energy storage: Challenges and the role of particle technology[J]. Particuology, 2014, 15:2-8.
[11] Ge Z , Ye F , Ding Y . Composite Materials for Thermal Energy Storage: Enhancing Performance through Microstructures[J]. ChemSusChem, 2014, 7(5):1318-1325.
[12] Regin A F , Solanki S C , Saini J S . Heat transfer characteristics of thermal energy storage system using PCM capsules: A review[J]. Renewable and Sustainable Energy Reviews, 2008, 12(9):2438-2458.
[13] Soares N, Costa J J, Gaspar A R, et al. Review of passive PCM latent heat thermal energy storage systems towards buildings' energy efficiency[J]. Energy & Buildings, 2013, 59:82-103.
[14] Tyagi V V, Buddhi D. Thermal cycle testing of calcium chloride hexahydrate as a possible PCM for latent heat storage [J]. Solar Energy Materials & Solar Cells, 2008, 92(8):891-899.
[15] Mekaddem N, Ali S B, Mazioud A, et al. A numerical study of latent thermal energy storage in a phase change material/carbon panel[C]// 2016.
[16] Touati B, Kerroumi N, Virgone J. Solar thermal energy discharging from a unit contains multiple phase change materials[C]// Renewable Energy Congress. 2017.
[17] Qiang S, Xing Y, Mu L, et al. Numerical research of solid-liquid phase change process in porous media[C]// International Conference on Materials for Renewable Energy & Environment. 2011.
[18] Al-Saadi S N, Zhai Z. Systematic evaluation of mathematical methods and numerical schemes for modeling PCM-enhanced building enclosure[J]. Energy & Buildings, 2015, 92:374-388.
[19] Alsaadi S N. Modeling and simulation of PCM-enhanced facade systems[J]. Dissertations & Theses - Gradworks, 2014.
[20] Touati B, Kerroumi N, Virgone J. Solar thermal energy discharging from a unit contains multiple phase change materials[C]// Renewable Energy Congress. 2017.

[21] Qiang S, Xing Y, Mu L, et al. Numerical research of solid-liquid phase change process in porous media[C]// International Conference on Materials for Renewable Energy & Environment. 2011.

Acknowledgments
This project is supported by the State Grid Science and Technology Cooperation Project "research and application of key technologies of optimal operation and collaborative control of phase-change heat storage stations for multi-scenario absorption of clean energy" (SGRIDLKJ2017-164) and the National Natural Science Foundation of China (51706124).

Journal of Physics: Conference Series **1419** (2019) 012042 doi:10.1088/1742-6596/1419/1/012042

Compared Analysis of Calculation and Simulation on Heat Transfer Characteristics in PCM Heat Exchange Unit, Part II: Thermal Release Process

Guizhi Xu[1], Xiao Hu[1], Shaoming Liu[1], Zhanfeng Deng[1], Qingqi Zhao[2], Meng Tang[2] and Changnian Chen[3,*]

[1]State Key Laboratory of Advanced Transmission Technology (Global Energy Interconnection Research Institute Co.,Ltd), Changping District,Beijing 102209, China
[2]State Grid Liaoning Electric Power Co., Ltd., Shenyang 110006, China
[3]School of Energy and Power Engineering, Shandong University, Jinan 250061, China

Email: chen.cn@sdu.edu.cn

Abstract. In this part, the thermal release processes of PCM phase change in the same energy storage unit are analyzed, and the PCM unit model is the same as in Part I: channel number is 39, the plate spacing is 0.01 m, the air channel is 40 with its plate spacing of 0.003 m, and the unit length is total of 1.62 m. MATLAB solution about the above PCM energy storage unit was performed according to the model and physical property data, and the numerical simulation was carried out by using Fluent software as well. The theoretical calculation and numerical simulation were compared and analyzed. For the release process, it also can be seen that only in the case of smaller or larger flow rate working conditions, the solution results of MATLAB for the changing rate of the average temperature of the phase change material inside the system is quite different from the FLUENT simulation result; the total results are very similar to thermal storage process. When the flow rate is higher, the difference between the two results is even smaller than the release process; the comparison result of the PCM liquid rate is different under the small flow rate condition, and this is obvious for the release process, especially under the condition of small flow rate. However, also the maximum error does not exceed 10%. Conclusion can be drown that the flow rate conditions in the model and working conditions are the important factors for affecting the compared results, but the calculation and simulation results can more accurately describe the thermal release characteristics of the PCM phase change energy storage unit, which was calculated in nearly 90% ranges of working conditions, which furtherly illustrates the effectiveness of analytical method used in this paper.

Nomenclature

C1	Solid specific heat, J/(kg K)
C2	Liquid specific heat, J/(kg K)
h	Enthalpy, J/kg
Δh	Enthalpy increase, J/kg
K1	Solid conductivity, W/(m K)
K2	Liquid conductivity, W/(m K)
q	External heat flux, W/m^2
T1	Solid temperature, K
T2	Liquid temperature, K
T_f	Fluid working temperature, K
V	Velocity vector, m/s
ρ	Density, kg/m^3
Subscripts	
M	Material
a	Air
0	Inlet state

1. Introduction

As illustrated in part I, energy storage technology and its application have become one of the key areas to make full use of energy and effectively utilize energy in recent years,. Among them, the application of phase change energy storage materials in energy storage has gradually attracted the attention of scholars all over the world. It has been widely used in water heaters, air conditioning, medical engineering, even military, Aerospace Engineering and other fields [1-5]. Phase change materials (PCMs) have the characteristics of small size, large energy storage per unit, friendly environment and easy control. Using PCMs, thermal energy can be stored in the material in the form of latent heat by phase change, and then converted into other forms of energy when used. The output temperature and energy are stable and the storage density is high. The storage heat per unit volume is 5-14 times that of sensible heat storage. Also, phase change energy storage materials have the advantages of environment friendliness, stable output, easy control and high heat storage density, which are incomparable with other energy storage methods [6-8].

In this part, the enthalpy method is also used to calculate and analyze the performance of PCM phase change energy storage unit by MATLAB, and the numerical simulation is carried out by Fluent 12.0 software too. The results of the two methods are compared, and the characteristics of the thermal release process of PCM phase change energy storage unit are analyzed, as well as the rationality and validity of the two simulation methods.

2. Physical and mathematical model analysis

2.1. Physical model of PCM heat exchange unit

In this part, the layout and physical model of the phase change energy storage device used is the same as in part I, as shown in Figure 1. The structure of the energy storage device is a kind of plate heat exchanger, in which the composite phase change material is encapsulated between staggered parallel plates. The heat transfer fluid (air) flows through the outer parallel channel and exchanges heat with the internal phase change material to achieve the purpose of thermal energy storage and release. In the heat storage process, the heat transfer fluid enters from the inlet above the device and flows out through the lower outlet; in the exothermic process, the flow direction of the heat transfer fluid is opposite to that of the heat storage process, and enters from the lower part and flows out from the upper part. The structure dimension of energy storage device is as follows: veneer area is 0.81 m^2, plate size is 1.6 m *0.505 m, steel plate is as thin as possible, i.e. about 1.0 mm, plate spacing is asymmetric channel, PCM channel is 0.010 mm, air channel is 0.003 mm, corresponding single channel cross-sectional area PCM channel is 0.00505 m, air channel is 0.001515 m^2. A summary of the unit structure parameters and material physical properties has been listed in Table 1 and Table 2.

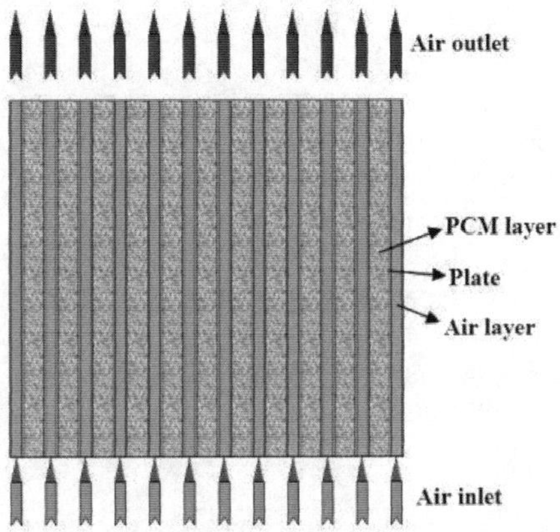

Figure 1. Structure of the PCM heat exchange unit.

In order to obtain more accurate simulation result, the following assumptions are made: 1) the heat conductivity and density of phase change media are constant and do not change with temperature; 2) the temperature of phase change is constant; 3) the heat transfer in phase change media is dominated by heat conduction, ignoring the effect of natural convection; 4) the initial temperature of phase change materials is close to the melting temperature [9-11].

Table 1. The PCM unit structure parameters.

No.	Structure parameters	Values
1	Air flow plate spacing/ m	0.003
2	PCM spacing/ m	0.01
3	Heat transfer area of veneer/ m^2	0.81
4	Cross section area of single channel/ m^2	0.001515
5	Section area of PCM in single channel/ m^2	0.00505
6	PCM channels number	39
7	Total heat transfer area/ m^2	63

Table 2. The PCM physical properties.

No.	Performance parameters	Values
1	Phase change temperature/ ℃	138
2	Energy storage density of materials/ kJ/kg	300
3	Density / kg/m^3	~1000
4	Thermal conductivity/ W/(mK)	~1
5	Specific heat / J/(kgK)	1.9
6	PCM quantity/ kg	200
7	Heat storage capacity /kWh	16.7
8	PCM volume / m^3	0.200
9	Thermal resistance / mK/W	0.005

2.2. Mathematical model of PCM heat exchange unit

For the plate energy storage unit shown in the figure above, because the heat transfer along its thickness direction is very small compared with the flow direction, its internal heat storage and discharge area can be simplified to a two-dimensional structure for simulation calculation [12-14]. In the process of simulation calculation, the average temperature of the PCM area and the outlet temperature of the heat transfer fluid are taken as the criteria, that is, when the average temperature of the PCM area is consistent with the temperature of the heat transfer fluid, or when the outlet temperature of the heat transfer fluid is consistent with the inlet temperature, the PCM inside the device is considered to be completely melted or solidified, other words, the heat energy storage or release are complete.

According to the air exothermic process, a coordinate system with the air inlet position as the origin and the x-axis direction parallel to the air flow direction has been established, as shown in Figure 2.

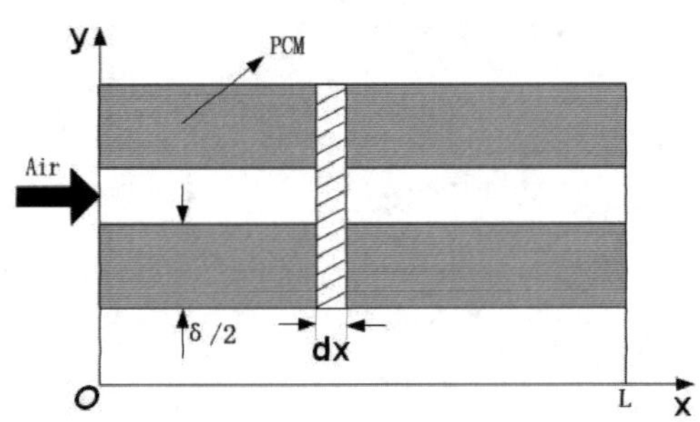

Figure 2. Coordinate system and infinitesimal analysis for the PCM unit.

Therefore, the following heat conduction equation can be established by analyzing the energy balance of dx volume from the origin of x.

$$S\rho_s c_{p,s}\partial T_s / \partial\tau = S\lambda_s \partial^2 T_s / \partial x^2 + KU(T_a - T_s) \tag{1}$$

The ideal gas equation of state is described as follows:

$$p=\rho_a R_M T_a \tag{2}$$

And the mass conservation equation is:

$$\rho_a u_a = \rho_{a,0} u_{a,0} \tag{3}$$

Energy equation:

$$\frac{\partial(\rho h)}{\partial t}+\mathrm{div}(\rho u h)=\mathrm{div}(\alpha\,\mathrm{grad}h)+S_h \quad (4)$$

$$\frac{\partial(\rho h)}{\partial t}=\lambda\nabla^2 T=\frac{\partial}{\partial t}\left(\lambda\frac{\partial T}{\partial x}\right)+\frac{\partial}{\partial t}\left(\lambda\frac{\partial T}{\partial y}\right) \quad (5)$$

The boundary and initial conditions are listed as bellows:
For boundary conditions,

$$T_a(0,\tau)=T_{a,0} \quad (6)$$

$$\partial T_s(0,\tau)/\partial x=\partial T_s(L,\tau)/\partial x=0 \quad (7)$$

For initial conditions,

$$T_s(x,0)=f(x) \quad (8)$$

Considering the non-steady phase change process for heat storage and discharge in device, it is difficult to obtain the analytical solution under this problem [15-16]. Therefore, numerical method is used to simulate the heat storage and discharge process of heat storage device, time and space are discretized in the solution area, and partial differential mathematical model equation is approximated by forward difference.

3. Results and discuss

3.1. Analysis of air temperature change rate

Hereby also taking the inlet velocity range of heat transfer fluid from 30% to 300% of the design value (3.14 m/s), i.e. 1-9 m/s, the MATLAB discrete solution results are compared with the simulation results of FLUENT under the same conditions. The air temperature rise curve is compared with the time change in Fig. 3. It can be seen that within the calculation range of heat transfer fluid velocity, the average temperature increase rate of phase change material side at the initial stage of overall heat storage process is greater than that at the later stage of heat storage. This is consistent with the results in thermal storage process. And the reason for this is also due to the large temperature difference; at the later stage of heat release, the heat transfer rate in the system gradually decreases, and the average temperature of phase change material decreases slowly.

Two points with different slope (heat discharging rate) on the air temperature distribution curve of Fig.3 are analyzed, and it showed the same law as described in part I. That is to say, the calculation methods are suitable for both thermal storage and release heat transfer processes based the model in this paper.

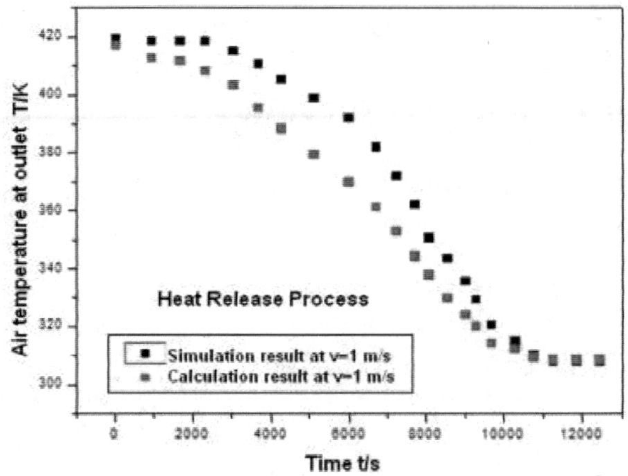
Air temperature at outlet T/K
Time t/s
Heat Release Process
Simulation result at v=1 m/s
Calculation result at v=1 m/s

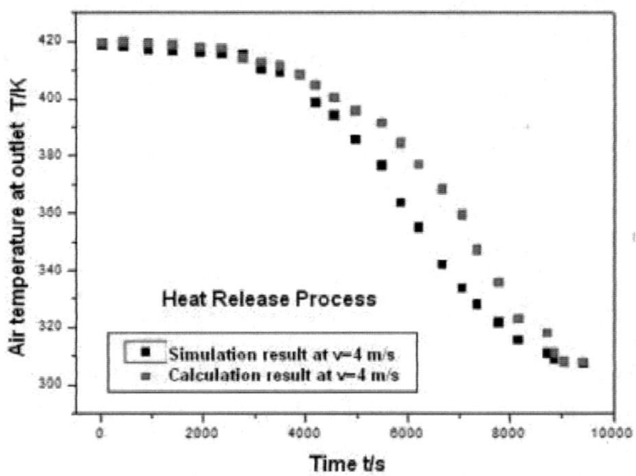
Air temperature at outlet T/K
Time t/s
Heat Release Process
Simulation result at v=4 m/s
Calculation result at v=4 m/s

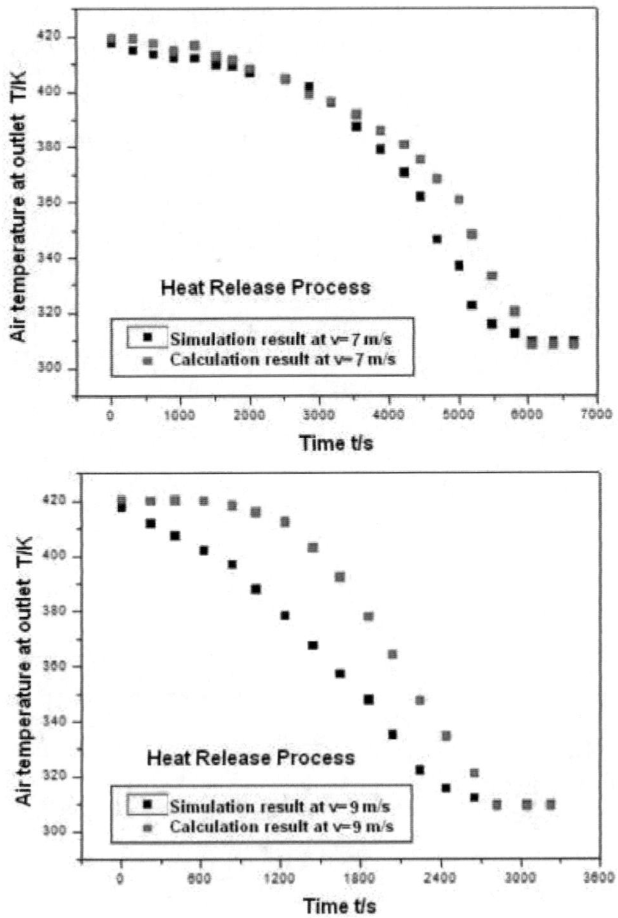

Figure 3. Air temperature change rate at outlet.

Also tt can be seen that the average temperature decrease rate of phase change material in the system decreases with the increase of the inlet velocity of heat transfer fluid. The results of MATLAB calculation show that the thermal release time is shorter than that of FLUENT simulation, especially when the flow rate is smaller and larger. This results are also consistent with the heat storage process. When the velocity of flow is 1 m/s, 4 m/s and 9 m/s, the results of MATLAB and FLUENT simulation show great difference as well. The average errors of calculation are 17.3%, 15.5% and 19.1% respectively. When the velocity of flow is 7 m/s, the calculated results are in good agreement with the simulation results, the curve coincidence is high, and the average error is reduced to less than 10%. This result is also consistent with the heat storage process.

3.2. Analysis of liquid phase ratio of PCM material

By calculating the liquid phase ratio of PCM material at a certain time, the simulation results and design results can be compared and analyzed. From Fig.4, it can be seen that the overall liquid phase change rate trend of the simulation results and the design results is the same, but comparing the four different flow rate conditions, it is found that the difference between the two results is smaller at higher flow rate, but larger at smaller flow rate, especially in the case of lower flow rate, the error of release process is larger, such as under 1 m/s flow rate. These trends are consistent with the heat storage process. It is noticed that under the

1 m/s flow rate the difference is more than 12%, but the appearance of this situation cannot explain the inappropriateness of the design results. From the comparison results of 4m/s and 9m/s, good consistency has been achieved.

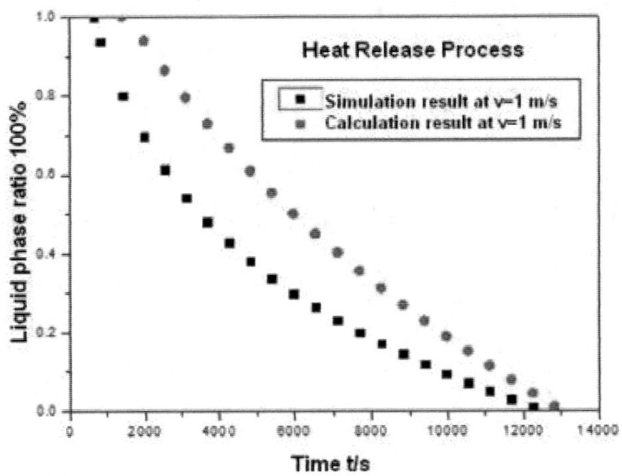

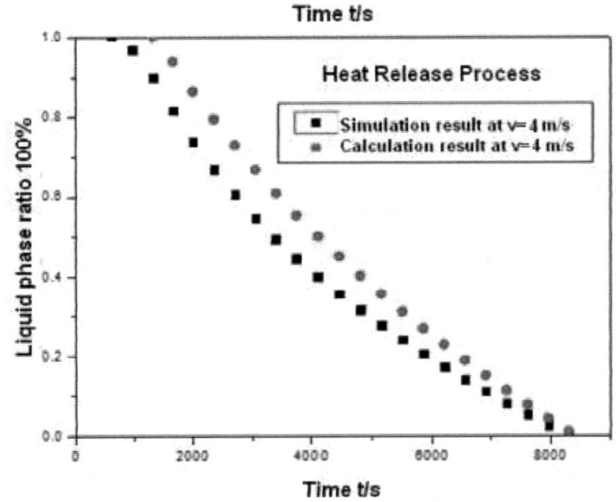

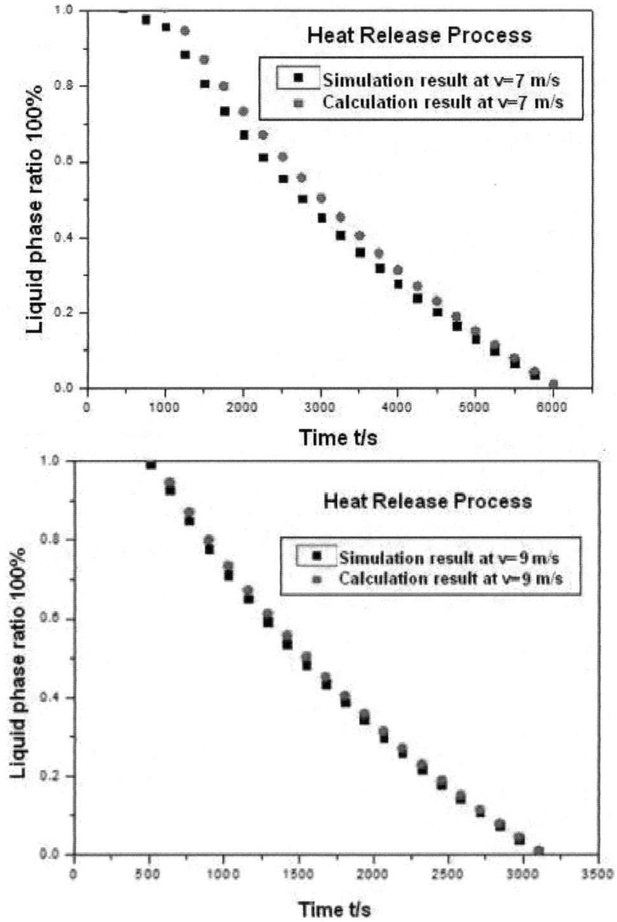

Figure 4. Liquid phase ratio of PCM material.

3.3. Heat release amount of PCM materials

For the total heat release amount of PCM, the residual heat release of PCM materials continue to change with the continuous heat transfer process. In this part, these changes also has been analyzed.

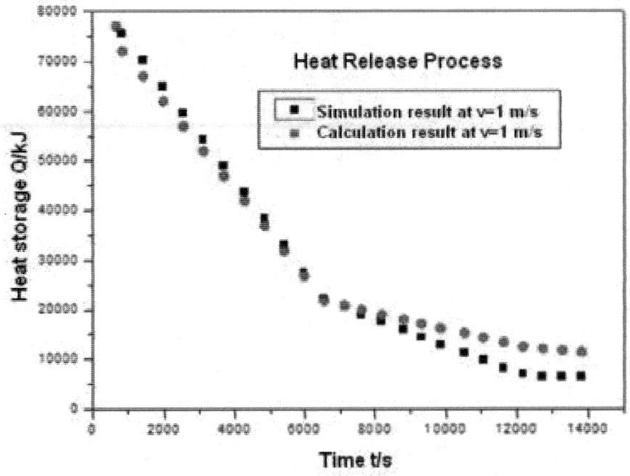
Heat Release Process
Simulation result at v=1 m/s
Calculation result at v=1 m/s
Heat storage Q/kJ
Time t/s

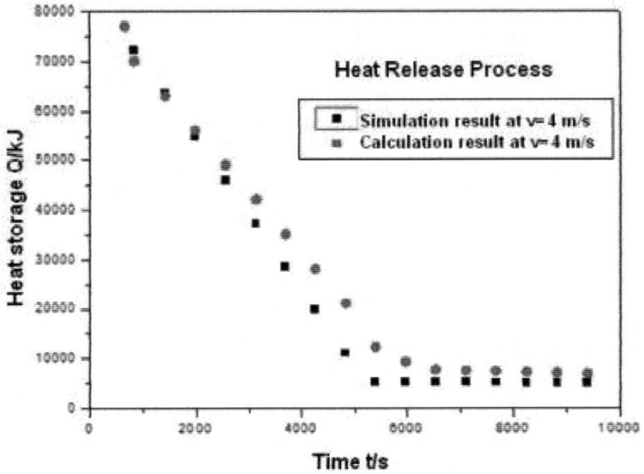
Heat Release Process
Simulation result at v=4 m/s
Calculation result at v=4 m/s
Heat storage Q/kJ
Time t/s

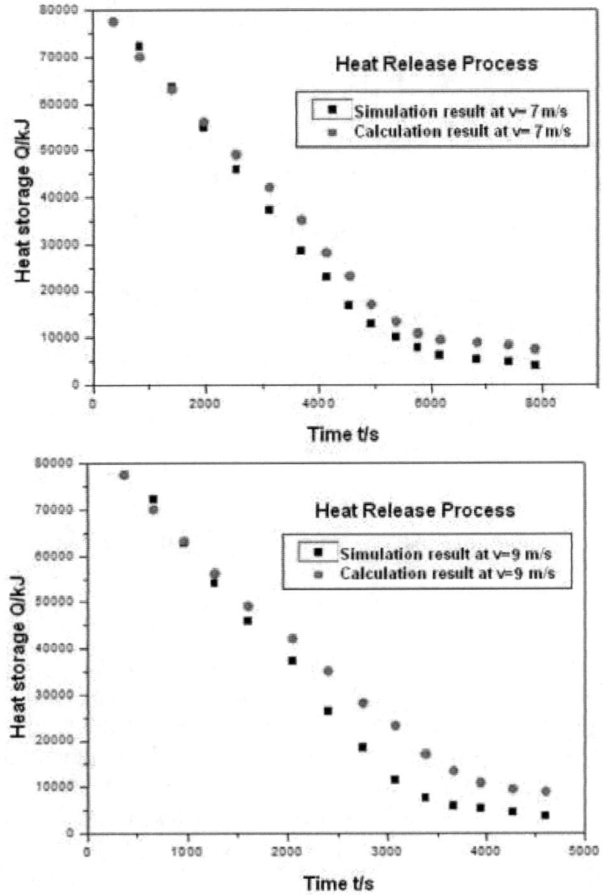

Figure 5. Heat storage amount during the phase change process.

From Fig. 5, it can be seen that the overall heat release change trend of simulation results and design results is consistent. Four comparisons show that the growth rate of heat release is relatively small at the beginning stage the following. When phase change occurs, it increases approximately linearly, and the growth rate is relatively large, these results are similar to the storage process. From the four comparison figures, it can draw the conclusion that the maximum check error is also not more than 10%, which shows the accuracy of the comparing results.

4. Conclusions

(1) By comparing and analyzing the change rate of air temperature, it is found that the similar results are for both thermal storage and release processes. And the results show that the errors of air temperature changing rate under different flow velocities are less than 10%.

(2) With the continuous heat transfer process, the liquid phase ratio of PCM material also changes continuously. The trend of the overall liquid phase change rate of the simulation results and the design results is the same, which is consistent with the results in part I.

(3) With the constant total heat amount of PCM and the continuous heat transfer process, the residual storage heat and release heat of PCM materials are also changing continuously. The analysis and

comparison show that the overall release heat trend of simulation results and design results is similar to storage process.

References

[1] Hammond M J. Reversible Liquid/Solid Phase Change Compositions [P]. US Pat.: 5785884, 1998−07−28.
[2] Y. L. Li, Y. Jin, Y. Huang, etc., Fundamentals of Heat Storage Technology (I) - Basic Principles of Heat Storage and New Trends of Research, Energy Storage Science and Technology, 2013 (01): 69-72.
[3] Feldman D, Shapiro M M, Banu D. Organic Phase Change Materials for Thermal Energy Storage [J]. Sol. Energy Mater, 1986, 13(1): 1−10.
[4] Syed M T, Kumar S, Moallem I M K, et al. Thermal Storage Using Form Stable Phase Change Materials [J]. ASHRAE J., 1997, 39(5): 45−50.
[5] Akiyama T, Yagi J. Encapsulation of Phase Change Materials for Storage of High Temperature Waste Heat [J]. High Temp. Mater. P., 2000, 19(4): 219−222.
[6] Akiyama T, Yagi J. Encapsulation of Phase Change Materials for Storage of High Temperature Waste Heat [J]. High Temp. Mater. P., 2000, 19(4): 219−222.
[7] Vigo T L, Frost C M. Temperature Adaptable Fibers [J]. Text. Res. J., 1986, 56(12): 737−740.
[8] Chen Xiao, Zhang Renyuan, Mao Lingbo. Progress in research and application of paraffin wax phase change material [J] .Materials Research and Application, 2008 (2):89-92.
[9] Soares N, Costa J J, Gaspar A R, et al. Review of passive PCM latent heat thermal energy storage systems towards buildings' energy efficiency[J]. Energy & Buildings, 2013, 59:82-103.
[10] Tyagi V V, Buddhi D. Thermal cycle testing of calcium chloride hexahydrate as a possible PCM for latent heat storage [J]. Solar Energy Materials & Solar Cells, 2008, 92(8):891-899.
[11] Mekaddem N, Ali S B, Mazioud A, et al. A numerical study of latent thermal energy storage in a phase change material/carbon panel[C]// 2016.
[12] Touati B, Kerroumi N, Virgone J. Solar thermal energy discharging from a unit contains multiple phase change materials[C]// Renewable Energy Congress. 2017.
[13] Qiang S, Xing Y, Mu L, et al. Numerical research of solid-liquid phase change process in porous media[C]// International Conference on Materials for Renewable Energy & Environment. 2011.
[14] Al-Saadi S N, Zhai Z. Systematic evaluation of mathematical methods and numerical schemes for modeling PCM-enhanced building enclosure[J]. Energy & Buildings, 2015, 92:374-388.
[15] Alsaadi S N. Modeling and simulation of PCM-enhanced facade systems[J]. Dissertations & Theses - Gradworks, 2014.
[16] Touati B, Kerroumi N, Virgone J. Solar thermal energy discharging from a unit contains multiple phase change materials[C]// Renewable Energy Congress. 2017.

Acknowledgments

This project is supported by the State Grid Science and Technology Cooperation Project "research and application of key technologies of optimal operation and collaborative control of phase-change heat storage stations for multi-scenario absorption of clean energy" (SGRIDLKJ2017-164) and the National Natural Science Foundation of China (51706124).

Emergency Evacuation: Dynamic Network Diversion

Yulu Yang[1], Jiawei Wang[2],Hongxia Zhao[2] and Xiangdong Liu[3]

[1]College of Economic, Jinan University, Guangzhou, Guangdong510632, China
[2]College of Cyber Security, Jinan University, Guangzhou, Guangdong510632, China
[3]Jinan University, Guangzhou, Guangdong510632, China

E-mail: ylu1906@stu2017.jnu.edu.cn, wangjiawei@stu2017.jnu.edu.cn, zhaohongxia@stu2017.jnu.edu.cn, tliuxd@jnu.edu.cn

Abstract. Under the dangerous conditions, rapid evacuation of large public spaces is critical to the safety of visitors. Taking the Louvre as an example, this paper comprehensively considers the characteristics of tourists, the internal structure of Louvre, the dangerous situation and the openness of some secret exits. At the same time, based on the improved Cellular Automaton model and Network Flow model, this paper constructs the Dynamic System Optimal model. By applying the Frank-Wolfe algorithm, this model is simulated in MATLAB to simulate the evacuation of visitors in the emergency situations. Through the simulation results, we can find out the potential bottleneck position that restricts the movement to the exits and make the system's optimal evacuation program. It can be verified that the evacuation efficiency of the Louvre has increased by 21.3% after using the system's optimal evacuation plan. In addition, our model also has a broad application prospect. It can be extended to the evacuation of people in other large buildings by changing the parameters of initialization to improve the evacuation efficiency and protect people's safety.

1. Background

An increasing number of accidents such as terrorist attacks in France arouse the need of effective emergency evacuation plans for many popular tourism destinations. When designing the evacuation plan for the Louvre in Paris, France, it is important to consider that the number and the diversity guests in the Louvre vary from day to day and from year to year. Multilingual groups of tourists and cases of disabled tourists make evacuation in emergencies an challenging task. At the same time, different security issues brought by different exits are also attracting our attention.

Located on the north bank of the River Seine in the heart of Paris, the Louvre is one of the largest and most visited art museums in the world, attracting more than 10.2 million visitors in 2018. In 2018, the number of foreign tourists increased significantly to nearly 7.7 million [1]. According to the Louvre official website, in 2017, shows that tourists from the United States accounted for 18.6% of the total number of the foreign visitors, China accounted for 11.4%, Brazil accounted for 5%, Britain 5%, Germany 4.6%, Spain 4% [2].There are five main entrances and exits and other non-main entrances with low levels of security and low public attention. As shown in the figure below, the red circle marks the main entrance and exit on the second floor of the underground, the blue is the main entrance and exit of the basement, and the rest is the main entrance on the ground.

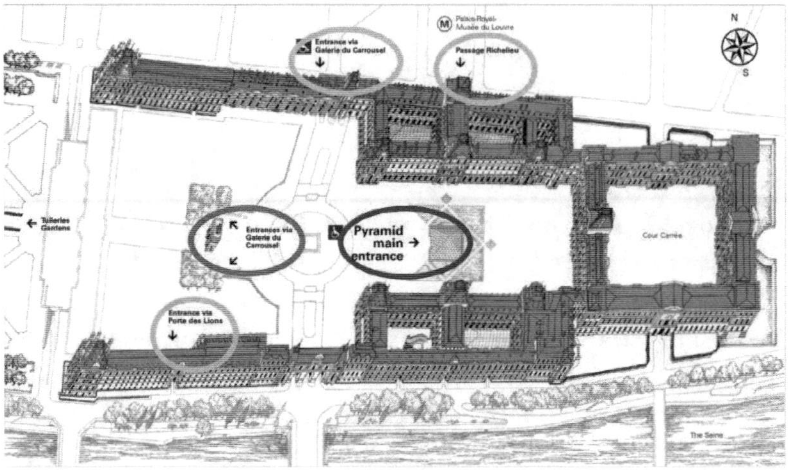

Figure 1. Map of the Louvre.

2. Introduction

This paper develops a model of visitors' attribute generation based on variable parameters, sets visitor's different attributes after considering the environment factor and physical condition. Assuming without guidance, all visitors will leave along the shortest path at the beginning of the evacuation. Through the simulation of our improved Cellular Automaton model simulation, we can determine the location of the bottleneck. After that, based on the system optimization model of Wardrop Principle, we construct the flow network of Louvre architecture. By applying the Frank-Wolfe algorithm, we can obtain the optimal capacity of each path and the congestion coefficient of the section at the minimum total evacuation time. Finally, we can obtain the specific fastest evacuation route.

At the end of the article, we propose the following evacuation plan: Firstly, dispatch officials to the nodes near the bottlenecks. Secondly, officials should guide visitors based on the fastest evacuation route. Thirdly, the open of secret exits depends on whether the overall congestion of the road network. Considering that model input relies heavily on real-time data, we develop a corresponding evacuation plan based on the number of visitors and adopt high-tech as a monitoring segment traffic tool.

The test of the improved Cellular Automaton simulation shows that the evacuation time was reduced by 21.3% after adopting our evacuation plan. In a word, after expanding the scope of application of the model and considering a wide range of factors, our model can effectively find the Louvre's fastest evacuation plan and can be applied to other large crowded occasions with high stability and high fault tolerance.

3. The Math Models

3.1. The Basic Models

3.1.1. Evacuation Path Model Based on Network Flow.

One of the most important data structures in dynamic human traffic distribution based on the Network Flow model [3] is the topology of the road network [4].The main elements of the composition are nodes and road segments, where the nodes include intersections and OD points. The road segments connect two nodes and have different geometric characteristics (such as length) and road segment characteristics (such as maximum capacity). Characteristics have dynamically variable properties (such as crowding, traffic) in the road segment.

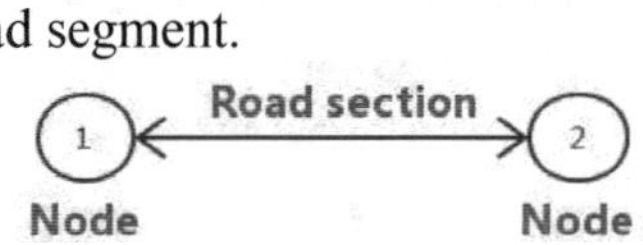

Figure 2. Road Network Structure.

(1) Dynamic system optimal model
When an emergency occurs in the Louvre, pedestrians will choose the path that is most beneficial to them to escape. Congestion often occurs in actual situations, which increases the overall evacuation time. According to the user optimal principle [5] all utilized paths have equal or less impedance, and the unused paths have equal or greater impedance. It is also known that the impedance encountered by a pedestrian on a connection is determined by the actual travel time spent.

In the system optimal model, pedestrians will choose to follow the guidance of the devised plan to maximize the overall efficiency of evacuation and minimize the evacuation time. At this time, the pedestrian escape route is mainly formulated by the relevant staff. During the evacuation process, pedestrians are given guidance to achieve the shortest total evacuation time. In designing the evacuation path, it is necessary to find the optimal solution for the total time that each region reaches the exit, which the following can be solved by the F-W algorithm.

$$\min\{z\} = \sum_{a} t_a(x)x_a \#(4.1)$$

s.t

$$\sum_{k} f_k^{rs} = q^{rs} \#(4.2)$$

$$x_a = \sum_{k}\sum_{s}\sum_{v} \delta_a^{rs} p_k^{rs} \qquad f_k^{rs} \geq 0, x_a \geq 0 \#(4.3)$$

$$h(G, k_v^0, k_v^n) = 0, \ v \in V \#(4.4)$$

Here q^{rs} is the traffic demand between OD pair $r(r \in R)$ and $s(s \in S)$; x_a is the flow on rode section a and f_k^{rs} is the flow on path k OD pair r and s; If the road section a is part of the path k between OD pair r and s $\delta_{a,k}^{rs} = 1$, otherwise $\delta_{a,k}^{rs} = 0$; k_v^o is the path that tourists v originally choose; k_v^n is the tourists' new path selection; G is the adopted boot plan.

Equation 4.2 represents the total flow conservation constraint between OD pair r and s; Equation 4.3 indicates that the traffic on segment a is equal to the sum of the number of visitors for all new paths through path a between all OD pairs; Equation 4.4 represents flow non-negative constraints; Equation 4.5 represents the induction scheme G for the vehicle path.

(2) Principle of the Frank-Wolfe algorithm

In the process of exploring the minimum evacuation time in the Louvre, the evacuation model can be transformed into the optimal solution to the population's arrival and leaving in each region. The general form of the constraint optimization problem based on the Frank-Wolfe algorithm is:

$$\min\{f(x)\} \#(4.5)$$

s.t

$$g_i(x) \geq 0, i = 1,2\ldots,n \#(4.6)$$

If $g_i(x)$ is a linear function, then the constraints of the linear programming are the same, only the objective function becomes nonlinear. The Taylor function is approximated by Taylor and linearized [6]. Expand f(x) at x_k, there is

$$f(x) \approx f(x_k) + \nabla f(x_k)T(x - x_k) \#(4.7)$$

Therefore, the original problem is similar to

$$\min\{f(x)\} \approx \nabla f(x_k)T(x - x_k) \qquad s.t \quad x \in m \#(4.8)$$

Let the optimal solution of this problem be y_k, then $p_k = y_k - x_k$, which is the feasible downward direction of the original problem. An iterative can be done by doing a one-dimensional search along this direction. In order to prevent the result of the one-dimensional search exceeding the feasible domain, the step size is $0 \leq \lambda \leq 1$.

At first a feasible point is initialized in the planned network flow area. When the feasible point $x \in g_i(x)$ is known, the optimal solution can be obtained by solving the feasible linear programming. If $\nabla f(x_k)T(y_x - x_k) = 0$, the K-T point in the region can be known from the F-W convergence theorem. Otherwise, it can be established in the feasible descent direction $p_k = y_k - x_k$ of $x \in g_i(x)$. Starting from point x_k, a one-dimensional linear search is performed along the direction p_k. That is, λ_k is obtained, so that

$$f(x_k + \lambda_k p_k) = \min\{f(x_k + \lambda p_k)\} \quad 0 \le \lambda \le 1\#(4.9)$$

Since x is a convex set, the next iteration point $x_k + 1 = x_k + \lambda_k p_k$ can be obtained. And thus iteratively iterative, a K-T point satisfying the planning condition can be obtained.

3.1.2. Simulation Model Based on Improved Cellular Automaton.
When an accident occurs in a large building that requires evacuation of the crowd, collisions between pedestrians, friction and panic often lead to differences in evacuation speed. Therefore, based on the traditional Cellular Automaton model, we lead into the social force model. We also take the factor of knowledgeable person into consideration, combining the situation of the Louvre during the crowd's moving. In other words, during the evacuation process, pedestrians can tell the people around them what they know.

So, we adopt an improved Cellular Automaton model. The Cellular Automaton model divides the two-dimensional space in the region into square grids of the same size and each grid represents a cell. In the traditional Cellular Automaton model, the cell can only move one grid per unit time, and the speed remains the same. Each grid has two states, one is occupied and the other is not occupied. The mobile type of the cell selects the Moore type field. In the same time, the cell has a total of 8 moving directions [7]. Cellular individuals can choose to stay in the same place or move to the surrounding 8 grids during the evacuation process by judging the surrounding environment.

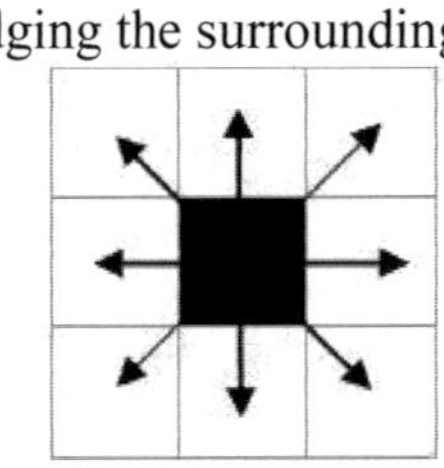

Figure 3. Cell diagram.

Under the improved Cellular Automaton model, the movement of the cell is affected by collision repulsive force (Details in 4.2.3), friction and panic emotions, which cause the pedestrian to change direction and speed during the evacuation process.

In order to make the simulation closer to the actual situation, we make the following improvements to the Cellular Automaton. First, the distance that pedestrians are allowed to walk in a unit time is not limited to one grid. And it can be changed depending on the actual conditions. Second, in the event of exclusion, pedestrian will slow down or turn around to find other paths in order to complete the evacuation in the shortest possible time. Third, when a pedestrian goes through a path and arrives at a place, he will tell the nearby pedestrian the situation of the path in order to find the shortest path.

3.2 Other Sub models

3.2.1. Speed Model
In the event of an accident, the evacuation of people from large buildings to safe areas is an important guarantee for reducing the number of casualties. According to V.M Predtechenskii and A.I Milinskii's study of the flow velocity in different situations, and taking the impact of personal physique into account, we consider different speed of people on the straight roads, walking along the stairs, and in the event of an accident [8].

(1) On the straight roads

On the straight roads, the average evacuation rate is a function of the density of people.

$$V_{L1} = 1.867(\rho_1 + \alpha)^4 - 6.333(\rho_1 + \alpha)^3 + 7.233(\rho_1 + \alpha)^2 - 3.617(\rho_1 + \alpha) + 1.502$$
$$-0.13 \le \alpha \le -0.09\#(4.10)$$

α is related to the attribute generation model, which will be explained in 4.2.2

$$\rho_1 = \frac{N * S}{W * L} \qquad 0.15 < \rho_1 < 0.95\#(4.11)$$

Here N is the total tourist number of the flow.

(2) When walking along the stairs

When the crowd walks along the stairs, the crowd speed needs to be modified.

$$V_{D1} = X_{down} V_{L1} \#(4.12)$$

$$X_{down} = 0.775 + 0.44e^{-0.92\rho_2} \sin(5.16\rho_2 - 0.224) \#(4.13)$$

(3) In the event of an accident

In the event of an accident, people will suddenly speed up due to the spread of panic in the crowd in the same situation.

$$V_{L2} = U_{down} V_{L1} \#(4.14)$$

$$V_{D2} = U_{down} V_{D1} \#(4.15)$$

Here V_{L2} is the evacuation speed under straight roads in the event of an accident; V_{D2} is the evacuation speed along the stairs in the event of an accident.

3.2.2. Attribute Generation Model

(1) Language attribute

Assuming that the number of native speakers of French, English and other languages in the Louvre is A, B, C respectively.

The total number of tourist M is get from the Louvre's report in the past two years. The number of tourists in the several countries with the most tourists is m_i $(0 \le i \le im)$. Assuming that the total population of the world is Np, the total population of all the above countries is np and the population of other country is n_i $(0 \le i \le im)$. The number of tourists whose mother tongue is French is n_{lan1}. The number of tourists whose mother tongue is English is n_{lan2}. Let the first language attribute is $m_i.lan$.When the native language of the tourist is French, $m_i.lan = 1$. When the native language of the visitor is English, $m_i.lan = 2$. When the native language of the visitor is other languages, $m_i.lan = 3$.

$$\begin{cases} A = \frac{1}{M} \sum_{k=0, m_k.lan=1}^{im} m_k + \frac{n_{lan1}}{Np - np} \\ B = \frac{1}{M} \sum_{k=0, m_k.lan=2}^{im} m_k + \frac{n_{lan2}}{Np - np} \\ C = 1 - A - B \end{cases} \#(4.16)$$

According to the above-mentioned total proportion of the whole year, we assume that the proportion of the native speakers of French, English and other languages in the Louvre at a certain moment are a, b, c respectively.

$$\begin{cases} a = A + x \\ b = B + y \\ c = 1 - a - b \end{cases} \#(4.17)$$

The values of x, y are subject to a uniform distribution of$(-\beta_1, \beta_2)$, $\beta_1 < 0.15$

(2) Physical condition attribute

In addition to the disabled, the physical condition index H of the tourists is subject to the normal distribution $N(\mu, \sigma^2)$. And we take the physical condition index as h_1. The proportion of tourists whose physical condition index is lower than h1 is Mh_1,

$$Mh_1 = \emptyset\left(\frac{h_1 - \mu}{\sigma}\right) \#(4.18)$$

The physical condition index of the disabled person is set to a fixed value of 0 and the total proportion Mh_2 is set to 0.5%. Because the proportion of disabled people is very low, we can consider Mh as the proportion of the total number of tourists whose physical condition index is higher than h_1 and lower than h_2. According to the above ratio, it can be assumed that the proportion of the people whose physical condition is lower than h_1 in the Louvre at a certain moment is mh_1 and the proportion of disabled persons is mh_2.

$$\begin{cases} mh_1 = Mh_1 + z_1 \\ mh_2 = Mh_2 + z_2 \end{cases} \#(4.19)$$

The value of z_1 is subject to a uniform distribution of$(-\beta_2,\beta_2)$, $\beta_2 < 0.15$
The value of z_2 is subject to a uniform distribution of$(-\beta_3,\beta_3)$, $\beta_3 < 0.05$

(3) Team attribute

Assuming that the number of groups whose population is i in the whole year is p_i $(0 \leq i \leq 10)$, and the group ratio is:

$$N_{p_i} = \frac{ip_i}{M} \#(4.20)$$

According to the above ratio, we can assume that the number of tourists in the Louvre at a certain moment is np_i.

$$np_i = Np_i + w\#(4.21)$$

The value of w is subject to a uniform distribution of$(-\beta_4,\beta_4)$, $\beta_4 < 0.1$

Under the above three models, consider the influence of these three attributes on the influence factor of the flow situation. The α value of tourists whose native language is French, English, and other languages increases in turn. The α value of tourists decreases with the body condition index increasing. The α value of tourists increases with the number of teams increasing. Besides, the lower the physical condition index, the greater the effect will be on the α value. The impact on the α value is small when the team size is large.

$$\alpha = 0.22\mathrm{In}(np + 1) * (0.98a - 0.15b + 0.33c) * \left(mh^2 + 1.02\right) - 0.097\#(4.22)$$

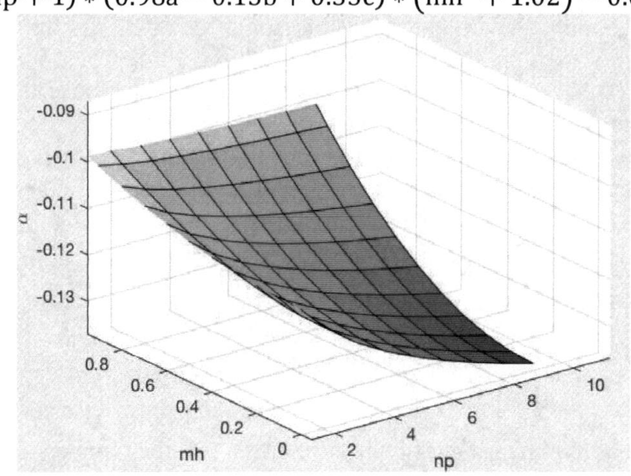

Figure 4. α with respect to np, mh at a=0.25, b=0.16.

3.2.3. Exclusion model

For the convenience of narrative and calculation, we assume that the coordinates of the cell at each position are $E_{ij}(i,j)$. When an unexpected situation requires evacuation, the pedestrian will choose the shortest path to leave the building. In the same floor, the location of the different exits is $E_{mn}(m,n)$.

At this time, if there is no obstacle, the minimum Euclidean distance from the pedestrian to the exit is recorded as

$$d_{min} = \min_n \left\{\min_n \sqrt{(m-i)^2 + (n-j)^2}\right\} \#(4.23)$$

If there is an obstacle, the minimum distance is calculated using the Dirichlet algorithm to allow the pedestrian to leave as soon as possible.

During the evacuation process, repulsive forces occur due to collisions between people and between people and the wall. In addition, when an accident such as a fire or a terrorist attack occurs, the tendency to reach a dangerous distance will cause the pedestrian to repel the source of the danger. According to the quantification of repulsive force by Weiguo Song [9],the Sigmoid function is used, and the neural network algorithm is used to define the rejection probability as

$$p = \frac{1 - e^{-kv}}{1 + e^{-kv}} = \frac{x_{ij} - x_{min}}{x_{max} - x_{min}} \#(4.24)$$

In the above formula, k is the hardness coefficient. x_{ij} is the Euclidean distance of the cell $E_{ij}(i,j)$ to the dangerous source. x_{max} is the distance farthest from the adjacent 8 cells from the danger source. And x_{min} is the nearest to the 8 cells from the dangerous source distance. The closer the source of danger is, the greater the repulsive force generated. V is the relative movement speed of pedestrians. When the collision between pedestrians is mutually exclusive, $V = 2v$. When the pedestrian and the stationary object repel, V = v.

4. Experimental

4.1. Parameter determination

(1) Speed parameter

According to the data mentioned in 2.1.2, among the foreign tourists who come to the Louvre, the native English speakers account for about 15.61% of all tourists, and native speakers of other languages account for 21.07% of all visitors. The remaining tourists from non-mentioned countries accounted for 38.82%. According to the current number of people in the world, we estimate that A is about 25.138%, B is about 16.248%, and C is about 58.614%.

According to the above-mentioned total proportion of the whole year, it is assumed that the proportion of the native speakers of French, English and other languages in the Louvre at a certain moment respectively is a = 0.25, b = 0.16, c = 0.59.

Take $H \sim N(0.5, 153^2)$, the proportion of disabled tourists is 0.005,

Then $mh = E(h) = 0.005 * 0 + 0.95 * \mu = 0.475$;

$np = \sum_{i=0}^{10} np_i = 2.5$;

U=1.2;

$\alpha = 0.22 In(np + 1) * (0.98a - 0.15b + 0.33c) * (mh^2 + 1.02) - 0.097 =- 0.118$

(2) Road network topology

According to the Louvre plan and the Network Flow model in 4.1.1, the doorway of the main exhibition area and the main staircase are used as nodes. The network structure of the Louvre can be established as follows:

Since the evacuation exits are concentrated on the ground floor, the evacuation plan for the ground floor is more critical and representative than that for the other layers. Considering the complexity of the Louvre building structure, the evacuation of the road sections will only be carried out. Solving the evacuation scheme for other layers can also be obtained by applying the model.

Journal of Physics: Conference Series **1419** (2019) 012043 doi:10.1088/1742-6596/1419/1/012043

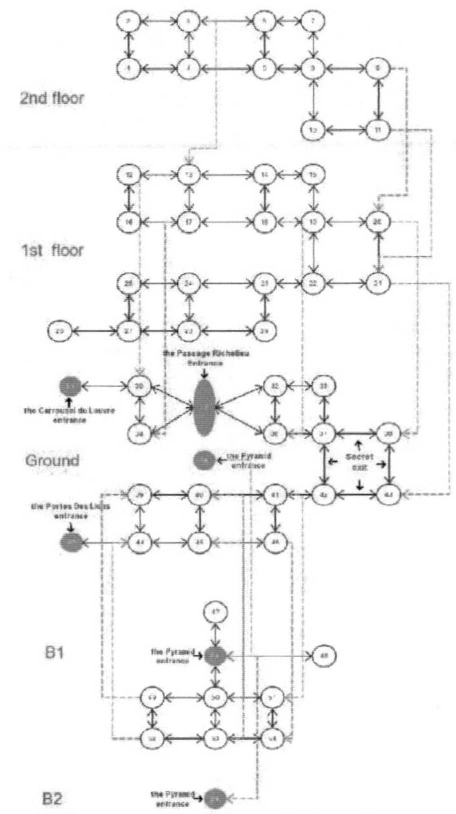

Figure 5. Path Network Diagram of the Louvre.

(3) Segment parameters

We assume that the capacity of the road section in the exhibition hall is 2.75 per/m^2. Then we divide the road section into main roads and branch roads and set the corresponding maximum road section capacity to 850 people and 390 people.

Combined with the analysis of the behavior of visitors to the Louvre Museum based on Yuji Yoshimura et al, the main visiting routes of the Louvre visitors and the visits can be known by using the data of Bluetooth data [10]. The popularity of the exhibition area determines the congestion of the path and stairs: the most congested route flow density ρ_1 was 0.8, and the congestion route ρ_1 was 0.6. Uncrowded route flow density ρ_1 has a value of 0.15. The route flow density ρ_2 of the stairs and the stairway has a value of 0.8. The congested route ρ_2 has a value of 0.6. And the uncongested route ρ_2 has a value of 0.2. According to the flow velocity model in 4.2.1, the average velocity of the flow of each direct pathway can be calculated. It can be seen from 1.1.2 that the total number of visitors to the Louvre in 2018 is about 10.2 million, so the total flow of people is assumed to be 27,945 per day. Of course, if you can monitor the flow of people in each section and perform dynamic input in real time, the accuracy of the output will be improved.

Finally, we can get the impedance value of each direct path.

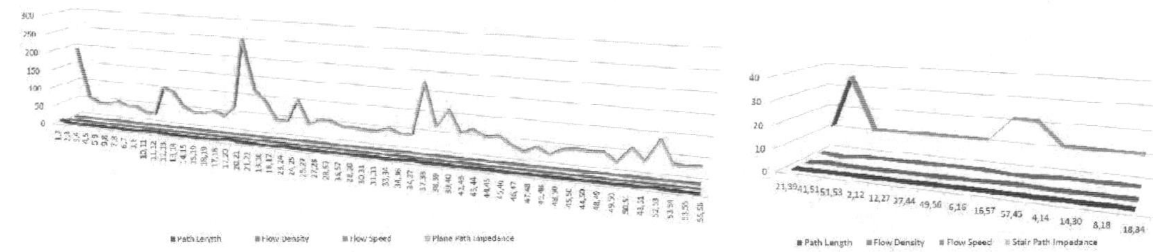

(a) Plane Path Impedance (b) Stair Path Impedance

Figure 6. Path Impedance.

4.2. Solution Process and Analysis

4.2.1. Determination of the Louvre Bottleneck Position

(1) Preliminary judgment of bottleneck position

Bottleneck locations often occur in restricted areas during crowd evacuation. The bottleneck position is characterized by the narrow position of the bottleneck itself, and the relatively wide space connecting the two ends of the bottle. In the process of advancing toward the bottleneck position, the flow of people is often blocked due to the sudden contraction of the space, and the pedestrian movement drastically slow down even stop [11].The actual bottleneck position usually exists in the restricted area of the crowd evacuation process, often near the doorway, underground passages, and so on. Therefore, when selecting nodes, the doorway of each pavilion shall prevail.

(2) Bottleneck position under simulation

Taking the ground layer as an example, the pedestrian simulation process is as shown in the following figure. In an emergency situation, the crowd usually speeds up and escapes from the building along the shortest path. After the evacuation begin, pedestrians gradually move to the exit. Initially, pedestrians left the doorway of each room according to the known escape routes. It can be observed that as some pedestrians become denser near the entrance, while the people lagging behind push those in ahead, resulting in collisions and fractions, and blockage at the door slows down the speed of moving sharply.

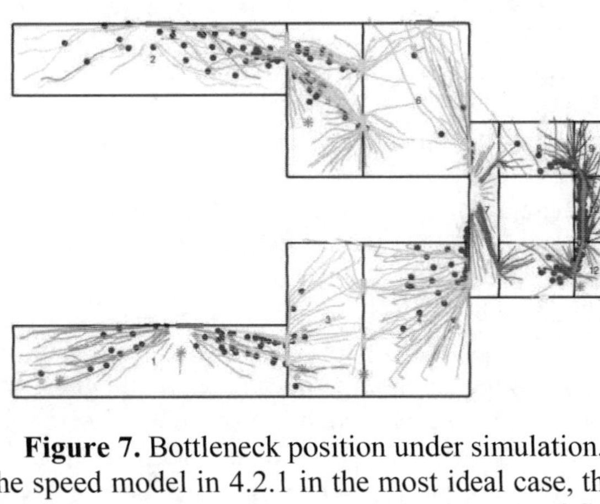

Figure 7. Bottleneck position under simulation.

From the solution of the speed model in 4.2.1 in the most ideal case, the speed of the crowd in the emergency state is 1.124m/s. The doorway of each room on the ground floor of the Louvre is divided into 12 locations as the following figure. When the blockage occurs, the speeds at 4, 4, 9, and 12 will drop to below 0.8m/s, in line with the above judgment on the bottleneck position

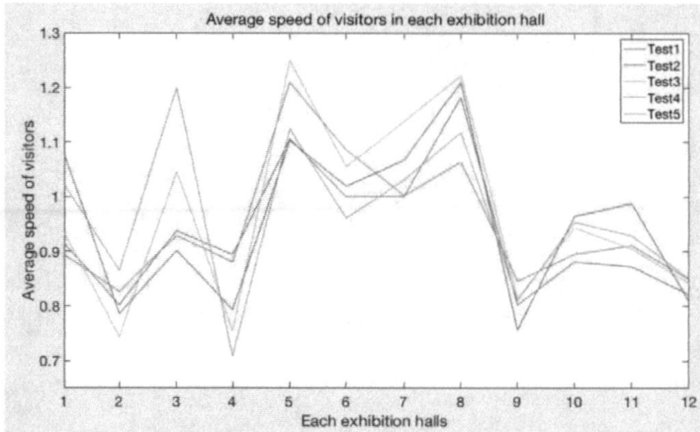

Figure 8. Average speed of visitor in each exhibition hall.

4.2.2. Determination of the Louvre Evacuation Plan

The following are the 12 routes with the largest expected traffic flow and the corresponding information of the road segment when the minimum evacuation time is reached.

Path type	Main road	Main road	Main road	Main road	Main road	Main road	Branch road	Main road	Main road	Main road	Branch road
Starting node	42	37	38	E3	37	43	41	41	40	40	36
Ending node	37	33	43	34	38	42	46	40	39	45	E3
Segment capacity	850	850	390	850	850	850	390	850	850	850	850
Flow density	0.6	0.15	0.15	0.15	0.15	0.15	0.15	0.6	0.6	0.15	0.15
Forward traffic flow	521.445	859.49	440.12	502.928	490.08	340.981	500.439	409.992	480.721	448.23	458.739
Reverse traffic flow	701.325	21.31	402.392	308.123	301.211	420.623	80.232	148.239	42.872	43.221	23.298
Total traffic flow	1222.77	880.8	842.512	811.051	791.291	761.604	580.671	558.231	523.593	491.451	482.037
Forward VC ratio	0.613	1.011	1.129	0.592	0.577	0.401	1.283	0.482	0.566	0.527	0.540
Reverse VC ratio	0.825	0.025	1.032	0.362	0.354	0.495	0.206	0.174	0.050	0.051	0.027
Average evacuation time/s	27	40	32	24	48	24	39	28	20	32	35

Figure 9. 12 Paths with the largest expected distribution of traffic.

VC ratio: The segment saturation (v/c) is the ratio of the segmented traffic to the maximum capacity of the segment, which reflects the congestion of the segment. From the above data we can formulate the following solutions:

(1) Reasonably deploy emergency personnel to indicate evacuation locations.

It is estimated that the roads with heavy flows of people are 37 →33, 42 → 43, 38 → 37, 37 →42 and they are concentrated in adjacent areas. Due to the limited number of emergency personnel, museum officials and emergency personnel should be dispatched to guard the nodes with the highest visitors flow rate, which are the nodes 37,38, 41, 42, 43, so as to maximize the effect of evacuation.

(2) Guide the division of people in different routes.

There are 4 roads for people at node 37 to choose from. The flow of people allocated to the segment 37→33 is the heaviest, so the emergency personnel should guide the visitor to select this road (although it is not the shortest way for some visitors). At node 37, the visitor should be guided to select the road 37 → 33. At node 38, the visitor should be guided to select the link 38→ 43. At node 41, the visitor should be guided to select the link 41→ 46. At node 42, the visitor should be guided to select the link 42 → 43. At node 43, the visitor should be guided to select the link 43 → 38.

(3) Rational use of secret exports.

Based on the literature [12] ,we can know the rating criteria of section load (VC): stable traffic (0.7 1), slow traffic (1-1.5), general congestion (1.5-2), and severe con-gestion (≥2). According to the degree of congestion in each section, the museum should determine the time to open the secret entrance. When 75% of the path is in slow transit, a secret exit should be opened. After a period of time, when the above limitation is reached again, other secret exits should be opened in order to reduce the load on the road section.

(4) Solve the impact of fixed hazard sources on optimized road sections.

When there exists a fix hazard source, some evacuation route might be changed or closed. In such cases, the plan of the evacuated human traffic should be modified and the different evacuation scheme should be re-allocated.

(5) Guide the evacuation of visitors on other floors according to the traffic flow of ground.

For the 1st and 2nd floors, considering that the stairway capacity is usually smaller than the inner passage of the building, visitors should be guided to evacuate using the stairs which is near the path allocating traffic in the ground floor. Officials should direct visitors to leave the stairway near the ground floor. For example, guide the visitors in first floor to use stairs 11 →21; 9 →20.

For B1, considering that the traffic on the ground floor is already large, visitors should be diverted from the second floor.

For B2, visitors should leave directly from the exit in the second floor.

4.2.3. Solution Verification

Repeatedly applying of the improved Cellular Automaton simulation shows that the average evacuation time of the ground floor is around 333.84 seconds While in 5.2.1, the evacuation of pedestrians along their shortest path takes about 424.16 seconds. Evacuation efficiency was increased by 21.3% after applying the system's optimal evacuation scheme.

5. Conclusions

5.1. Policy and procedural recommendations for emergency management of the Louvre

In order to make the evacuation speed the fastest in the Louvre when an unexpected situation happen, it is necessary to consider the diversion of the visitors near the bottleneck and the path and time of emergency personnel entry.

(1) In order to reduce the congestion of the bottleneck position and decrease the overall evacuation time, we encourage the Louvre to develop the following measures:

- Take into account the impact of high-tech science and technology. We can set up Bluetooth monitors in various sections to monitor the traffic of each section in real time to help us determine the evacuation plan in case of an emergency.
- At the intersections where visitors leave the Louvre, a sign based on text or arrow content should be set to guide visitors in order to achieve the shortest evacuation time.
- After several simulations and actual rehearsals, the location of the bottleneck in the Louvre can be determined. The Louvre officials should be set up on the road leading to the main branch of the bottleneck to make a proportional diversion of the people in an emergency. According to the guidance of 5.2.2, all the people can move to the exit at the fastest speed to achieve the shortest total evacuation time.

(2) In order to minimize the time for emergency personnel to enter the Louvre and ease traffic pressure, we encourage the Louvre to develop the following measures:

- Take into account the low security level at the secret channel, we should try not to open the secret exits. Based on the relationship between real time traffic and optimal current capacity, we can determine the number of secret exits that should be open to allow emergency personnel to enter the Louvre.
- We should arrange emergency personnel to be in a position to the secret exit and other exits, and to inform the tourists the path leading to secret exits to ease the traffic pressure of other exits.

(3) In order to allow slower visitors to leave the Louvre as soon as possible, we encourage the Louvre to develop the following measures:

- Considering the different attributes of tourists, the elderly, the disabled and the children are slower. Therefore, when the emergency personnel are guiding, they do not divert such groups of people. Such groups of people should escape through the fastest path to minimize the overall evacuation time

Journal of Physics: Conference Series **1419** (2019) 012043 doi:10.1088/1742-6596/1419/1/012043

5.2. Other large crowded structural adjustment and implementation model

Apart from museums,, we also consider two other types of large congestion structures for evacuation in the event of an emergency.

(1) Transportation hubs (subway interchanges, train interchanges, etc.)

The evacuation plan for the subway interchange is different from that of the Louvre. It is necessary to take into account that convection would not occur when people move to different exits, because they often walk in a single direction. Therefore, we made the following improvements to the evacuation model.

- In the improved Cellular Automaton model, the case in subway interchange does not need to consider the repulsion caused by factors such as collisions between pedestrians. However, it is necessary to consider the friction between people due to crowding.
- In the process of abstracting the Louvre into a Network Flow model, the connection between nodes is considered to be two-way. While in the subway interchange, the connection between the nodes should be abstracted as one-way communication, which is more realistic.

(2) High-rise buildings (high-rise residential buildings, high-rise office buildings, etc.)

- The high-rise building evacuation plans are different from the Louvre evacuation model. Pedestrians need to reach the lowest level of exit to escape. Therefore, in the evacuation process in high-rise buildings, the Network Flow model should be abstracted, and the connections between the floors should be connected in one direction to determine the shortest evacuation time.

References

[1] 2018. http://presse.louvre.fr/10-2-million-visitors-to-the-louvre-in-2018/
[2] 2017. http://presse.louvre.fr/8-1-million-visitors-to-the-louvre-in-2017/
[3] Yunhan Li. 2015. Research on Design Methods and Evaluation of Evacuation Plan in Urban Rail transit Transfer hub Based on Simulation. [D]. Beijing: Beijing Jiaotong University. 22-24
[4] Hongliang Ma. 2015. Research on the Emergency Rescue and Evacuation System Based on Dynamic Tr
[5] affic Assignment[D]. Beijing: Tsinghua University.
[6] Wardrop J G. 1952. Some theoretical aspects of road traffic research[J]. Proceedings of the Institution of Civil Engineers, 1(2):325-378
[7] Shengjia Xue. 2002.The Method of Finding the Optimal Solution Set of Linear Fractional Programming[J]. Journal of Applied Mathematics and Computational Mathematics. 16(1):90-96.
[8] Xinlong Zhang, Xiuwan Chen, Huaiyu Li and Fei Li. 2017.An Improved Cellular Automata Model for Simulating Pedestrian Evacuation[J]. Geomatics and Information Science of Wuhan University. (49):1130-1136
[9] V.M.Predtechenskii, A.I.Milinskii. 1978. Planning for foot traffic in buildings [M]. New Delhi: Amerind Publishing Co. New Delhi.
[10] Weiguo Song, Yanfei Yu, Weicheng Fan. 2005. A person evacuation Cellular Automata model considering friction and repulsion[J]. Science in China Ser.E Engineering & Materials Science. 35(7):725-736
[11] Yuji Yoshimura, Alexander Amini, Stanislav Sobolevsky, Josep Blat and Carlo Ratti. 2017.Analysis of pedestrian behaviors through non-invasive Bluetooth monitoring[J]. Applied Geography. 43-51
[12] Weichen Liao. 2017. Experimental and Simulation Research on Pedestrian Dynamics in Bottlenecks[D]. Beijing: Beijing University of Chemical Technology. 5-6.
[13] Yongquan Liao, Xiaoliang Sun, Limin Jia, Honghui Dong and Qinghua Zhang. 2012. Evaluating Traffic Status of Urban Expressway in Beijing Based on Road Vehicle Capacity[J]. Logistics Technology. 31(2):87

Research and Simulation Analysis of Systematic Voltage Fluctuation in DC Commutation Failure

Lei Yang[1], Xin He[1], Shenghong Ma[1], Chuan Xiang[1], Can Xiao[2], Chao Xing[1], Shoudong Xu[1] and Chang Chang[2]

[1]Yunnan Power Grid Co., Ltd. Electric Power Research Institute, Kunming, China, 650217
[2]School of Electrical Engineering, Southwest Jiaotong University, Chengdu, China, 610031

Abstract: The failure of high-power DC commutation leads to large fluctuations in the voltage of the transmitting end grid, which brings great harm to the safe and stable operation of the transmitting end grid. In this paper, the detailed analysis of PMU data is carried out for the voltage fluctuation events of the transmitting end caused by the frequent DC commutation failure in recent years. By constructing a simulation model in BPA and RTDS, the fault back-tracking of the voltage fluctuation of the power supply with the commutation failure is carried out. The simulation results show that the main cause of the fluctuation is caused by the timing coordination and the delay of the rectification side AC filter. This is of great significance for reducing voltage fluctuations in the case of subsequent DC system commutation failure, and it also has great significance for maintaining safe and stable operation of the power grid.

1. Introduction

In order to reduce the risk of large-scale power outages in the power grid and enhance the controllability of the impact of large-scale power grid accidents, in 2016, China Southern Power Grid proposed to use Yunnan Power Grid as a separate transmitting power grid in the medium- and long-term grid planning, and implemented the main network of China Southern Power Grid. Asynchronous networking operation [1-3]. The problem of DC transmission line after asynchronous networking has become a major problem that plagues the safe and stable operation of China's power grid.

In recent years, voltage fluctuations have often added a lot of trouble to industrial production, scientific research and daily life, sometimes damaging equipment and causing accidents. With the rapid development of modern science and technology, the increasing popularity of electronic computers and various electronic devices, the quality of supply voltages in factories, mines, scientific research, post and telecommunications, and hospitals are getting higher and higher. However, due to the large amount of impact load, intermittent load, and various short-circuit faults in the power supply system, attempts to cause the system voltage to change rapidly and rapidly, that is, voltage fluctuations. DC voltage fluctuation is one of the difficulties that plague the safe and stable operation of the DC transmission system of the Southern Power Grid. Through a large amount of research and practice [4-6], the operation and maintenance unit has gradually mastered the abnormal causes and possible effects, not only taking the action. Effective improvement measures to avoid the adverse effects of DC voltage fluctuations on the operation of the system as much as possible, and also to develop corresponding maintenance plans. A large number of practices have proved that these overhauls can quickly and accurately identify and eliminate fault points under the conditions of system operation, which is of great significance for maintaining safe and stable operation of the power grid.

Literature [4-6] has confirmed the root cause of the abnormal fluctuation of high-voltage DC voltage caused by the failure of the photoelectric sensor in the high-voltage DC voltage measurement loop through a lot of research and practice. The literature [4,7] carried out simulation analysis on the abnormal voltage fluctuations of Tianguang and Niu Cong from the DC transmission system, and analyzed the causes of abnormal voltage fluctuations. In the literature [8], a DC voltage fluctuation suppression method suitable for modular multilevel converter HVDC transmission system (MMC-HVDC) is proposed for the defects of the DC voltage fluctuation suppression algorithm in flexible DC transmission systems. Thus, the MMC AC side current remains symmetrically operated while the DC side voltage, current and power remain constant. Literature [9] analyzed the phenomenon of several outages caused by the action of the valve protective triggering function caused by the high voltage DC voltage fluctuation of the rectifier side of a DC transmission system. It is proposed that the DC line voltage measurement system needs to be modified to solve this problem. In [10], the mechanism of voltage fluctuation was studied and verified by constructing a high-voltage DC actual control protection device in RTDS. The change in internal resistance at the time of sensor failure is the direct cause of DC voltage fluctuations.

In August 2018, a DC system of Yunnan Power Grid failed to commutate, which led to fluctuations in the voltage of Yunnan Power Grid. This incident caused great concern of relevant personnel. In this paper, the voltage fluctuations after this commutation failure are simulated and analyzed in RTDS and BPA respectively, and compared with the PMU observation data. The main reason leading to this voltage fluctuation is the timing matching and the rectifier side AC filter triggered by the delay.

2. Voltage fluctuation

The ideal state of the power supply system is that the three-phase AC power supply should be symmetrical, the voltage rms value is constant, and the load characteristics are independent of the system voltage level. This requires ensuring that the user's load is distributed in three-phase balance, and that the power is absorbed at a constant power. At the same time, it is necessary to ensure that the common connection point has an infinite short-circuit capacity, and the equivalent reactance of the system is zero. However, in the actual operation process, it is impossible to meet such conditions. Generally, the voltage of the power supply system changes from time to time. Therefore, the voltage rms value cannot be kept constant [11], or the actual voltage and system. The phenomenon in which the nominal voltage deviates is collectively referred to as voltage fluctuation. Its value d is often expressed as a percentage of the ratio of the difference ΔU between the two adjacent extreme values U_{max} and U_{min} (which are the root mean square values) to the nominal value U_N. which is

$$d = \frac{\Delta U}{U_N} \times 100\% \tag{1}$$

The number of voltage fluctuations per unit time is usually called frequency, expressed in r, and the unit is min^{-1} or s^{-1}. There are various reasons for voltage fluctuations. Short circuit faults in power systems or switching of large rectifier devices and reactive power compensation devices may cause fluctuations in plant voltage.

Since the volatility load frequently uses the rapidly changing electric energy from the power distribution system during the operation, the impact power change occurs, causing the voltage of the common connection point to fluctuate rapidly in a short time and obviously deviate from the nominal voltage. value.

Voltage fluctuations are an important indicator of power quality. Rapid voltage fluctuations can degrade the performance of electrical equipment, the length of time that automatic remote devices and electronic equipment work, the inferior quality of products, and the long-term flashing of lighting. With the development of production automation and intelligence, the level of electrification continues to increase, and voltage fluctuations have caused widespread concern.

3. Data Analysis

This paper builds a schematic diagram of the grid structure of a DC system in Yunnan Power Grid as shown in the Figure 1.

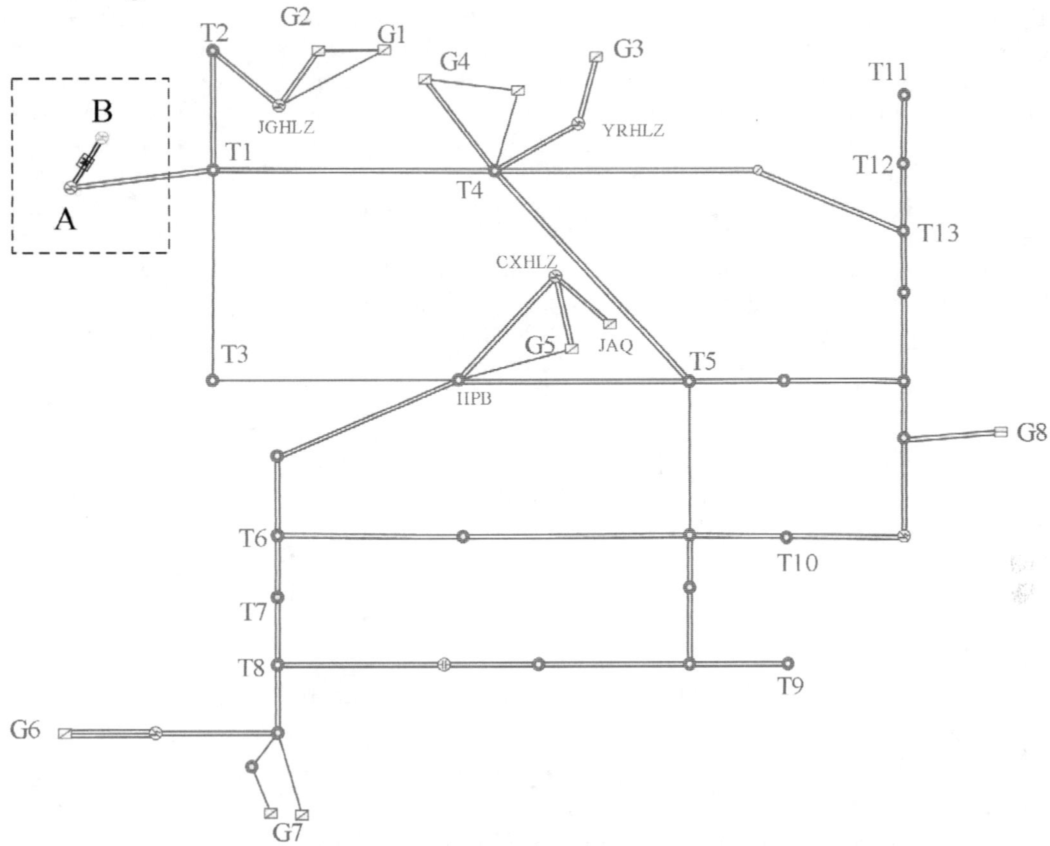

Figure 1. Yunnan power grid structure diagram.

Among them, G1-G8 is a power plant, and T1-T13 is a substation. This paper compares the voltage changes of these stations and the subsequent analysis of the voltage fluctuations of these plants in the case of commutation failure through BPA and RTDS. The main reason for voltage fluctuations is obtained.

3.1. Power Plant Voltage Fluctuation Analysis

By reading the PMU data, the voltage fluctuations of the DC power supplies are as follows.

Table 1. Power plant voltage fluctuations.

Measuring point	Us（kV）	U_{min}	T_{min}（s）	U_{max}	T_{max}	ΔU
G1	310.30	285.8	29.39	331.8	29.45	46.0
G2	309.95	281.7	29.39	333.3	29.45	51.6
G2	312.82	294.3	29.39	324.2	29.45	29.9
G4	303.26	288.2	29.38	317.2	29.44	29.0
G5	311.64	295.6	29.38	318.3	29.44	22.7
G6	309.79	302.6	29.38	316.0	29.98	13.4
G7	308.26	302.6	30.38	313.3	30.95	10.7
G8	310.09	299.5	29.37	316.3	29.43	16.8

The voltage fluctuation Δ U is the difference Us between U_{max} and U_{min} is the steady state value of the voltage, the unit is kV, U_{min} is the voltage minimum value, U_{max} is the voltage maximum value, T_{min} is the voltage minimum time, and T_{max} is the voltage maximum time, the unit both are seconds, and the starting time of voltage fluctuations of each power plant is basically the same, between 209.34-209.37.

It can be seen from the above table: (1) It can be seen from the table that the power point close to the DC line has a significant voltage amplitude, and the voltage fluctuation amplitude is relatively small from the power point far from the DC line. (2) The voltage reaches a minimum value after 0.02 seconds (corresponding to time 29.38), and reaches a maximum value after 0.06 seconds (corresponding to time 29.44), indicating that the voltage adjustment is basically synchronized. The distance of the power plant is shown in Figure 1. The time lag of the maximum G6 and G7 power plants far from the DC line reaches the maximum at 29.98 and 29.95 (minus 1 second timing error). Figure 2 is a comparison of the voltage fluctuation curves of G2 and G6. It can be seen from the figure that the voltage trends of the front end of the two power plants are the same, so the possible reason for reaching the maximum value after the G6 and G7 is that the voltage fluctuation of the near-end unit of the DC line is clear. Clear, and the remote unit has a small adjustment range. With the reactive power distribution and voltage change, the voltage regulation will reach the maximum value after the disturbance.

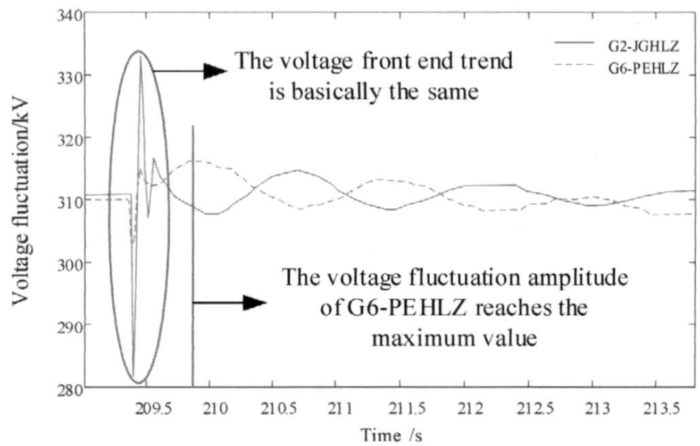

Figure 2. Comparative Analysis of Voltage Fluctuation Curves between G2 and G6.

(3) The voltage fluctuation of the DC near-end unit is the largest, and the voltage fluctuation of the unit farther from the DC line is smaller. The voltage fluctuations from large to small are: G2, G1, G3, G4, G5, G8, G6, G7.

3.2. Analysis of substation voltage fluctuation

Table 2. Substation voltage fluctuations.

Measuring point	Us(kV)	U_{min}	T_{min}(s)	U_{max}	T_{max}	ΔU
T1	308.007	252.0	29.38	344.8	29.44	92.8
T2	307.292	263.3	29.37	338.4	29.43	75.1
T3	288.68	278.8	29.37	288.7	29.4	9.9
T4	306.212	284.6	29.38	322.2	29.44	37.6
T5	310.145	294.7	29.37	318.7	29.43	24.0
T6	308.624	298.0	29.38	314.2	29.44	16.2
T7	310.149	302.5	29.37	315.4	29.44	12.9
T8	309.97	304.1	29.38	314.8	29.44	10.8
T9	309.894	305.9	29.39	315.6	29.45	9.7
T10	310.976	300.2	29.38	316.8	29.44	16.6
T11	311.316	303.1	29.38	315.1	29.8	12.0
T12	312.813	303.0	29.38	316.0	29.44	12.9
T13	313.156	300.5	29.38	318.6	29.44	18.1

Parameter meaning as described above and the voltage fluctuation start time of each substation is basically the same, between 209.35-209.37.

It can be seen from the above table: (1) The voltage reaches a minimum value after 0.02 seconds (corresponding to time 29.38), and reaches a maximum value after 0.06 seconds (corresponding to time 29.44), indicating that the voltage adjustment is basically synchronized. (2) The time lag of T11 which is far from the DC line reaches the maximum value, and reaches the maximum value at 29.8. Figure 3 is a comparison of the voltage fluctuation curves of T1 and T11. It can be seen from the figure that the voltage changes of the front end of the two substations are consistent. Therefore, the possible reason for reaching the maximum value after the T11 is that the voltage fluctuation of the near-end substation of the DC line is clear and clear. However, as the remote substation changes with the reactive power and voltage, the voltage regulation will reach the maximum value after the disturbance.

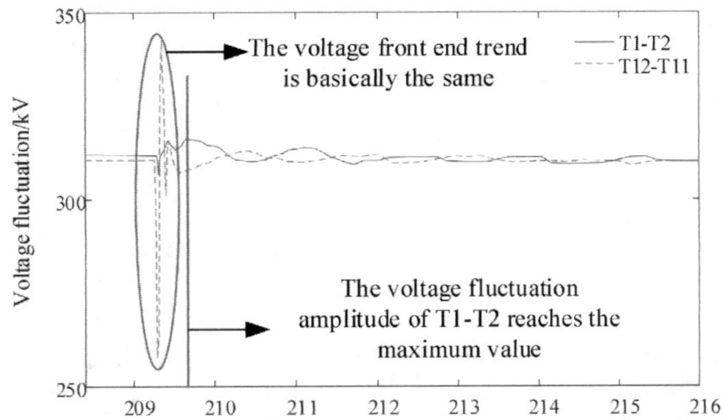

Figure 3. Comparative Analysis of Voltage Fluctuation Curves between T1 and T11.

(3) The voltage fluctuation of the near-end substation of the DC line is the largest, and the voltage fluctuation of the substation is farther away from the DC line. The voltage fluctuations from large to small are: T1, T2, T4, T5, T13, T10, T6, T7, T12, T11, T8, T3. It should be noted that the steady-state voltage of T3 itself is low, only 288.68kV. During the voltage fluctuation process, the voltage only has a falling process and there is no rising process. Therefore, it is not considered in the comparison of voltage fluctuations.

It can be seen from the table that the magnitude of the voltage fluctuation depends mainly on the distance of the plant. The closer to the DC line, the greater the voltage fluctuation of the plant station. It can be seen from the voltage fluctuation diagram that after about 4-7 cycles of fluctuation, the voltage fluctuation subsides and gradually disappears.

3.3. Effect of Commutation Failure on Reactive Power and Voltage of Rectifier Station

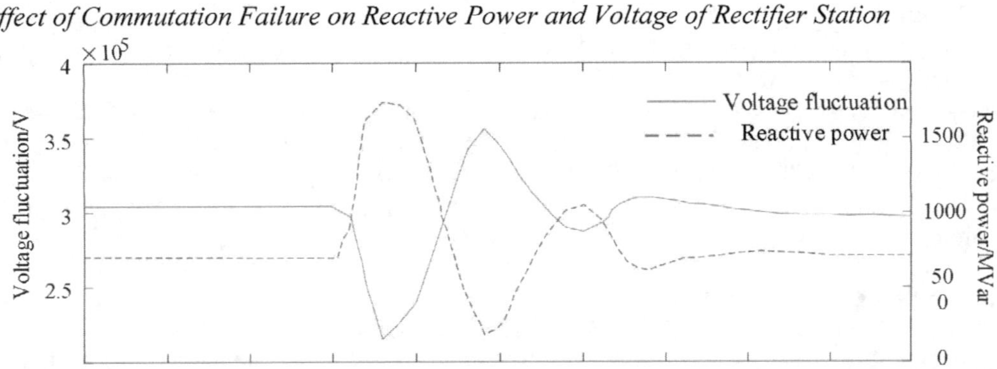

Figure 4. Switching voltage and reactive power of A Converter Station.

Figure 4 shows that when the commutation voltage is reduced, the reactive power is absorbed from the system, and when the commutation voltage is increased, the reactive power is delivered to the system.

From the view of the reactive power of the Xinping double line, it also shows that the DC commutation fails, resulting in a voltage drop on the rectifier side. The system fills the reactive power through the Xinping double line; after the DC power drops to 0, the commutation voltage rises. High, the reactive power is delivered to the system side; the commutation failure ends, the DC voltage is restored, and the system and the commutation voltage are restored. The power of each station is positive as it flows into the plant, and the flow station is negative.

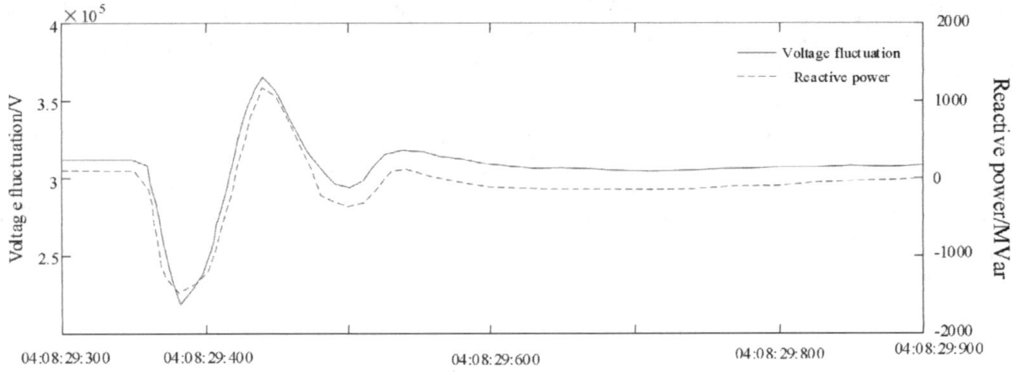

Figure 5. Xinping A line voltage and reactive power observed from A Converter Station.

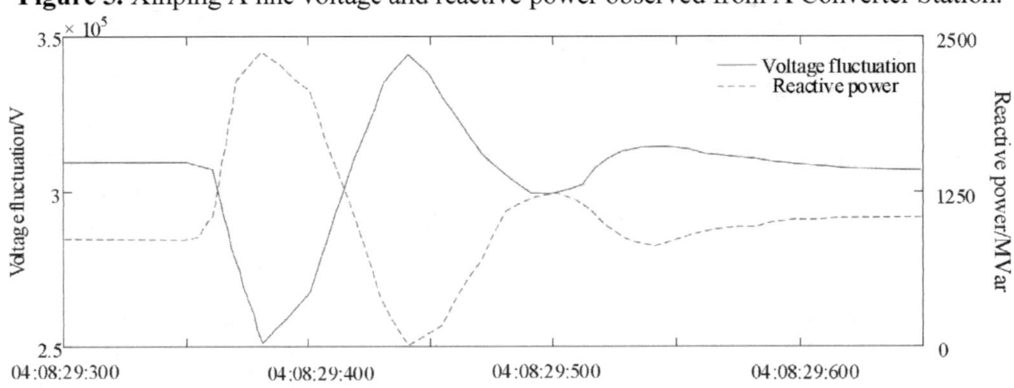

Figure 6. Xinping A line voltage and reactive power observed from T1.

It can be seen from Figures 5 and 6 that after the commutation failure, the DC voltage is reduced, the reactive power is changed from T1 to A, and the new Songs voltage rebounds in the later period, and the reactive power flows from A to T1.

Judging from the voltage fluctuations of other stations in the whole network, the basic and the failure of the commutation are basically the same, and the voltage fluctuation of the whole network is basically judged by the failure of the DC commutation. In addition, the voltage fluctuation caused by the commutation failure is generally unidirectional, and there is a positive or negative offset in the new loose voltage fluctuation.

4. RTDS and BPA simulation analysis

It can be seen from the table that the magnitude of the voltage fluctuation depends mainly on the distance of the plant. The closer to the DC line, the greater the voltage fluctuation of the plant station. As shown in the figure, the darker the color, the closer the voltage is to the DC terminal, and the greater the voltage fluctuation generated.

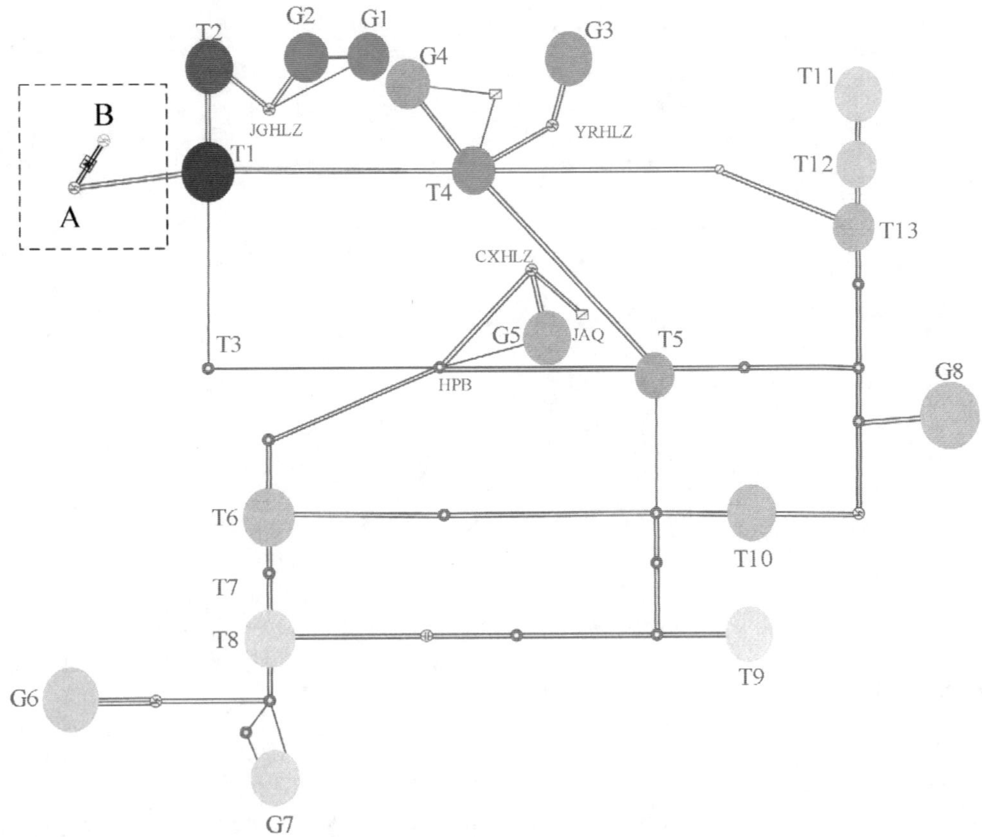

Figure 7. Voltage fluctuation intensity map.

(1) BPA-based system simulation

In the BPA, the actual operating conditions are established and simulated, and the waveform is shown in Figure 7.

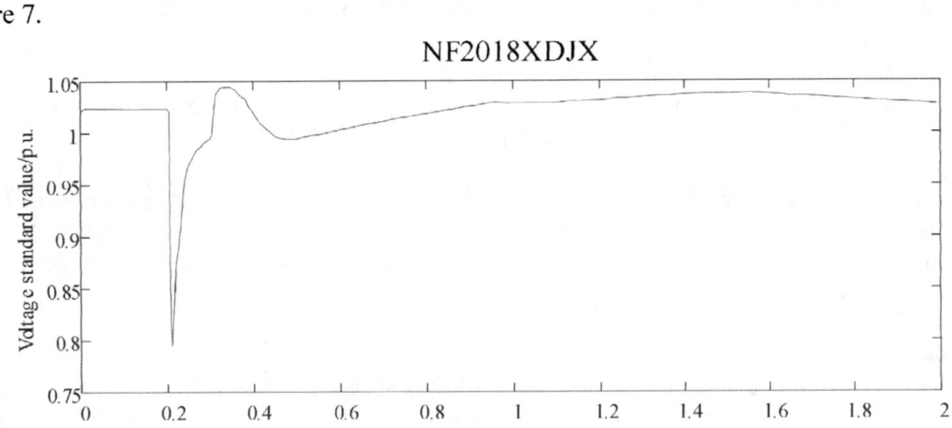

Figure 8. BPA-based DC commutation failure recurrence.

The simulation results show that the BPA-based electromechanical transient simulation can not reproduce the systematic voltage fluctuation, and can only characterize the voltage and current when the DC commutation fails. The lowest point voltage is 0.8pu, which is approximately 100kV, which cannot reflect the DC power. Electromagnetic transient voltage fluctuation process caused by electronic devices.

(2) Closed-loop simulation of DC control and protection based on RTDS

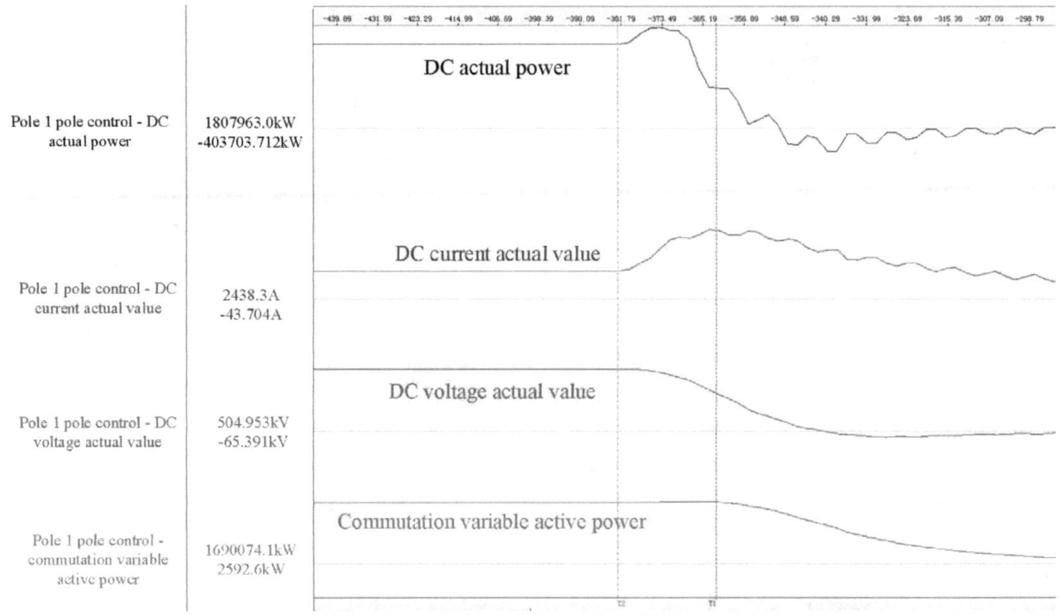

Figure 9. DCDS-based closed-loop simulation of RT control.

The closed-loop simulation result of DC control based on RTDS is shown in Fig. 8. At the time of T1, the DC system is reduced due to the failure of the AC system, and the DC commutation fails. However, the DC current rises first and then decreases, and the actual DC power is raised first. Decrease, at time T2, the DC current reaches a maximum value, and the time difference from T1 to T2 is approximately 20 ms.

The above simulation results show that there is a time difference between the attenuation of DC power and DC current and the attenuation of DC voltage. In the initial stage of commutation failure, the voltage drops, and the DC transmission power does not fall and rises due to the increase of DC current. It needs to absorb more from the system. The reactive power supports DC operation. At this time, the DC is characterized by absorbing more reactive power into the system. At the same time, the reactive power always flows from the point where the voltage is high to the point where the voltage is low, which intensifies the filling of the reactive power of the system. Due to the increase of reactive power between DC and system switching, the AC filter input command is issued, and the actual input is made after the delay. After T2, the DC current and DC current are rapidly attenuated to 0, the DC reactive power demand is reduced to 0, the AC filter is out of time, and a large amount of reactive power is output to the system. The AC filter with delayed input will increase the overflow of reactive power. After the commutation failure ends, the DC power is restored to the original value, the reactive power demand is restored to the original value, and the reactive power and voltage fluctuation processes are completed.

5. Conclusion

(1) The magnitude of the fluctuation of the voltage of the transmitting end grid caused by the failure of DC commutation depends mainly on the distance between the plant and the station. The closer to the DC line, the greater the voltage fluctuation of the plant station.

(2) The voltage fluctuation of the transmitting end grid caused by the failure of DC commutation is large, and the magnitude of the voltage fluctuation may cause the off-net new energy to be disconnected. It is necessary to take corresponding measures to avoid the high-low penetration of the near-end power electronic equipment which caused a bigger grid accident.

(3) The voltage fluctuation of the transmitting end grid caused by the DC commutation failure is millisecond, and the electromechanical simulation cannot trace back the fault process.

(4) The main reasons for voltage fluctuation: First, in the process of DC commutation failure, DC current and DC power rise first and then fall, and the timing lags behind the decrease of DC voltage,

causing the reactive power required for DC to rise first and then fall; Second, the reactive power exchange with the system increases, and the long-term commutation failure may cause the AC filter to operate, and the switching delay further aggravates the reactive power and voltage fluctuation.
This paper lists the actual operation cases, analyzes the system voltage fluctuations of the transmission network caused by DC commutation failure, and uses the high-fidelity simulation method to analyze the fault backtracking and the cause, which is the next phase failure. The induced power grid voltage fluctuation prevention and control and the high-low penetration setting of the DC near-end new energy plant station provide ideas to effectively ensure the safe and stable operation of the power grid.

References

[1] ZHANG Donghui, HONG Chao, ZHOU Baorong, YAO Wenfeng, et al. Asynchronous Interconnection System Scheme for Yunnan Power Grid and the Main Grid of China Southern Power Grid .Southern Power System Technology. 2014,8(06):1-6.
[2] FU Chao, ZHANG Dan, LIU Yongjun, HONG ChaoSystem, et al. Risk Analysis on Implementation Stage of Asynchronous Interconnection of Yunnan Power Grid and Main Grid of CSG. Southern Power System Technology ,2016,10(07):24-28.
[3] FU Chao, LIU Yongjun, TU Liang, LI Peng, et al. Experiment and Analysis on Asynchronously Interconnected System of Yunnan Power Grid and Main Grid of China Southern Power Grid. Southern Power System Technology. 2016,10(07):24-28.
[4] ZHANG Haifeng, ZHU Taoxi.Causes and Impact Analysis of Extreme I Voltage Abnormal Fluctuation in Tianguang HVDC Transmission System. Automation of Electric Power Systems. 2007(24):102-104.
[5] Chen Hongzhi, Gao Qi, Shi Yangwei. Analysis and Solution to DC Voltage Pole II Photoelectric Transformer Failure in Gaozhao HVDC Transmission System. Modern Electric Pow er 2009,26(05):24-26.
[6] WU Zehui, ZHANG Peng, ZUO Ganqing. Analysis and Treatment of Voltage Fluctuation in High-voltage DC System. Automation of Electric Power Systems. 2008(05):104-107.
[7] LI Guogen, SONG Jiahan, PI Jie, et al. Analysis on HVDC voltage abnormal fluctuation of Niuzhai-Conghua DC transmission system. Electrical Measurement & Instrumentation: 1-7.
[8] ZHOU Yuebin, JIANG Daozhou, HU Pengfei,et al. A New Approach for Suppressing DC Voltage Ripples of MMC-HVDC. Proceedings of the CSEE. 2013,33(27):36-43+7.
[9] LIN Rui. The Affection of DC Voltage Abnormal Fluctuation on Rectifier Side and Its Improvement Advice. Southern Power System Technology. 2012,6(02):43-46.
[10] HAO Zhijie, ZHANG Jianshe. RTDS-based Research on the Principle of HVDC Voltage Fluctuation . Southern Power System Technology. 2008,2(06):52-55.
[11] M. M. Moreos, J. C. Gomez. Flicker Sourees and Mitigation [J]. IEEE Power Engineering Review. 2002, (2): 5- 11.

Overall setting and cooperation method for power protection of large hydropower plants

Lei Yang[1], Chuan Xiang[1], Xin He[1], Chao Xing[1], Xinze Xi[1], Shixian Sheng[2], Can Xiao[2], Jiasi Liao[2]

[1]Electric Power Research Institute of Yunan, Kunming ,Yunan Province, China
[2]Southwest Jiaotong University, School of Electrical Engineering, Chengdu, Sichuan Province, China

E-mail: 1373591113@qq.com

Abstract: Aiming at solving the operation risk problems caused by the improper protection settings, this paper proposed a new method for the protection setting of the auxiliary power of large hydropower plants. This method applied the idea of setting from LV grade to HV grade, where each type of protection sets the value based on the setting principles and calibrates their sensitivity with their own protection range. This method handled the insufficient sensitivities of protection at the price of sacrificing the lower levels of the protection. Then based on the settings from the dispatch and control center, it vertically conducted the verification of protection setting coordination, preventing the override tripping and expanded scope of accidents. The practical applications indicate that this method could improve the protection setting accuracy of large hydropower plants and limit the operation risks of the auxiliary power within a low level.

1.Introduction

Yunnan is rich in water resources. In recent years, large hydropower plants connected to the Yunnan 500kV main network have: A power plant in Zhaotong area with a total installed capacity of 9×700MW; and B power plant in Pu'er area with a total installed capacity of 9×650MW; Lijiang area C power plant, the total installed capacity of 5×360MW; Lijiang D power plant, with a total installed capacity of 5×360MW; Lijiang area E power plant, the total installed capacity of 5×400MW; In addition, there are F power plants and G power plants. Yunnan Power Grid not only plays an important role in power supply in the South Network, but also has a large amount of power transmission to neighboring countries in the southwest. Therefore, ensuring the accuracy of relay protection setting of the large hydropower plants in Yunnan is one of the important links to ensure the stability of the power supply of the South Network and neighboring countries.

Comparatively speaking, the operation mode of the plant power consumption of large hydropower plants and the requirements for the accuracy of protection settings have their own characteristics. First of all, the impact of large hydropower plant failures on grid stability is large. The analysis of "Yunnan Power Grid Operation Mode" shows that the fault of the A power plant in the flood season will cause 220kV I line overload, and the grid frequency will drop by about 0.3Hz. The fault of the B power plant will cause the grid frequency to drop 0.3Hz, both have exceeded the frequency fluctuation range of the South Network. Secondly, large hydropower plants have large single-unit capacity and a large number of units. In order to ensure the reliability and flexibility of the power supply of the plant, many plant power supplies and power supply points are often set up. The wiring and operation modes of the plant power system are complicated. It brings great difficulties to relay protection setting and protection value matching. If the setting and coordination are not proper, it will lead to the chain reaction of the plant power system, causing multiple units to trip, and even affect the entire power plant. In addition, due to the repeatability of the large hydropower plants, the structure and setting of the power consumption of the plant often adopt uniform principles and setting methods. If the method is improper, it will usually result in the error of consistency of multiple units, leaving great security risks to the stability of the power grid. The stability calculation criterion of Yunnan Power Grid requires that any generator set in the system trip or lose its magnetism, and the protection and circuit breakers operate correctly, which can maintain the stable operation of the system and the normal power supply of the power grid. However, the setting calculation and coordination errors of the plant power supply usually cause multiple units to trip at the same time. Therefore, it is urgent to sort out the protection and setting methods of power consumption in large hydropower plants, verify whether all levels of protection cooperate with each other, ensure the power supply reliability of large hydropower plants, and avoid the occurrence of over-grade tripping that leads to the disalignment of multiple units [1]~[13].

In recent years, many scholars have proposed ways to improve the rationality and accuracy of power plant setting values for large power plants. Reference [10] proposes to shorten the time difference to 0.2s to ensure the quickness of the main protection. Reference [11] proposes to improve the wiring of the system to make the operation mode of the plant power system clearer and more reasonable. Reference [12] pointed out that the selectivity of over-current protection can be realized according to the step-by-step matching principle of over-current protection time limit, which will make the fault current increase and the damage degree of equipment more serious when the fault occurs closer to the power side. It is suggested to reasonably determine the timing matching grade difference. Reference [13] proposes to design and develop a set of engineering and practical tuning software platform that can meet different protection configurations of different manufacturers.

The existing literature mainly tends to improve some aspects of the plant power setting, and lacks the analysis of the existing problems of the existing large-scale hydropower plant protection settings, and does not sort out the setting process and method. This paper proposes a holistic approach and method for plant power setting by analyzing the actual over-the-counter trip of plant power in Yunnan Power Grid. The practical application shows that the proposed method is of great significance to ensure the quality and accuracy of the fixed value setting of large-scale hydropower plants.

2.Power operation risk of large hydropower plants

2.1. Current status of power plant setting in large hydropower plants

The existing power plant setting of large hydropower plants mainly faces the following problems:

1)The operation mode is complicated. With the increase of power plant units and the increase of installed capacity, the system wiring of large-scale hydropower plants is complicated and the operation modes are numerous. Any omission of operation mode may lead to incorrect operation of protection.

2)There is a contradiction between the sensitivity and the step-by-step coordination of protection. In order to satisfy the stepwise coordination of protection, the upper relay protection may be required to have a higher overcurrent setting value, but this value cannot meet the requirements of the upper relay protection's sensitivity, causing a contradiction.

3)The backup protection value of the high-voltage transformer used in the factory is given by the central dispatch. The protection of other parts of the power system of the plant is self-tuned by the power plant. If the two lack effective communication and coordination with the protection value, it may cause a leap-level trip of the protection.

2.2. Large-scale hydropower plant power consumption risk

The following two typical plant power over-the-counter trip events occurred in Yunnan Power Grid are used to detail the operational risks of existing plant power.

Accident 1: In 2012, the high-variable backup protection action of a No. 3 generating unit of a power station jumped off the 500kV two-side circuit breaker, the #3 generator outlet short-circuiter, and the #4 generator outlet short-circuiter, causing the two units to be shut down. The two overhaul drain pumps of the 400V fourth group of common electric B sections are started at the same time, and the starting current and starting time exceed the fixed value of the high-variable over-current section II of the No. 3 plant, resulting in protection action.

Accident 2: In 2012, the protection of generators and transformers of No. 4 and No. 5 of a power plant "protection action of the high-voltage transformer over-current II section of the plant" caused the shutdown of Units 4 and 5. The cause of the accident was that the 400V bridge machine was approaching during the completion of the power transmission. The friction between the B and C phase cables on the power supply side of the slip wire and the steel structure bracket causes a phase-to-phase short circuit. 0.4, 10.5kV protection failed to work because: the setting value of the main body protection of the power switch of the bridge machine inlet line, the setting value of the main body protection of the main body protection of the b section of the first group of 400V public electricity and the setting value of the power transformer protection device of the b section of the first group of public electricity is too large; and this short circuit is short-circuited by the transition resistor so that the short-circuit current is too small. The 10.5kV protection timing calculation is low in sensitivity and improperly matched with high plant overcurrent protection, which eventually leads to the No.5 high plant overcurrent II protection action, and the 4th and 5th sections of the plant use the power reserve action, the 4th and 5th sections of the plant to be closed by the tie switch. Then the maintenance personnel see that the power transmission voltage is abnormal. Manually disconnect the 400V power supply of the bridge machine, and the fault point is cut off. Otherwise, the 5th and 6th sections are prepared for self-injection, and the #6 unit will also be cut off.

Based on the above two accidents, it can be seen that the operational risks caused by the improper setting of the protection value of the existing large hydropower plants are as follows:

1)There is no step-by-step coordination of plant power protection (including time coordination and reference volume coordination), which is one of the operational risks of existing hydropower plants. The main reason for causing accident 1 and accident 2 is that the fixed value of the plant power protection is not coordinated step by step, resulting in over level tripping in the event of a fault, loss of selectivity, and an increase in the scope of the accident.

2)There is no uniform setting principle and mutual cooperation between the plant protection and the same level, especially through the parts that are connected by self-injection. Accident 2 reflects that the 400V bus-coupled circuit breaker cannot reflect the fault on the connected busbar, resulting in a chain reaction, and the fault is difficult to remove, which greatly expands the scope of the accident. At the same time, the plant change backup protection action should make the self-injection action lock and prevent the switch from being connected to the faulty line.

3)The sensitivity of protection is lacking. Next level of protection is less sensitive, resulting in over level tripping; the sensitivity of the same-stage bus-coupled short-circuit protection is low, resulting in the transmission of faults.

In summary, the setting of the power protection rating of the plant needs to be improved from the depth and breadth of the propagation.

3. Setting of power protection value for large hydropower plants

The process of setting up electricity for hydropower plants is shown in Figure 1. The setting of the power system of large hydropower plants mainly includes the following parts: calculation of plant power system parameters, calculation of 10.5kV short-circuit current, calculation of 400V short-circuit current, calculation of 400V protection setting, calculation of 10.5kV protection setting, and inspection of protection coordination.

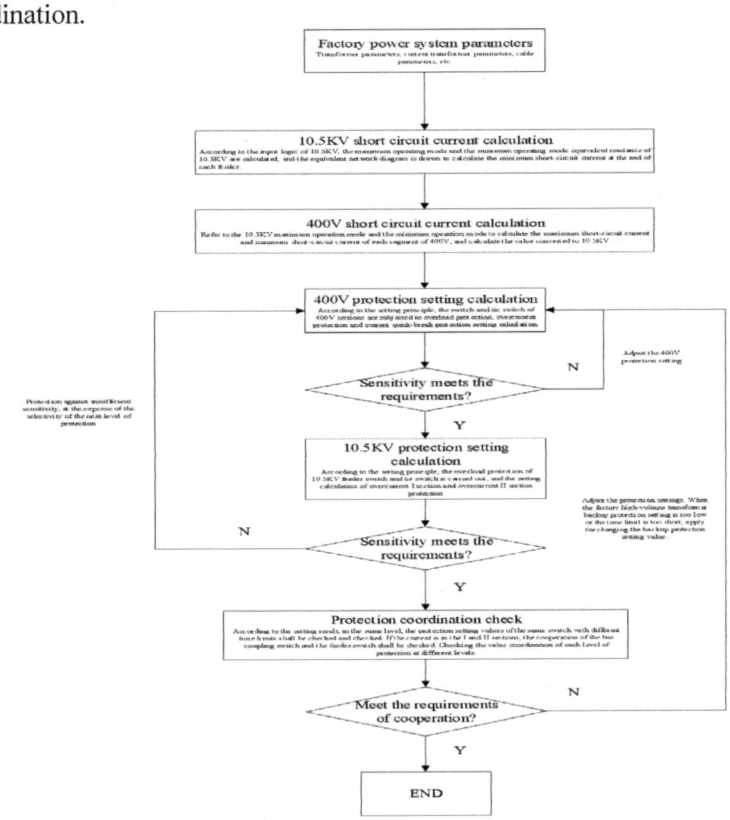

Figure 1. Flow chart of power plant power plant.

3.1. Short circuit current calculation
Assume that the upper part of the transformer used in the factory is an infinite system, $S=\infty$, $X_j^*=0$,generally take the reference power: $S_j=100MVA$, the reference voltage is the average voltage $V_j=V_p$. The maximum operating mode and the minimum operating mode are selected from all possible operating modes including the self-injection. 10.5kV needs to calculate the minimum short-circuit current at the end of each segment of the feeder $I_{d\_10.5kV}^{\min}$, for the 10.5kV feeder switch and the bus-coupled switch over-current I segment sensitivity check. Similarly, 400V needs to calculate the maximum short-circuit current $I_{d\_400V}^{\max}$, and minimum short-circuit current $I_{d\_400V}^{\min}$, and calculate the value converted to 10.5kV $I_{d\_10.5kV}^{eq\ \max}$ and $I_{d\_10.5kV}^{eq\ \min}$.

3.2. 400V protection setting calculation
1) Overload protection (L)

Since the overcurrent II segment avoids the maximum starting current of the motor connected to the 0.4kV bus, the overload protection acts only as a severe overload protection for the transformer and acts as a trip.

Setting principle: Avoid the normal overload current and allowable running time of the low-voltage factory transformer[14].

$$\begin{aligned} I_{set.I} &= K_{rel} I_e / K_{re} / I_n \\ &= K_{rel} S_n / (\sqrt{3} U_n) / K_{re} / I_n \end{aligned} \quad (1)$$

In the formula: K_{rel} ——the reliability factor, taken as 1.2;

K_{re} ——return coefficient, taken as 0.85;

I_e ——rated current of low voltage transformer used in factory;

I_n ——The rated current of the circuit breaker.

2) Time-limited overcurrent II segment protection (S)

Since the protection object belongs to the end of the load, it is only necessary to avoid the maximum load current and the motor is self-starting [14].

The maximum motor self-starting current is: $I_{ss.\max} = \sum K_{ss} I$;

Setting value: $I_{set}^{II} = K_{rel} \times I_{ss.\max}$

$n = I_{set}^{II} / I_n$

Among them, I_n is the rated current of the switch, K_{rel} is the reliability factor, which is 1.2.

Sensitivity check:

$$K_{sen} = I_{d\_400V}^{\min} / (nI_n) > 2 \quad (2)$$

Where: $I_{d\_400V}^{\min}$ ——400V minimum short-circuit current, calculated from short-circuit current.

3) Current quick-break protection (I)

Two-phase short circuit with 400V busbar has 2 times sensitivity calculation $I_{set}^{I} = I_{d\_400V}^{\min} / 2$

Setting value $n = I_{rel}^{I} / I_n$

I_n is the rated current of the switch.

400V factory power contact switch protection setting calculation

According to the operation mode of the 400 V system, that is, during normal operation, each segment runs independently. When one of the power failures occurs, the communication switch is closed by the self-injection device, and the power supply continues to be supplied to the power failure segment. According to the protection tuning selectivity principle, when a short-circuit fault occurs in the power failure section, the contact switch should be tripped first. Therefore, the protection setting calculation of the tie switch is set according to the minimum short-circuit current of each segment.

3.3. 10.5kV protection setting calculation

1) Overcurrent I

The overcurrent I segment, quick-break, is calculated according to the following principles [14]:

a) Maximum short-circuit current flowing through the protective installation when avoiding an external short circuit:

$$I_{set}^{I} = K_{rel} I_{d\_400V}^{\max} / n_{TA} \quad (3)$$

Where: K_{rel} - the reliability factor, taken as 1.1;

$I_{d\_400V}^{\max}$ ——In the maximum mode, the short-circuit current of the three-phase short circuit occurs on the low-voltage side of the low-voltage transformer of the factory;

n_{TA} - CT ratio.

b) Avoid the magnetizing inrush current of the transformer:

$$I_{set}^{I} = (3 \sim 5) I_n / n_{TA} \quad (4)$$

In the formula: I_n - the rated current of the high voltage side of the low voltage transformer used in the factory $I_n = S_n / (\sqrt{3} \times U_n)$.

c)Sensitivity check:

$$K_{sen} = I^{\min}_{d\_10.5kV} / (n_{TA} I^{I}_{set}) > 2 \quad (5)$$

Where: $I^{\min}_{d\_10.5kV}$ - Minimum short-circuit current of 10.5kV bus.

2) Overcurrent section II

a) The overcurrent II section , quick-break, is matched with the maximum starting current of the motor that needs to be self-starting on the low-voltage transformer used in the factory, and cooperates with the 400V incoming timing definite current protection [14]. Available:

$$I^{II}_{set} = K_{rel} I^{II}_{set.400V} \quad (6)$$

Where: K_{rel} - the coefficient of cooperation, take 1.1;

$I^{II}_{set.400V}$ -400V incoming line timing limit overcurrent protection setting.

b) The action delay matches the overcurrent protection of the low side branch.

c) Sensitivity check

$$K_{sen} = I^{\mathrm{eq}\ \min}_{d\_10.5kV} / I^{II}_{set} > 2 \quad (7)$$

In the formula: $I^{\mathrm{eq}\ \min}_{d\_10.5kV}$ -The minimum short-circuit current of the two phases occurs on the low-voltage and low-voltage side busbars of the plant (returned to the 10.5kV side).

3) Overload protection

Operating current $I_{set.l} = K_{rel} I_{nH2} / K_{re}$

In the formula: K_{rel} - the reliability factor, generally takes 1.05 when acting on the signal;

K_{re} - return coefficient, generally 0.95

I_{nH2} - Maximum load current, $I_{nH2} = S_n / (\sqrt{3} \times U_n) / n_{TA}$.

4)Low-voltage and low-voltage side zero-sequence overcurrent protection

Calculated by avoiding the normal maximum unbalance current:

Operating current $I_{set.0} = 25\% I_{nL2} / n_{TA0}$

In the formula: I_{nL} - rated current of low voltage side of low voltage transformer used in factory $I_{nL} = S_n / (\sqrt{3} \times U_{nL})$;

n_{TA0} - Low-voltage side zero-sequence CT ratio.

3.4. Protection coordination check

According to the setting result, in the same level, the protection setting values of the same switch with different time limits are checked; the matching of the bus coupling switch and the incoming line switch and the feeder switch is checked. In different levels, check the matching of the protection of each level of protection. The inspection of the protection cooperation mainly includes the following points:

1) The sensitivity of 10.5kv feeder switch over current I section, 10.5kv feeder switch over current II section, 400V inlet switch speed break protection, 400V inlet switch over current II section is greater than or equal to 2;

2) Check the coordination of 10.5kv feeder switch over current section I, II and overload; Check the 400V current break, over current section II and over load.

3) 10.5kV feeder switch over-current I section and 400V incoming line switch quick-break protection cooperate with each other, the former is larger than the latter;

4) 10.5kV feeder switch overcurrent II segment and 400V inlet switch overcurrent II cooperate with each other, the former is greater than the latter;

5) 10.5kV feeder switch overload and 400V line switch overload cooperate with each other, the former is greater than the latter;

6) The protection coordination time difference satisfies Δt=0.2~0.5 seconds;

7) The 10.5kV feeder switch overcurrent I and II sections must also be matched with the factory high voltage transformer backup protection.

4. Instance verification

In order to verify the feasibility of the above-mentioned plant power setting method, the method is used to set the power setting value of a power plant in Yunnan. The setting results and the coordination check of protection are shown in Table 1.

Obviously, using the tuning method of this paper, the accuracy of the setting from the depth and the breadth is effectively ensured, and the protection of the leap-level trip situation is avoided. The actual engineering application shows that the improved setting value by using this method greatly improves the reliability of the plant power protection action and is conducive to the safe and stable operation of the power grid.

Table 1. Consolidation results and coordination check of protection.

Circuit breaker name	Remarks	10kV incoming feeder switch Overcurrent I section (T=0.05S)	10kV incoming line feeder switch overcurrent I section sensitivity check	10kv incoming feeder switch overload protection (T=20S, alarm)	Low-voltag e side zero-seque nce overcurren t (T=1S, trip)	400kV incoming current fast-break Idmin/2 (sensitivity is 2)
011	Unit 2 self-use b segment	7.49	2.93	0.51	2.26	3.35
012	The first group of public electricity a segment	6.96	2.36	0.48	2.84	4.03
021	Unit 3 self-use electricity a segment + 5 machine self-use electricity b segment	7.53	2.91	0.51	3.01	3.35
022	Unit 4 self-use electricity a segment +6 machine self-use electricity b segment	10.52	2.09	0.51	3.01	4.45
023	The first group of public electric b segments	7.07	2.33	0.48	2.84	4.06
026	The second group of public electricity a segment	7.07	2.33	0.48	2.84	4.06
027	Unit 1 self-use a segment	8.23	2	1.22	7.22	6.91
031	Unit 2 self-use a segment	7.57	5.71	0.51	2.26	3.81
032	Unit 6 self-use electricity a segment + 4 machine self-use electricity b segment	7.57	5.71	0.51	3.01	3.81

033	Unit 5 self-use electricity a section + 3 machine self-use electricity b section	10.56	4.09	0.51	3.01	5.28
089	Unit 1 self-use electricity b segment	8.85	2	1.62	7.22	6.17
045	The second group of public electric b segments	6.63	2	0.48	2.84	3.8

Continued Table 1. Consolidation results and coordination check of protection.

Circuit breaker name	**Quick break test**		**400kV incoming line overcurrent II segment setting (T=0.5)**		**400kV incoming line overcurrent II section sensitivity**		**10kV overcurrent II section (T=0.7S)**	**10kV overcurrent II segment sensitivity**	**Factory high change backup test (T=3.5S)**		**400V overload setting**
	10kV incoming line feeder switch overcurrent I section current value	400kV incoming current quick-break current value (10kV side)							10kV incoming line feeder switch overcurrent II section current	10kV incoming line feeder switch overcurrent I section current	
011	449.4	159.5238		2		3.35	1.75	3.03	105	449.4	0.82
012	556.8	191.9048		2.5		3.22	1.64	2.93	131.2	556.8	1.03
021	451.8	159.5238	2	1.5	3.35	4.46	3.06	1.74	183.6	451.8	0.82
022	631.2	211.9048	1.5	1.5	5.93	5.93	2.62	2.7	157.2	631.2	0.82
023	565.6	193.3333		2.5		3.25	1.64	2.94	131.2	565.6	1.03
026	565.6	193.3333		2.5		3.25	1.64	2.94	131.2	565.6	1.03
027	658.4	329.0476		5.5		2.51	3.6	2.58	288	658.4	2.61
031	454.2	181.4286		2		3.81	1.75	3.45	105	454.2	0.82
032	454.2	181.4286	2	1.5	3.81	5.07	3.06	1.97	183.6	454.2	0.82
033	633.6	251.4286	1.5	1.5	7.05	7.05	2.62	3.2	157.2	633.6	0.82
089	531	293.8095		3		4.11	2.62	4.22	157.2	531	2.61

045	530.4	180.9524		2		3.8	1.31	3.45	104.8	530.4	1.03

5. Conclusion

This paper describes in detail a calculation method for the power setting of large hydropower plants. The system adopts the calculation method of the sub-module, and points out that it should cooperate with the logic of the self-injection, list all possible operation modes, select the maximum operation mode and the minimum operation mode to prevent the protection setting value from being incorrect due to the omission of the operation mode. . For the case where the sensitivity requirements and the protection coordination requirements conflict, a treatment method that satisfies the compatibility requirements by sacrificing the selectivity of the lower protection is proposed. The problem of protection coordination is checked from both depth and breadth, which effectively avoids the occurrence of protection overstepping trips. The power-setting method proposed in this paper is of great significance to the stable operation of the power plant. It is also conducive to the establishment of the plant power setting and professional inspection system, laying the foundation for the programmatic realization of the plant power setting and protection coordination check.

References

[1] Zhang Binghai and Liu Jian 2006 Discussion on the problem of fixed value matching of power system protection in low voltage power plant of thermal power plant *Power System Protection And Control* vol 40(06) pp 70-6

[2] Li Baozhong, Zhang Liwen and Song Shuyong 2009 Misoperation analysis and preventive measures for power plant protection *Shanxi Electric Power* vol 37(06) pp 66-9

[3] Liu Ling and Liang Wei 2008 Analysis and selection of grounding faults in power system of 380V low voltage power plant *China Electric Power Education* vol 20(S3) pp 33-7

[4] Yang Ming and Lu Guowei 2007 Cause analysis and preventive measures of a relay protection action *Hubei Electric Power* vol 24(06) pp 54-7

[5] Yue Disong 2007 Coordination analysis of multi-power relay protection setting *Electrical application* vol 24(06) pp 54-7

[6] Shan Baotao, Liu Zhe and Guan Fenghua 2007 Research on relay protection setting and comprehensive analysis management system of power plant *China Electric Power Education* vol 35(S1) pp 65-9

[7] Zhao Min and Wang Suhua 2009 Discussion on the cooperation between the main transformer backup protection and the 10kV side outlet protection in 220kV substation *Power System Protection And Control* vol 23(07) pp 34-8

[8] Yunnan Power Grid Corporation 2013 *Yunnan Power Grid Operation Mode in 2014*

[9] Cui Jiapei 1993 *Power System Relay Protection and Safety Automatic Device Setting Calculation*(Beijing: China Electric Power Press)

[10] Li Xianshan, Wang Zhufeng, Wang Xiaojian and Yu Ming 2011 Study on the coordination value of power plant protection for complex power plants *Hydropower Automation And Dam Detection* vol 35

[11] Peng Jun 2013 Analysis of measures to improve the reliability of power plant's power system *Zhejiang Electric Power* vol 8

[12] Gao Daquan 2007 Analysis and improvement of protection performance of power plant system *Relay* vol 35

[13] Cai Zhenghua 2008 Research on protection setting calculation and software of large hydropower plants(Huazhong University of Science and Technology)

[14] *Design Guidelines for Relay Protection of Hydropower Plants*

Journal of Physics: Conference Series **1419** (2019) 012046 doi:10.1088/1742-6596/1419/1/012046

Simulation application of virtual reality technology in legal education

Kunyi Jian[1], Peng Nai[1,3] and Tao Yu[2]

[1]Law School, Yunnan University of Finance and Economics, Yunnan Province, People's Republic of China

[2]Law School, Yunnan Minzu University, Yunnan Province, People's Republic of China

[3]E-mail: 863193534@qq.com

Abstract. The virtual simulation teaching of law is a systematic practice course of judicial case handling. This course introduces VR technology, enabling testers to play the role of a judicial officer in a virtual environment to handle cases. This design is realized through the intersection of the teaching design by the arts and sciences, the construction of virtual models, and man-machine interaction. Through this design, students can improve their ability to handle cases, develop students' thinking, and enhance students' ability to use their knowledge comprehensively. In order to ensure better experimental results, it is necessary to provide technical support from scene simulation, case simulation, mechanism simulation and so on.

Legal education is related to the construction of a country under the rule of law and a society ruled by law. Its purpose is to train qualified legal talents for the country [1]. Chairman Xi Jinping pointed out: "The discipline of law is a highly practical discipline. Legal education should handle the relationship between knowledge teaching and practical teaching. It is necessary to break the institutional barrier between universities and society and introduce the high-quality practical teaching resources of practical work departments to colleges and universities. Strengthen the exchange of legal education, legal researchers and practitioners of the rule of law." The Ministry of Education's "Opinions on the Use of a Combination of Rule of Law Education and Moral Education to Train Excellent Legal Talents 2.0" (hereinafter referred to as "2.0 Opinions") also states: "It is the key to strengthening law education to emphasize practice." At present, legal education attaches great importance to knowledge and despise practical ability. It attaches importance to the cultivation of academic ability and despise the training of applied ability. The legal knowledge is shelved and cannot be used [2]. In view of this, we are trying to innovate in systems and technologies to introduce the teaching methods of natural science laboratories into law teaching. Through VR technology, we will use simulation experiments to enable students to enter the virtual environment to act as judicial officers and enhance students' judicial practice.

1. Research background

Since the promulgation of the "2.0 Opinions" and the "National Standards for Undergraduate Professional Teaching Quality in Colleges and Universities", it has become urgent to change the teaching methods of law practice teaching and improve the teaching system of law practice teaching [4]. It should be noted that China will adopt a profound and systematic reform of legal education.

1.1. Status and classifications of current teaching contents of legal education

Legal education is a very practical subject. The realization of fairness and justice requires both theoretical development and case handling. Judging from the supply and demand of labor in the society, the legal talents required by the society are mainly concentrated in the substantive departments. In order to prove this argument, we have established an analytical model. Taking Kunming city in Yunnan Province of China as an example, we conducted statistics on the personnel engaged in legal research in public universities in this city, and compared the number of court staff in this area. (See Table 1, Table 2)

Table 1. Number of people engaged in legal research in public universities in Kunming city of Yunnan province.

Universities	**Number**
Yunnan University	81
Kunming University of Science and Technology	64
Yunnan Normal University	80
Yunnan University of Finance and Economics	43
Yunnan Minzu University	33
Southwest Forestry University	32
Yunnan Police Academy	30
Total	363

Note: the data comes from the official website of the universities (last accessed on September 24, 2019).

Table 2. Statistics on the number of judges in the city of Kunming

Cities	**Number**
Yunnan Higher People's Court	178
Kunming Intermediate People's Court	144
Kunming Wuhua District People's Court	54
Kunming Xishan District People's Court	48
Kunming Guandu District People's Court	49
Kunming Panlong District People's Court	46
Kunming chenggong District People's Court	20
Total	539

According to the standards of China's judicial system reform, judges account for 34% of the total number of court staff. Therefore, based on the above data, it can be estimated that the total number of court staff in Kunming is about 1,585. It can be seen that in Kunming, the number of court staff is more than four times higher than that of university legal researchers. This is calculated without considering the types of other practical work such as prosecutors and lawyers. Therefore, it is not difficult for us to conclude that the demand for talents in the legal profession is mainly concentrated in the legal practice area, and the training of law students needs to be oriented to practice.

The contents of current legal practice teaching can be divided into the following categories: mock court, case teaching method, and holiday internship. Although the above several teaching models have practical forms, they lack the essence of practice. Most of the models return to the path of theoretical study, and it is difficult to improve the students' practical ability.

The first is the mock court. Although this teaching method restores a certain court trial site, it pays too much attention to the form of performance and neglects the training of students' ability to handle cases. Teaching presents the shortcomings of fragmentation and lack of simulation. The so-called fragmentation means that the cases used by students often only have basic cases, excluding evidence, procedures, etc. The disputes of students in the courts are only related to the application of the law, but the complete case handling procedure was split. Insufficient simulation refers to the lack of simulation in the mock court. Essentially, students are allowed to put on the robe for qualitative discussion of the case, because the case that the student contacts is a case that is cut into pieces, and the only part of the legal dispute is there are a lot of opinions on the Internet, so in the teaching, we will see many students go online to get the script of the court speech, and then put on the robe to interpret it, which greatly weakens the teaching function of the mock court, more like a stage play.

The second is the case teaching method. This teaching method was originally a research method of scholars. For example, Chen Xingliang from Peking University used this method to write teaching materials and gained wide acclaim [12]. Case teaching method is different from "Follows the precedent". It is not a method of judges' referee cases, but a teaching method. It uses a specific case to help students understand the use of abstract knowledge. Therefore, the essence of the case teaching method is theoretical teaching, but it is only teaching by means of cases. Cases can help students understand knowledge, but they do not allow students to understand the process of handling cases, so the role of improving students' practical ability becomes limited.

Again, it is a holiday internship. Take the Yunnan University of Finance and Economics Law School as an example. The student's holiday internship mainly includes regular and irregular (see Table 3). During regular internships, the school will arrange holidays (usually two weeks) to conduct legal practice training for students. The main contents include inviting staff from the legal practice department to communicate with students in the school and leading students to visit the courts and law firms. During irregular internships, students use the summer vacation to go to the courts, procuratorates, law firms and other legal work departments for internships, and then submit certificates and reports to the school, according to which the school provides credits to students.

Table 3. Yunnan University of Finance and Economics Law School Student Internship Mode

Regular internship internships (two weeks)	Campus internship	Interview with staff in the legal work department
	Off-campus internship	Visit the legal work department
During irregular internships	Off-campus internship	legalwork department internship

At present, there are three shortcomings in the model of holiday internship: first, the organized internship time is too short. It is difficult for students to handle a complete case in two weeks, whether it is interviews with legal staff or visiting legal workdepartment, students are only onlookers. They can only obtain a kind of intuition through observation, and it is difficult to have a real experience in legal work. Second, although the irregular internships are long-term, there is a lack of school organization, and students' participation in internships lacks guidance and institutional guarantees. The staff of legal work department is unwilling or difficult to hand over the work of assisting the case to a student whose legal status is unknown. Judging from the student's internship report, most of the students' internships are trivial, and there are not many opportunities to get in touch with the case. From the 2018 student internship report of Yunnan University of Finance and Economics Law School, almost all the student internship spent a lot of time introducing them to learn to bind the files and cleaning, but they are not deeply aware of the judicial cases. Over time, students regard the internship as a task for graduation, as long as they can get an internship certificate to solve this trouble. Third, students have no sense of role in the process of internship, and it is difficult to form a real feeling of handling the case. As a practical operation, it is difficult to form an experience and grasp by listening alone. Once engaged in specific work, the students also do not know how to proceed. This is like driving and surgery,listening to other people's operating experience is not likely to make you a driver or a surgeon directly.

1.2. Construction of virtual simulation teaching model of legal education

The crux of the problem lies in the fact that the legal employment market requires students to have the practical ability of law, but students do not have the legal status to handle the case, it is difficult for students to acquire this ability through training. The inconsistency between job market demand and teaching methods has led to an imbalance in legal education (see Figure 1).

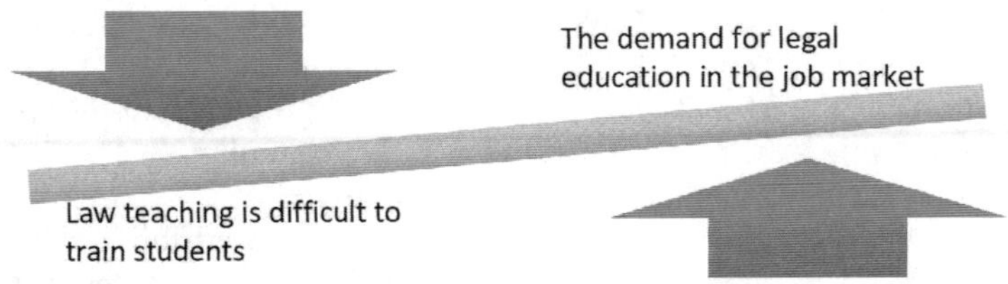

Figure 1. Teaching leverage imbalance.

Therefore, we need to re-construct a teaching model to cope with this challenge. On the one hand, we must consider the ability of students to handle cases, and on the other hand, the reality that students do not have legal status. To this end, we try to introduce VR technology into the teaching of law practice, and construct a legal practice teaching model of virtual simulation. The purpose of practical teaching is to develop students' vocational skills [5]. Therefore, the teaching model needs to be compatible with the characteristics of the legal profession, and then the following design principles are required.

Principle 1: the virtual simulation teaching of law needs to simulate the real legal profession, so that students can experience the case in an immersive way. A major feature of the legal profession is the personal experience of the judiciary. This can also be confirmed from the goal of the state's reform of the "judicial case responsibility system". The judges need to be responsible for their own judgments, which will be the future legal profession trend of our country. Regardless of the role played in the legal profession, personal handling of the case will be a basic requirement. Therefore, the first purpose of virtual simulation teaching is to let students actually enter the case and personally review the evidence of the case to understand the content of the legal profession.

Principle 2: virtual simulation teaching should enable students to experience the complete procedures for handling cases. The content of the legal profession is to handle the case, and the realization of all fairness and justice will ultimately be reflected in the specific case. As Chairman Xi Jinping pointed out, it is necessary for the people to feel fair and just in every judicial case. Practice teaching can't be separated from case-handling training, students will not handle cases, and instilling more fair and just ideas will fail. Only by making students experience the entire process of the case, the knowledge learned in the classroom can truly be associated with practice and become knowledge that can be used.

Principle 3: virtual simulation teaching should realize the simulation of professional roles, so that students can experience professional content in virtual roles. Each profession has its own ethics, methods and skills, which are the wealth gained through reflection on work. Bystanders can evaluate a profession, but cannot experience a profession. Theory becomes a profession, and people's actions are required to transform knowledge into their own experience. As some scholars have pointed out, the teaching of law depends on the personal experience of students [6]. Therefore, the best way for students to understand a profession is to let them go through this profession and gradually acquire the essence of knowledge in the course of action.

Focusing on the above experimental principles, we introduce VR technology into the teaching of law practice. In the past, VR technology was mainly limited to the research scope of science and engineering. Its advantage is that when the experimental target and experimental conditions conflict, VR technology forms the sensory simulation in the human brain through audio and video output, thus making people feel the feeling of entering the real scene. Engineers can experiment in virtual reality. For example, some engineers have built an interactive Mars rover virtual training system through Unity3D virtual reality software to simulate the Mars environment trainers operating space vehicles [7]. The conclusion that VR technology is applied to teaching can effectively save teaching costs has also been confirmed in the area of science and engineering. For example, in the field of mechanical teaching, domestic colleges and universities have overcome the difficulties by introducing VR teaching methods in response to many problems such as short practice time and tight equipment resources [8]. Another example is Zhu Wenhua and other scholars took the cylindrical gear reducer as the research object, and developed a set of teaching software based on Unity3D software to improve the interaction and immersion of practical teaching through virtual technology [9]. Whether it is experiment or teaching, the application of VR technology in science and engineering neighborhoods is relatively mature, but it is not common to introduce this technology in liberal arts teaching. Even if liberal arts teaching transplant VR software into the classroom, it only creates a virtual simulation of audio and video effects, and is never a teaching system built using liberal arts logic. Therefore, we try to establish a VR system based on the needs of law practice teaching. The core program of the system is a complete case-handling process that provides students with a scenario of each judicial workflow in a virtual simulation manner, which in turn allows students to advance the proceedings in this environment.

We envision the first phase of the experiment starting with a criminal case. We will restore the cases that have been completed by the court to the computer system, so that students can access the virtual evidence materials. The litigation process is promoted from the discovery of cases, analysis of the case, construction of the evidence system, and response to trials. We will group students, including the investigation team, the public prosecution team, the defense team, the trial team, etc. Some students will act as defendants, victims and witnesses. Then, within the prescribed time, students are required to carry out all aspects of the case in the virtual environment. We will follow the criminal procedure to form a VR system design with different tester groups (See Figure 2, Figure 3, Figure 4, Figure 5, Figure 6).

Testers in the investigation team should learn how to obtain and fix evidence. We will use VR technology to carry out virtual crime scene restoration. The tester will conduct on-the-spot investigation through the VR helmet to the crime scene, obtain evidence and fix the evidence, and conduct virtual investigation experiments when necessary. At the same time, for the acquisition of verbal evidence, we consider designing two test paths. In path 1, we connect the tester's VR helmet to the VR helmet that acts as a suspect and witness, and directly implement the simulation interrogation or inquiry in the form of identity play; in path 2, according to the dossier material, the engineer virtual suspects and witnesses in the computer, allowing the virtual electronic robot to accept the tester's question or inquiry. The former test is more conducive to improving the tester's ability to communicate with real people, while the latter test is more conducive to improving the tester's ability to deal with real cases.

The Public Prosecution Team testers need to train the skills to review the case. Similarly, VR technology can be used for on-site exploration and investigation experiments; in addition, VR technology will simulate the scene of suspects being interrogated, and test whether the tester can effectively protect the legitimate rights and interests of the suspects. If necessary, the virtual defendant will redressing injustices, let the tester judge whether it is necessary to initiate a special investigation procedure; at the same time, as a prosecution, it is necessary to communicate with the testers of the defense team through the VR system to listen to the opinions of lawyers and parties.

Testers in the defense team need to train defense skills. Through the environment of the VR system virtual law firm, the tester needs to meet the parties and the family members of the parties to examine the tester's negotiation skills; through the VR system to restore the case environment, the lawyer must find evidence that the defendant is not guilty or worthy of forgiveness at the virtual scene of the case; talk to role-playing or virtual witnesses to gain testimony; communicate with judges and prosecutors through VR helmets.

Testers in the trial team need to train trial skills. Through the interconnection function of VR technology and information system, the court trial site is simulated to examine the evidence submitted by the public prosecutor and the defender to the court, and listen to their opinions, organize the debate in the virtual environment, and finally announce the judgment of the judge. The test group testers need to judge the case according to the logic of "evidence-fact-law application", so as to obtain a real case-handling experience. At this time, the instructor can also enter the VR virtual court as an observer and watch the final performance of each group of testers.

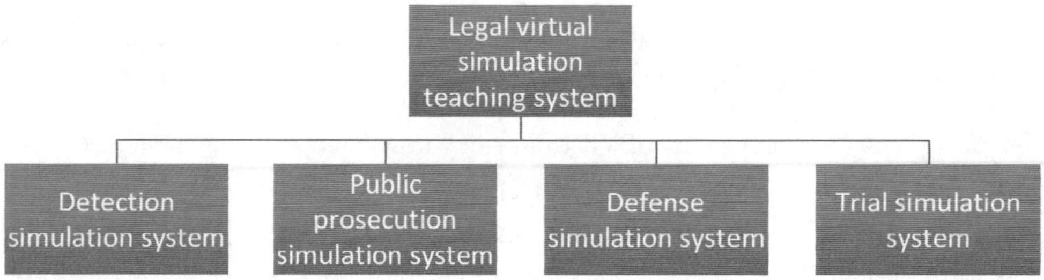

Figure 2. Virtual Simulation Teaching System Structure.

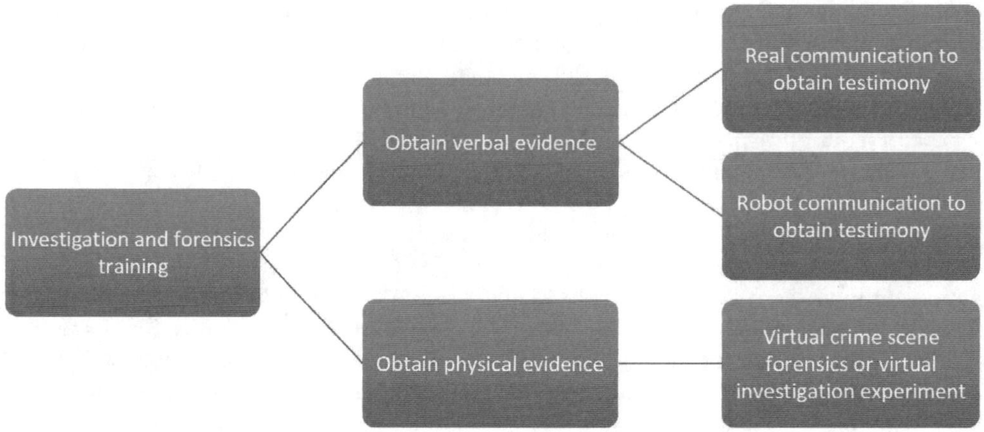

Figure 3. Detection simulation system.

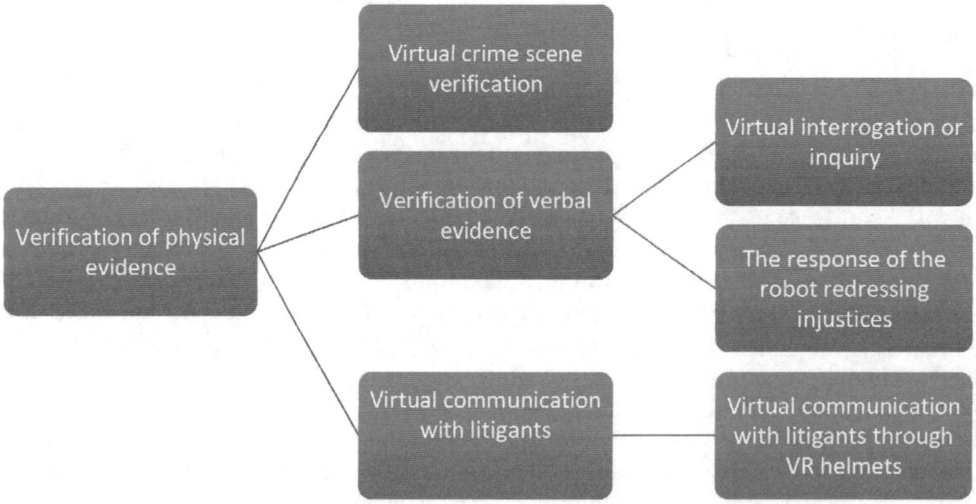

Figure 4. Public prosecution simulation system.

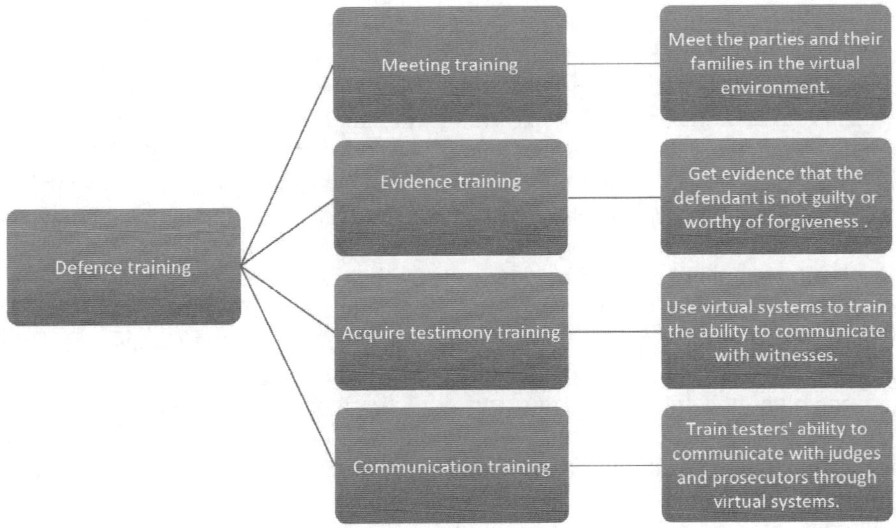

Figure 5. Defence simulation system.

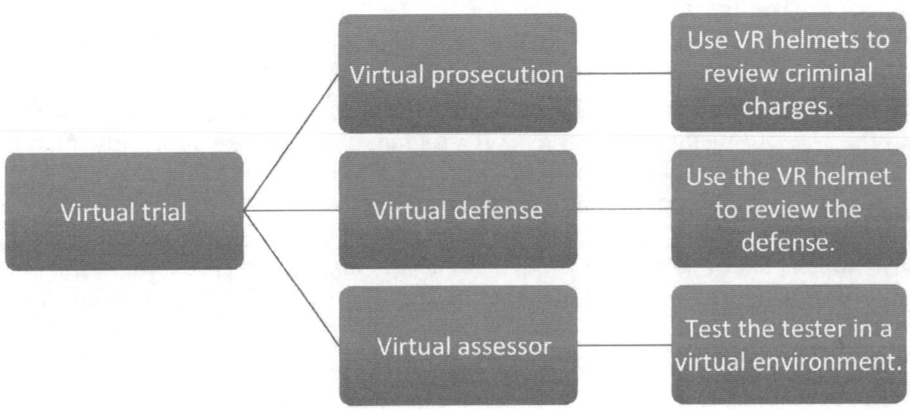

Figure 6. Trial simulation system.

It should be noted that law is a complex academic system composed of multiple legal departments. For each kind of legal department, the virtual simulation teaching system needs to be designed according to the characteristics of its own disciplines. Here we only try criminal law as an example. If we achieve good results, we will promote the technology of other legal departments. Improve the entire virtual teaching system of law.

2. Research methods

Generally speaking, this teaching design uses the idea of introducing VR technology into law teaching, forming a model virtual teaching design in detail, and then promoting the simulation teaching experiment process through information technology, so that students can get the real case handling experience.

2.1. The teaching design of the intersection of arts and sciences

In the past, the simulation of law teaching was insufficient, and it was difficult for students to have a real experience of handling cases. However, relying solely on liberal arts teaching methods has to face the obstacles that students do not have legal status. Teaching requires students to handle cases like judges, but the law prohibits those who do not have legal qualifications to handle cases. This creates a paradox that cannot be resolved in the liberal arts teaching mechanism. The VR technology of science and engineering can solve this problem very well. On the one hand, the virtual simulation of VR technology enables students to obtain the opportunity to handle cases without having legal status; on the other hand, VR virtual simulation technology provides students with virtual case handling scenarios allow students to directly enter the environment of case handling.

Although the application of VR technology to teaching has been practiced by many scholars, the system is still limited to science thinking only from the perspective of assisted teaching. If VR technology is not programmed based on the key to the litigation process, it is only the effect of audio and video. Students can be brought into the scene in a certain teaching session, but it is difficult to be brought into the proceedings. VR technology becomes a simulation game machine in the classroom, and it is difficult to improve the ability of students to handle cases.

Our design combines the liberal arts thinking with the technology of science, using liberal arts thinking as the basis for programming, and using science technology to break through the experimental barriers of the liberal arts area. Then in the virtual world to build a social relationship and social collaboration that meets the needs of liberal arts teaching, testers can gain a sense of social role through experimentation.

2.2. Construction of virtual teaching model

This design attempts to form the basis of modeling the VR system through litigation procedures. Specifically, it converts the legal language of the litigation procedure into a computer language, turning the real-world litigation procedure into a computer world litigation procedure. This allows the tester to feel the true existence of the proceedings in the virtual world. We mainly implement computer modeling in terms of both the type of litigation and the type of tester. First, the model is classified according to the type of litigation, so that the VR program matches the various litigation processes in the real world. Secondly, the model is classified according to the type of litigation participants, so that the VR program matches the realworld litigation role.

2.3. The combination of networked interaction and human-computer interaction

This design will establish two kinds of interactive training: VR helmet networking interaction and human-computer interaction. If the test requires different litigation participants and the evaluation of the test results is more subjective, we tend to adopt the VR helmet networking interaction method, so that the tester can obtain the virtual world role to carry out virtual communication. A typical example is the virtual trial. If the test needs to be operated in accordance with the litigation procedure, the evaluation of the test results is more objective. We tend to adopt a human-computer interaction method. If the tester can complete the operation according to the preset procedure, the corresponding score can be obtained. The VR helmet networking interactive test needs to be scored by the teacher, while the human-computer interaction test only needs to be evaluated by the computer.

3. Research significance and promotion

Legal Virtual Simulation VR teaching is a teaching innovation for the lack of practical ability of law students. The advantage of this system is that it can enable students to obtain the real experience of handling cases in a virtual environment, thus truly improving students' legal practice ability.

First of all, the virtual simulation experiment of law helps to improve students' ability to handle cases. In the past, the cases in which students studied in the law classroom actually removed the facts and evidence, leaving only pure legal application. The student's debate is based on existing facts and evidence. This led to the case being artificially separated, and the student was only learning a piece of the case, even the smallest piece. Therefore, it is very limited to improve students' ability to handle cases. Once the VR system is adopted, a real case and case handling process can be restored in the virtual world, which can effectively improve students' comprehensive analysis ability of the case [9]. Students discover the facts of the case based on the evidence, use evidence to prove their claims, and finally debate and judge based on the evidence, the theory can be realistic, and the knowledge can be flexible use.

Secondly, the VR simulation teaching of law has opened up the academic vision of students. A contradiction in law teaching: students need to acquire systematic legal knowledge, establish a methodology to analyze problems and solve problems, and they need to receive systematic training in schools; however, the relatively closed environment of schools also causes students to have narrow academic horizons. Students are not paying enough attention to legal practice issues and staying in textbooks. It is difficult to form criticism and reflection on knowledge. Through the VR simulation teaching of law, students can truly participate in the handling of real cases, which is to enable students to comprehensively apply knowledge, and thus enhance students' ability to identify cases, apply laws, explain laws, and choose legal methods [10].

Thirdly, the virtual simulation teaching of law is conducive to improving students' ability to use knowledge comprehensively. In the past, law teaching often introduced knowledge to students through scattered departmental laws. The links between the courses were not close, and students' ability to comprehensively use knowledge was insufficient. The departmental law curriculum is relatively isolated. Even if students use a specific case to study, they only discuss the legal application of the case and do not pay attention to the evidence of the case. The training that students can get is still at the theoretical level, and it is difficult to form a comprehensive judgment training system for evidence, facts and law. However, in dealing with cases, the knowledge of a single legal department is often difficult to deal with, which requires students to have the ability to use legal knowledge comprehensively. Virtual simulation teaching takes the case as the main teaching content, and the separation between knowledge is repaired based on the purpose of handling the case. In the process of simulating the case, students will not clearly feel the boundaries between knowledge. Students will apply knowledge according to their own needs, and improve their ability to analyze and solve problems. At this time, students have become the center of teaching, knowledge serves the needs of students, and students' ability to use knowledge comprehensively has been improved.

4. Conclusion: prerequisites and safeguard measures

In fact, simulation teaching has long been used in training rule of law in many countries. For example, in Germany, candidates want to serve as judges and prosecutors. He or she has to study four years of law and pass the first national exam before completing. Two years of internship and passed the second national exam [11]. It can be seen that in Germany, practical teaching of law is as important as theoretical teaching. Practice teaching in Germany shows a simulation: candidates are asked to work like judges and prosecutors. This kind of simulation training is to provide a training mechanism by legally constructing a social mechanism. And what we are thinking about is creating this condition in the virtual world. Therefore, the virtual training conditions of the computer need to be as perfect as the real world mechanism, so as to ensure that the VR simulation teaching can obtain the same training effect as the real world. In order to achieve this goal, we believe that the design of the VR virtual teaching system requires the following conditions: scene simulation, case simulation, and mechanism simulation.

The first is the scene simulation condition. The scene of the case is not just the court, but every space environment needed to handle the case. As a case-handler, personally experiencing the space environment of a case is the basic condition for obtaining facts and making judgments. The teaching of virtual simulation is not only to establish a virtual court, but to restore the space environment of the entire case in the virtual world. The trainer must obtain verbal evidence in a virtual environment, be able to conduct on-site investigations of the case, and be able to meet with the client at a specific place like a lawyer. These spatial and environmental conditions enable students to gain a deeper knowledge. Therefore, when designing the VR virtual teaching system, engineers and legal experts are required to establish a separate database for the scenarios required by the proceedings, as the basis material of the VR virtual teaching system.

The second is the conditions for case simulation. The reason why the previous practical teaching did not achieve the effect of teaching simulation is that the case for training is not complete and true. Although the case itself may have happened in the real world, this truth can only be called the truth of the matter, and it cannot be called the truth of the case. The truth of the matter is a simply example to provide material for the debate. However, as a case, in addition to the facts of a case, evidence of the case is also required. If students are involved in a case, it is not enough to see the facts of the case. Just as every citizen can get the case matter through the Internet and present his own analysis, but we will not say that this is the handling the case, if the law school students It is only an analysis of the case matter and it cannot be regarded as handling the case.

Therefore, providing students with evidence-supported cases becomes the key to the success of VR virtual teaching. Each experimental project of the VR teaching system must be designed around real cases, which has to consider the problems of cooperation and risk management between the university and thejudicial organs. On the one hand, colleges and universities need to obtain dossiers from the judicial organs to provide students with real cases. This requires action to break the institutional barriers between universities and the judicial organs to establish a true cooperative relationship. Turning the use of the judicial organsdossiers into an institutionalized operational process. On the other hand, universities and the judicial organs need to consider the issue of case data storage and confidentiality risks. In addition to considering the educational value of the case, the choice of case must also consider the confidentiality of the case. The content of information in the evidence materials involving national and trade secrets and personal privacy needs to be technically handled by professional to prevent information leakage, and a complete monitoring mechanism and accountability mechanism should be formed in the process of using the dossiers.

The teaching use mechanism of the judicial organs'dossiers is a complicated process, and it takes a huge technical cost for engineers to design a test program based on the dossiers. So our idea is to build a library of teaching cases to provide material for engineer programming. Yunnan University of Finance and Economics is negotiating with the Yunnan Provincial People's Procuratorate to build this cooperation platform. We consider forming a team of experts to focus on selecting a batch of cases with teaching value, focusing on the technical processing of secret information, and building a case library for simulation teaching.

The third is the mechanism simulation condition. Judicial justice is first manifested through procedures, so virtual simulation teaching can not be separated from the justice of the program to discuss the justice of the case. When conducting virtual simulation teaching, a set of procedures that are strictly in accordance with the rules of evidence and the rules of litigation are required. Due to the savings in teaching time, the time for the proceedings can be appropriately shortened. However, virtual teaching should be accompanied by a systematic litigation rule: for example, a virtual prosecutor must sue in court within one week, and a virtual judge should make a judgment within one week. These mechanisms must be developed in advance and entered into a simulation system to foster students' awareness of the procedures for handling cases. At the same time, in addition to relying on computers, teaching evaluation requires the participation of expert groups. It is necessary to form a professional experimental evaluation team (which can be composed of legal researchers, judges, prosecutors, and lawyers) to score testers' performance and provide practical guidance outside the VR system.

References

[1] Zhang W X 2011 Selected Works of Law *Legal Education* **10** p 7

[2] Zhu J W 2016 *China Renmin University China Law Development Report* Beijing: China Renmin University Press p 6

[3] He J H 2008 *Comparative study of procuratorial systems* Beijing: China Procuratorate Press p 166-167

[4] Fang S K 2018 Problems and Countermeasures in the Teaching of Law in China *People's Rule of Law* **16** p 79-81

[5] Zhang J 2019 Research on the Reform of Law Practice Teaching under the Background of Cultivating Excellent Rule of Law Talents *J. of Liupanshui Normal University* **4** p 118-119

[6] Hu M 2018 Innovative Law Education Model, Cultivating High-Quality Legal Personnel with Morality and Law *Chinese Higher Education* **9** p 27-28

[7] Zhang J, Sun S L, Wu Y F 2019 Research on Virtual Simulation System of RV Reducer Based on Unity3D *Combined Machine Tools and Automated Machining Technology* **9** p 120

[8] Liu H M, Qin X P, Huang J J 2016 Research on Virtual Prototype Simulation and Transmission Accuracy of RV Reducer *Mechanical Transmission* **5** p 55-60

[9] Zhang L X, Wang L J, Xu W K 2009 Research on Real-time Control System of High Speed Material Handling Robot *Mechanical Science and Technology* **4** p 32-34
[10] Zeng L J 2018 PracticalistJurisprudence Research Paradigm *J. of Zhejiang University (Humanities and Social Sciences Edition)* **4** p 43-45
[11] Zhang J 2006 Research on the Design of Teaching Methods of Legal Ability Improvement Clinic *J. of Hulunbeier College* **6** p 102-105
[12] Chen X L 2017 *Case of Criminal Law* Beijing: China Renmin University Press

Journal of Physics: Conference Series **1419** (2019) 012047 doi:10.1088/1742-6596/1419/1/012047

Simulation test of driving assistance system based on virtual scene

Xing Meng[1], Haiyun Gan[1,*] and Ying Yan[1,2]

[1]National and Local Joint Engineering Center for Intelligent Vehicle-Road Collaboration and Safety Technology, Tianjin University of Technology and Education, China

[2]E-mail: yanying@tute.edu.cn

Abstract. Assisted driving technology has received increasing attention. In order to verify the function of the assisted driving technology, this paper takes the automatic emergency braking system as an example and uses CarMaker simulation software to construct a standardized test scenario. Build a vehicle model and add sensors to detect the environment around the vehicle. Set up the Simulink control model. Use the PID control to brake the vehicle. The simulation test of the automatic emergency braking system was completed and the dynamic characteristics of the test vehicle were analyzed. Finally, through simulation experiments, it is verified that the simulation test can achieve extremely dangerous working conditions and reduce the possibility of accidents. Simulation tests can improve the efficiency of the test, and can repeat the test multiple times, reducing the cost of testing.

1. Introduction

With the development of the automotive industry, assisted driving and automatic driving technology has become a development trend. The assisted driving technology can not only reduce traffic accidents caused by driver human factors, but also reduce traffic congestion [1]. The core issue of this technology is functional reliability. As shown in Table 1, from 2016 to 2018, there have been nearly 10 traffic accidents caused by autonomous driving in the world [2]. Therefore, assisted driving and autonomous driving technologies require a large number of functional tests to improve safety. Therefore, assisted driving and autonomous driving techniques require a large number of functional tests to improve safety [3]. RAND Corporation said: to prove that autonomous vehicles can reduce traffic accidents by 20% compared to human driving, it requires about 5 billion miles of public road testing, using a fleet of 100 vehicles. It is tested at an average speed of 25 mph 24 hours a day, 365 days a year, and will take about 225 years [4].

In the process of car V process development, testing and verifying the control system is a very important step [5]. At present, commonly used driving assistance systems include adaptive cruise control system (ACC), front collision warning system (FCWS), automatic emergency control system (AEB), lane departure warning (LDW), lane keeping assist system (LKAS) and so on. This paper selects the automatic emergency braking system (AEB) for simulation test for functional verification [6].

Table 1. Global autopilot traffic accidents in 2016-2018.

Time	Company	Autopilot level	Accident Cause	Casualties
January 2016	Tesla	L2	The driver did not operate as required by the system	1 death
February 2016	Waymo	L2	Automatic driving design imperfect, predicting rear vehicle behavior errors	No casualties, vehicle collision
May 2016	Tesla	L2	Automatic driving design imperfect, unable to identify large objects docked	1 death
March 2017	Uber	L3	Ordinary vehicle fault	No casualties vehicle rollover
August 2017	Waymo	L4	Automatic driving safety officer intervention fault	No casualties vehicle collision
March 2018	Uber	L4	Automatic driving system software does not recognize pedestrians	1 death
March 2018	Tesla	L2	The driver did not operate as required by the system	1 death
May 2018	Waymo	L4	Ordinary vehicle fault	1 minor injury

2. Automatic emergency control system

The task of the Automatic Emergency Brake (AEB) system is to decelerate the vehicle to the speed of the target object. To do this, the system compares the time-to-collision t_{tc} with a time-threshold-brake t_{tb} to decide if a braking intervention is required

For a stationary or a very slow moving target object, the time-to-collision t_{tc} is calculated as follows:

$$t_{tc} = \frac{d}{v_{rel}} \tag{1}$$

with relative distance d and relative velocity v_{rel} between the ego vehicle and the target object.

If the target object is decelerating, the time-to-collision is calculated as follows:

$$t_{tc} = \frac{\sqrt{v_{rel}^{2} - 2 \cdot \mathrm{d} \cdot D_{\mathrm{rel}}} - v_{rel}}{D_{rel}} \tag{2}$$

with relative deceleration D_{rel}.

The time-threshold-brake t_{tb} for a stationary target object is calculated in following way:

$$t_{tb} = \tau_B + \frac{v_{rel}}{2 \cdot D_{\max}} \tag{3}$$

with brake loss time τ_B.

For a non-stationary target object the time-threshold-brake is calculated with:

$$t_{tb} = \tau_B + \frac{v_{rel} + D_{rel} \cdot \tau_B}{2 \cdot (D_{\max} - D_{obs})} \tag{4}$$

with maximum allowed deceleration of ego vehicle D_{max} and the actual target object deceleration D_{obs}.

If $t_{tc} < t_{tb}$, the AEB system sets as target acceleration for the controller the maximum allowed deceleration D_{max}.

3. Simulation text

3.1. Simulation tool

CarMaker is a passenger vehicle dynamics simulation software for vehicle component development, controller development and vehicle development. CarMaker is mainly composed of two parts that one is CarMaker Interface Toolbox which mainly includes Graphic User Interface that provides all the functions of CarMaker; virtual instrument which displays information such as pedal and vehicle status; IPGControl is a visualization tool that can observe selected output variables such as speed acceleration etc; IPGMovie used to display 3D animations in real time during simulation. The other part is the Virtual Vehicle Environment which is the core of the entire simulation including CarMaker executable files control software such as CarMaker for Simulink and some external models.

3.2. Testing scenario

China Automotive Technology and Research Center has carried out extensive domestic and international technical exchanges and practical tests based on the in-depth study and analysis of foreign Euro-NCAP [7] by combining China's automobile standard regulations, road traffic conditions and vehicle characteristics. Finally, China Automotive Technology and Research Center confirmed the test and scoring rules for C-NCAP for China. This paper selects one of the typical AEB test scenarios in the C-NCAP management rules [8].

The description of scene two is as follows:

Site requirements for scenario two: sunny weather, there is no precipitation and snowfall, the wind speed is no more than 10m/s, the road surface is dry and level, the peak braking force coefficient is greater than 0.9, the test area is 30m ahead of the test end point within 6m from the driver's side of the test center line and 4m on the passenger side. There are no other vehicles obstacles or other protrusions inside.

The movement of pedestrians in scene two: the path of the pedestrian is perpendicular to the path of the test vehicle. The lateral distance between the pedestrian and the test vehicle is 4m. The pedestrian accelerates from the static state to 5km/h from the left side of the test vehicle, and the acceleration distance is 1m. The pedestrian walks in a straight line at a speed of 5km/h. In the absence of braking measures, the test vehicle collides with the pedestrian crossing, with the collision location at 25% of the transverse section of the vehicle itself.

3.3. Vehicle model

The target vehicle constructed is a three-seat five-seat passenger car. The weight of the car is 1714kg, 4.44m long, 1.93m wide and 1.24m high. Vehicle centroid position coordinates is (2.275, 0, 0.5). The suspension spring has a stiffness of 35000 N/m. Add an object sensor. Sensor position coordinates is (4, 0, 0.5), field of view angle is 60°, and maximum sensing range is 15m, as shown in Table 2.

Table 2. Vehicle parameters.

Serial Number	Parameter	Definition
1	vehicle weight	1714kg
2	Vehicle length	4.44m
3	Vehicle width	1.93m
4	Vehicle height	1.24m
5	Centroid position coordinates	(2.275, 0, 0.5)
6	Suspension spring stiffness	35000N/m
7	Sensor position coordinates	(4, 0, 0.5)
8	Sensor field Angle	60°
9	The maximum sensing range of the sensor	15m

3.4. Driver model

The cruising speed of the driver model is 30km/h. The driver's foot is stepped from the accelerator pedal to the brake pedal for 0.5s. The maximum longitudinal acceleration is 3m/s^2, the maximum longitudinal deceleration is 4m/s^2, and the maximum lateral acceleration is 4m/s^2, as shown in Table 3.

Journal of Physics: Conference Series **1419** (2019) 012047 doi:10.1088/1742-6596/1419/1/012047

Table 3. Drive parameters.

Serial Number	Parameter	Value
1	Cruising speed	30km/h
2	The time the driver's foot stepped on the brake pedal from the accelerator pedal	0.5s
3	Maximum longitudinal acceleration	$3m/s^2$
4	maximum longitudinal deceleration	$4\ m/s^2$
5	maximum lateral acceleration	$4\ m/s^2$

3.5. Control model

The test vehicle continuously acquires the road information ahead through the sensor during driving. When an obstacle is detected in front of the vehicle, the test vehicle will feed back the speed and position information of the obstacle acquired by the sensor to the vehicle brake system. The braking system adjusts the opening of the vehicle throttle and pedal by the distance and relative speed of the obstacle, thereby achieving the braking of the test vehicle. The Simulink block diagram of the brake system of this experimental car is shown in Fig.1. In addition, the control algorithm of this braking system is PID control, as shown in Fig.2. The accelerator pedal is used as the input of the controller to output the wheel torque of the front and rear wheels of the vehicle to realize the vehicle speed control.

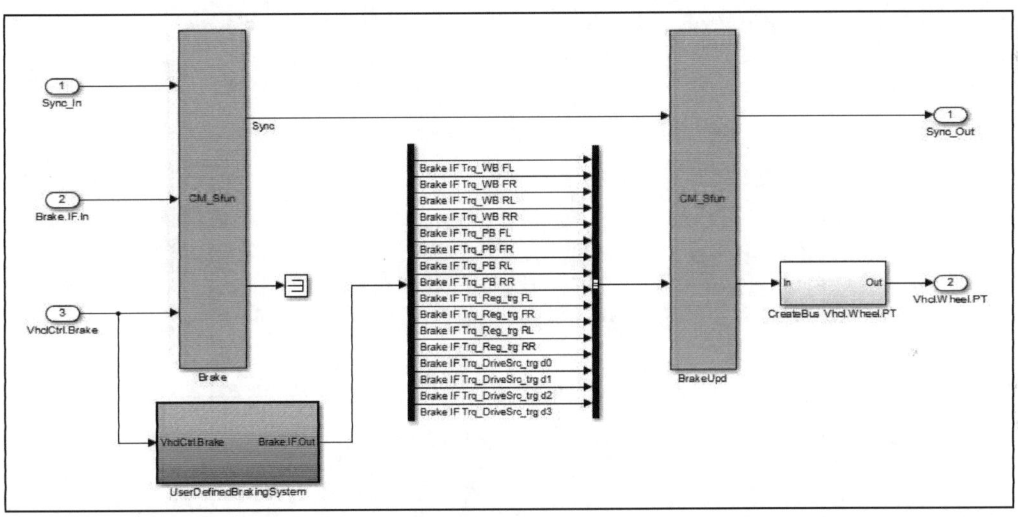

Figure 1. Brake model.

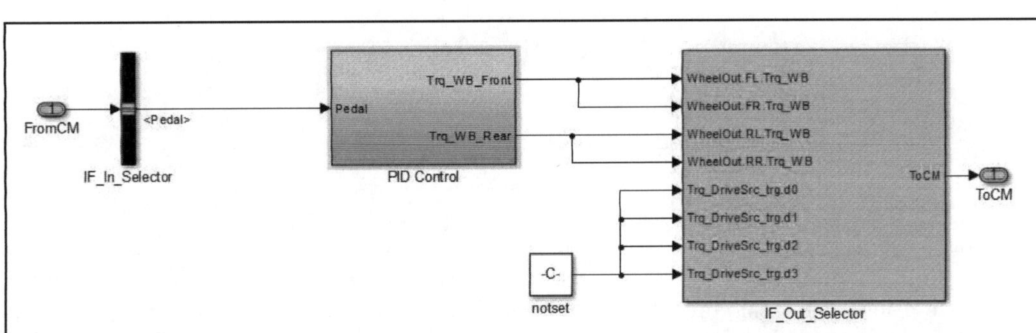

Figure 2. Control module.

4. Analysis of results

Figure 3 is a simulated animation of the scene. Multiple result information can be obtained through the IPGControl subroutine.

Figure 3. Simulation animation.

Figure 4 is a velocity-time image of the test vehicle. It can be analyzed from the figure that the entire driving time of the test vehicle is about 7s. The initial speed of the test vehicle is approximately 30 km/h. When the current party detects a pedestrian, the test vehicle begins to brake until it stops.

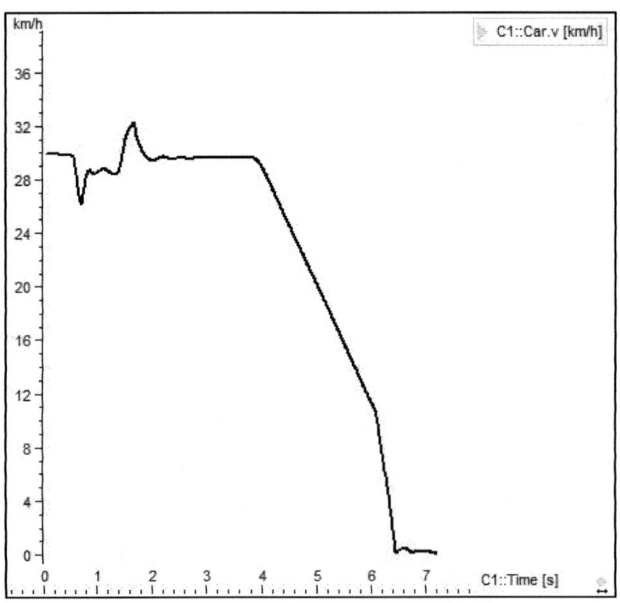

Figure 4. Test vehicle v-t image.

Of the four tires, two tires act as a brake. Figure 5 is the left front wheel braking torque image. Small fluctuations occurred when the vehicle started to drive. When the sensor detects a pedestrian in front, the braking torque increases. From the 6s to the 7s, the braking torque reaches the maximum value. When the vehicle is completely stopped, the braking torque is zero. Figure 6 is the right rear wheel braking torque, the principle of action is similar to the left front wheel.

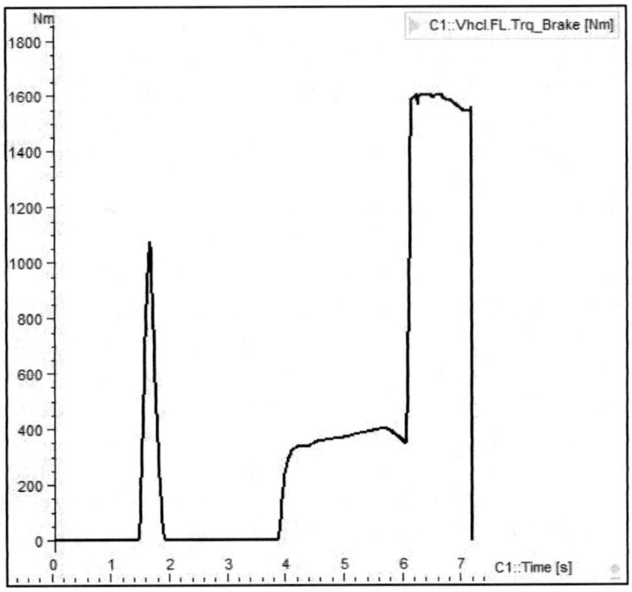

Figure 5. Left front wheel braking torque.

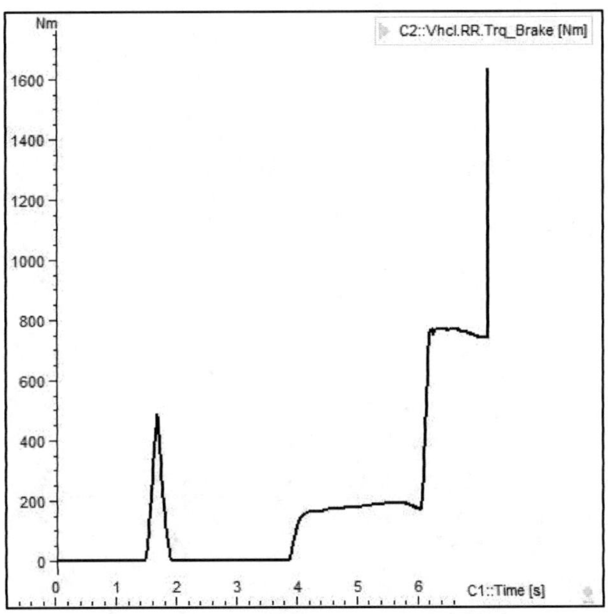

Figure 6. Rear right wheel braking torque.

5. Conclusion

This paper completed a virtual road scene ADAS function test based on CarMaker software, and used an AEB scene in the C-NCAP standardized test scenario as an example to verify the simulation. Through simulation experiments, the efficiency and accuracy of ADAS simulation are improved, the test cost is saved, and the test cycle is reduced, which brings great convenience to vehicle development.

Acknowledgments

This work was finally supported by the National Science Foundation of Tianjin (18JCTPJC68500).

Journal of Physics: Conference Series **1419** (2019) 012047 doi:10.1088/1742-6596/1419/1/012047

References

[1] Felipe Jiménez, José Eugenio Naranjo, José Javier Anaya, Fernando García, Aurelio Ponz and José María Armingol 2016 Advanced Driver Assistance System for road environments to improve safety and efficiency *Transportation Research Procedia* **14** pp 2245-2254
[2] Marius Feilhauer, Juergen Haering 2016 Current Approaches in HIL-Based ADAS Testing *SAE International* (09): 63-69
[3] Maria Staubach 2009 Factors Correlated with Traffic Accidents as a Basis for Evaluating Advanced Driver Assistance Systems *Accident Analysis and Prevention* (06): 1025-1033
[4] Daza I G, Bergasa L M and Bronte S 2014 Fusion of Optimized Indicators from Advanced Driver Assistance System (ADAS) for Driver Drowsiness Detection *Sensors* **14** (1): 1106-1131
[5] Lijun Jiang, Jinpeng He and Weiguo Liu 2014 Research on Automatic Emergency Braking System Test Scenario *Automotive Technology* (01): 39-43
[6] Pengfei Li, Ziqi Zhan 2018 Research on ADAS Simulation Test Based on Scene Standardization *Conf. of China Automotive Engineering Society* pp 463-468
[7] Chinese Automotive Technology and Research Center 2018 *C-NCAP Management Rules* [Z]
[8] Yu J, Lee G and Lee H 2016 Validation of Automotive Body ECU Using Hardware-in-the-Loop Simulation *SAE 2016 World Congress and Exhibition*

Research on simulation fixture for seat anchorage strength test

Changjiang Du[1], Yunpeng Cong[1,2], Gang Li[1] and Zeyang Tian[1]

[1]China Automotive Technology and Research Center Co., Ltd. No.68, Xianfeng East Road, Dongli District, Tianjin, P.R. China.

[2]E-mail: congyunpeng@catarc.ac.cn

Abstract. The fixture is often used to the test of automotive component parts. This paper states the research ideas of a new type simulation fixture for the seat anchorage strength test. The effectiveness of the simulation fixture is verified by FEA (Finite Element Analysis) and experiments. The new simulation fixture has higher strength and better applicability, which can simulate preferably the position of the belt anchorage in the actual vehicle.

1. Background

Automotive is a highly integrated industrial product. An ordinary car usually contains more than 10,000 component parts. So many parts make up a machine that can run at high speed, and it often needs to work in harsh working condition [1]. It is necessary to strictly control the quality during the design and production process of the automotive. Therefore, there are numerous test projects in the automotive industry, and automotive component parts are the main object of these test projects [2]. For example, car seats, seat belts, air bags, etc. The whole vehicle or the body in white is usually not available during the component test. It is necessary to make some fixtures that simulate the state in which the components are installed in the whole vehicle [3,4]. It is generally necessary to make a simulation fixture to realize the installation of the seat belt on the vehicle body during the seat anchorage strength test.

The seat anchorage strength test is described in detail in GB 14167-2013, GB 15083-2006 and ECE R17. The test conditions for the seats with three-point belt installed on the M1 and N1 models are as follows: the upper and lower body modules are fixed to the seats by seat belts. A test load of 13500 ± 200 N is applied to the upper and lower body modules respectively, and a load equivalent to 20 times the mass of the seat assembly is applied to the seat [5]. It is not difficult to find that the load of the seat test is large. Therefore, the fixture for the test should have a certain strength. The original simulation fixture of the laboratory has been damaged during the high-frequency test, and local areas may be deformed or even broken. In addition, the original fixture cannot adjust the position of its belt anchorage, and it is difficult to meet increasingly stringent test conditions. Therefore, we summarize the experience gained in the test, design and manufacture the new type simulation fixture for the seat anchorage strength test. This paper will elaborate the structural design, simulation and verification of fixture.

2. The design of simulation fixture

2.1. The thinking of simulation fixture

At the beginning of the design, the goal is to increase strength and enhance applicability. Based on the advantages of CATARC (China Automotive Technology and Research Center), the survey and statistics of the position of belt anchorage commonly used in the market were carried out, and the statistics results were applied to the design. In addition, the laboratory undertakes the seat tests of more than 100 companies every year, and produces fixtures according to different test items to assist the test. It has accumulated valuable experience in this work. Combining extensive research and rich experience, engineers repeatedly research and demonstrate in the design stage of the project, so that the new fixture has the advantages of high strength and adjustable belt anchorage position. The design process is shown in Figure 1.

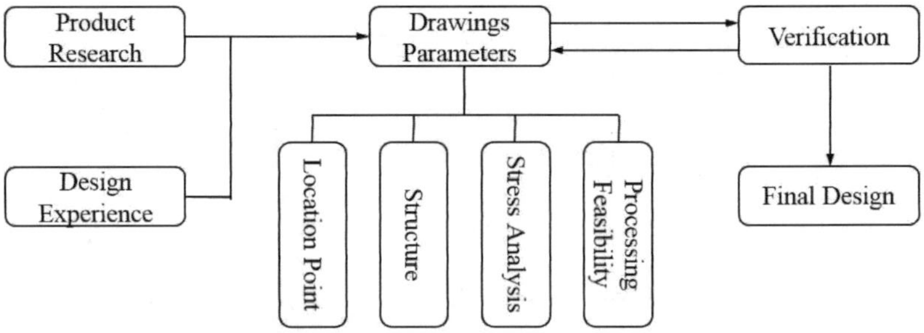

Figure 1. Design process of simulation fixture.

2.2. *The design scheme of simulation fixture*

2.2.1. Introduction. The simulation fixture is divided into upper and lower parts. Each of the upper and lower parts includes structural module, backing plate and convex plate. Structural modules have support and connection functions. There are a lot of through holes on the back plate to fix the convex plate which is connected to the lock ring to simulate the belt anchorage.

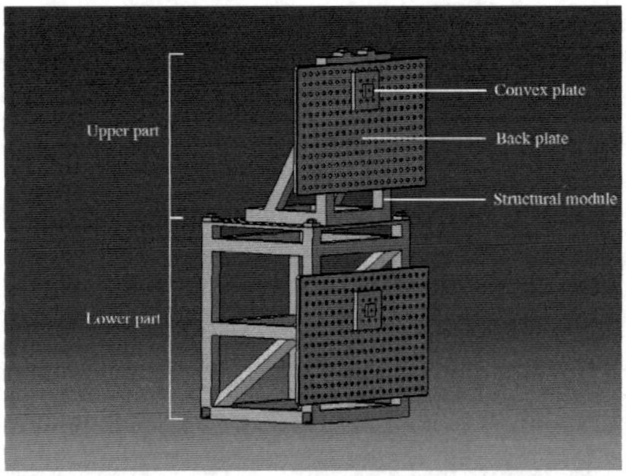

Figure 2. The stereogram of simulation fixture.

2.2.2. Characteristics

(1) High strength. The load of the seat test is large. Therefore, the fixture for the test should have a certain strength and needs to have no effect on the seat test. In view of the precedent of the original fixture deformation, the strength of the new one is optimized by the structure and material selection. The original one adopts angle steel welding design, and the new one has firmly frame structure. No.45 steels which is widely used in industry is selected as materials [6].

(2) Independent design of upper and lower structure. The simulation fixture adopts an independent design of the upper and lower parts, and they are connected through the positioning holes at the bottom of the upper module and the top of the lower module, so that the position can be adjusted.

Modern automotive design is streamlined, and there are a lot of surface shapes in the vehicle. This type of melding not only exists in the sheet metal structure but also in the body in white structural parts. The B-pillar and C-pillar of the vehicle are usually not a vertical cylinder, so the upper and lower anchorage of the seat belt are not in a vertical section. The up and down independent design of the fixture can simulate this situation and makes the test conditions closer to the vehicle situation.

(3) Matrix design of convex plate mounting holes. There are a lot of holes on the back plate of the upper and lower parts, and thus form a matrix for mounting the convex plate modules. Therefore, the position of belt anchorage can be adjusted up and down, left and right in the cross section of the back plate. Combined with the up and down independent adjustment function of the fixture, the belt anchorage can be adjusted on the x, y, z axis. Simulation fixture can simulate the belt anchorage position of most vehicle in the market. The applicability enhanced, the efficiency is improved.

3. Finite element analysis of simulation fixture

Finite element analysis simulates real physical systems (geometry and load cases) by mathematical approximations [7]. With simple and interacting elements (units), a finite number of unknowns can be used to approximate an infinitely unknown real system. FEA not only has high calculation accuracy, but also can adapt to various complex shapes. It has become an effective engineering analysis method.

3.1. Stress analysis

In the seat test, the upper and lower body modules are respectively subjected to a pulling force that is perpendicular to the plane of the seat. The pulling force are loaded on the anchorage of the simulation fixture and the seat by the transmission of the seat belt. The convex plates on the fixture are subject to tensile and shear forces. The strength and stiffness of the fixture are analysed by FEA.

3.2. Establishment and analysis of finite element model

When the finite element model is built to simulate the actual problem, it is necessary to restore the force of the actual problem as much as possible, and the calculation accuracy and calculation efficiency should also be taken into consideration. The type of unit is selected by test object and calculation accuracy. The fixture is a typical solid structure, and the convex plates of the upper and lower modules are subjected to tensile and shear forces. The solid No.95 cell grid is selected to divide the model in the simulation process. The influence of some small structures on the force situation is neglected in the division process [8]. The model is divided into 121500 nodes and 512657 grids. The meshing and constraints of the model are shown in Figure 3.

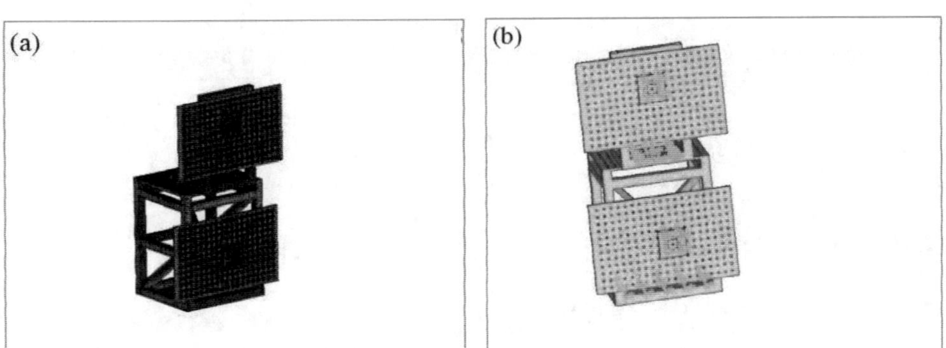

Figure 3. Meshing (a) and constraint (b).

The upper and lower parts are separately simulated, and the force and displacement of are calculated. The analysis of the force is to ensure that the fixture has sufficient strength. The analysis of the displacement is to ensure that the fixture should not has significant displacement due to the test load. The force of the upper and lower modules is shown in Figure 4. (a) and (b), respectively. The maximum stress is 49.6 Mpa and 15.7 Mpa respectively. This is far less than the yield strength of the material. The displacement of the upper and lower modules is shown in Figure 4. (c) and (d), respectively. The maximum displacement is 1.19×10^{-3} mm and 1.08×10^{-3} mm respectively.

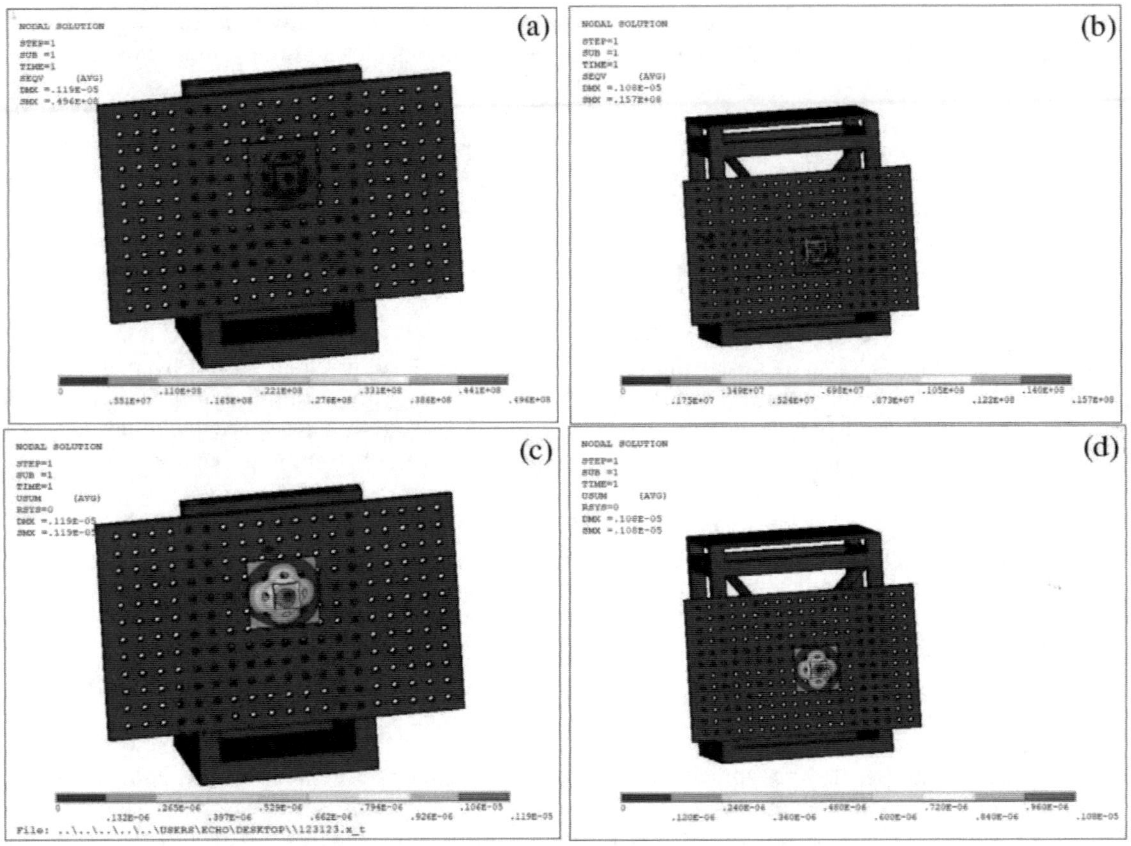

Figure 4. Simulation of force and displacement.

4. Processing and verification

4.1. Production requirements

(1) There should be no scratches, scratches, etc. on the surface of the part.
(2) The machining of machined parts must comply with the product drawings, technical procedures and the provisions of this standard.
(3) The defects must be completely removed before welding, and the groove surface should be smooth, and no sharp corners should exist.
(4) The sand, oil, water, rust and other dirt within 20mm around the weld zone and the groove must be cleaned.
(5) The oxide layer on the part should be removed.
(6) Not specified square steel size is 5mm.
(7) The surface of the part should be painted.

4.2 Test verification

Simulation fixture was applied to the seat test. A load was applied to the seat according to GB 15083-2006. The displacement of the fixture in the vertical position of the test was measured by a dial gauge. The test load settings are shown in Table 1, the test picture is shown in Figure 5, and the actual load is shown in Table 2.

Table 1. The load setting of seat test.

Channels \ Items	Initial load	Test load	Loading time	Hold time
Chest module	500N	13500N/16200N	30s/10s	10s/1s
Hip module	500N	13500N/16200N	30s/10s	10s/1s
Centroid	500N	4161N/4993N	30s/10s	10s/1s

Figure 5. The picture of seat anchorage strength test.

Table 2. The actual load of seat test.

Channels \ Items	Maximum displacement	Maximum load	Loading time	Hold time
Chest module	160.6mm/175.8mm	13835N/16347N	29.75s/9.81s	10.04s/1.00s
Hip module	129.0mm/142.0mm	13714N/16324N	29.79s/9.40s	10.00s/1.41s
Centroid	61.1mm/66.6mm	3605N/4241N	29.54s/9.46s	10.25s/1.35s

During the verification test, the displacement of the fixture in the vertical direction was measured by a dial gauge. In the case where the chest module, the buttocks module, and the seat centroid position are simultaneously loaded, the vertical displacements of the upper and lower modules of the simulated tooling are 0.13mm and 0.09mm, respectively. Although the measured conditions are somewhat different from those obtained by FEA, such displacement does not have a significant impact on the seat test. It shows that the new fixture has a significant improvement in strength and rigidity compared to the old one. At present, the new one has been used in the test for half a year, and no abnormal situation has occurred during the period.

5. Conclusion

CATARC (China Automotive Technology and Research Center) is committed to becoming the No.1 think tank in the automotive industry and continuously improving its capabilities. In order to better serve the test, the new simulation fixture with higher strength and wider applicability is designed and manufactured. The new fixture overthrew the old design and adopted the innovative split design of the upper and lower parts. The back plate which has the hole matrix is respectively welded on the upper and lower parts, thereby realizing the adjustment function of the upper and lower fixed points in six directions in space. The upper and lower anchorages can be adjusted in x, y, z axes in space. New fixture can more accurately simulate the actual situation of seat belt installation. The feasibility and effectiveness has been proved by FEA and test.

References

[1] Editorial Department of China Journal of Highways 2017 Summary of China Automotive Engineering Academic Research 2017 *Chinese Journal of Highways* **30** (06): 1-197

[2] Li Songyang 2019 Research on Test Methods of Auto Parts *Modern Manufacturing Technology*

and Equipment (05): 121-122
[3] Wang Yanqing 2018 On the development of a steering wheel assembly auxiliary tooling *Technology and Market* **25** (08): 160
[4] Ye Peng, Bo Shi and Wang Sen 2018 Design and application of engine trial auxiliary tooling *Automotive practical technology* (02): 167-170
[5] Huang Yuqiang 2011 Safety design and optimization of car seat structure [D] Jilin University
[6] Wu Hailong 2013 Research on the weld ability of No. 45 steel *Science and Technology Innovation and Application* (25): 130
[7] Wang Wanlin, Xu Congchang, Wang Zhenhu, Wang Zheyang and Li Luoxing 2019 Research on Finite Element Analysis and Structure Evaluation Method of Aluminum Alloy Car Body *Automotive Engineering* **41** (06): 607-614+640
[8] Zhou Xiaojun, Yu Ji 2018 Numerical simulation analysis of oil-water separator based on Fluent *Industrial Control Computer* **31** (05): 78-79

AUTHOR INDEX

AUTHOR INDEX

Institute of Physics
Dirac House, Temple Back
Bristol BS1 6BE UK

ISSN: 1742-6588
ISBN 978-1-7138-1324-8

9 781713 813248

CURRAN ASSOCIATES INC.
proceedings.com

IV International Conference of Computational Methods in Engineering Science (CMES'19)

IOP Conference Series: Materials Science and Engineering
Volume 710

Kazimierz Dolny, Poland
21 - 23 November 2019

IOP
Institute of Physics